T0297758

# The Health of Populations
Beyond Medicine

# The Health of Populations
## Beyond Medicine

**Jack E. James**
Reykjavík University, Reykjavík, Iceland
National University of Ireland, Galway, Galway, Ireland

AMSTERDAM • BOSTON • HEIDELBERG • LONDON
NEW YORK • OXFORD • PARIS • SAN DIEGO
SAN FRANCISCO • SINGAPORE • SYDNEY • TOKYO
Academic Press is an imprint of Elsevier

Academic Press is an imprint of Elsevier
125 London Wall, London, EC2Y 5AS, UK
525 B Street, Suite 1800, San Diego, CA 92101-4495, USA
225 Wyman Street, Waltham, MA 02451, USA
The Boulevard, Langford Lane, Kidlington, Oxford OX5 1GB, UK

**Notices**
Knowledge and best practice in this field are constantly changing. As new research and experience
broaden our understanding, changes in research methods, professional practices, or medical
treatment may become necessary.

Practitioners and researchers must always rely on their own experience and knowledge in evaluating
and using any information, methods, compounds, or experiments described herein. In using such
information or methods they should be mindful of their own safety and the safety of others, including
parties for whom they have a professional responsibility.

To the fullest extent of the law, neither the Publisher nor the authors, contributors, or editors, assume
any liability for any injury and/or damage to persons or property as a matter of products liability,
negligence or otherwise, or from any use or operation of any methods, products, instructions, or
ideas contained in the material herein.

**Library of Congress Cataloging-in-Publication Data**
A catalog record for this book is available from the Library of Congress

**British Library Cataloguing in Publication Data**
A catalogue record for this book is available from the British Library

ISBN: 978-0-12-802812-4

Printed and bound in the United States of America

For information on all Academic Press publications
visit our website at http://store.elsevier.com/

This book has been manufactured using Print On Demand technology. Each copy is produced
to order and is limited to black ink. The online version of this book will show color figures
where appropriate.

Working together
to grow libraries in
developing countries

www.elsevier.com • www.bookaid.org

# Contents

# Foreword

*Reverence for human suffering and human life, for the smallest and most insignificant, must be the inviolable law to rule the world from now on. In so doing, we do not replace old slogans with new ones and imagine that some good may come out of high-sounding speeches and pronouncements. We must recognize that only a deep-seated change of heart, spreading from one man to another, can achieve such a thing in this world.*

(Albert Schweitzer, Reverence for Life, 1969, New York: Harper and Row, p. 104)

For more than 50 years Albert Schweitzer lived and labored at the hospital he founded at Lambaréné, in Gabon, equatorial Africa. His basic philosophy: meet patients on their own terms. Tend the needs of individuals, but within the context of their community and of the broader society. It is an ethic to which all physicians aspire.

When Jack James approached me about writing this foreword, I was initially puzzled. Why ask a practicing emergency physician to introduce a book that disputes the utility of modern biomedicine?

Emergency medicine is grounded in the French principle of triage, first promulgated by Napoleon's surgeon-in-chief, Dominique Jean Larrey. When vital supplies are limited, justice demands their distribution for efficacy. Triage is practiced daily in emergency wards the world over. Patients are prioritized for treatment based on the severity and acuity of their illness or injury. In large disasters, emergency medical providers use colored tags to indicate the likelihood of survival: green for walking wounded; yellow for urgent; red for emergent; black for non-survivable. Precious resources are preferentially allocated to the red-tagged patients, then to those with yellow tags. Patients who are judged non-survivable receive comfort measures if supplies permit. When resources are insufficient for everyone, impartial triage drives their distribution in a way that benefits the majority, even at the expense of individuals.

Physicians and surgeons have come a long way since the fields of Waterloo. Modern medicine would be unrecognizable to Drs. Larrey and Schweitzer, in many beautiful and positive ways. But there is an economic dark side. Doctors have become carnival barkers on the Great Medical Midway. Our sales pitch? With enough money and unlimited access, everyone everywhere can live infinitely productive lives, if sufficiently bolstered with enough personalized pills to guarantee *jeunesse dorée* until the end of our protracted and consummately healthy natural lives.

This is fantasy.

Biomedicine has not significantly altered the global burden of disease, global health inequality, and the global rise of noncommunicable diseases, as Dr. James demonstrates using scrupulous scientific analysis. His facts are plain and stubborn. They do not support the genomic and informational technologies upon which personalized medicine is built, nor do they support, in justice, its global expense. Modern medicine is a reckless juggernaut, grinding population health under its personalized platinum wheels.

And yet, physicians have an intimate and deeply personal relationship with their individual patients. For more than 30 years I have treasured this as a doctor. As a sick patient, I have depended on it. Therein lies the paradox.

*The Health of Populations: Beyond Medicine* is a wonderfully articulate, incisive, comprehensive review of the challenges facing global health. It is also a call to battle. Dr. James offers incontrovertible evidence to support the triage of limited resources, away from efforts to control the progression of severe chronic disease in individuals, toward population-wide promotion of health and prevention of disease. Biomedicine is not a magic wand; it should be used as a complement to behavioral, legislative, social, and other preventive means to achieve personal and global health. For physicians, healthcare focus must shift to risk-factor reduction, from pre-birth to end of life, in a way that balances the needs of individual patients and those of society. Reverence for human suffering and human life are the foundation of that argument, and of this book.

It will take a deep-seated change of heart to achieve such a thing in this world. *The Health of Populations: Beyond Medicine* is our roadmap.

**Mary Claire O'Brien, MD**
Professor
Department of Emergency Medicine
Department of Social Sciences and Health Policy
Wake Forest School of Medicine
Winston-Salem, North Carolina, USA

# About the Author

Dr. Jack James was born and raised in Sydney, Australia. He studied psychology, biology, and social sciences, trained and practiced as a clinical psychologist, and completed a PhD in experimental clinical psychology. He has been on the faculty of several universities in Australia and Europe as a professor of psychology with particular interests in health, and has been principal investigator of health-related research projects funded by the European Union and public bodies in Australia, Ireland, and Iceland.

# Preface

Health is a dynamic state, guaranteed by no one thing. Rather, health derives from myriad aspects of living that comprise the habits and habitats of individuals and populations. Health may be optimized, but not perfected. While it is never too late to strive to recover compromised health, recovery is usually partial rather than complete. Thus, optimal health is founded on ways of living that favor prevention, and prevention rests on minimizing exposure to disease and injury risk factors.

Authorities, national and global, fret about ways to expand biomedical healthcare to avert catastrophe from the escalating global burden of chronic noncommunicable diseases. However, history and current evidence show that biomedical healthcare has contributed comparatively little to the health of populations past and present. Moreover, despite occasional promises of imminent transformational discoveries, there is no realistic prospect of biomedicine succeeding in optimizing the health of current or future populations. Instead of solutions, biomedicine has contributed to the current crisis by way of medical harm to patients, which is now a leading global cause of mortality and morbidity. The evidence, both historical and contemporary, is incontrovertible: Susceptibility to disease and injury is determined more by behavioral and social determinants associated with ways of living than by any other factors. Therefore, the proper role for biomedicine is as an adjunct to risk factor reduction throughout the lifecourse.

In a world characterized by an aging demographic and an unprecedented global burden of disease, individuals can do much to optimize personal health by minimizing individual exposure to risk factors. Ultimately, however, optimizing health requires collective action to transform healthcare by repositioning biomedicine from a dominant to an adjunctive role. Change is required at all levels from the individual to the societal, and from the local to the global. The alternative course, maintenance of the status quo and business as usual, will bring disaster. Continued reliance on biomedical healthcare—with its immense inherent lack of safety, modest efficacy, disappointing effectiveness, and unsustainable cost-effectiveness—will exacerbate the already worsening global epidemic of noncommunicable diseases.

I gratefully acknowledge the many students and colleagues who contributed in diverse ways to the evolution of this book. I am especially grateful to Valmai Gendle, Drífa Harðardóttir, and Janet McQueen for their reading of drafts of the work. I thank the staff of Reykjavík University Library, and especially Unnur Valgeirsdóttir, for the expert bibliographic support they supplied. I also thank the publishers, Elsevier, the production team, and especially Senior Acquisitions Editor, Kristine Jones, for their unwavering encouragement and support. Above all, I am grateful to Drífa for enduring the inordinate hours of my time that this task consumed.

**Jack E. James, PhD**
**Reykjavík, Iceland**

# Part 1

# The Science of Health

*Medical science and services are misdirected, and society's investment in health is not well used, because they rest on an erroneous assumption about the basis of health [that has] led to indifference to the external influences and personal behavior which are the predominant determinants.*

(Thomas McKeown, 1979, pp. xv-xvi)

Human health varies greatly within and between individuals as well as populations. At any given time, illness afflicts some people and not others, and among those afflicted some are more stricken and others less so. Modern biomedical healthcare has given rise to strong beliefs about the role of biology in explaining variability in health. However, the science of health shows that variation in personal and population health is explained more by psychosocial variables than by biology. In short, biology is not *the* cause of health and ill-health. Biology is mostly a mediator in causal chains that begin and end in the interactions between people and the environments they inhabit.

Considering the evidence, biomedical dominance of healthcare is difficult to comprehend or to justify. While medical research, innovation, and practice make positive contributions to health, the totality of benefit is comparatively modest. Worse, harm caused by medical interventions, discussed in detail in Part 2, is widespread and of shocking proportions. A profit-driven ethos, which shows no signs of abating, exacerbates much medical harm. Optimizing personal and population health, wellbeing, and quality of life requires the reordering of current healthcare priorities (discussed in Part 3), which need to be directed away from predominantly biomedical preoccupations toward comprehensive recognition that human health is founded on ways of living.

Health and illness are due to innumerable interactive influences that can be thought of collectively as comprising the habits and habitats of human populations. ***Habitat***[1] is familiar in ecology and refers to a given species' characteristic milieu, inclusive of the physical and social features of that environment. ***Habits*** can be considered broadly as comprising the usual or persistent patterns of behavior of individuals and groups. Though rarely used in discussions about human health, the composite phrase *habits and habitats* is used in the pages that follow to encompass myriad *biological*, *psychological*, and *social* processes that collectively determine human health and wellbeing.

---

1. Terms in bold italics are defined in the Glossary.

# Chapter 1

# The Origins of Health

*The health of nations is more important than the wealth of nations.*
(William James Duran, 1885–1981, American writer, historian, and philosopher)

## Contents

It is curious that much informed discussion about health is actually not about health but about disease. The *disease perspective*, widespread in biomedical science and practice, assumes that health is the default state defined by the absence of disturbances in physiology. Belief in the preeminence of biological explanations of health is part of a continuing yet anachronistic philosophical tradition in biomedicine, traceable to the seventeenth century origins of modern science and the *mind-body problem* addressed in the philosophy of René Descartes (see Box 1.1). Disinclination to consider nonbiological variables is evident in the healthcare that patients routinely receive, which is dominated by interventions that target biological processes.

The need to consider patients' ways of living, their *habits and habitats*, including health-related behavior (e.g., diet, physical activity, tobacco use, and use of substances such as alcohol), family life, cultural norms, and socioeconomic circumstances, is often acknowledged in biomedical healthcare but generally does not progress much beyond lip service. Reverence for biology is

The Health of Populations. http://dx.doi.org/10.1016/B978-0-12-802812-4.00001-1

**BOX 1.1  Descartes and Mind-Body Dualism**

René Descartes (1596–1650) *(Source: http://upload.wikimedia.org/wikipedia/commons/ 7/73/Frans_Hals_-_Portret_van_René_Descartes.jpg)*

Seventeenth-century Europe was marked by intense conflict between the teachings of the Roman Catholic Church on one hand, and the development of reasoning and science on the other. Because the Church regarded humans as spiritual beings controlled by supernatural forces, any systematic study of the body, for example, by means of dissection to learn about anatomy, was perceived as a threat to scripture and was prohibited. René Descartes, French philosopher and mathematician, proposed a solution to the impasse. He argued that mind (soul) and body consist of two distinct "substances." The mind, being spiritual and moral, is the province of the Church, while the body, being physical and mechanical, can rightly be subjected to scientific enquiry.

Descartes (Internet Encyclopedia of Philosophy, 1996) did not, as is sometimes claimed, argue the complete separation of mind and body. On the contrary, he argued that the two existed in unity by means of interaction in the brain. Although he was less than precise in describing how such interaction occurred, he did venture to speculate that mind-body interaction was seated in the pineal gland, a small organ resembling a tiny pine cone (hence, its name) located near the center of the brain. The functions of the pineal gland, now known to include regulation of sleep/wake cycles, were unknown to Descartes and his contemporaries. Unlike other brain structures, the pineal gland is a single organ not duplicated in the two hemispheres of the brain. This may have encouraged Descartes to intuit it as a likely candidate for the unique role of mind-body communication. In any event, Descartes' account provided the foundation for what has come to be known as *mind-body dualism*. The idea appeased seventeenth-century clerics and paved the way for the future advancement of science, including medical science.

**BOX 1.1  Descartes and Mind-Body Dualism—cont'd**

Although Descartes argued that mind and body exist in "real distinction" from one another and are *separable*, his speculations about the role of the pineal gland suggest that he may not have intended mind and body to be thought of as being entirely *separate*. The trouble is that separate is indeed how the person (mind) and body commonly came to be regarded, and biomedicine is a legacy of that traditional way of thinking. Modern biomedical healthcare represents the clearest embodiment of the pretense that physiology is separate from subjective, psychological, and behavioral processes. Biomedicine's dualistic approach to health is evidenced by its preoccupation with interventions aimed at remedying physiological disturbances in the body without much regard to individual circumstances or wider social influences. Today, mind-body dualism underpins much of the critical discourse about medicinal research and practice. That critique is frequently used in support of a variety of *complementary and alternative therapies*, including *mind-body medicine*.

Mind-body practice is claimed to take account of the needs of the whole person by addressing both mind and body. However, mind-body practice is itself open to two major criticisms. First, by arguing the need to take account of *both* mind and body, mind-body practice can be accused of perpetuating the very thing (i.e., separation of mind and body) that it seeks to denounce. Secondly, whether considered as two separate entities or as a unified whole, "mind-body" obviously pertains to individuals. Consequently, mind-body practice tends to have an individualistic focus, a feature it shares with biological medicine. Health, however, cannot be appreciated fully by considering the individual alone, even when the individual is treated as a mind-body whole. As argued throughout this book, understanding health requires a population perspective that takes account of the habits and habitats that define ways of living. Only then is it possible to begin to appreciate the biological, psychological, and social complexity of the true determinants of health.

implicit in most contemporary healthcare, and biological *exceptionalism* (the belief that biological understanding is uniquely important) is sometimes asserted explicitly. When outlining his vision of what is needed to improve the health of current and future populations, the eminent American geneticist Muin Khoury claimed that "the most effective way to improve health is to understand normal biology ... and its perturbations" (Khoury et al., 2012, p. 34). In fact, the contention that biomedical understanding is the foundation for optimizing human health is demonstrably false. The history of medical achievement, discussed in the present chapter, and the contemporary practice of biomedical healthcare, discussed in later chapters, shows that the health of populations has not, is not, and indeed cannot be optimized through advances in biological science alone.

## 1.1 WHAT IS HEALTH?

With its focus on biology and disease, biomedicine's slender approach to health does not reconcile easily with traditional or contemporary lay perspectives. The word "health" derives from Old English hælþ, of Germanic origin, perhaps influenced by Old Norse *heill*, meaning "whole, sound, or well" (http://dictionary.reference.com/browse/health). Echoes of these ancient origins can be found in brief everyday encounters between people exchanging pleasantries. When asked how they are feeling, the average healthy person typically answers with idioms such as "fine," "great," "pretty good," "fit as a fiddle," "couldn't be better," and the like. Though nebulous, these phrases allude to positive states of wellbeing without reference to the absence of ailments or disabilities.

### 1.1.1 Definition of Health: World Health Organization

A notable exception to the inclination of healthcare authorities to equate human health with the absence of disease is the view promulgated by the peak international authority on health, namely, the World Health Organization (WHO). In 1948, the year of its inception, the WHO proposed that health is:

> a state of complete physical, mental and social wellbeing and not merely the absence of disease or infirmity (p. 100).

In its emphasis on multiple factors (mental and social, as well as physical/biological), the WHO definition of health conflicts with the biomedical view, and this may explain much of the considerable criticism that the definition has received over the years. Most criticism has centered on the definition's use of the word "complete." The concern is that, at best, complete health is experienced by few—and even then, possibly only fleetingly. As one wit opined, if health depended on complete physical, psychological, and social wellbeing, it would be achieved only at the point of simultaneous orgasm (Smith, 2002).

Long before WHO's inception, the idea of complete or perfect health must have occurred to the Nobel Laureate in Literature (1925), Shaw (1909), who wrote in the preface to *The Doctor's Dilemma*, "there is no such thing as perfect health [and therefore] nobody is ever really well." This possibly reflects what is generally believed intuitively. Health is a relative state of being that ranges across a continuum that rarely includes perfection. Individually, we are inclined to use comparison as the main method for judging ourselves to be more or less (though rarely completely) healthy. Typically, we judge ourselves to be healthy or not compared to how we have felt in the past, while also comparing ourselves with how healthy we perceive others to be, especially those in our cohort of similar age, occupation, and social group.

Although the WHO definition of health continues to be criticized as unrealistic, such criticism is possibly not merely exaggerated but inaccurate because it is typically made without reference to other key WHO provisions.

One particularly relevant provision is the WHO's (1948) position on health as "one of the fundamental human rights of every human being" (p. 100). In that context, it is notable that whereas the idealized *state* of health in the WHO definition comprises "complete" wellbeing, the *right* to health is limited to "the highest attainable standard." In that respect, the distinction between *access* to the highest standard of health attainable and the *state* of complete health mentioned in the definition parallels the distinction that is sometimes made in moral philosophy between means and ends. In the study of ethics, *means* are actions and *ends* are goals, with the latter presumed to possess greater intrinsic worth than the former. Immanuel Kant, for example, argued that it is immoral to use a person merely as a means to achieving ends because to be human is an end of greater intrinsic worth than means to ends (Johnson, 2014).

Moreover, just as Kant argued that people should be considered both means and ends (and not *merely* means), the WHO asserts that health is "a resource for everyday life" (WHO, 1986, p. 329) and "a major resource for social, economic and personal development" (p. 330). Therefore, the WHO holds that health is an inherent right (an end) as well as a means for achieving individual and collective ends. Accordingly, if health is a *right*, and it seems reasonable to consider it as such, it necessarily follows that health involves *responsibilities*. If so, who bears those responsibilities? As with other fundamental rights, health must, by definition, be a responsibility of collectives (groups, communities, and governments). On the other hand, as with other fundamental rights, it is inconceivable that the collective should bear total responsibility for the health of everyone at all times. Consequently, the right to health also involves individuals assuming responsibility for their own health to the extent that they are able. Thus, health is an idealized state of complete physical, mental, and social wellbeing, and access to the highest attainable standard is a right involving personal and collective responsibilities.

Despite the passage of several decades since the WHO definition of health was ratified as a universal declaration, there is little evidence of its impact on the healthcare that is available to most people. Healthcare remains dominated by biological preoccupations, and most healthcare services consist of what physicians do to people whose biology is deemed to be perturbed. Given that comparatively little healthcare is aimed at protecting health in the healthy, it could be said that health largely is treated differently from other fundamental rights, such as the rights to freedom of movement, freedom of thought, freedom of expression, and freedom of assembly. All such rights are deserving of protection at all times, not only after the fact when a violation has occurred. Health, as a fundamental right, is deserving of similar protections. However, contemporary healthcare is substantially devoid of efforts to protect health, generally being deployed only after health has been compromised. In that regard, contemporary *healthcare* might more reasonably be referred to as *disease care*. Furthermore, as discussed in later chapters, the fundamental human right to health is characterized by widespread inequality within and between countries.

In the context of the WHO declaration that health is a fundamental human right to wellbeing that transcends mere absence of disease, the vast, elaborate, and immensely costly structures of contemporary biomedical healthcare are shown to be glaringly deficient. For the benefit of anyone tempted to think of *health* and *biomedicine* as essentially synonymous, the WHO definition serves notice of the profound limitations of most contemporary healthcare. Though often criticized and sometimes maligned, the WHO definition of health has remained unchanged for more than six decades. Illuminating past and continuing shortcomings, the WHO definition of health is a beacon to future possibilities. It is a persistent reminder of the radical transformations in healthcare that are needed to address current and looming crises in global health, including continuing widespread inequalities.

## 1.2   WHAT MAKES HUMANS HEALTHY?

Guided by the broad framework of health conceived by the WHO, we may venture to ask: what makes humans healthy? It would be natural to think that this question would be a preoccupation of healthcare professionals. However, in the vein suggested above, rather than addressing questions of health, medical dominance of healthcare has resulted in attention being mostly directed at the opposite question, namely, what causes disease? Even if we ignore mental and social wellbeing and focus only on biological health, which the WHO definition implores us *not* to do, it is perverse to try to account for health solely in terms of what causes illness. Whereas the number and complexity of ways of becoming ill are numerous, the fundamental causes of biological health are comparatively few. Moreover, understanding what makes humans healthy does not require in-depth study. On the contrary, the main causes of health are prosaic. They involve the familiar habits and habitats of everyday human existence.

Despite the priorities of current biomedical healthcare, history shows that the health of populations has never much depended on the arcane knowledge and practices of physicians, hospitals, and biomedical research institutes. In relatively recent historical time, many parts of the world experienced two major transformations in population health and the ***burden of disease:***

- Average life expectancy increased; and
- patterns of mortality (death) and morbidity (sickness and disability) changed from being caused by predominantly *acute communicable* diseases to *chronic noncommunicable* diseases.

Understanding those transformations provides major insights into what makes (and keeps) humans healthy.

## 1.2.1   The McKeown Thesis

Thomas McKeown, British physician and medical historian, has possibly contributed more than anyone to understanding the main determinants of human health. In his book, *The Role of Medicine: Dream, Mirage or Nemesis* (1979), and other writings (McKeown and Brown, 1955; McKeown et al., 1972, 1975), McKeown reported major historical trends in patterns of population health and disease. His most influential work concerns analyses of historical records for England and Wales covering a period of approximately 200 years from about the mid-eighteenth century. He aimed to explain the rapid growth in population size and especially the dramatic increases in average life expectancy for the period. The overall trend for the latter can be seen in Figure 1.1, which charts a period of about 300 years beginning a little before and ending a little after the period studied by McKeown. Life expectancy at birth in England and Wales, while not always following an unrelentingly smooth trajectory, increased dramatically throughout the nineteenth and twentieth centuries.

During the first decades after the beginning of the Industrial Revolution in the mid-eighteenth century there appears to have been little or no improvement in life expectancy, and possibly a slight decrease. However, by about the beginning of the nineteenth century, life expectancy began to increase to an extent unprecedented in human history. Average life expectancy, it should be said, is not merely a measure of longevity, but also an indicator of the overall health of populations. Obviously, without life there can be no health, and intuitively we sense, correctly, that an increase in the average length of life of a population portends an increase in average health. Because average life expectancy

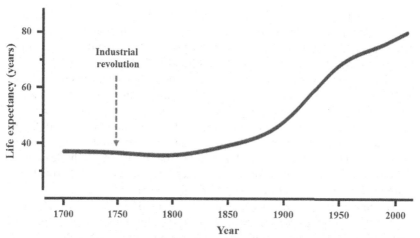

**FIGURE 1.1**   Life expectancy at birth in England and Wales from 1700 to 2009. *(Adapted from Wrigley and Schofield (1981) for the period 1700–1850, Kinsella (1992) for the period 1850–1950, and the Human Mortality Database (2012) for the period 1990–2009.)*

measured at birth is positively correlated with most key indicators of human health, it is often used as a surrogate measurement for comparing the general health of different populations or the same population over time.

The period that McKeown examined coincided with profound change in science and technology. Consequently, it may have been reasonable to assume, as many did, that the increased life expectancy and associated increased population health that were the topics of McKeown's investigations were due to greater biomedical understanding and improvements in clinical practice. However, McKeown's analyses showed that biomedicine was not the main cause of the observed improvements in health. Indeed, compared to other factors, the contribution of medical advances to population health was marginal.

For most of the period covered by McKeown's analyses, illness and death were caused mainly by infectious diseases. McKeown charted the recorded death rates for a wide range of infectious diseases, including tuberculosis (TB), cholera, typhus, typhoid, and a number of childhood infectious diseases, including scarlet fever, diphtheria, and whooping cough. Without exception, deaths from those infections decreased steadily throughout the period. Remarkably, however, McKeown's analysis showed that most of the decrease in death rate occurred *before* practical medical interventions had been introduced, and therefore the improvements he observed could not have been due to those initiatives. Of the several diseases studied, the historical records showed that reductions in the rate of death from TB contributed most to the massive gains in health that were achieved.

## 1.2.2 Tuberculosis

In recent historical time, TB was *the* main cause of death in England and Wales—and in other comparable countries up to and including the early twentieth century. There was, however, a relentless decline in the **incidence** of the disease (and in other common infectious diseases) that continued well into the twentieth century. The decline, which is likely to have begun sometime before records began, is depicted in Figure 1.2. Comparing Figures 1.1 and 1.2, it can be seen that the increase in life expectancy (Figure 1.1) mirrored the decrease in TB mortality (Figure 1.2), illustrating the vital contribution of TB mortality to population health. It is important to ask: what brought about the decrease in deaths from TB and the associated increase in life expectancy? The seemingly obvious answer, *medicine*, cannot be correct because TB had almost disappeared before the advent of relevant medical interventions.

The first real advance in medical understanding of TB came in 1882 when the bacterium responsible for the disease, *Mycobacterium tuberculosis* (then known as *tubercle bacillus* from which the abbreviation, TB, derives), was isolated by the German physician Robert Koch (1843–1910). The first public announcement of that breakthrough work was as part of an inspirational lecture

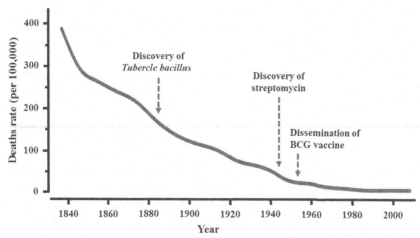

**FIGURE 1.2**   Respiratory tuberculosis mortality in England and Wales (1838–2008). *(Adapted from McKeown (1979, p. 92) for the period 1838–1900 and the Health Protection Agency (2012) for the period 1913–2008.)*

Koch gave on 24 March 1882 to the Physiological Society of Berlin (Daniel, 2006). Koch began by recounting the terrifying statistics,

> *one in seven of all human beings dies from tuberculosis, and [of] the productive middle-age groups, tuberculosis carries away one-third and often more.*
>
> (Murray, 2004, p. 1086)

Koch enthralled his audience with a methodical account of his invention of a novel staining method for viewing the bacterium under a microscope, success in isolating the bacterium by growing it in a culture medium, and use of the isolated organism to infect previously healthy laboratory animals with the disease. In due course, his research into TB earned Koch the Nobel Prize in Physiology or Medicine, awarded in 1905. Some consider Koch's lecture of March 1882 to be the most important in medical history, and one person who was in the audience, Paul Ehrlich, himself a Nobel Laureate in Physiology or Medicine (1908), came to describe it as "my single greatest scientific experience" (Gradmann, 2006).

In 1890, Koch announced that he had discovered a "cure" for TB, a development that promised to eclipse even his earlier highly celebrated success. In practice, however, the alleged cure, *tuberculin*, which he extracted from TB cultures, failed "spectacularly" (Gradmann, 2006). Notwithstanding his personal lack of success in directly finding a cure, it is instructive to consider the consequences of Koch's work in the context of the life-saving impact it is sometimes assumed to have had in laying the foundation for subsequent work by others. To begin with, it is important to note that although the toll of death was appalling, as duly noted by Koch at the time of his initial breakthrough discoveries, TB mortality was actually falling rapidly. Figure 1.2 shows that the annual rate of TB deaths had approximately halved from when recording began to when Koch's work was conducted. Obviously, none of the decline in incidence of TB *before* that work can be attributed to any life-saving interventions that may have come *after*.

In fact, more than 60 years passed before the discovery, in 1943, of streptomycin, the first effective **antibiotic** treatment for TB infection. By the end of that decade, clinical trials confirmed the general efficacy of the new drug, and from about the middle of the century it quickly came into widespread use as a treatment for TB. Nevertheless, more so even than Koch's original discoveries, the advent of streptomycin came long after the incidence of TB began its relentless decline. McKeown estimated that for the whole of the period from 1838, when TB deaths were first recorded, to 1971, streptomycin probably contributed only about 3% to the total number of lives saved. Even then, the drug was not free of complications, including side effects such as hearing loss, vertigo, and tinnitus. Moreover, rather than delivering cure in all instances, the use of antibiotics in the treatment of TB has in more recent times contributed to the spread of new strains of the disease that are resistant to all currently available treatments. Without diminishing the importance of the lives that were saved, it is obvious that greater importance attaches to the conditions (discussed below) that were responsible for the millions of lives that were spared in which streptomycin and other drugs had no role.

Work on a vaccine for TB had been under way since early in the twentieth century, but human trials involving the use of Bacille Calmette-Guérin (BCG) vaccine (named after French researchers, Albert Calmette and Camille Guérin) did not take place until the 1920s, and attempts at widespread immunization in the form of population vaccination programs did not commence until the 1950s. By that time, almost the entire decline in TB infection that could occur, from the high rates of earlier centuries to the relatively low rates of recent times, had already taken place. The fact is that the transformation of TB from being the leading cause of death to no longer being a major threat happened well before the advent of any effective medical response to the problem. On the simple premise that an effect cannot precede its cause, medical intervention could not have been responsible for the precipitous falls in TB mortality that occurred in many countries over the past 200 years nor the consequential dramatic contributions of decreased TB mortality to increased population life expectancy.

### 1.2.2.1  TB Today

Despite the comparatively small contribution of TB to current levels of mortality in high-income countries, it continues to be a leading cause of mortality from infectious disease globally, being second only to HIV/AIDS as a cause of death due to infection. Worldwide, there are more than 9 million new cases of TB infection annually and more than 1.5 million deaths (Ottenhoff and Kaufmann, 2012). TB is spread from person to person through the air, and although capable of attacking any part of the body it typically infects the lungs. Neonatal vaccination with BCG has long been thought to provide relatively good protection against childhood manifestations of the disease (Trunz et al., 2006). However, protection is limited, and although estimates vary, the period of cover appears to be in the range of 10-20 years (Aronson et al., 2004; Newton et al., 2008; Sterne et al., 1998). That timeframe coincides with the age-related distribution of the disease in the population, with adults generally showing a higher incidence.

Thus, adult pulmonary TB appears to be the main form of the disease, being responsible for most new diagnoses and contributing most to the global burden of TB-related mortality and morbidity. On the other hand, recent evidence suggests that the incidence of TB infection in childhood may previously have been substantially underestimated due to widespread underreporting of pediatric cases in many countries (Dodd et al., 2014). Notwithstanding possible underreporting, the current global burden of TB infection remains high for all age groups, with by far the largest burden being concentrated in low-income countries. Indeed, this is one aspect of the problem about which there is no disagreement. The countries with the highest rates of TB are those in which poverty and malnutrition are prevalent. Even in countries where population *prevalence* of the disease is low, the incidence of TB infection is consistently found to be highest among the poor, marginalized, and socially disadvantaged (de Vries et al., 2014).

### 1.2.2.2   TB Immunization

More than 3 billion people are estimated to have received BCG vaccine, making it the most widely used vaccine in history (Andersen and Doherty, 2005). The first mass BCG vaccinations occurred in countries where population-wide infection rates had already reached relatively low levels (i.e., Britain and some Western European countries). However, BCG vaccine policies and practices vary markedly between countries (Zwerling et al., 2011).[1] That there is variation in policy among countries that have well-established public health systems tends to suggest that the vaccine may have limited efficacy. Were vaccination truly effective, attempts at mass immunization are likely to have been more uniform. In reality, despite policies of population-wide vaccination in some countries, the practice has not been shown to be effective in producing and maintaining low rates of TB infection.

For countries where the risk of infection had already fallen to low levels prior to the adoption of mass vaccination, subsequent additional falls in the incidence of disease cannot alone be taken as evidence of the efficacy of vaccination. In those instances, whatever was responsible for lowering infection risk in

---

1. An interactive Internet database of global BCG vaccination practices can be found at www.bcgatlas.org.

the first place could also have been responsible for lowering infection rates still further after mass vaccination was adopted. Some countries, such as those of Western Europe, Britain, Australia, and New Zealand, which have low rates of infection, adopted mass vaccination, whereas other countries, such as the Netherlands and the United States, which also have low infection rates, did not adopt mass vaccination. Rather than attempting to immunize the entire population, some countries, notably the United States, have favored selective vaccination of "at risk" groups (e.g., healthcare workers and infants living in high-risk settings). Today, mass vaccination is mostly practiced in developing countries[2] where TB infection risk is high. However, the very fact that infection rates have often remained persistently high in those countries, despite adopting policies of mass vaccination, provides further reason for doubting the efficacy of such programs.

Nevertheless, confidence in modern biomedical innovation and practice appears to run high, judging from the many who evidently believe in the possibility of a fast-track to low population levels of TB for developing countries. That confidence is evidenced by repeated calls from the biomedical community for resources to be committed to the development of new vaccines (e.g., Kaufmann et al., 2010). Conversely, the purported need for such investment merely confirms the limited efficacy of past and present biomedical efforts to confer mass immunity. Had available interventions been effective in achieving near-eradication of TB in those countries where risk of TB infection is low, new vaccines would not be needed. Indeed, it is curious how mass vaccination for TB has been transported from high-income countries to developing countries despite no clear evidence of success in either. Nevertheless, belief in a biomedical solution to TB persists at the highest levels within the biomedical research community (e.g., Sizemore et al., 2012).

The *Global Plan to Stop TB, 2006–2015* (Stop TB Partnership, 2006) seeks to eliminate the disease "as a global public health problem by 2050" at an estimated cost of USD56 billion. Part of the plan includes the development and implementation of new TB vaccines for mass immunization. While it is a truism that mass immunization with a *genuinely* effective vaccine could eradicate TB, experience to date does not encourage confidence that any such outcome will be achieved. Decreases in TB mortality and associated increases in life expectancy seen in high-income countries have been real and spectacular. Those transformations, however, occurred *despite* mass vaccination programs. In an apparent repetition of that pattern, there is evidence of similar transformations from high- to low-infection rates in progress today in countries that are currently experiencing rapid economic development and social change (WHO, 2012), notably, China (Wang et al., 2014).

---

2. Consistent with World Bank usage, *developing* denotes low- and middle-income countries. It does not imply that all economies in the group are experiencing similar levels of development or that other economies have reached a preferred or final stage of development.

### 1.2.2.3    The Key to Successful TB Infection Control Lies in the Centuries-Old Past

Medical interventions for TB, including antibiotic treatment of cases and mass vaccination of populations, have been widespread for more than half a century. Yet, the evidence concerning TB infection rates suggests that, where infection is common, medical intervention has had little success in containing the spread of disease to below epidemic levels, let alone achieving eradication. The current situation, moreover, is exacerbated by the emergence of drug resistance, including *multidrug-resistant* and *extensively drug-resistant* strains (Mingote et al., 2015; WHO, 2013), and the more recent and essentially untreatable, *totally drug-resistant* forms (Rowland, 2012). It is no small irony that medical intervention is known to be the critical causal factor responsible for the increased prevalence of such strains, due largely to the use, overuse, and misuse of antibiotics intended to cure infection.

A recent South African study reported poor long-term outcomes for patients with extensively drug-resistant TB, for whom there are no further options once all existing interventions have been tried and failed (Pietersen et al., 2014). Many of these patients are discharged due to unavailability of long-term treatment facilities, causing the disease to be transmitted to the community. In those circumstances, young children are at highest risk of severe disease and death once infected, but unfortunately until recently the global incidence of multidrug-resistant TB in children had not been quantified. Notably, a recent comprehensive assessment of total global incidence of child cases for all forms of TB yielded estimates 2- to 3-times that of previous official estimates (Jenkins et al., 2014). Almost one million new cases of childhood TB were identified for a single year, including more than 30,000 multidrug-resistant cases, most of which were not officially notified.

In a heartfelt protest to the continued emphasis on biomedical approaches to TB-infection control, Khan and Coker (2014) recently outlined "five easy steps" that may be taken to *hinder* TB control. The first step is to continue with current incentives that encourage those working in national TB programs to hide facts and avoid difficult problems. Because stated national goals for infection control are often infused with political capital, healthcare personnel are sometimes reluctant to reveal the true scale of infection rates or intractable problems (e.g., multidrug resistance) for fear of being perceived publicly to have failed. Second, there is the converse step of rushing to address urgent problems (e.g., multidrug resistance), typically by adopting medical or technological solutions of limited efficacy, that divert resources from interventions that will produce better results (e.g., the larger numbers of patients whose infection is of a type that is likely to respond to existing and cheaper **antimicrobials**). Third, purchasing medical or technological solutions (e.g., for treating multidrug resistance) when the clinical infrastructure (e.g., healthcare centers in remote areas) to disseminate those interventions does not exist. Fourth, encouraging widespread misuse of antimicrobials by allowing them to be dispensed by the unregulated private sector, including private doctors, pharmacies, and local grocery stores in many developing countries. Finally, permitting sudden revisions in level of funding such as may accompany changes in foreign-aid policy that disrupt local infrastructure and impede TB infection control.

To summarize the record of TB infection and control, the relentless fall in the incidence of TB that occurred over long spans of time in widely geographically dispersed economically-developed countries has been largely independent of advances in biomedicine. Biomedical intervention, long the focus of global effort to contain TB, has contributed only marginally to limiting infection rates, while causing substantial harm by contributing to the emergence of new strains of the disease that are resistant to intervention. Despite decades of extensive medical intervention, including mass vaccination, a high prevalence of TB persists in many regions of the world. Moreover, after many years of concerted research and development in TB diagnostics, drugs, and vaccines, the WHO (2014) in its most recent global report on TB estimated that an additional USD2 billion per annum is needed to continue that effort. Results thus far from such research include new drugs currently being tested that show "encouraging prospects" (WHO, 2014, p. 15), although a vaccine to prevent TB in adults "remains elusive" (p. 17).

Notwithstanding the substantial global commitment to biomedical approaches to control TB infection, extensive evidence shows that the incidence of TB is closely related to economic development. Consistent with centuries-old trends in countries that were the first to industrialize, present-day developing countries are also experiencing decreases in rates of TB infection. If dramatic benefits of the past and continuing benefits of the present are not attributable to biomedical healthcare, what is responsible? More than 60 years ago, René Dubos, a microbiologist by training, asserted that TB is a "social disease" (Dubos and Dubos, 1952). He believed that its occurrence is associated with human economic and social conditions, and that its decline in many countries was due to improvements in those conditions. That understanding was confirmed and expanded by McKeown (1979) in his analyses of trends for TB and other infectious-disease mortality in England and Wales during the period in history when Britain industrialized.

### 1.2.3 The Broad Universality of the McKeown Thesis

The decline in mortality from TB and other major infectious diseases revealed by McKeown's analyses of records for England and Wales was not a unique episode in history. Essentially the same pattern has been found to have occurred widely throughout Europe (McKeown et al., 1972) and high-income countries elsewhere (McKinlay and McKinlay, 1977). In the United States, total mortality from all causes decreased more than threefold between 1900 (the earliest date for which reliable national data are available) and 1973, with more than 90% of that decrease occurring before 1950 (McKinlay and McKinlay, 1977). In 1900,

11 major infectious diseases (typhoid, smallpox, scarlet fever, measles, whooping cough, diphtheria, influenza, TB, pneumonia, infections of the digestive system, and poliomyelitis) were responsible for about 40% of deaths in the United States. By 1973, the same diseases accounted for just 6% of deaths. While not conducting as extensive analyses as those conducted by McKeown (1979) for England and Wales, McKinlay and McKinlay (1977) arrived at essentially identical conclusions for the United States. Investigating the "specific and supposedly effective" medical interventions for the 11 major infectious diseases of the period, McKinlay & McKinlay concluded that biomedical healthcare had contributed little to the overall pronounced decrease in mortality. Specifically, they estimated that no more than 3.5% of the total decrease in mortality that occurred in the 70 years from 1900 could be attributed to medical intervention.

The McKinlay and McKinlay (1977) analyses revealed a further intriguing detail concerning annual expenditure on medical care. In the quarter century preceding the mid-1950s, medical expenditure, measured as a percentage of gross national product (*GNP*), rose from about 3.5 to 4.5% of GNP, but less than 20 years later it had almost doubled to more than 8.5% of GNP. Whatever the reason for the sudden increase in expenditure, it was not rational accounting. As in comparable economies elsewhere, the health of the American population improved greatly during the first half of the twentieth century. However, McKinlay and McKinlay found that, like no other nation, Americans saw fit to massively increase their expenditure on medical care during the third quarter of the century, despite biomedicine actually having contributed little to improvements in the national's health in the half century before.

Given the consistently observed nexus between the "modernization" of national economies and decreased incidence of infectious diseases (accompanied by extension of life and improvement in general health), it is moot what might have been the population impact had the medical treatments of the modern era been available two centuries earlier. Would, for example, the decline in TB mortality have been faster and more complete, and the increase in life expectancy greater? The evidence indicates that the pattern of improvement in population health would probably have been little different. While crediting a degree of clinical efficacy to streptomycin, McKeown (1979) attributed little or no benefit to BCG vaccine, citing the Netherlands, which never adopted a national program of BCG vaccination, as having the lowest rate of TB mortality in Europe. The present-day global pattern of TB mortality supports McKeown's conclusion. Biomedical healthcare was largely irrelevant to the massive saving of life in countries where TB mortality fell sharply in the wake of economic growth and social reforms precipitated by the Industrial Revolution, and the same appears to be true in present-day developing countries that are currently experiencing rapid declines in mortality from TB and other major infectious diseases.

## 1.3 HISTORICAL CAUSES OF INCREASED LIFE EXPECTANCY

The importance of the Industrial Revolution in human affairs cannot be overstated. As explained by American economist, Robert Emerson Lucas, Jr, recipient of the Nobel Prize in Economic Sciences (1995), for the first time in history, the living standards of masses of people underwent sustained *accelerated* growth (Lucas, 2004). Economic prosperity transformed patterns of death and disease, and presaged unprecedented improvements in population health. Immediately after the beginning of the Industrial Revolution, around the mid-eighteenth century, life expectancy changed little for several decades, possibly due to the large influx of people into major population centers that led to overcrowding and worsened living conditions for many. This, however, was followed by a steep and largely unrelenting increase in health throughout the nineteenth and twentieth centuries accompanied by pronounced increases in average life expectancy.

McKeown identified three main causes for the sharp decline in infectious disease and the consequential marked improvements in health and life expectancy that occurred in Britain and Wales. The most important cause, according to McKeown, was improved *nutrition*. The Industrial Revolution brought with it improved methods of agriculture, including improvements in large-scale food production, storage, and distribution. Increased reliability in the quality and supply of food meant that populations were generally better nourished and less subject to assaults on health from intermittent famine. Consequently, there were major population increases in host resistance to disease leading to lower rates of infection. With a lower infection rate in the population in general, any given individual, whether or not in a state of compromised health, is less exposed to infection and therefore at lower risk than previously of becoming infected. Moreover, improved general health due to better nourishment contributes to an improved rate of survival if infected.

The emphasis McKeown placed on improvements in host (i.e., individual) immunity due to improvements in nutrition resonates with current understanding of infectious disease in general and recent observations concerning the incidence of TB in particular. The worldwide appearance of HIV in the 1980s was followed by increases in the incidence of TB infection. Compromised immune function due to HIV infection dramatically undermines host resistance to infection. Thus, with the spread of HIV, there was a resurgence of TB in countries where its incidence had previously fallen to low levels. TB was found to be concentrated among formerly healthy young adults in whom the disease had all but disappeared, but whose sexual activity during an era of rising HIV infection led to markedly increased susceptibility to TB infection.

Secondly, McKeown identified improvements in population standards of *sanitation*, especially in relation to the provision of clean water and the disposal of sewage. Those changes were critical for the control of water- and food-borne diseases such as cholera and typhoid. Thirdly, McKeown argued that *social innovation* made possible by increased affluence contributed to changes in individual behavior and environmental conditions that contributed to improvements in population health. More specifically, declining birth rate accompanied by reduced infant mortality contributed to reduced poverty and increased personal life expectation. Improvements in public education and literacy fostered ongoing improvements in standards of public sanitation and personal hygiene, and large-scale slum clearance and urban renewal projects helped to eradicate sites where infectious disease was most concentrated.

In summary, McKeown found that the main historic causes of increased life expectancy and associated increases in population health were improved nutrition, access to clean water, slum clearance, public sanitation, and personal hygiene. These aspects of life, taken largely for granted in present-day high-income economies, were consequences of the scientific, technological, and social developments that accompanied economic growth precipitated by the Industrial Revolution. Looking only at the surface of things, it is understandable that commentators tended to assume that medical innovation was responsible for the historic improvements in health that characterized the period. McKeown's work, however, revealed that the main scientific innovations responsible for improvements in health concerned agriculture and engineering whereas developments in biomedicine were largely irrelevant. Importantly, too, the period was characterized by the emergence of progressive social movements that advocated reforms such as slum clearance, public sanitation, general education, and the adoption of personal hygiene (see Box 1.2).

## BOX 1.2 Soap and the Advent of Personal Hygiene

Box of Amigo del Obrero (Worker's Friend) soap, Museo del Objeto del Objeto, Mexico City. (Source: http://commons.wikimedia.org/wiki/File:MODOAmigo.jpg)

Continued

**BOX 1.2 Soap and the Advent of Personal Hygiene—cont'd**

The importance of personal hygiene to health should not be underestimated. According to Curtis (2007), hygiene may be defined as "the set of behaviors [used] to avoid infection" (p. 660). Curtis believes that most animals exhibit hygiene behavior, which has an ancient evolutionary history. Certainly, personal cleanliness in humans pre-dates recorded history. Water, it may be assumed, has been used throughout the millennia to cleanse the body of obvious signs of dirt, and it is now known that washing with water lowers the risk of infection from microorganisms. Ancient Egyptians, Greeks, and Romans bathed regularly, although cleaning the body with soap appears not to have been common until much later. The first recorded use of soap is from Babylon, where it appears to have been used primarily for cleaning animal skins. Other early soap-like compounds also appear to have been used for cleaning clothes, as hair-styling agents, and as lotions for treating skin diseases.

By the time of the fall of the ancient Roman Empire in about the middle of the first millennium, the habit of bathing all but disappeared in Europe, although it continued elsewhere. For example, in Japan, regular bathing appears to have remained customary throughout the medieval period (fifth to the fifteenth centuries). It was not until the seventeenth century that bathing returned to fashion in Europe. Increasingly, cleanliness came to be regarded as important not only for social distinction but also for health. As bathing increased in popularity, the use of soap for washing the body also became fashionable and demand for soap quickly grew (Geels, 2005). Today, there is extensive scientific evidence for the usefulness of soap. Cleansing the hands is especially effective against the spread of diarrheal organisms that contaminate food and water (Biran et al., 2012).

Handwashing has been particularly well researched in low-income countries, where washing with water alone has been found to reduce diarrhea by about 30% and by about 45% when soap is used (Curtis et al., 2011). Handwashing also reduces neonatal mortality, trachoma (a contagious eye disease), parasitic worm infections, and lower-respiratory-tract infections. In one study, children younger than 5 years in households that received plain soap and handwashing promotion that included information and literature about personal hygiene had a 50% lower incidence of pneumonia than control families that did not receive the intervention (Luby et al., 2005). In an era smitten by technology, it is notable that the most cost-effective single intervention for reducing the global burden of infectious disease may be something as uncomplicated as handwashing with soap.

Town planning and urban renewal were pursued to an unprecedented level. Although often initiated for the express purpose of eradicating disease and improving health and wellbeing, the initial impetus for these reformist movements was not drawn from advances in the medical sciences. For example, although false, the theory that disease epidemics are caused by miasma (poisonous vapor or "bad air" from decomposed organic matter) underpinned many of the sanitary innovations and reforms of the nineteenth century. Improved sanitation indeed contributed to improved population health, even if later understanding of microbiology led to *miasma theory* being replaced by *germ theory*. Of the main nineteenth century humanitarians and social reformers, Dubos (1959) wrote:

> *Their doctrine...was scientifically naïve but proved highly effective in dealing with the most important health problems of their age (p. 23).*

Notwithstanding popular beliefs about the importance of advances in medical science, the reality is that socially inspired applications of the physical sciences (notably, civil engineering) have contributed immensely more to past improvements in human health. Similarly transformational improvements in health are ongoing, as infrastructure for providing clean water and disposing of sewage is brought to ever-larger numbers of people in developing countries.

Apart from being well supported by the historical record, informal confirmation of the McKeown thesis is to be found in media reports of calamities of every description that devastate the lives and wellbeing of communities around the globe, especially in low-income countries. When a country or region experiences destructive natural events, gross governmental neglect, or widespread conflict, relief efforts typically focus on the everyday "necessities" that also underpin health during times of peace and tranquility: namely, nutrition, clean water, shelter, and sanitation. Emergency medical assistance is often also part of the humanitarian effort mounted in response to disasters. However, whereas medical assistance can assist the survival of individuals suffering acute illness or injury, the saving of life on a large scale in emergency-relief settings almost always depends crucially on the provision of food, clean water, shelter, and sanitation, not clinical medicine.

Similar realities seem also to apply even in relation to obvious "medical" disasters such as the 2014 West Africa outbreak of Ebola, so much in the news at the time this text was written. A vital feature of attempts to contain the disease is physical isolation of infected persons, both living and deceased, from those not infected (Fauci, 2014; Fletcher et al., 2014; Frieden et al., 2014). Understandably, much attention is being given to development of a cure, not least because of social and cultural resistance in some communities to healthcare workers' attempts to isolate infected persons. However, even if a cure for Ebola were found, contact isolation (a behavioral, social, and environmental intervention), though not free from formidable challenges, would (as it is now) still be the most effective first line of defense against the spread of the disease.

## 1.3.1 The McKeown Thesis: Criticism, Challenge, and Vindication

The dramatic improvements in population health described by McKeown have come to be referred to as the *epidemiologic transition*, wherein the incidence of fatal infectious (communicable) diseases decreased dramatically to be largely replaced by *noncommunicable diseases* (*common complex diseases*), such as cardiovascular disease and cancer. Mortality declined, leading to the historically recent dramatic increases in average life-expectancy at birth in high-income countries. The epidemiologic transition, further details of which are discussed in later chapters, is accepted historical fact. Although McKeown's explanation of the phenomenon is also largely accepted, particular details of his account have been challenged. One charge is that McKeown did not adequately consider differences in experience between urban and rural dwellers. At times, conditions for workers in the most intensively industrialized urban centers deteriorated to appalling levels that remained long after overall conditions had improved. At other times, changes in agricultural production displaced rural laborers, forcing them to endure conditions rivaling those of impoverished urban laborers. Another charge is that McKeown did not adequately appreciate the relative importance of the public health benefits from social reforms in Britain during the nineteenth century compared to improvements in other factors that he emphasized, notably, nutrition. Although such criticisms have merit, they do not challenge McKeown's main conclusions. Indeed, considering the quality of the data he had to work with, McKeown is almost certain to have misconstrued some aspects of detail, a likelihood he acknowledged repeatedly in his writings.

Taking account of imperfections in data and occasional possible errors of interpretation, the hyperbole surrounding some criticisms of the McKeown thesis has been less than edifying. For example, Colgrove (2002) asserted that the McKeown thesis had been "overturned" and his (McKeown's) conclusions "largely discredited." Those opinions were mostly based on McKeown's purported misunderstanding of demographic details, especially birth rate and population growth (discussed in Chapter 2), and the seemingly minor charge that he had overestimated the importance of nutrition while underestimating the benefits of clean water supply. However, even if valid, those criticisms do not alter McKeown's conclusions concerning the relative unimportance of biomedical healthcare to the major gains in life expectancy and health that occurred during the two centuries preceding the late-twentieth century. Indeed, despite his harsh assessment of McKeown, Colgrove (2002) concluded:

> The consensus among most historians about the McKeown thesis [is] that curative medical measures played little role in mortality decline (p. 728) [and a] large and growing body of research suggesting that broad social conditions must be addressed…has validated the underlying premise of McKeown's inquiries (p. 729).

If these are the conclusions of a harsh critic, then perhaps we can be confident in accepting the broad sweep of McKeown's thesis.

Even while arguing the wide generalizability of his conclusions, there are exceptions pertaining to individual diseases that McKeown was careful to acknowledge. In particular, *poliomyelitis*, which appears to have been rare before the late-nineteenth century, occurred in epidemic proportions in many countries during the early twentieth century. However, rates fell sharply with the advent of a vaccine in the 1950s. The disease has been almost eliminated from countries that have had mass vaccination while remaining common in countries that have not had mass vaccination. This apparent success with polio-myelitis was confirmed by McKinlay and McKinlay (1977) in their analysis of infectious diseases in the United States during the first three-quarters of the twentieth century, where poliomyelitis was found to be alone in showing a "noticeably changed" rate as a consequence of medical intervention. Neverthe-less, despite extremely disabling effects, the number of deaths from polio, even during periods of epidemic, was comparatively small. Therefore, while contrib-uting substantially to a reduced prevalence of disability, the eradication of polio contributed little to the overall rate of death. Thus, we are repeatedly drawn to the same conclusion. The long life expectancy characteristic of contemporary high-income countries has comparatively little to do with biomedical healthcare and a great deal to do with advances in non-medical science and technology, changes in individual behavior, and improvements in economic and social conditions.

In addition to its implications for the promotion of health, McKeown's work has also been examined for its implications for economic and political ideology. In particular, McKeown's findings have been used to support a laissez-faire stance toward health and welfare, which argues that gains in these areas are the inevitable and welcomed "by-product of economic growth" (Szreter, 1988). This is reminiscent of the idea expressed by Smith (1776) in the *Wealth of Nations* that industry, though entirely self-serving, may be

*led by an invisible hand to promote an end which was no part of [the] intention [but which nevertheless promoted] the public good (p. 386).*

However, Smith did not assert that public good was an *inevitable* consequence of industry, merely a *possible* (if frequent) outcome.

In that vein, McKeown's findings could, in theory, be taken to imply that governments should avoid actions that disrupt free-market forces because improved population health depends on economic success. Closer consider-ation, however, shows that any succor given to a laissez-faire economic inter-pretation of the McKeown thesis is undermined by the fact that much of the improvement in health that accompanied nineteenth-century economic devel-opment was due to successful social-reform and public-health initiatives which were led by government, both nationally and locally (Szreter, 2003). Moreover, many such initiatives, including new building codes and town-planning regula-tions, were expressly aimed at remedying harm caused by free-market industri-alization, such as the squalid slums that emerged in rapidly developing industrial regions. Thus, McKeown's thesis implies that health benefits accom-panying economic development are likely to be greater, and may even depend upon, government intervention, regulation, and control of private commercial enterprise.

## REFERENCES

Andersen, P., Doherty, T.M., 2005. The success and failure of BCG: implications for a novel tuberculosis vaccine. Nat. Rev. Microbiol. 3, 656–662. http://dx.doi.org/10.1038/nrmicro1211.

Aronson, N.E., Santosham, M., Comstock, G.W., et al., 2004. Long-term efficacy of BCG vaccine in American Indians and Alaska Natives: a 60-year follow-up study. J. Am. Med. Assoc. 291, 2086–2091.

Biran, A., Curtis, V., Gautam, O.P., et al., 2012. Background Paper on Measuring WASH and Food Hygiene Practices: Definition of Goals to Be Tackled Post 2015 by the Joint Monitoring Program. London School of Hygiene and Tropical Medicine, London, UK.

Colgrove, J., 2002. The McKeown thesis: a historical controversy and its enduring influence. Am. J. Public Health 92, 725–729.

Curtis, V.A., 2007. Dirt, disgust and disease: a natural history of hygiene. J. Epidemiol. Community Health 61, 660–664.

Curtis, V., Schmidt, W., Luby, S., Florez, R., et al., 2011. Hygiene: new hopes, new horizons. Lancet Infect. Dis. 11, 312–321.

Daniel, T.M., 2006. The history of tuberculosis. Respir. Med. 100, 1862–1870.

de Vries, G., Aldridge, R.W., Caylà, J.A., et al., 2014. Epidemiology of tuberculosis in big cities of the European union and European economic area countries. Euro Surveill. 19, 42–54.

Dodd, P.J., Gardiner, E., Coghlan, R., Seddon, J.A., 2014. Burden of childhood tuberculosis in 22 high-burden countries: a mathematical modeling study. Lancet Glob. Health 2, e453–e459.

Dubos, R.J., 1959. Mirage of Health: Utopias, Progress, and Biological Change. Harper and Brothers, New York.

Dubos, R.J., Dubos, J., 1952. The White Plague: Tuberculosis, Man, and Society. Little, Brown, Boston, MA.

Fauci, A.S., 2014. Ebola: underscoring the global disparities in health care resources. N. Engl. J. Med. 371, 1084–1086.

Fletcher, T.E., Brooks, T.J., Beeching, N.J., 2014. Ebola and other viral haemorrhagic fevers. Br. Med. J. 349, 1–2. http://dx.doi.org/10.1136/bmj.g5079.

Frieden, T.R., Damon, I., Bell, B.P., Kenyon, T., Nichol, S., 2014. Ebola 2014: new challenges, new global response and responsibility. N. Engl. J. Med. 371, 1177–1180.

Geels, F., 2005. Co-evolution of technology and society: the transition in water supply and personal hygiene in the Netherlands (1850–1930): a case study in multi-level perspective. Technol. Soc. 27, 363–397.

Gradmann, C., 2006. Robert Koch and the white death: from tuberculosis to tuberculin. Microbes Infect. 8, 294–301.

Health Protection Agency, 2012. Tuberculosis mortality and mortality rate, England and Wales, 1913-2008. Retrieved 29 October 2012 from http://www.hpa.org.uk/Topics/InfectiousDiseases/InfectionsAZ/Tuberculosis/TBUKSurveillanceData/TuberculosisMortality/TBMortality01trend/.

Internet Encyclopedia of Philosophy, 1996. Meditations on First Philosophy. The 1911 edition of The Philosophical Works of Descartes (Cambridge University Press), translated by Elizabeth S. Haldane. Retrieved on 29 October 2013 from http://www.sacred-texts.com/phi/desc/med.txt.

Jenkins, H.E., Tolman, A.W., Yuen, C.M., et al., 2014. Incidence of multidrug-resistant tuberculosis disease in children: systematic review and global estimates. Lancet 383, 1572–1579.

Johnson, R., 2014. Kant's moral philosophy. In: Zalta, E.N. (Ed.), The Stanford Encyclopedia of Philosophy. Retrieved 21 January 2015 from http://plato.stanford.edu/archives/sum2014/entries/kant-moral/.

Kaufmann, S.H., Hussey, G., Lambert, P.H., 2010. New vaccines for tuberculosis. Lancet 375, 2110–2119.

Khan, M.S., Coker, R.J., 2014. How to hinder tuberculosis control: five easy steps. Lancet 384, 646–648.

Khoury, M.J., Gwin, M.L., Glasgow, R.E., Kramer, B.S., 2012. A population approach to precision medicine. Am. J. Prev. Med. 42, 639–645.

Kinsella, K.G., 1992. Changes in life expectancy 1900-1990. Am. J. Clin. Nutr. 55, 1196S–1202S.

Luby, S.P., Agboatwalla, M., Feikin, D.R., et al., 2005. Effect of hand washing on child health: a randomized controlled trial. Lancet 366, 225–233.

Lucas Jr., R.E., 2004. The industrial revolution: past and future. Econ. Educ. Bull. 44, 1–8.

McKeown, T., 1979. The role of Medicine: Dream, Mirage or Nemesis? Basil Blackwell, Oxford, UK.

McKeown, T., Brown, R.G., 1955. Medical evidence related to English population changes in the eighteenth century. Popul. Stud. 9, 119–141.

McKeown, T., Brown, R.G., Record, R.G., 1972. An interpretation of the modern rise of population in Europe. Popul. Stud. 26, 345–382.

McKeown, T., Record, R.G., Turner, R.D., 1975. An interpretation of the decline of mortality in England and Wales during the twentieth century. Popul. Stud. 29, 391–422.

McKinlay, J.B., McKinlay, S.M., 1977. The questionable contribution of medical measures to the decline of mortality in the United States in the twentieth century. Milbank Mem. Fund Q. Health Soc 55, 405–428.

Mingote, L.R., Namutamba, D., Apina, F., et al., 2015. The use of bedaquiline in regimens to treat drug-resistant and drug-susceptible tuberculosis: a perspective from tuberculosis-affected communities. Lancet 385, 477–479. http://dx.doi.org/10.1016/S0140-6736(14)60523-7.

Human Mortality Database, 2012. England and wales total population. Retrieved 26 June 2012 from http://www.mortality.org/.

Murray, J.F., 2004. Mycobacterium tuberculosis and the cause of consumption: from discovery to fact. Am. J. Respir. Crit. Care Med. 169, 1086–1088.

Newton, S.M., Brent, A.J., Anderson, S., et al., 2008. Pediatric tuberculosis. Lancet Infect. Dis. 8, 498–510.

Ottenhoff, T.H., Kaufmann, S.H., 2012. Vaccines against tuberculosis: where are we and where do we need to go? PLoS Pathog. 8, 1–12. http://dx.doi.org/10.1371/journal.ppat.1002607.

Pietersen, E., Ignatius, E., Streicher, E.M., et al., 2014. Long-term outcomes of patients with extensively drug-resistant tuberculosis in south Africa: a cohort study. Lancet 383, 1230–1239.

Rowland, K., 2012. Totally drug-resistant TB emerges in India. Nature News and Comment. Retrieved 23 March 2014 from http://www.nature.com/news/totally-drug-resistant-tb-emerges-in-india-1.9797.

Shaw, G. B., 1909. The Doctor's Dilemma: preface on doctors. Retrieved from 24 October 2013 from http://www.online-literature.com/george_bernard_shaw/doctors-dilemma/0/.

Sizemore, C.F., Schleif, A.C., Bernstein, J.B., Heilman, C.A., 2012. The role of biomedical research in global tuberculosis control: gaps and challenges. Emerg. Microbes Infect. 1, 1–6. http://dx.doi.org/10.1038/emi.2012.21.

Smith, A., 1776. An inquiry into the nature and causes of the wealth of nations. In: Manis, Jim (Ed.), Electronic Classics Series. Pennsylvania State University, Hazleton, PA.

Smith, R., 2002. In search of "non-disease" Br. Med. J. 324, 883–885.

Sterne, J.A., Rodrigues, L.C., Guedes, I.N., 1998. Does the efficacy of BCG decline with time since vaccination? Int. J. Tuberc. Lung Dis. 2, 200–207.

Stop TB Partnership, 2006. The global plan to stop TB, 2006–2015. Actions for life: towards a world free of tuberculosis. Int. J. Tuberc. Lung Dis. 10, 240–241.

Szreter, S., 1988. The importance of social intervention in Britain's mortality decline c. 1850-1914: a re-interpretation of the role of public health. Soc. Hist. Med. 1, 1–37.

Szreter, S., 2003. The population health approach in historical perspective. Am. J. Public Health 93, 421–431.

Trunz, B.B., Fine, P.E.M., Dye, C., 2006. Effect of BCG vaccination on childhood tuberculous meningitis and miliary tuberculosis worldwide: a meta-analysis and assessment of cost-effectiveness. Lancet 367, 1173–1180.

Wang, L., Zhang, H., Ruan, Y., et al., 2014. Tuberculosis prevalence in China, 1990–2010: a longitudinal analysis of national survey data. Lancet 383, 2057–2064.

WHO, 1948. Preamble to the constitution of the world health organization. In: Adopted by the International Health Conference, New York, 19-22 June, 1946; signed on 22 July 1946 by the Representatives of 61 States (Official Records of the World Health Organization, no. 2, p. 100), and Entered into Force on 7 April 1948.

WHO, 1986. Ottawa charter for health promotion. World Health Organization, Geneva. Retrieved 2 January 2013 from, http://www.who.int/healthpromotion/conferences/hpr_special%20issue.pdf.

WHO, 2012. Global tuberculosis report 2012. World Health Organization, Geneva. Retrieved 19 October 2013 from, http://www.who.int/tb/publications/global_report/en/.

WHO, 2013. Global tuberculosis report 2013. World Health Organization, Geneva, Switzerland. Retrieved 21 March 2014 from, http://apps.who.int/iris/bitstream/10665/91355/1/9789241564656_eng.pdf?ua=1.

WHO, 2014. Global tuberculosis report 2014. World Health Organization, Geneva, Switzerland. Retrieved 19 December 2014 from, http://apps.who.int/iris/bitstream/10665/137094/1/9789241564809_eng.pdf?ua=1.

Wrigley, E.A., Schofield, R., 1981. The Population History of England, 1541-1871: A Reconstruction. Harvard University Press, Cambridge, MA, Table 7.15, p. 230.

Zwerling, A., Behr, M.A., Verma, A., et al., 2011. The BCG world atlas: a database of global BCG vaccination policies and practices. PLoS Med. 8, e1001012.

Chapter 2

# Current Patterns of Death and Disease

*Socially we are people of the 21st century, but genetically we remain citizens of the Paleolithic era.*

(O'Keefe and Cordain, 2004, p. 101)

## Contents

Accepting the consensus that the McKeown thesis is broadly correct and biomedicine's contribution to past improvements in health was marginal, it is relevant to ask whether what is true historically is also true for patterns of health that characterize the contemporary world. With the epidemiologic transition, the population burden of disease shifted from high incidence of early mortality due to acute infectious disease to postponed mortality (increased life expectancy) and extended morbidity due to chronic noncommunicable disease. Given the contrasting patterns of health that prevailed before and after the epidemiologic transition, we might wonder what significance in contemporary times should be attached to the fact that improvements in the health of former generations were largely unrelated to medical intervention.

In the period following World War II, there was much optimism regarding the power of biomedical healthcare to eradicate major illnesses. In part, that

The Health of Populations. http://dx.doi.org/10.1016/B978-0-12-802812-4.00002-3

optimism stemmed from the false belief that medical science and innovation were responsible for containment of infectious diseases that had previously been common. The fact that biomedicine contributed little to past gains in the health of populations gives pause for thought, but does not necessarily mean that its contribution to enhancing the health of current and future generations is similarly fated to be minor. We may grant that the radical reductions in communicable diseases that characterized the epidemiologic transition depended on transformations in ways of living unrelated to clinical medicine. It is important, however, to consider as separate questions the contribution of biomedical healthcare to the *maintenance* of the improved standards of health that followed the epidemiologic transition and the potential for further biomedical *enhancement* of health for future populations.

The *noncommunicable diseases*, which, as the name implies, are noninfectious and nontransmissible between persons, include cardiovascular diseases (primarily, coronary heart disease and stroke), cancers, chronic respiratory diseases, and diabetes. These are typified by slow progression (incubation) and long duration. It might be noted, however, that the frequent depiction of noncommunicable diseases (sometimes referred to as *common complex diseases*) as *chronic* and communicable diseases as *acute* is not strictly true in all instances. For example, in the comparatively infrequent and tragic instance of sudden cardiac death in a young apparently healthy person, the cause of death, though noncommunicable, would ordinarily be regarded an acute event in the sense of being without significant prior indicators. Conversely, HIV/AIDS, characteristically a chronic disease of potentially long duration, is due to communicable infection.

Additionally, communicable and noncommunicable diseases are not wholly distinguishable due to the fact that some of the latter are linked to infectious diseases. For example, cervical and liver cancers are linked with the human papilloma and hepatitis viruses, respectively (Schiller and Lowy, 2014). Furthermore, in some major systems of disease classification, such as used by the World Health Organization (WHO), the term *noncommunicable diseases* includes injuries, both unintentional (accident) and intentional (e.g., suicide and violence), which are indeed noncommunicable but are neither chronic nor diseases in the usual sense. Nevertheless, while being aware of exceptions both minor and major, it is useful to retain the terms *communicable*, to refer to infectious diseases that are transmitted between persons and mostly have acute outcomes, and *noncommunicable*, to refer to chronic diseases characterized by slow progression and long duration.

## 2.1 CURRENT MAIN CAUSES OF DEATH

The WHO publishes health information at regular intervals for every country and region of the world, and Table 2.1 summarizes the most recent estimates for the 10 leading causes of death (WHO, 2014). The table reports deaths separately for high-income and developing countries, and it can be seen that about four-fifths of the global population of approximately 7 billion people reside in

**TABLE 2.1** The 10 leading causes of death worldwide and separately for high-income and developing countries[a]

| Cause of death | Worldwide (population: 7,075,456,000) | | High-income countries[b] (population: 1,293,593,000) | | Developing countries[c] (population: 5,781,863,000) | |
|---|---|---|---|---|---|---|
| | Deaths (000s) | Percent | Deaths (000s) | Percent | Deaths (000s) | Percent |
| Cardiovascular diseases (incl. ischemic heart disease, stroke, etc.) | 17,519 | 31 | 4438 | 38 | 13,082 | 30 |
| Cancers (incl. lung, liver, stomach, colon, breast, etc.) | 8206 | 15 | 2894 | 25 | 5312 | 12 |
| Infectious and parasitic diseases (incl. diarrheal diseases, HIV/AIDs, childhood-cluster diseases, tuberculosis, etc.) | 6432 | 12 | 303 | 3 | 6128 | 14 |
| Injuries (incl. unintentional and intentional) | 5144 | 9 | 747 | 6 | 4398 | 10 |
| Chronic respiratory diseases (incl. chronic obstructive pulmonary disease, asthma, etc.) | 4042 | 7 | 645 | 6 | 3397 | 8 |
| Respiratory infections | 3061 | 6 | 396 | 3 | 2665 | 6 |
| Neonatal conditions | 2475 | 4 | 40 | 0 | 2435 | 6 |
| Digestive diseases (incl. | 2264 | 4 | 515 | 4 | 1749 | 4 |

*Continued*

**TABLE 2.1** The 10 leading causes of death worldwide and separately for high-income and developing countries—cont'd

| Cause of death | Worldwide (population: 7,075,456,000) | | High-income countries (population: 1,293,593,000) | | Developing countries (population: 5,781,863,000) | |
|---|---|---|---|---|---|---|
| | Deaths (000s) | Percent | Deaths (000s) | Percent | Deaths (000s) | Percent |
| cirrhosis of the liver) | | | | | | |
| Diabetes mellitus | 1497 | 3 | 254 | 2 | 1244 | 3 |
| Neurological conditions (incl. Alzheimer's disease, Parkinson's disease, epilepsy, etc.) | 1420 | 3 | 732 | 6 | 688 | 2 |
| Other causes | 3798 | 5 | 708 | 6 | 3090 | 7 |
| Total (all causes) | 55,859 | | 11,671 | | 44,187 | |

[a]*Summary estimates of mortality for the year 2012, based on analyses of the latest available national information on levels of mortality and causes as at the end of August 2013. Details may be accessed at http://www.who.int/healthinfo/global_burden_disease/en/.*
[b]*For fiscal year 2015, high-income economies are defined as those with a **gross national income** (GNI) per capita of USD12,746 or more. Details may be accessed at http://data.worldbank.org/about/country-and-lending-groups.*
[c]*Consistent with World Bank usage, developing denotes low- and middle-income countries with GNI per capita of less than USD12,746 for fiscal year 2015.*
Derived from WHO (2014)

developing countries. It is notable that cardiovascular diseases are the leading cause of death for both high-income and developing countries, which is evidence that the epidemiologic transition from communicable to noncommunicable diseases has spread to include most of humanity.

That the transition from communicable to noncommunicable diseases is continuing is also evident in Table 2.1. In particular, it can be seen from the columns showing number and percent of deaths that infectious and parasitic diseases are the second most common cause of death in developing countries, having been displaced in the recent past by cardiovascular diseases as the most common cause of death. By comparison, infectious and parasitic diseases rank as the eighth most common cause of death in high-income countries. Today, infectious and parasitic diseases continue to be the leading cause of death in only a relatively small number of low-income countries, mostly in sub-Saharan Africa. In general, they are countries where economic development and social initiatives are insufficient to guarantee the foundations for health that McKeown (1979) identified as responsible for the epidemiologic transition of Western Europe more than 150 years earlier.

The global predominance of noncommunicable diseases is illustrated in Figure 2.1, which shows that, of approximately 56 million deaths worldwide, almost 38 million (just over two-thirds) are due to noncommunicable diseases. In turn, Figure 2.2 shows that almost 80% of deaths from noncommunicable

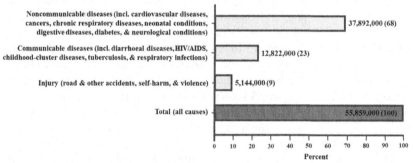

**FIGURE 2.1**   Deaths (%) worldwide from noncommunicable diseases, communicable diseases, injury, and total combined (all causes). *(Derived from WHO (2014).)*

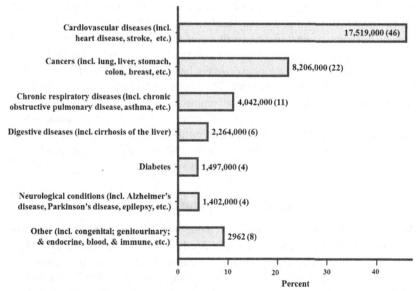

**FIGURE 2.2**   Worldwide noncommunicable disease deaths (% of total of 37,892,000). *(Derived from WHO (2014).)*

diseases are due to just three diseases, namely, cardiovascular diseases (accounting for almost half of all deaths from noncommunicable diseases), cancers (almost a quarter), and chronic respiratory diseases (at about one-tenth) (WHO, 2014).

## 2.2 HABITS AND HABITATS ANCIENT AND MODERN

If the McKeown thesis is helpful in understanding the decline in infectious disease that is characteristic of the epidemiologic transition, what explains the comparatively sudden rise in prevalence of noncommunicable fatal diseases? Much progress in understanding the rise of the currently predominant common complex diseases has been made by applying general principles of Darwinian evolution to what is known about how human habits and habitats have changed since the origin of our species. At the most rudimentary level, we can apply what might loosely be referred to as the "caveman" heuristic, which entails trying to imagine the nature of the conditions that prevailed during human evolutionary history. Natural selection ensures the fitness of a species for the *ecological niche* it occupies. A legacy of natural selection is that each generation of a given species is genetically adapted to the environment of preceding generations. Hence, when habitats change, the fitness of a species for the new environment is challenged, possibly threatening that species' viability.

During stable epochs there is evolutionary concordance, wherein selective pressure favors the *status quo*, including the maintenance of traits, biological and behavioral, that had been optimal during a species' evolutionary past. Conversely, when habitat change is long term or permanent, new selection pressures alter the average **genome** in favor of characteristics better suited to changes in the environment. In time, this may lead to the evolution of a new species possessing different characteristics to those of the ancestor; characteristics that establish the fitness of the new species to the changed environment. Obviously, present-day humans have adopted different habits and occupy different habitats to those of our Paleolithic ancestors. The caveman heuristic involves the exercise (largely intellectual due to lack of empirical evidence) of trying to identify which aspects of present-day habits and habitats are neutral or advantageous to the health of our species and which are harmful. Of the many ways in which the habits and habitats of Paleolithic and modern humans differ, the most obvious pertain to diet, physical activity, and social structure.

### 2.2.1 Nutritional Habits and Habitat

Among the differences that distinguish modern human habits and habitats from those of our ancient forebears, changed diet has received most attention. More than 70% of the energy supplied in the contemporary Western diet, summarized in Table 2.2, is believed to belong to food groups that were largely unavailable to preagricultural humans (Cordain et al., 2005). Moreover, the postagricultural advent of new food groups appears to have contributed to a substantial

**TABLE 2.2** Comparison between the contemporary Western diet and the diet of hunter-gatherers

| Food group | Value (% of energy/level) |
| --- | --- |
| **Western food types generally unavailable to preagricultural humans[a]** | |
| Dairy products | 11 |
| Cereal grains | 24 |
| Refined sugars | 19 |
| Refined vegetable oils | 18 |
| Alcohol | 1 |
| Total | 72 |
| Salt (added) | 10 g/day |
| **Hunter-gatherer diet[b]** | |
| Protein | 27 |
| Carbohydrates (no refined grains) | 31 |
| Total fat | 38 |
| Saturated fat | Moderate |
| Monounsaturated fat | High |
| Polyunsaturated fat | Moderate |
| Omega-3 fat | High |
| Total fiber | High |
| Fruits and vegetables | High |
| Nuts and seeds | Moderate |
| Salt | Low |
| Refined sugars (seasonally available honey) | Low |
| Glycemic load | Low |

[a]Adapted from Cordain et al. (2005).
[b]Adapted from O'Keefe and Cordain (2004).

narrowing of the range of foodstuffs generally present in the human diet (Trevathan, 2007). Humans are natural omnivores, and hunter-gatherer foraging is believed to have delivered a greater diversity of foodstuffs and nutrients than is typical of contemporary human diets.

With specific reference to the typical American diet, the diet of Paleolithic humans is estimated to have contained two to three times more protein and fiber, about twice the amount of polyunsaturated and monounsaturated fats, four times more Omega-3 fats, while also containing four to five times less salt and 60-70% less saturated fat (O'Keefe and Cordain, 2004). Additionally, whereas sugar-sweetened, calorie-dense, low-nutritious sodas are the predominant beverage consumed in some countries, Paleolithic humans drank water almost exclusively. It would be surprising if such radical shifts in diet could be endured without implications for health.

## 2.2.2   Physical Activity Habits and Habitat

The nexus between food acquisition (i.e., foraging) and consumption in hunter-gatherer societies meant that individual physical exertion was obligatory. However, the physical effort of hunting and gathering, now no longer required to eat, has been largely replaced by sedentary habits. Estimating ancestral physical activity depends largely on examination of excavated Paleolithic human remains and on studies of recent hunter-gatherers who are assumed to share the habits of their ancestors. Estimates suggest that average daily energy expenditure for preagricultural humans was probably two to three times that of modern humans (Eaton and Eaton, 2003). Moreover, although many people engage in sports and recreational activities that involve physical exertion, such activity may often lack the diversity of that which was required of people who needed to forage for food.

The physical activity involved in gathering, hunting, digging, carrying, shaping of tools, rendering of food (e.g., butchering and cleaning game), constructing and reconstructing temporary shelter, and other tasks required of subsistence-living hunter-gatherers is akin to what today might be called "cross-training" (Ruff, 2000). Such activity is likely to contribute more to a combination of endurance, strength, and agility than many of the physical activity regimens of modern humans, which often are based primarily on a single activity such as running, throwing, hitting, or lifting weights. Apart from the possible absence of variety in a range of activities undertaken by people who qualify as *active*, overall levels of physical activity are generally low and falling. Recent large-scale international research has provided, for the first time, reliable measurements of physical activity levels across most of the globe (Hallal et al., 2012). The results show that physical activity levels for approximately one-third of the adult population worldwide satisfy the formal definition of being physically *inactive*, and that approximately 80% of children in their mid-teens are less active than current guidelines recommend.

## 2.2.3   Social Structure

Social structure is an aspect of habitat that has changed profoundly. Although challenging to infer in detail, it is known that during the Paleolithic period people were widely dispersed in kinship groups or bands generally believed to consist of up to a few hundred, occasionally a few thousand, and rarely as many as 10,000 (Layton et al., 2012). There were no cities, since these emerged only after the transition to agriculture. The subsequent protracted process of global urbanization accelerated markedly in the twentieth century (Cohen, 2003). In 1800, 2% of people lived in cities; by 1900, it was 12%; and by 2010 more people were urban dwellers than rural dwellers. It is anticipated that, by 2050, 70% of the world's population will live in cities. In 1900, no cities had 10 million people or more, but today there are as many as 40 cities of that size and larger (United Nations, 2012). Such change has undoubtedly affected dietary patterns and the level of physical activity, but probably also myriad other aspects of life that influence health and wellbeing (e.g., access to services and entertainment, as well as stress and alienation).

However, beyond speculation, little is known about the implications for health and wellbeing of being settled in permanent structures in massive cities versus temporary dwellings in natural settings, or of extensive impersonal interaction with countless anonymous others versus the intimacy afforded by living primarily among kin. Similarly, little is known about the implications for health and wellbeing of the daily use of mechanized devices, electronics, telecommunications, and services characteristic of modern living versus the absence of those "conveniences" in hunter-gatherer societies. Although remnants of hunter-gatherer cultures exist among contemporary native peoples, probably none are untouched by modern technology and communication (Marlowe, 2005). Consequently, comparisons between the effects of Paleolithic and contemporary social structures, even more than the habits and habitats of nutrition and physical activity, are probably beyond extensive systematic study.

## 2.3   DISCORDANCE BETWEEN PALEOLITHIC AND MODERN HABITS AND HABITATS

Human habits and habitats appear to have remained relatively stable throughout the long span of human evolutionary history that predates the *Agricultural Revolution*, when our Paleolithic ancestors transitioned from hunting and gathering to settled agriculture. The adaptive pressures that existed throughout prehuman and human evolution influenced the selection of genes that influence the nature of the cardiorespiratory, musculoskeletal, and metabolic character of our species, as well as our psychological traits and patterns of social organization. Conversely, too little time has elapsed for natural selection to produce major adaptive changes in the human genome in response to the monumental behavioral and environmental transitions of the Agricultural Revolution. Thus, in the vein of the caveman heuristic mentioned above, scientists have argued that substantial threats to human population health and wellbeing may be expected (Eaton and Eaton, 2003) due to what has been described as *evolutionary collision* and pervasive *evolutionary discordance* between ancient human genomic inheritance and modern human habits and habitats (Cordain et al., 2005).

Hunter-gatherers are believed to have been generally lean, fit, and largely free from the world's current major chronic diseases. From this, it is inferred that hunting and gathering are in accord with human genetic predispositions, and that the current global epidemic of common complex diseases is largely the result of evolutionary discordance due to the changed habits and habitats that characterize modern humans. As a possible illustration, the highest recorded levels for prevalence of obesity and diabetes have been observed in former hunter-gatherer populations of Australian Aborigines (Daniel et al., 1999) and Alaskan Eskimos (Ebbesson et al., 1998), whose traditional ways of living were supplanted within relatively short periods of time by modern habits and habitats.

The apparent craving among modern humans for calorie-dense foods, such as fats, sweets, and starches is believed by some to be a legacy of our Paleolithic ancestry. Those foods are thought to have been preferred, and were consumed whenever possible, because of the survival advantage they may have conferred in habitats characterized by intermittent food scarcity. To explain the process, American geneticist James Neel (1962) proposed the *thrifty gene hypothesis*, wherein thrifty genes equip those who possess them to be particularly efficient at metabolizing and storing food. It is speculated that the habit of overeating calorie-dense foods when such foods were readily available created caloric excess, which, in individuals possessing thrifty genes, was stored as intra-abdominal fat. This capacity to fatten during times of plenty may have been advantageous for hunter-gatherers, especially child-bearing women, because the energy stored in body fat increased the chances of survival when food was scarce. However, when the habit of caloric excess persists unabated in habitats free of food scarcity, the formerly adaptive "thrifty" genotype becomes a liability, offering a plausible hypothesis for much of the current burden of obesity-related disease, including cardiovascular disease, hypertension, and diabetes.

Notwithstanding intuitive appeal and wide acceptance, the alleged genetic foundation of diet-related evolutionary discordance has been questioned (Turner and Thompson, 2013). To begin with, more than half-a-century since it was hypothesized, the "thrifty gene" has still to be identified. Moreover, the putative human predilection for energy-rich, sweet, and fatty foods attributed to genes that have been selected for their adaptive advantage in the Paleolithic habitat of intermittent food scarcity suggests that over-consumption of such foods is inevitable in the modern habitat of perpetual abundance. That presumed inevitability, however, ignores human behavioral flexibility and potential to adapt to changes in habitats. For example, there is good evidence that early taste *experiences* can have an enduring influence on later taste *preferences*. In particular, early exposure to processed "junk foods" rich in energy, fat, sugar, and salt, but deprived of essential nutrients, has been found not merely to be important but to be more important than genetic disposition in the development of later food preferences and eating habits that compete with healthier food choices (e.g., Eertmans et al., 2001; Greene et al., 1975; Rozin and Millman, 1987).

In a study of laboratory rats, comparisons were made of the offspring of mothers fed either a regular rat diet alone or regular diet plus junk food (Bayol et al., 2007). Compared to mature male and female rats whose mothers had a regular diet, those whose mothers had consumed junk food showed a preference for fatty, sugary, and salty foods at the expense of protein-rich foods. In addition, offspring exposed to junk food during gestation and suckling had increased body weight and body mass index, suggestive of susceptibility to obesity. These propensities were not due to any *genetic* difference among the animals, but solely to differences in prenatal and infant dietary *experience*.

Less controlled observations of human populations suggest similar processes. Whereas it has long been accepted that diet in adulthood can influence susceptibility to noncommunicable diseases, it appears that early nutrition also has a key role in determining susceptibility to such diseases in later life (Barker, 2012; Gluckman et al., 2005; Lillycrop and Burdge, 2012). In longitudinal studies of birth size involving 25,000 men and women in the United Kingdom, impaired fetal growth and development not attributable to prematurity were found to be predictive of higher rates of coronary heart disease, high blood pressure, high cholesterol concentrations, and compromised glucose-insulin metabolism in adulthood (Godfrey and Barker, 2000). A key conclusion of the study was that the quality of fetal nutrition, a function of mothers' food choices during pregnancy, was a major factor contributing to disease susceptibility of offspring in adulthood.

Some behavioral patterns appear fixed and largely determined by genetic heritage, whereas others are not the direct manifestations of genes. The consumption of food is, in the most general sense, a genetically-determined habit involving the ingestion of nutrients required to support life-sustaining metabolic processes. However, for humans in particular, the specific foods ingested; the timing, length, and frequency of feeding; and the artifacts and paraphernalia of eating are largely the products of learning. They are cultural habits that are transmitted socially. Social and cultural influences include the plethora of low-cost palatable yet unhealthy foods promulgated by multinational food industries whose marketing strategies have the capacity to instill poor food choices and harmful eating habits in successive generations of consumers.

Engagement in frequent physical activity is no longer required for purposes of hunting and foraging, thereby contributing to the evolutionary discordance of modern life characterized by sedentary activities, including nonmanual occupations and passive entertainments such as those that involve "screen use" (e.g., television and computer gaming). Although many cultural traditions encourage early participation in physically active games and the incorporation of physical activity as part of formal educational practice, encouragement of life-long participation in diverse and effortful physical activity is a major global challenge that must be addressed if diseases of inactivity are to be avoided.

It would be a mistake, however, to suggest that reinstatement of Paleolithic habits and habitats is the sole means available for countering the many potentially harmful influences of contemporary life. Suggestions of that kind are not a realistic response to current challenges, and they hint at romanticized desire for a return to an imagined earlier golden age of human history and the biocentric view that gives priority to all that is "natural." World population density alone suggests that any large-scale reinstatement of ancient habitats is unlikely, if only because many such habitats have been lost, probably forever, due to eons of human intervention. It is more realistic and more consistent with current human economic and technological capabilities to counter harmful influences by employing social and cultural strategies that encourage habits and habitats consistent with those of Paleolithic ancestry without demanding wholesale reinstatement of ancient ways of living. Achievable congruence between past and present ways of living includes a diet rich in plant material and lean protein, and physical activity levels substantially above those characteristic of a large and increasing proportion of the global population.

Similarly, it is a mistake to interpret genetic processes as bequeathing an immutable heritage, and therefore it is mistaken to assume that evolutionary discordance is inevitable. An antiquated genome does not preclude innovation and change in habits and habitats. Among all animals, humans possess an unsurpassed capacity to shape the environment in ways congruent with our genetic heritage. Intervention with one generation to encourage healthier habits (e.g., sound food choices) has the potential to instill healthier proclivities (e.g., taste preferences) in the next generation that may leave them less susceptible to disease-creating influences (e.g., marketing by junk-food peddlers). That is, contrary to exclusively genetic explanations of human inheritance, the *habits* of one generation can affect long-term food choices, body mass, and healthy longevity versus long-term morbidity and premature death in the next generation. Given those possibilities, it may be the *behavior* of one generation rather than *genes* that is the critical factor in determining the health of successive generations.

### 2.3.1 Niche Construction Theory

The social and cultural transmission of attributes from one generation to the next looks like *evolution*, which it may reasonably be considered to be, but it is not synonymous with Darwinian evolution. Much current theorizing about evolution concerns the role of processes responsible for social and cultural evolution and such processes can often be distinguished from those that define standard evolution theory. One such proposed process is *niche construction*, whereby the habits of species are seen to bring about changes in habitat and consequential changes in selection pressures that influence the future evolutionary course of that species.

The term *natural selection* was coined by Darwin (1859) to differentiate the processes he wished to delineate from *artificial selection* (selective breeding) in which humans intervene to manipulate the breeding process of animals with the intention of selecting human-preferred traits in those animals. Natural selection, which is nonteleological (i.e., not governed by an ultimate purpose), describes the process whereby natural variations in phenotype (observable traits) are likely to lead to some traits being better adapted adventitiously to particular features of the environment (ecological niche). Those adaptive traits confer a reproductive advantage for the individuals possessing that phenotype. Greater reproductive success is likely to lead, in turn, to a gradual increase in the population frequency of the genotype (genetic complement) associated with the successful phenotype. Over time, numerous successive adaptations may lead to the emergence of new species.

Niche construction, on the other hand, occurs when the organism's own behavior results in persistent change to the environment. Such *ecosystem engineering* bequeaths an *ecological inheritance* to subsequent generations, whereby the altered selection pressures of the inherited ecology may be capable of producing evolutionary change (see Box 2.1). In this way, reciprocal feedback between the habits of organisms and the habitats they occupy is seen as an intrinsic feature of niche construction. Both standard evolution theory and niche construction theory identify environmental selection pressure as the key mechanism of adaption, but the difference between the two is in relation to the role posited for the environment. In standard evolution theory, the environment may be seen as exogenous in that it is a preexisting state to which the organism adapts and evolution ensues as a consequence of selection pressures imposed by the environment. By contrast, in niche construction theory, the environment may be seen as endogenous in that its modification by the organism is held to be an intrinsic part of the evolutionary chain.

The idea that the habits of animals can influence habitat and thereby affect the selection pressures that underpin evolution may seem fairly self-evident. However, the possibility appears to have been less than obvious to some, including the American biologist George C. Williams, a major proponent of the *gene-centered view* of evolution popularized by British biologist Richard Dawkins (1976, 1982). Williams (1992) wrote: "Adaptation is always asymmetrical; organisms adapt to their environments, never vice versa" (p. 484). However, the idea that the relationship between genotypes and phenotypes is unidirectional (i.e., that genotype alone determines phenotype and not vice versa) appears certain to be wrong. The modern study of genetics, including epigenetics (discussed in later chapters), suggests that genotypes and phenotypes coevolve through processes of reciprocal influence.

It is obvious that humans are responsible for immense changes to the environment of the planet. Much of that change, if viewed as niche construction, has often amounted to environmental degradation, whereby deforestation, industrial development, and urban expansion have destroyed habitats for many species, including, at times, our own. Whether the totality of human niche construction proves ultimately to be to the benefit or detriment of our species (not to mention the planet) has yet to be gaged. Notwithstanding, human ecological inheritance is linked to human social and cultural proclivities, leading biologists and social scientists alike to suggest niche construction as a mechanism by which social and

**BOX 2.1 Evolutionary Niche Construction: Dam-building Beavers**

*(http://www.flickr.com/photos/31563480@N00/2452702213.)*

Beaver dams consist of vertical poles, crisscrossed with horizontal branches, and a combination of weeds and mud painstakingly positioned to fill the gaps. The resulting artificial lake provides protection from predators, serves as a food store, and serves as a safe site for the nest (or "lodge"). Dams have a major effect on the surrounding ecosystem, creating wetlands that provide habitat for a diverse wildlife.

It is reasonable to assume that the habit of dam building did not appear fully intact. Rather, it must have evolved over many generations during which feebler and more rudimentary practices by beaver ancestors entrapped small pools of water that benefited the survival of those ancestors. In time, the selection pressures arising from advantages created by those ancestral practices evolved into the complex and highly integrated behavior of dam building that is characteristic of modern beaver populations.

Thus, in addition to evolution due to natural selection of traits suited for survival in a given habitat, long-term habitat change caused by the behavior of a given species can similarly lead to selection of traits suited to the new habitat. In a loose sense, by changing its environment, "the beaver" can be seen to have affected the course of its own evolution, not to mention the evolution of other species that share its habitat. These include those that thrive on the resources provided by the lake, but also would-be predators in need of alternative foraging opportunities due to the protection afforded the beaver by its dam-building habits.

cultural habits (and not biology solely) have the potential to influence human evolution (e.g., Laland et al., 2014; Odling-Smee et al., 2013).

A frequently cited example of cultural adaptation is the evolution of adult lactose absorption in populations that adopted dairy farming following the Agricultural Revolution. Culturally determined farming of cattle for milk is believed to have created selection pressures in pastoralist populations favoring the gene variant (*allele*) that enables lactose consumed in milk to be absorbed from the intestine (McCracken, 1971; Simoons, 1970). Examples such as this have led some theorists to conclude that gene-culture coevolution is a general feature of human evolution, and possibly even the dominant form (Laland et al., 2014). There is even the suggestion that human niche construction could be "used" to speed up or impede human biological evolution by way of niche engineering for that purpose. Thus, although traditional evolutionary theory eschews a teleological dimension, it is difficult for niche construction theory to do likewise. Just as selective breeding practice with farm animals and pets demonstrates the fact of purposeful manipulation of evolution, niche construction theory draws attention to such possibilities arising from human culture. Nevertheless, rather than a radical departure from traditional evolution theory, as some proponents argue, the main contribution of niche construction theory may be in suggesting new emphases. Competition for limited resources within and between species underpins natural selection. Accordingly, the idea that the evolutionary paths of species can be affected by the ecological impact of competitors is inherent in natural selection. Thus, it seems a minimal rather than radical reconceptualization to emphasize that the ecological inheritance of organisms can affect the evolutionary path of their own (niche construction theory) as well as the evolutionary paths of competitor species (traditional evolution theory).

Admittedly, in the context of human evolution, niche construction acquires particular significance because of the emergence of human culture as an adaption to habitat challenges. For example, farming is a cultural adaption to deal with the problem of food supply. As outlined above, the adoption of cattle farming presaged genetic evolution in the form of lactose absorption in human populations. Humans have an unparalleled capacity for cultural adaption. So much so, that it has been suggested that cultural adaptation has the potential to replace genetic adaptation (Laland and Brown, 2006). Despite the genome of our hunter-gatherer ancestors being little changed, cultural adaption, including engineering and technology, has allowed humans to thrive in every kind of habitat from the equator to the arctic. Thus, compared with the more biocentric perspective of standard evolution theory, niche construction theory draws attention to alternative routes of evolutionary adaptation involving interaction between species habits and habitats with particular emphasis on the reputed adaptive potential of cultural inheritance.

### 2.3.2  Learning and Ecosystem Engineering

*Inside an Eskimo's anorak there is a tropical climate and inside an igloo there is the same (anonymous).*

Human social and cultural habits are founded on the human capacity for learning. Processes of learning common to humans and nonhumans have been the subject of intense study by psychologists for more than a century. Recognizing that learning is an evolved capability (Catania, 1978; Skinner, 1984), psychologists have long conceptualized learning as encompassing processes by which organisms adapt to their environment. Theories of learning describe forms of habit-habitat adaptation that are *isomorphic* to the theory of evolution by natural selection. Whereas evolution is a process of species adaptation spanning many generations, learning is a process of behavioral adaptation spanning the lifetime of the individual. Just as evolution involves stepwise change, so also does learning. Broadly, as individuals interact with their physical and social environment, behaviors that produce outcomes that are adaptive tend to be acquired (learned) while those that are not adaptive tend to be excluded (extinguished) from the individual's cumulative behavioral repertoire.

The linkage between behavior, environment, and outcomes/consequences is referred to as a learning *contingency*, and due to having a unique history of *reinforcing* contingencies each individual comes to possess a unique repertoire of learned behavior. American psychologist, Burrhus Frederic Skinner, the person most strongly identified with current understanding of human and nonhuman learning (especially, *operant conditioning*), defined culture as:

> *the contingencies of social reinforcement maintained by a group.*
>
> (Skinner, 1984, p. 221)

The quotation at the beginning of this section is a reference to Inuit ingenuity, wherein learned adaptation to an environment of extreme cold has been transferred from one generation to the next through culture. Evolution and learning are *homologues* comprising complementary and analogous mechanisms. Whereas the *genome* is the product of intergenerational transmission of biology, *culture* is the product of intergenerational transmission of learning.

Whether, and to what extent, niche construction can influence the course of human evolution is moot. Irrespective of such influence, the human capacity to engineer the ecosystem is a reminder that, in healthcare, it is important to take account of changed habits and habitats, both ancient and recent. Ignoring evolutionary discordance due to changes in human habits and habitats over the millennia is tantamount to ignoring likely sources of harm to human health. Conversely, interventions that engineer ecosystems in ways that succeed in minimizing discordance are likely to be neutral or beneficial to health. Assuming that wholesale reinstatement of ancient habits and habitats is impossible, the extent to which discordance can be minimized will depend on human ingenuity. As forms of ecosystem engineering, healthcare practices that are sensitive to threats to health from evolutionary discordance are likely to be most successful in optimizing personal and population health.

## 2.4   WHAT IS HEALTHCARE FOR? THE PREEMINENCE OF BEHAVIORAL OUTCOMES

Past President of the American Psychological Association Division of Health Psychology, Robert M. Kaplan, has proposed that healthcare has two main purposes: (1) to prolong life and (2) to improve the quality of life prior to inevitable death (Kaplan, 1990). When, as seems reasonable, prolonging life and improving the quality of life are identified as the core essentials of healthcare, certain logical inferences follow that may be unfamiliar to most people and possibly surprising to many. Kaplan's argument holds that the outcomes that matter most in health and healthcare are *behavioral*, and that biological preoccupations are a distraction that can sometimes be harmful. According to Kaplan, because behavior is the central outcome of healthcare, it is behavior not biology that should be the prime concern of healthcare policy, planning, and delivery.

All disease has environmental, behavioral, and/or biological *causes* of varying relative contribution. To illustrate, we may think of an individual contracting influenza. First, the influenza virus must be present in the *environment*; typically, this would mean that there are infected persons in the population. Secondly, some action (i.e., *behavior*) of the individual who succumbs causes them to come into contact with the virus. For example, the victim takes a public bus and inhales the virus which is present in the air due to the presence of other passengers who have also exercised choice (a behavior) to ride on the bus, and whose coughing and sneezing has dispersed the virus which is then inhaled by fellow passengers. The *biology* of infection is such that, within 1 or 2 days, newly infected persons may show symptoms, including fever, headaches, and fatigue. Thus, the individual instances of influenza infection have been *caused* by a combination of environmental, behavioral, and biological factors, but what is the critical *outcome* of the infection?

The physical symptoms of infection may or may not respond to medical treatment. With or without effective medical intervention, what is it about the affliction that is of principal concern to those infected? Typically, their principal concerns relate to the *behavioral outcomes* of being infected, including personal discomfort (reduced quality of life), loss of productivity (e.g., absenteeism from work and school), and impaired social function (e.g., inability to fulfill social obligations or disinclination to participate in group entertainments). In all instances of compromised health, whether the condition suffered is treated medically or left to run its natural course, the desired eventual outcomes are the same: restoration of behavioral wellbeing evidenced by participation in everyday life.

By way of another example we can consider HIV infection. Undiagnosed presymptomatic HIV infection has no immediate outcomes, either behavioral or medical. However, with the onset of infection, a biological process is initiated that is likely to have major consequences, both behavioral and medical. Nevertheless, it is not biological outcomes per se that are paramount, but the biologically mediated destruction of behavioral and social functions that are the tragedy of HIV/AIDS. Compared to no intervention, medical intervention is capable of improving quality of life and prolonging longevity, but again, it is the behavioral outcomes of quality of life and longevity that are the true measures of success. Moreover, it might be noted that the most effective strategy against HIV/AIDS is not medical but behavioral (e.g., prevention of infection through adoption of safe-sex practices). Thus, it is *behavioral* outcomes pertaining to quality of life and longevity, not *biological* mediators, that are (or should be) of most concern and the main consideration in healthcare.

## 2.4.1   Mortality

Many would agree that postponement of premature death is an obvious priority of healthcare, and many possibly also regard death as an essentially biological state. Death, however, is not solely a biological state, and it is not obvious why biological features seem often to be the dominant consideration. Rather, in common with all states of health, death is quintessentially a behavioral outcome. The fact that death is traditionally defined as the cessation of *biological* functions reflects arbitrary convention. Indeed, the biological dividing line between life and death has to date defied precise definition. Although historical attempts to define death, especially for legal purposes, have focussed on biological functions, the particular functions that comprise definitions have changed with the passage of time.

Death was once defined as cessation of the heartbeat. However, the advent of cardiopulmonary resuscitation and defibrillation rendered that definition obsolete. Today, *brain death* is often used as the defining condition, although there is difference of opinion as to what constitutes the critical features. Moreover, patients maintained on mechanical life support for extended periods, and who satisfy some definitions of brain death, have been observed to exhibit an array of biological functions in common with the living, including circulation, respiration, temperature control, waste excretion, wound healing, infection fighting, and fetal gestation in instances of pregnant brain-dead women (Miller, 2009). Such instances confound biological definitions of death, revealing intrinsic limitations that can be overcome only by considering behavior.

Behavioral outcomes are not merely necessary, they may even be sufficient, for defining death. Dead persons do not exhibit functional behavior, a fact that agrees with common sense. When, at a personal level, we recall persons who have died, it is their prior functional (i.e., behavioral) existence, rather than their biology, that dominates our memories of them. Granted, in memory, we may recall the stature and physical appearance of the deceased person, but so doing does not usually extend to recollections of aspects of their physiology, such as heart, liver, or kidney functions. Rather, our memories are comprised of what the person *did* in life: their mannerisms; the things they said; their achievements; their preferences and dislikes; their pastimes; and their social roles in life as parents, siblings, friends, and members of a community. It is the absence of all these (and many more) behavioral features that confirms the death of the person we knew.

Any satisfactory definition of death must include reference to *doing* (i.e., behavior). Death is *the irreversible cessation of meaningful behavioral function*. Although reference to biological function may help to illuminate correlates and mediators of extinguished behavior, definitions of death based on biological function alone are necessarily incomplete. For example, the presence of extensive brain damage may help to confirm irreversibility of lost behavioral function. Similarly, what constitutes *meaningful* behavior may include biological considerations, as well as a myriad of cultural and legal considerations. Ultimately, however, a satisfactory definition of death must contain reference to meaningful behavior having ceased.

## 2.4.2 Morbidity

It is customary to define morbidity (nonfatal outcomes) biologically, but to so do makes even less sense than attempts to define death that way. Threats to health acquire significance in exact proportion to the extent to which they affect behavior in the form of shortened life, or impaired present or future quality of life. Biology is important insofar as pathophysiology is correlated with longevity and present or future quality of life. To illustrate, although the biological state of raised blood pressure can be experienced asymptomatically, wherein the afflicted person is unaware of its presence and suffers no apparent decrease in function, we should not conclude that asymptomatic hypertension is unimportant. Persistent high blood pressure is associated with *future* pathological change in the cardiovascular system (e.g., atherosclerosis), which in turn is associated with reduced quality of life due to impaired function and decreased longevity. What matters ultimately is not biological states *per se* (e.g., hypertension or atherosclerosis), but the behavioral outcomes, such as inability to perform daily activities, that may be associated with biological states. Again, the relevance of biological function to health and wellbeing is not as an end in itself but as a mediator of behavioral function.

The relative importance of diseases of every description, ranging from the common cold to terminal cancer, is best assessed not on a scale of biological severity but on a scale of behavioral severity. Indeed, biological severity is essentially without meaning if account is not also taken of behavioral outcomes. The differentiation between the two is not merely semantic. There are practical advantages in keeping the focus in healthcare on behavioral rather than on biological outcomes. For example, prostate cancer is a leading cancer diagnosis in men, and debate has long persisted in relation to how the condition should be managed. A biological focus has encouraged intervention to treat the existing pathophysiology, including surgical removal of the diseased prostate. However, when combined with advanced age, functional decline, and coexisting medical conditions (e.g., diabetes and cardiovascular disease), aggressive medical intervention for prostate cancer can result in decreased quality of life and increased mortality (Hoffman, 2012).

Prostate cancer not infrequently progresses at a slow enough rate as not to cause significant discomfort in the lifetime of the patient who ultimately dies of other causes. Focussing on the behavioral outcomes of prospective quality of life and longevity encourages a wider range of individually tailored options for patients. Clinical management that prioritizes active surveillance rather than "cure" increases the likelihood of the condition being neither under- nor over-treated, thereby conferring maximum benefit for the behavioral outcomes of longevity and quality of life (Raldow et al., 2012).

Improving biological function in the absence of discernible benefit for present or future behavioral outcomes is pointless and potentially harmful, whereas improving behavioral outcomes is always worthwhile and beneficial. Accordingly, whereas it is reasonable for healthcare to seek to improve both biological and behavioral outcomes, the ultimate goal of healthcare, and the ultimate test of healthcare effectiveness, should always be improved behavioral outcomes because efforts to improve biological outcomes make sense *only* to the extent that they improve behavioral outcomes. As a final illustration, angioplasty may be performed to restore blood supply to the heart (biological outcome), which in turn improves the patient's functional status and engagement in daily activities (behavioral outcomes). However, without improvement in behavioral function the biological "improvements" would be meaningless. In other words, biological outcomes are adjunctive to behavioral outcomes, and biomedical healthcare is (or should be seen to be) adjunctive to efforts to improve behavioral outcomes.

## REFERENCES

Barker, D.J.P., 2012. Developmental origins of chronic disease. Public Health 126, 185–189.

Bayol, S.A., Farrington, S.J., Stickland, N.C., 2007. A maternal "junk food" diet in pregnancy and lactation promotes an exacerbated taste for "junk food" and a greater propensity for obesity in rat offspring. Br. J. Nutr. 98, 843–851.

Catania, A.C., 1978. The psychology of learning: some lessons from the Darwinian revolution. Ann. N. Y. Acad. Sci. 309, 18–28.

Cohen, J.E., 2003. Human population: the next half century. Science 302, 1172–1175.

Cordain, L., Eaton, S.B., Sebastian, A., et al., 2005. Origins and evolution of the western diet: health implications for the 21st century. Am. J. Clin. Nutr. 81, 341–354.

Daniel, M., Rowley, K.G., McDermott, R., et al., 1999. Diabetes incidence in an Australian aboriginal population: an 8-year follow-up study. Diabetes Care 22, 1993–1998.

Darwin, C., 1859. On the Origin of Species. John Murray, London.

Dawkins, R., 1976. The Selfish Gene. Oxford University Press, Oxford.

Dawkins, R., 1982. The Extended Phenotype. Oxford University Press, Oxford.

Eaton, S.B., Eaton, S.B., 2003. An evolutionary perspective on human physical activity: implications for health. Comp. Biochem. Physiol. A 136, 153–159.

Ebbesson, S.O., Schraer, C.D., Risica, P.M., et al., 1998. Diabetes and impaired glucose tolerance in three Alaskan Eskimo populations: the Alaska-Siberia project. Diabetes Care 21, 563–569.

Eertmans, A., Baeyens, F., Van den Bergh, O., 2001. Food likes and their relative importance in human eating behavior: review and preliminary suggestions for health promotion. Health Educ. Res. Theory Pract. 16, 443–456.

Gluckman, P.D., Cutfield, W., Hofman, P., et al., 2005. The fetal, neonatal, and infant environments—the long-term consequences for disease risk. Early Hum. Dev. 81, 51–59.

Godfrey, K.M., Barker, D.J., 2000. Fetal nutrition and adult disease. Am. J. Clin. Nutr. 71 (Suppl. 5), S1344–S1352.

Greene, L.S., Desor, J.A., Maller, O., 1975. Heredity and experience: their relative importance in the development of taste preference in man. J. Comp. Physiol. Psychol. 89, 279–284.

Hallal, P.C., Andersen, L.B., Bull, F.C., et al., 2012. Global physical activity levels: surveillance progress, pitfalls, and prospects. Lancet 380, 247–257.

Hoffman, K.E., 2012. Management of older men with clinically localized prostate cancer: the significance of advanced age and comorbidity. Semin. Radiat. Oncol. 22, 284–294.

Kaplan, R.M., 1990. Behavior as the central outcome in health care. Am. Psychol. 45, 1211–1220.

Laland, K.N., Brown, G.R., 2006. Niche construction, human behavior, and the adaptive lag hypothesis. Evol. Anthr. 15, 95–104.

Laland, K.N., Boogert, N., Evans, C., 2014. Niche construction, innovation and complexity. Environ. Innov. Soc. Trans. 8, 11, 71–86. http://dx.doi.org/10.1016/j.eist.2013.08.003.

Layton, R., O'Hara, S., Bilsborough, A., 2012. Antiquity and social functions of multilevel social organization among human hunter-gatherers. Int. J. Primatol. 33, 1215–1245.

Lillycrop, K.A., Burdge, G.C., 2012. Epigenetic mechanisms linking early nutrition to long term health. Best Pract. Res. Clin. Endocrinol. Metab. 26, 667–676.

Marlowe, F.W., 2005. Hunter-gatherers and human evolution. Evol. Anthropol. 14, 54–67.

McCracken, R.D., 1971. Lactase deficiency: an example of dietary evolution. Curr. Anthropol. 12, 479–517.

McKeown, T., 1979. The Role of Medicine: Dream, Mirage or Nemesis? Basil Blackwell, Oxford, UK.

Miller, F.G., 2009. Death and organ donation: back to the future. J. Med. Ethics 35, 616–620.

Neel, J.V., 1962. Diabetes mellitus: a "thrifty" genotype rendered detrimental by "progress"? Am. J. Hum. Genet. 14, 353–362.

O'Keefe, J.H., Cordain, L., 2004. Cardiovascular disease resulting from a diet and lifestyle at odds with our Paleolithic genome: how to become a 21st-century hunter-gatherer. Mayo Clin. Proc. 79, 101–108.

Odling-Smee, J., Erwin, D.H., Palkovacs, E.P., et al., 2013. Niche construction theory: a practical guide for ecologists. Q. Rev. Biol. 88, 3–28.

Raldow, A.C., Presley, C.J., Yu, J.B., et al., 2012. The relationship between clinical benefit and receipt of curative therapy for prostate cancer. Arch. Intern. Med. 172, 362–363.

Rozin, P., Millman, L., 1987. Family environment, not heredity, accounts for family resemblances in food preferences and attitudes: a twin study. Appetite 8, 125–134.

Ruff, C.B., 2000. Body mass prediction from skeletal frame size in elite athletes. Am. J. Phys. Anthropol. 113, 507–517.

Schiller, J.T., Lowy, D.R., 2014. Virus infection and human cancer: an overview. Recent Results Cancer Res. 193, 1–10.

Simoons, F.J., 1970. Primary adult lactose intolerance and the milking habit: a problem in biologic and cultural interrelations. Am. J. Dig. Dis. 15, 695–710.

Skinner, B.F., 1984. The evolution of behavior. J. Exp. Anal. Behav. 41, 217–221.

Trevathan, W.R., 2007. Evolutionary medicine. Annu. Rev. Anthropol. 36, 139–154.

Turner, B.L., Thompson, A.L., 2013. Beyond the Paleolithic prescription: incorporating diversity and flexibility in the study of human diet evolution. Nutr. Rev. 71, 501–510.

United Nations, 2012. World Urbanization Prospects: The 2011 Revision. United Nations, Department of Economic and Social Affairs, Population Division, New York.

WHO, 2014. Global Health Estimates 2014 Summary Tables: Deaths by Cause, Age and Sex, by World Bank Region, 2000-2012. World Health Organization, Geneva, Switzerland. Retrieved 29 August 2014 from, http://www.who.int/healthinfo/global_burden_disease/en/.

Williams, G.C., 1992. Gaia, nature worship and biocentric fallacies. Q. Rev. Biol. 67, 479–486.

Chapter 3

# Twelve Millennia of Changing Human Habits and Habitats

## Contents

The historical period of approximately two centuries that concerned McKeown and others is of vital importance to current understanding of the reasons for human epidemiologic transition. Although a brief epoch in the overall span of human history, the period (and geography) coincident with the Industrial Revolution was the foundation for momentous improvements in personal and population health. One indicator of those improvements is that an unusually large proportion of contemporary humanity has extended life compared to people living during earlier epochs. The transition towards extended life expectancy is of such magnitude as to cause some to speak of a current *longevity revolution*. Whereas it now seems natural that a large number of people will live to old age, the increase in life expectancy on which that anticipation is founded is of recent origin (Kaneko et al., 2011). Almost all of the increase in average human longevity during the whole of recorded history is believed to have occurred within the past approximately 200 years.

*The Health of Populations.* http://dx.doi.org/10.1016/B978-0-12-802812-4.00003-5

Anatomically modern humans (*Homo sapiens*) are estimated to have evolved about 200,000 years ago during the *Paleolithic Period* (*Old Stone Age*) from an earlier species of the *Homo* genus. The habits of early humans included foraging in small groups, and subsistence depended on gathering plants and hunting or scavenging wild animals in forest and savannah habitats. By the beginning of the *Upper Paleolithic* (*Late Stone Age*), about 50,000 years ago, specialized stone implements, language, and art had developed, but otherwise the habits and habitats of humans had not changed greatly, with people still subsisting in small hunter-gatherer groups.

It was not until approximately 12,000 years ago, at the beginning of the *Neolithic Period*, that the first major shift in human habits and habitats occurred in the form of the *Agricultural Revolution* (also called the *Neolithic Revolution* and the *Neolithic Demographic Transition*), which transformed the structure of human societies (Harper and Armelagos, 2010). However, as we have seen, it was not until the Industrial Revolution of the mid-eighteenth century that something particularly extraordinary began to happen in relation to human population longevity; it began to extend appreciably. By way of analogy, if the 200,000-year history of modern humans is condensed into 24 h, little may be said to have happened in terms of human longevity until a little after 2 min to midnight.

Consideration of what has been happening to the health of populations in those last "couple of minutes" of human existence raises questions that concern epidemiology and demography. Both disciplines involve the study of populations, but with differing emphases. Epidemiology is the study of the *population distribution of diseases and their causes*, whereas demography is the *study of population dynamics*, including size, growth, and density. While overlapping, the two fields of study draw attention to different aspects of the broad currents of change responsible for recent improvements in population health and the accompanying longevity revolution. To explain these changes, proponents of both disciplines have offered different *models of transition*, which are examined in the sections that follow.

It should be noted, however, that the respective models of epidemiologic and **demographic transition** are hypothetical accounts. Moreover, due largely to their speculative rather than factual nature, models of transition are themselves subject to transition due to ongoing theorizing, new information, and new speculation. Consequently, in the sections that follow, the reader may find it hard to escape the feeling that we are sinking ever deeper into a morass of endless speculation. Nevertheless, models of transition are useful for illuminating, however dimly, historical events responsible for major improvements in population health. Therefore, despite being imperfect and frustratingly transient, some discussion of the main models of epidemiologic and demographic transition is important for understanding the present and likely future health of populations.

## 3.1  EPIDEMIOLOGIC TRANSITION

Referring to population patterns of mortality and disease throughout human history, the Egyptian-born demographer and health scientist, Omran (1971), noted that, in present-day high-income countries, a transition had occurred wherein "degenerative and man-made" diseases came to displace pandemics of infection that had been the main causes of mortality and morbidity prior to "modernization." Omran characterized this process of epidemiologic transition as comprising three main "ages" (summarized in Box 3.1): pestilence and famine, receding pandemics, and degenerative and man-made diseases. Whereas McKeown's (1979) analyses, as discussed in Chapter 1, were primarily aimed at identifying factors responsible for the transformative improvements in health specific to postindustrialized England and Wales, Omran's (1971) aim was to identify the main features of a generalizable model that would be applicable universally. The inquiries of both led them to the same broad conclusion that economic development and social innovation are the catalysts for epidemiologic transition.

In Omran's account, the *Age of Pestilence and Famine*, the first of three ages covering the periods before, during, and after transition, is characterized by a high incidence of death from infectious and parasitic diseases and chronic malnutrition, especially among children. Due to a relatively small proportion of the population surviving the high mortality of youth, life expectancy at birth is low. Even with high fertility, high rate of mortality means that the population grows little if at all. Whereas biological features such as length of gestation impose natural limits on frequency of births, the potential rate of mortality in a given population is essentially unconstrained so long as there are any people remaining. Consequently, even as the fertility of a population approaches its biological

---

**BOX 3.1  Overview of the epidemiologic transition**

*Age of Pestilence and Famine.* Mortality is high and fluctuating, precluding sustained population growth. During this age, which is believed to have prevailed throughout the period of almost 12,000 years between the Agricultural and Industrial Revolutions, average life expectancy at birth is low.

*Age of Receding Pandemics.* "Modernization" triggers a decrease in mortality, which is initially gradual but accelerates as epidemic peaks become less frequent or disappear. Average life expectancy at birth increases steadily, and the resulting population growth is not only sustained but begins to increase dramatically (see Figure 3.1).

*Age of Degenerative and Man-Made Diseases.* Mortality continues to decrease then approaches stability at a relatively low level. Average life expectancy at birth continues to increase, reaching levels in the latter twentieth century that are unprecedented in human history.

*Adapted from Omran (1971).*

upper limit, population growth may not occur, and there may be depopulation due to deaths from epidemics, famines, and war exceeding the maximum achievable rate of birth. Overall, however, with rates of births and deaths both at high levels and often more or less in balance, the Age of Pestilence and Famine is characterized by an approximate equilibrium wherein the population experiences little growth for extended periods.

As described by Omran, modernization precipitates a marked downward trend in mortality, and this did not happen in any region of the world until the late-eighteenth century. The first occurrence was in Britain with the arrival of the Industrial Revolution, which precipitated increases in average life expectancy (as described by McKeown and illustrated in Figure 1.1) and a surge in population growth. Then, as modernization spread to other parts of the globe, so too did the epidemiologic transition. This, Omran called the *Age of Receding Pandemics*, characterized by a marked decline in deaths from infectious diseases, as occurred first in Britain, then throughout Western Europe and industrialized countries further afield.

Finally, in the *Age of Degenerative and Man-Made Diseases*, infectious disease—no longer a major cause of death—has been replaced by chronic noncommunicable diseases. Life expectancy is greatly extended compared to earlier ages and birth rate declines sharply, contributing to significant aging of the population. This period broadly corresponds with the first half of the twentieth century in Europe, North America, and other high-income countries. Omran (1971) referred to the epidemiologic transition experienced in those countries as the *classical* or *Western model*. He claimed that a model of *accelerated transition* occurred in a number of countries, notably Japan, where the transition started later but progressed more rapidly. Lastly, he cited other countries, mostly in Africa, Asia, and Latin America, as illustrative of a *contemporary* or *delayed model*, wherein transition is in progress or has yet to begin.

There is a sense in which Omran's use of the phrase "man-made diseases" is mistaken, and it is important to draw attention to the error. By man-made, he was making the point that all of the major noncommunicable diseases (e.g., cardiovascular diseases and cancers) are substantially caused by behavior and ways of living (e.g., smoking, poor diet, and physical inactivity), and the point is well taken. However, use of the term in this context implies that what came before was not man-made, and the error lies therein. High rates of infectious disease in human populations from the Agricultural Revolution to the present day are very much attributable to ways of living (e.g., high population density and poor sanitation), and therefore may be regarded as substantially "man-made" (to retain Omran's gender-specific terminology). Epidemiologic transition occurs because one set of human habits and habitats that encourage high rates of communicable diseases is replaced by another set of habits and habitats that encourage high rates of noncommunicable diseases. Thus, the Age of Degenerative and Man-Made Diseases could just as easily be referred to as the *Age of Noncommunicable Diseases*, and so doing would be more accurate.

Omran (1971) gave considerable attention to demographic and socioeconomic changes that, he argued, are the features of modernization that contributed most to epidemiologic transition. He emphasized what he described as the tendency for improved infant and childhood survival to contribute to a lower birth rate during the middle and subsequent stages of epidemiologic transition, and identified three categories of factors he regarded as most important: biological, psychosocial, and economic. Regarding biology, he argued that the increased chance of a live birth surviving infancy and early childhood contributed to prolonged lactation, tending in turn to lengthen the mother's postpartum period of natural protection against conception. The consequential lengthening of birth intervals, especially among young women who are highly fecund (fertile) and of low parity (referring to number of previous births), leads to fewer births overall. Thus, in time, a falling birth rate tends to stem the rate of population growth from a declining death rate.

Omran reckoned that improved infant and childhood survival also alter the dynamics of the psychosocial and economic factors that influence family size. The incentive to "replace" children lost to early death is reduced with enhanced survival, and there is a commensurate increase in psychological and economic investment in individual children. The children of smaller families tend to be provided better protection, care, and education than would be possible in families of larger size. Whereas many children may once have been an economic asset contributing to household income, large family size due to improved rate of survival of children at some point becomes an economic liability. Thus, as couples become aware of the increased certainty that their offspring will survive them, they are more likely to actively limit family size.

## 3.2  DEMOGRAPHIC TRANSITION

The decline in mortality that is a fundamental feature of epidemiologic transition contributes to a widening of the *demographic gap* between rates of births and deaths. With a widening of the gap between falling death rate and stable high birth rate, the population necessarily grows. The estimates of global population size and birth rate over the millennia summarized in Figure 3.1 show that, for most of human history, global population growth was modest. Opinion differs as to whether modest population growth for Paleolithic hunter-gatherer communities was due primarily to a high mortality rate or a relatively low birth rate. Prolonged breastfeeding could have inhibited postnatal fertility, resulting in a reduced rate of fertility (Ramos et al., 1996) which—coupled with a moderate rate of mortality—would have resulted in little population growth. Alternatively, there would also have been little growth in population if both birth rate and mortality rate were high. Despite uncertainty about the mechanism responsible for the balance between mortality and fertility during the Paleolithic period, birth rate is generally agreed to have been high for settled agricultural communities. Population growth in those communities was

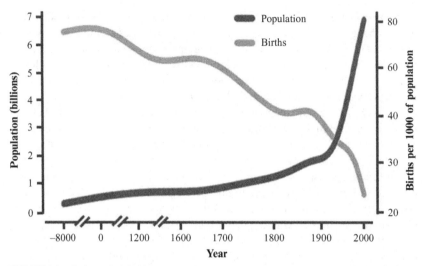

**FIGURE 3.1** Estimated global human population size and annual birth rate over recent millennia. Note that estimates before modern times are guestimates. *(Derived from Haub (2011).)*

limited by high mortality for the whole of the period from the Agricultural Revolution to the Industrial Revolution.

As illustrated in Figure 3.2, *Stage 1* of the model of demographic transition depicts populations as being comparatively small throughout human history prior to the Industrial Revolution. As early as 1798, Malthus (1798) in his famous work, *An Essay on the Population Principle*, described what he believed to be the relevant dynamics, the broad features of which are accepted by modern scholars (Lee, 2003). According to Malthus, any sharp rise in population would elicit the "positive" check of a rise in premature death due to famine, disease, or war, and/or the "preventive" check of a fall in birth rate. With modernization, population equilibrium is altered in *Stage 2* of the model of demographic transition due to a marked decrease in death rate. Birth rate, however, remains high and largely unaltered initially, leading to rapid population growth. Following population growth, *Stage 3* is characterized by a low mortality rate combined with a decrease in birth rate, which, in time, results in a return to population equilibrium, but in the context of an enlarged population compared to the pretransition era. Whereas Stages 2 and 3 characterize countries in varying stages of modernization, *Stage 4* typifies present-day high-income countries, where populations are large and stable with historically low rates for both births and deaths.

To summarize, demographic transition consists of an initial period of rapid population growth due to a decrease in formerly high mortality rates coupled with an unchanged high birth rate, followed by a period of decreased birth rate and unchanged mortality, leading to a return to stable, but large, population size. In addition to transformation from comparatively small to large population, the demographic transition is characterized by a marked increase in average age.

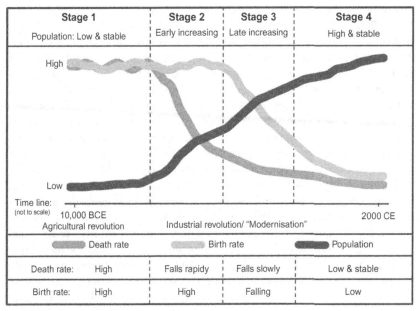

| Stage 1 | Stage 2 | Stage 3 | Stage 4 |
|---|---|---|---|
| Population: Low & stable | Early increasing | Late increasing | High & stable |

FIGURE 3.2    Model of demographic transition.

Aging demographic has already occurred in the countries that were first to modernize and is spreading globally.


## 3.3   TRANSITIONS IN TRANSITION

Over time, both models of epidemiologic and demographic transition have been infused with embellishments that emphasize one or other epidemiologic or demographic process deemed to be of particular interest to a particular time or place, including economic development, poverty, social disadvantage, wealth distribution and income inequality, contraceptive practices, and the role of women and children in society and economics. The processes that the models seek to explain relate to complex and interdependent events spanning broad swaths of time and geography. As mentioned above, mortality has been a main emphasis of the model of epidemiologic transition, with the transformations that the model seeks to explain sometimes also being referred to as *health transition* or *mortality transition*. On the other hand, the model of demographic transition tends to emphasize changes in birth rate to explain the dynamics of population growth.

Among embellishments of the model of epidemiologic transition, a fourth "age" has been suggested to follow the Age of Degenerative and Man-Made Diseases. This added *Age of Delayed Degenerative Diseases* has been inferred on the basis of evidence indicating postponement in some countries of the chronological age at which degenerative diseases tend to be fatal (Olshansky and Ault, 1986). Recent decreases in rates of cardiovascular deaths in several high-income countries (discussed in Chapter 4) is illustrative of the suggestion that age at time of death from degenerative disease may be extending. Whether that trend will continue or stall is a question of current debate, as is the question of whether recent gains in cardiovascular mortality in some countries will be lost to negative trends in other degenerative conditions, such as obesity and diabetes.

Taking a longer view of history, some accounts refer not merely to *one* epidemiologic transition triggered by industrialization, but to a total of *three* distinct epidemiologic transitions, each triggered by different but correspondingly momentous changes in human habits and habitats. By that account, the *first epidemiologic transition* occurred about 12,000 years ago when, as outlined above, humans transitioned from hunting and gathering in small nomadic groups to farming and husbandry in settled communities. The proliferation of human settlement that accompanied the Agricultural Revolution is believed to have had devastating consequences for the health of populations. Aggregation into settled communities encouraged the epidemic spread of infectious diseases responsible for the high rates of mortality that characterized human existence for the next 10-12 millennia following the Agricultural Revolution. Thus, the *second epidemiologic transition* refers to the account given by Omran (1971) of the positive impact on population health of the economic and industrial modernization that accompanied the Industrial Revolution.

The *third epidemiologic transition*, alleged by some to be imminent, refers to potential devastation from possible future resurgence of infectious diseases, including the spread of new antibiotic-resistant pathogens (Barrett et al., 1998; Harper and Armelagos, 2010). As discussed in Chapter 6, antibiotic resistance is a substantial present threat that could escalate with disastrous consequences for global health. Thus, of three epidemiologic transitions, only the second describes positive transformative change in personal and population health. The prospect, however, of a third transition having a transformative negative impact is a source of ongoing debate and speculation. For present purposes, unless indicated otherwise, mention of epidemiologic transition in the following pages will refer to the second and positive transition described by McKeown, Omran, and others.

With reference to demographic transition, a possible extension of the four-stage model shown in Figure 3.2 to include a fifth stage has been suggested to take account of the fact that the rate of fertility in several countries, especially in Europe (e.g., Austria, Denmark, Germany, and Sweden), has dropped to below the replacement level of 2.1 births per woman (i.e., there are fewer births than deaths). Under those circumstances, only inward migration can prevent population shrinkage and further aging of the population. Current trends in reduced fertility have been associated with changing personal priorities concerning marriage and family characterized by postponement of marriage, increased divorce, and increased nonmarital cohabitation. Thus, although the four-stage model of demographic transition sometimes elicits fears of global population growth exceeding sustainable levels, a possible fifth stage foreshadows different outcomes. The threat of population shrinkage due to declining fertility has led to speculation that long-term population decline in Europe, in particular, may be inevitable (van de Kaa, 1987), a phenomenon that the course followed by earlier epidemiologic and demographic transitions suggests could spread globally.

## 3.4  WHAT LONGEVITY REVOLUTION?

The survival of species across eons requires either of two strategies: individual immortality or reproduction coupled with mortality. Evolution mostly chose the latter. Reproduction and mortality are the twin necessities of the continuance of complex life forms.[1] The length of life necessary for successful reproduction is sometimes termed the *essential life span*. Notably, the span of human life typically exceeds the essential life span by a substantial margin, raising questions about the selective evolutionary pressures that endowed our species with a peculiar abundance of postreproductive life. A key question is, if life forms exist to self-perpetuate, what species-specific evolutionary advantage can there be in the survival of individuals who continue to consume resources that are in limited supply (food, clothing, housing, etc.) despite possessing little or no remaining reproductive potential? The answer, it seems, concerns human intelligence and capacity to learn, and the caregiving needs that those potentialities demand. Compared to other animals, including our closest primate relatives, the human brain is distinguished not merely by its disproportionately large size and greater cognitive potential. Additionally, the brains of human young require many years—an inordinately long time compared to other species—of active engagement with their surroundings before behavioral, intellectual, linguistic, and social maturity are conferred. The comparatively slow process of human maturation is accompanied by a commensurately long period of dependence on adults for subsistence and acculturation.

Caring for human young imposes heavy long-term burdens involving competition and potential conflict between the appetitive needs of caregivers and those of offspring. In short, caring for young consumes caregiver time and effort that could be spent on food gathering, hunting, shelter-building, and other self-preserving activities. Human infancy, childhood, juvenility, and adolescence are stages of life all characterized by slow development and low productivity. In that context, although incurring costs, postreproductive longevity offers advantages for collective productivity and survival. By exceeding the essential

---

1. Functional immortality reputedly exists among certain "simple" organisms such as hydra, tardigrades, planarian flatworms, and other species. Interest in the ingenuity of those species in being able to postpone or avoid mortality has spawned much speculation and scientific research. To date, however, results have fallen far short of fulfilling hopes of achieving anything approximating immortality for humans.

life span, human longevity creates *grand parenting* opportunities that alleviate the extraordinary caregiving demands imposed on younger more productive parents of slow-developing offspring (Bjorklund et al., 2002; Coall and Hertwig, 2011). Given the fact that human life span (defined as the observed duration of life from birth to death) typically exceeds the essential life span (required to reproduce), what then might we expect the human life span to be under optimal conditions of living? Curiously, the answer to that question not only lacks agreement among scholars, but opinion is widely divergent, ranging from an estimated few score years to immortality.

The current *maximum* human life span is 122 years, a record held by Jeanne Louise Calment. Born in February 1875, Calment died in August 1997, having lived the whole of her life in Arles, France. In 1988, the centenary of Vincent van Gogh's visit to Arles, Calment, still lucid at the age 113 years, recalled that van Gogh visited her father's fabric shop to buy canvas. Calment's advanced age and lucidity, as well as that of other *supercentenarians* (people aged 110 years or more) has encouraged speculation about the limits of human aging, but current knowledge does not permit firm conclusions. Surprisingly, the absence of a known upper limit has sometimes been interpreted as implying *no* limit. For instance, Aubrey de Grey and others at the California-based biogerontology SENS Research Foundation imagine a suite of biomedical interventions for achieving "engineered negligible senescence." If successful, these strategies would abolish *senescence*, the process of biological aging wherein death is presaged by cumulative time-related "wear and tear" to molecular and cellular structures that promote progressive deterioration and eventual failure of multiple vital biological functions. SENS researchers contend that purportedly soon-to-be-available biomedical technology will eliminate aging, enabling humans to become functionally immortal (Zealley and de Grey, 2012).

### 3.4.1 The Quest for Immortality

Belief in the attainability of human immortality is predicated on the assumption that all age-related deterioration can be identified, classified, and modified. Based on argument that the relevant categories of age-related deterioration have been delineated, it has been claimed that the achievement of immortality is imminent (Zealley and de Grey, 2012). There are allegedly seven such categories of senescence, including three that relate to functioning whole cells; two are intracellular, and two are extracellular. On the not insubstantial assumption that all time-related instances of biological degeneration can be repaired at rates fast enough to ensure that vital functions are never irreversibly compromised, comprehensive elimination of senescence can, in theory, be achieved. Repair, it is claimed, will be achieved using "rejuvenation biotechnology" currently under development, mostly comprised of envisioned new pharmaceuticals, tissue engineering, and organ replacement.

Though undeniably creative, speculations about bioengineered immortality are perversely optimistic. It is claimed that senescence-abolishing technologies, which do not currently exist except in imagination, will soon be developed as extensions of existing interventions. However, as discussed throughout this book, especially in Part 2, existing biomedical interventions are themselves extensively imperfect and limited even for dealing with considerably lesser everyday health problems than the problem of rendering mortality obsolete. Consequently, claims that similar interventions currently under development will soon succeed in eliminating mortality are implausible. The incredulity of one commentator was expressed in the following terms:

> Aging has an extremely long evolutionary history, and the anatomical structure and physiology of animals is directly related to their finite lifespan. The anti-aging movement proposes in a few decades to reverse what has been the result of millions of years of evolution.

Holliday (2009, p. 223)

Eternal youth and immortality have been the stuff of speculation, experimentation, and myth throughout the course of human history. The danger is that promises of success with imagined future technologies could encourage greater general confidence in biomedical solutions than past performance or likely future success warrant. Imaginings are one thing and realities another. Current realities of individual and population health demand strategies guided by results rather than the heroic optimism of advocates of engineered human longevity. The facts show that biomedicine's past contributions to population health have been modest. There is nothing in antiaging musings to suggest a different future.

Notwithstanding the possible appeal of immortality as myth on one hand, thoughts of any such actuality, on the other hand, may be uncomfortable to many. Current human population growth already poses many unanswered moral and practical questions, all of which would appear to be exacerbated were humans to live many times longer than at present. How could the planet sustain an ever-burgeoning population due to an indefinite life span being conferred on those currently living? Would there not be intolerable tensions arising from competition for resources? Birth rate would have to be controlled, but how? Could humanity justify withholding opportunity for life from the not-born to satisfy the vanity of those who wish to be immortal? In the fictional tale, *Death by Intervals*, Jóse Saramago, recipient of the Nobel Prize in Literature (1998), tells of a country in which death suddenly and mysteriously ceases (Saramago, 2008). Celebrations are brief, as the new order quickly proves intolerable. Death continues to exist everywhere else, and before long citizens of the deathless country find clandestine ways of crossing borders into neighboring lands for the sole purpose of dying. The resulting international tension escalates to the brink of war.

Zealley and de Grey (2012), however, are not prepared to countenance *any* ill effects of immortality, whether fictional or real. They simply dismiss all concerns as necessarily subordinate "side effects" to a higher ethical imperative.

Estimating a current global rate of roughly 100,000 age-related deaths per day, they assert that the advent of immortality is ethically necessary, because immortality is the only way of stopping the "slaughter" due to aging. Furthermore, arguing that current biomedical interventions succeed only in prolonging life at the expense of failing health, Zealley and de Grey (2012) propose immortality as some kind of solution to current problems of resource allocation. Achieving immortality by curing senescence, they argue, is the solution to the problem of resources being "increasingly devoted to the support of the frail."

Notably, Zealley and de Grey apparently see no contradiction in portraying nonexistent rejuvenation biomedicine as impeccably efficacious and cost-effective while characterizing extant clinical medicine (from which rejuvenation medicine is derived) as unequivocally ineffectual and bankrupting. Possibly the most fanciful aspect of engineered negligible senescence as a proposed solution to the "disease" of aging derives from its unabashed biological exceptionalism. The approach takes no heed of the effects of ways of living on the body or on health—be it smoking, poor diet, substance abuse, sedentary habits, over-exposure to ultraviolet sunlight, the quality of social relationships, socioeconomic position, or any of the myriad known mediators of senescence. Apparently, all will succumb to the power of a limited menu of would-be rejuvenation biotechnologies. In that respect, engineered negligible senescence is supreme among immortality myths: it promises life without limit and living without restraint.

### 3.4.2 Extraordinary Life Extension

Predictions of immortality occupy one end of the spectrum of human longevity predictions. At a more modest point on the continuum, a number of mathematical and demographic models converge to suggest a maximum human life span of about 125 years (Weon and Je, 2009), just slightly longer than the record already held by Jeanne Calment. Elsewhere on the continuum of human longevity predictions, claim and counterclaim about the likelihood or not of extraordinary extension of the human life span abound and are strenuously contested (Carnes, 2011; Carnes et al., 2010; Oeppen and Vaupel, 2002; Olshansky and Carnes, 2012; Olshansky et al., 1990, 2001; Vaupel, 2010; Vaupel and Rau, 2012). Much of the speculation about marked extension of human life, wherein most people born are predicted to become centenarians and supercentenarians, derives from statistics showing impressive improvements in life expectancy over the course of recent history. One such prediction is that most children born in high-income countries in the year 2000 will still be alive in the year 2100 (Christensen et al., 2009). Predictions of that kind, however, frequently confuse life expectancy and life span.

As discussed in Chapter 1, most recent gains in life expectancy at birth were due to dramatic decreases in mortality from infectious diseases, especially during childhood and early adulthood. The sparing of young lives contributes greatly to increased average life expectancy, while potentially having little or no effect on life span. In just two centuries, following many centuries of relative stability, life expectancy at birth in England and Wales doubled from less than 40 years to nearly 80 years (Figure 1.1), and similar progress was made in Japan in approximately half that time (Ikeda et al., 2011). The same overall pattern is evident for high-income countries in general, all of which experienced major increases in average *life expectancy at birth* but relatively modest increases in average *life span* (Mathers et al., 2014). The point is illustrated in Table 3.1, which summarizes life expectancy for different age cohorts in the United States from 1850 to 2011. For males, life expectancy at birth increased by 38.0 years between 1850 and 2011, and for females there was an increase of 40.5 years. By comparison, the increase in life expectancy at age 80 years was small, 2.3 years (from 5.9 to 8.2) and 3.3 years (6.4 to 9.7) for males and females, respectively. This shows that, despite impressive increases in life expectancy at birth, average human life span increased comparatively little.

In other words, living to old age is not the modern phenomenon it is sometimes portrayed as being. There have always been old people among human populations, there just happen to be proportionately more of them today. Moreover, it is of interest to know when the improved rate of survival (indicated by increased life expectancy) in older age occurred. Table 3.1 gives a hint, where it can be seen that, in the United States for the 100 years from 1850, life expectancy at 80 was more or less stable for both men and women. The increase that occurred in the United States sometime during the past half-century is consistent with recent findings for other high-income countries (Mathers et al., 2014). It appears that older-age mortality remained essentially static throughout the period of the epidemiologic transition, during the period when younger-age life expectancy improved dramatically. However, beginning about 30 years ago, mortality in older age began to fall in high-income countries, and the improvements were similar to those summarized in Table 3.1 for the United States. Specifically, between 1980 and 2011, life expectancy at 80 years in high-income countries increased on average 2.5 and 3.3 years for men and women, respectively (Mathers et al., 2014).

**TABLE 3.1** Life expectancy by age in the United States, 1850-2011

| | Males (age in years) | | | Females (age in years) | | |
|---|---|---|---|---|---|---|
| Year | 0 | 40 | 80 | 0 | 40 | 80 |
| 1850 | 38.3 | 27.9 | 5.9 | 40.5 | 29.8 | 6.4 |
| 1900 | 48.2 | 27.7 | 5.1 | 51.1 | 29.2 | 5.5 |
| 1950 | 66.3 | 31.2 | 5.9 | 72.0 | 35.6 | 6.6 |
| 2011 | 76.3 | 38.6 | 8.2 | 81.0 | 42.5 | 9.7 |

Adapted from Information Please Database, Pearson Education (2014).

Thus, evidence is lacking of any pronounced extension in the human life span. Rather, "population aging," which has been appreciable in many countries, is due primarily to increases in rate of survival into older age. Population aging is also encouraged by declining birth rate, which has produced historically atypical proportions of older-to-younger aged groups in many countries. In that respect, the longevity revolution might be more appropriately called a *longevity bulge*. Nevertheless, growth in the proportion of older people whose health and welfare needs depend on support from a declining workforce has the potential to create economic stresses. Most commentators expect these trends to continue for the foreseeable future (e.g., Bloom et al., 2014). However, some have speculated about a possible reversal of trends. In particular, it has been suggested that obesity will, if unchecked, have a negative effect on life expectancy (Olshansky et al., 2005). Examining trends for the United States, Olshansky et al. see potential major threats from recent substantial increases in the prevalence of obesity and possible additional threats from increased prevalence of infectious diseases. Overall, they anticipate that these threats could halt or even reverse current trends in life expectancy at birth and at older ages.

Notwithstanding speculation about possible impending declines in life expectancy at all ages, the increase in older age life expectancy that has occurred in recent decades is possibly more modest than might have been expected. The sparing of early mortality from infectious disease due to improvements in standards of nutrition, sanitation, housing, education, and hygiene has created communities that are relatively free of infectious disease. This, in turn, could be expected to benefit older people as they become increasingly vulnerable to infection due to the frailties of health that accompany advanced age. Yet, despite the benefit of a healthier environment, human life span, as distinct from life expectancy at birth, has remained largely unaffected. Overall, then, the available evidence suggests that postreproductive human longevity is relatively inflexible and not by any means easily extended. Popular beliefs about clinical medicine having delivered major improvements in length of life are incorrect. Ironically, with respect to infectious disease, as discussed in later chapters, the biggest contemporary threat of infection confronting older people is that which is contracted during the course of medical intervention itself, especially hospital-based care.

Some indication of the relative stability of the human life span can be gleaned from verified records of maximum life span. The oldest person currently living is a Japanese woman, Misao Okawa, aged 117 years at the time of writing. If, in 2020, Ms Okawa is still living, it will have taken a quarter-century for Jeanne Calment's milestone of human longevity to be equaled[2]. That fact contradicts popular belief that humans are currently navigating an inexorable trajectory of rapidly increasing life span. Humans, indeed, may be considered fortunate in

2. Expectations of Jeanne Calment's record being equaled and surpassed in the near future receded during the printing of this book. Ms Okawa died in April 2015, followed by the deaths of the next oldest persons, Gertrude Weaver (April 2015) and Jeralean Talley (June 2015), both of the United States.

possessing, compared to other animals, the long postreproductive life span that evolution has endowed. However, predictions of functional immortality are for the time being, at least, ornate conjecture. With regard to less rarefied predictions about longevity extension, it may also be concluded that there is little evidence, past or present, supporting speculation, both lay and scientific, about prospects for anything approximating what might be regarded as extraordinary extension of the human life span. Populations today are longer lived primarily in the sense that more people than in previous centuries achieve older age by surviving childhood and early adulthood. Additionally, global "population aging" is currently being augmented by a markedly falling birth rate in an increasing number of countries.

## 3.5 ARE POPULATIONS HEALTHIER NOW AS WELL AS LONGER LIVED?

It is worth noting that in much of biomedicine, and especially in the medical subfields of genetics, epidemiology, and to a lesser extent, public health, anything of nonbiological origin believed to affect health is often referred to as an "environmental" risk. In the literature of those disciplines, cigarette smoking, for example, is sometimes said to be an environmental hazard. Such usage, however, is unacceptably imprecise. Whereas airborne cigarette smoke poses an *environmental* threat to smokers and nonsmokers, the act of smoking is *behavior*. Efforts to reduce the incidence of smoking behavior are not helped by referring to that behavior as something it obviously is not. It is more sensible to think of problems of smoking in general as behavior (something people do) that affects the environment (e.g., pollutes the air people breathe) while also being affected by the environment (e.g., rates of smoking behavior are greatly influenced by aspects of the social environment comprised, for example, of the approving or disapproving behavior of others).

By referring to "external influences *and* personal behavior" (emphasis added) in the quote that appears at the beginning of Part 1, McKeown was careful to avoid the error of lumping everything nonbiological into an all-encompassing "environment." The strongest defense that could be made for using *environment* as an all-embracing term is that it serves merely as convenient shorthand. However, a better and no less convenient abbreviation is *habits and habitats*, which neatly encompasses the totality of environmental and biopsychosocial influences relevant to health. Given that changes in habits and habitats have led to populations in high-income regions being longer lived on average now than in previous centuries, should we conclude that they are also healthier? To answer that question, we must first consider how health is measured.

### 3.5.1 Quantifying the Burden of Disease: Summary Measures of Health

On the basis of population-wide measurements of life expectancy at birth, we may be tempted to conclude that over the course of recent generations there has been extraordinary improvement in population health in many regions.

However, although life expectancy at birth is widely accepted as a reasonable surrogate measurement of the general health of populations, it is crucial to acknowledge that longevity ignores a lot of potentially important detail revealed by other metrics. The **burden of disease** concept serves to reveal some of the shortcomings of relying on life expectancy as the sole indicator of health. Disease burden refers to the accumulation of negative health-related outcomes, inclusive of mortality *as well as* morbidity and disability. The burden of disease concept can be applied to single specific disease outcomes, multiple diseases and disease categories, and *all causes* collectively. Although mortality rate and average life-expectancy continue to be used as proxy indicators of the health of populations, it has become increasingly necessary to take account of morbidity rate; the rate of nonfatal states of impaired health (Molla et al., 2003).

The rising prevalence of noncommunicable diseases, which often result in long periods of impaired health, is one reason death rate has become less useful as a proxy measurement of health. The need for more comprehensive measurements has led to the development of a variety of indices that aim to quantify population health status by *combining* information about mortality and morbidity into composite indices. These *summary measures of population health* have undergone substantial development over the past 3 decades to become important and widely used indicators of differences in the burden of disease between populations and changes within the same population measured at different times. By way of illustration, it may be recalled that biomedical success in curtailing the poliomyelitis epidemic was discounted by McKeown (1979) as an important medical contribution to rising standards of population health for the reason that polio is relatively infrequently fatal. Therefore, lives saved due to success in halting polio contributed little to the overall declining rates of mortality and increased life expectancy that were McKeown's principal interests. However, had summary measures of health been available and been used in McKeown's analyses, stemming the polio epidemic would have been shown as having had a greater impact on reducing the population burden of disease than was revealed by examining only mortality and life expectancy.

The main summary measures of health that have been developed may be divided into two broad groups: health expectancy and health gap measurements (Donev et al., 2010; Gold et al., 2002; Pinheiro et al., 2011). The first category, measurements of health *expectancy*, extend the life-expectancy concept by adding information about disability, quality of life, or other non-fatal indicators of morbidity, to information about mortality. There are several such indices, including health-adjusted life expectancy (*HALE*), quality-adjusted life expectancy (*QALE*), and disability-adjusted life expectancy (*DALE*), all of which estimate the number of years of life lived in full health (i.e., healthy life expectancy) *plus* years lived in a reduced state of health, diminished quality of life, or with disability.

Health *gap* measurements, on the other hand, focus not on life lived but on *lost* years of full health compared to an accepted standard of longevity and standard of health. The chosen standard for longevity may vary to reflect, for example, the observed (i.e., empirical) average life expectancy for the population of interest, or it may be based on an external point of reference, such as the country with the longest life expectancy (currently, Japan). Moreover, a different standard may be chosen to reflect the difference in longevity between men and women, reflecting the fact that women generally live longer than men. Health gap measurements include health-adjusted life years (HALY), quality-adjusted life years (QALY), and disability-adjusted life years (DALY).

Both health expectancy and health gap measurements require value judgments to be made in relation to inferred norms of health for a given population. This is necessary in order to assign values or weights to life lived in varying states of diminished quality of life or with disability of varying severity. In short, the underlying principle is that life lived in diminished states of health is valued less than life lived in full health. The weights are determined on a scale from 0 to 1. Health quality represents gain to be maximized, whereas disability represents loss to be minimized. Thus, consistent with intuition (though not necessarily always without some confusion), health quality and disability are weighted oppositely to one another. Zero quality indicates death, and "1" represents maximum quality, whereas zero disability represents full function and "1" indicates death.

The judgments used to derive scale gradations are obtained by surveying opinion of either lay persons or experts, or both. When the opinions of large numbers of people are averaged, the resulting measurement scales can have good measurement properties, including high reliability and validity. This is contrary to what had been thought might be the case: namely, that variability in value judgments between people would lead to unacceptably high measurement instability. In particular, it had long been hypothesized that people from divergent populations and cultures would be likely to attribute quite different values to different diseases and disabilities. However, a recent comprehensive study involving 187 countries reported a high level of consistency between cultures in the average disability weights assigned to 220 unique health states, and measurement stability was observed for both physical and mental health (Salomon et al., 2012a). Table 3.2 contains a sample of about 40% of the health states that were surveyed. Given that mental health generally receives substantially less healthcare attention than physical health, a noteworthy feature of Table 3.2 is the relatively high weights given to mental, behavioral, and substance-use disorders.

Summary measures of health are a boon to rational health planning, especially in relation to utilitarian decisions that seek to optimize benefit, for example, from the distribution of limited healthcare resources. From a utilitarian perspective, a health gain for many may, for example, be deemed a greater good for society than a larger gain that benefits only a few. Although the quantification afforded by summary measures of health should not be the sole arbiter of all such difficult decisions, having quantifiable indicators is a good starting point.

For example, having access to a limited resource, how should that resource be used given a choice between, say, extending life by 5 years for

**TABLE 3.2** Brief description and morbidity weights sampled from a list of 220 unique health states quantified using summary measures of health

**Infectious disease**

| | | | |
|---|---|---|---|
| Diarrhea: mild | 0.061 | Diarrhea: moderate | 0.202 |
| Diarrhea: severe | 0.281 | Ear pain | 0.018 |
| HIV: symptomatic, pre-AIDS | 0.221 | HIV/AIDS: receiving antiretroviral treatment | 0.053 |
| AIDS: not receiving antiretroviral treatment | 0.547 | Intestinal nematode infections: symptomatic | 0.030 |
| Tuberculosis: without HIV infection | 0.331 | Tuberculosis: with HIV infection | 0.399 |

**Cancer**

| | | | |
|---|---|---|---|
| Cancer: diagnosis & primary therapy | 0.294 | Cancer: metastatic | 0.484 |
| Mastectomy | 0.038 | Terminal phase: with medication | 0.508 |
| Terminal phase: without medication | 0.519 | | |

**Cardiovascular and circulatory disease**

| | | | |
|---|---|---|---|
| Heart failure: mild | 0.037 | Heart failure: moderate | 0.070 |
| Heart failure: severe | 0.186 | Stroke: long-term consequences, mild | 0.021 |
| Stroke: long-term consequences, moderate | 0.076 | Stroke: long-term consequences, severe | 0.539 |

**Diabetes, digestive, and genitourinary disease**

| | | | |
|---|---|---|---|
| Diabetic foot | 0.023 | Chronic kidney disease (stage IV) | 0.105 |
| End-stage renal disease: with kidney transplant | 0.027 | End-stage renal disease: on dialysis | 0.573 |
| Gastric bleeding | 0.323 | Urinary incontinence | 0.142 |
| Impotence | 0.019 | Infertility: primary | 0.011 |
| Infertility: secondary | 0.006 | | |

**Chronic respiratory diseases**

| | | | |
|---|---|---|---|
| Asthma: controlled | 0.009 | Asthma: partially controlled | 0.027 |
| Asthma: uncontrolled | 0.132 | COPD and other chronic respiratory diseases: mild | 0.015 |

**TABLE 3.2**  Brief description and morbidity weights sampled from a list of 220 unique health states quantified using summary measures of health—cont'd

| | | | |
|---|---|---|---|
| COPD and other chronic respiratory diseases: moderate | 0.192 | COPD and other chronic respiratory diseases: severe | 0.383 |
| **Neurological disorders** | | | |
| Dementia: mild | 0.082 | Dementia: moderate | 0.346 |
| Dementia: severe | 0.438 | Multiple sclerosis: mild | 0.198 |
| Multiple sclerosis: moderate | 0.445 | Multiple sclerosis: severe | 0.707 |
| Parkinson's disease: mild | 0.011 | Parkinson's disease: moderate | 0.263 |
| Parkinson's disease: severe | 0.549 | | |
| **Mental, behavioral, and substance-use disorders** | | | |
| Alcohol-use disorder: mild | 0.259 | Alcohol-use disorder: moderate | 0.388 |
| Alcohol-use disorder: severe | 0.549 | Cocaine dependence | 0.376 |
| Heroin and other opioid dependence | 0.641 | Anxiety disorders: mild | 0.030 |
| Anxiety disorders: moderate | 0.149 | Anxiety disorders: severe | 0.523 |
| Major depressive disorder: mild episode | 0.159 | Major depressive disorder: moderate episode | 0.406 |
| Major depressive disorder: severe episode | 0.655 | Bipolar disorder: manic episode | 0.480 |
| Bipolar disorder: residual state | 0.035 | Schizophrenia: acute state | 0.756 |
| Schizophrenia: residual state | 0.576 | Anorexia nervosa | 0.223 |
| Asperger's syndrome | 0.110 | Autism | 0.259 |
| **Hearing and vision loss** | | | |
| Hearing loss: mild | 0.005 | Hearing loss: moderate | 0.023 |
| Hearing loss: severe | 0.032 | Distance vision: mild impairment | 0.004 |
| Distance vision: moderate impairment | 0.033 | Distance vision: severe impairment | 0.191 |
| Near vision impairment | 0.013 | | |
| **Musculoskeletal disorders** | | | |
| Low back pain: acute, without leg pain | 0.269 | Low back pain: acute, with leg pain | 0.322 |

*Continued*

**TABLE 3.2** Brief description and morbidity weights sampled from a list of 220 unique health states quantified using summary measures of health—cont'd

| | | | |
|---|---|---|---|
| Low back pain: chronic, without leg pain | 0.366 | Low back pain: chronic, with leg pain | 0.374 |
| Gout: acute | | | |
| **Injuries** | | | |
| Amputation of one arm: with or without treatment | 0.130 | Amputation of both arms: with treatment | 0.044 |
| Amputation of both arms: without treatment | 0.359 | Amputation of one leg: without treatment | 0.164 |
| Amputation of both legs: with treatment | 0.051 | Amputation of both legs: without treatment | 0.494 |
| Fracture of skull | 0.073 | Traumatic brain injury: minor | 0.106 |
| Traumatic brain injury: moderate | 0.224 | Traumatic brain injury: severe | 0.625 |
| Spinal cord lesion below neck: treated | 0.047 | Spinal cord lesion below neck: untreated | 0.440 |
| Spinal cord lesion at neck: treated | 0.369 | Spinal cord lesion at neck: untreated | 0.673 |

Adapted from Salomon et al. (2012a).

100 people—all of whom have a QALY of 1.0 (i.e., there is no disability following intervention)—or extending life by 5 years for 500 people, all of whom have a QALY of 0.2 (i.e., there is substantial disability following intervention)? In both instances, the quantitative benefit equals 500 years of life in full health: 5 years × 100 people with a QALY of 1.0 versus 5 years × 500 people with a QALY of 0.2. Given the quantitative equality of these two outcomes, a decision about how to deploy the resource might necessitate other (possibly more subjective) considerations.

Conversely, we might imagine similar circumstances with the exception that the same 100 people in the first group are expected to benefit by 6 (not 5) years (to yield 600 years of life in full health). The estimated larger benefit (600 vs. 500 life years) could be the crucial factor in arriving at a decision to deploy an intervention that restores full function for longer for a smaller group compared to a smaller overall benefit for a larger group. As a further illustration, we might again imagine similar circumstances with the exception that the same 500 people in the second group are expected to have a QALY of 0.4 following intervention, indicative of less serious disability than in the initial illustration above. This would yield a total benefit of 1000 life years (5 years × 500 people with a QALY of 0.4), in which case the decision might be made to deploy the limited resource for their benefit.

Decisions based on quantitative analysis are not devoid of subjectivity and should not be made to the exclusion of other considerations. However, in general, rational and transparent decision-making is likely to be assisted by use of an appropriate summary measure of health to explicate and to quantify key outcomes. In that vein, there is reason to question much of what happens in biomedical healthcare. From the perspective of the good of society, no sane person would try to defend a *small gain for a small number of people* being valued over a *large gain for a large number* of people. Yet, those are exactly the priorities reflected in the actions of much contemporary biomedical healthcare.

### 3.5.1.1   Differentiating Between the Urgent and the Important

Biomedical healthcare often expends extraordinary levels of limited resources for the marginal benefit of a comparatively small number of ill patients, when similar effort and expenditure would deliver a large benefit to many more people who are not (yet) sick. One of the most difficult challenges in healthcare is said to be ensuring "that the urgent does not crowd out the important" (McGinnis and Foege, 2004). At the policy level, development of biomedical healthcare of marginal benefit frequently takes precedence over preventive interventions capable of benefiting personal and population health far more. In part, the dominance of subjectivity (the perceived urgency of the *immediate*) over rationality (addressing the seemingly less urgent but more *important* challenges) is attributable to understandable compassion for those who are presently sick. Such concerns, however, should not cause us to shy away from confronting evident anomalies, including irrational priorities. Summary measures of health have an important role in encouraging more rational healthcare policy and planning.

### 3.5.2   Do Summary Measures of Health Measure Health?

Summary measures of health are important for extending the characterization and quantification of the health of populations beyond mortality alone to include key dimensions of morbidity. However, they have important limitations, as revealed when viewed from the perspective of habits and habitats. Consistent with traditions in biomedicine, summary measures of health tend to focus on disease, wherein health is the assumed default condition in the absence of diagnosed medical conditions or disability. However, viewing health as a phenomenon of habits and habitats suggests the importance of also measuring variables other than disease, especially indicators predictive of positive health and not merely disease absence. Measuring variables that foster health would lead to better understanding of ways to optimize health and forestall disease. Alas, the measurement of *positive* health is in its infancy.

By way of illustration, commentators have contrasted the extensive efforts made by national governments worldwide to measure economic outcomes while simultaneously ignoring indicators of subjective wellbeing. Economic indicators, it has been argued, overlook much of what is most important to people, namely, their evaluations and feelings about their lives (Diener and Seligman, 2004). An exception to the prevailing preoccupation with economic performance to the exclusion of other considerations of human welfare is the small Himalayan Kingdom of Bhutan, which, for more than 40 years, has pursued a policy of emphasizing *gross national happiness* over *gross national product*. As outlined in Box 3.2, Bhutan has sought to apply a range of integrated indices of habits and habitats that measure diverse aspects of the natural

**BOX 3.2  Gross National Happiness in the Kingdom of Bhutan**

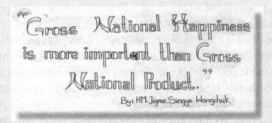

Most countries employ a range of indicators of "progress," and the indicator of progress that is most widely used is *gross domestic product* (*GDP*). Developed in the United States to manage the Great Depression and the war-time economy of World War II that followed, GDP is the sum of the economic value of all the goods and services produced in an economy in a given year. As a measure of national economic output, GDP was not designed to measure personal wellbeing, although it has come to be widely viewed as serving that purpose (Braun, 2009). However, despite some countries experiencing extended periods of prosperity as indicated by rising GDP, there is little evidence of associated improvement in subjective wellbeing. In the United States, for example, until recently, there had been several decades of steeply rising increases in economic output as measured by GDP. Yet, for the same period, there was no rise in life satisfaction, and substantial increases in depression and distrust (Diener and Seligman, 2004). Joseph Stiglitz, Nobel Laureate in Economic Sciences (2001), has described as *fetishism* the longstanding preoccupation with GDP as a measure of progress (Stiglitz, 2009).

In a courageous step, the small Himalayan Kingdom of Bhutan, with a population of approximately 750,000, has defied orthodox economics by adopting a more holistic approach to development and progress. Specifically, Bhutan has adopted the *gross national happiness* (*GNH*) index, rather than GDP, to guide development of the country's

**BOX 3.2 Gross National Happiness in the Kingdom of Bhutan—cont'd**

institutions and policies. The concept, first introduced in 1972 by Jigme Singye Wang-chuck, the Fourth King of Bhutan, has been revised periodically, and is currently measured by a questionnaire comprising 36 indicators covering nine domains, as follows:

- Community vitality
- Cultural diversity and resilience
- Ecological diversity and resilience
- Education
- Good governance
- Health
- Living standard
- Psychological wellbeing
- Time use

There is evidence that Bhutan's approach is having success (Brooks, 2013). Bhutan ranks high in global analyses of life satisfaction. Cultural and ecological heritage have been largely preserved, and there have been gains in standard of living and transition to democracy. GNH is not a panacea for society's ills nor does it provide a blueprint to be adopted uncritically by other nations. However, there is growing realization that material wealth alone is a poor indicator of fulfillment in life, and increased recognition of the importance of measuring subjective wellbeing (Beaglehole and Bonita, 2015).

The French national statistics office has sought to develop more holistic measures of wellbeing to guide policy, and the British, German, Canadian, and Chinese governments have begun to adopt similar measures (Braun, 2009; Brooks, 2013). In addition, the General Assembly of the United Nations (2012) resolution document, *Happiness: Towards a Holistic Approach to Development*, explicitly states that GDP is an inadequate indicator of human happiness and wellbeing. The document also states that the pursuit of happiness is a fundamental human goal, and invites Member States to develop alternative measurements of happiness and wellbeing for use in framing public policies.

environment, physical health, psychological wellbeing, social harmony, public infrastructure, cultural life, and governance, and in so doing appears to have had success in promoting the wellbeing of its citizens.

There is considerable variation in the habits and habitats of human societies, and better understanding of associated variation in health would be invaluable to the development of interventions for promoting health. Many details of the habits and habitats of ancient peoples are probably unknowable, and therefore precise comparisons between modern and prehistoric humans may never be possible. However, there has been little attempt to develop standard indices of contemporary healthy habits and habitats akin to the summary measures of health (so-called) that now serve to standardize the measurement of disease absence. Although the recently developed summary measures of health are a substantial improvement on measurements that focus solely on mortality, they nevertheless may be seen as perpetuating a disease focus over wellbeing.

## 3.6  MORBIDITY: COMPRESSION, EXPANSION, OR EQUILIBRIUM?

Despite their limitations, summary measures of health are important for understanding key implications of epidemiologic and demographic transition, especially in relation to population aging. However, summary measures of health do not alone answer the question of whether populations are healthier now as well as longer lived. On the contrary, there is intense speculation about the implications of population aging for society overall. On one hand, the longevity revolution of recent history is sometimes celebrated as a great achievement of the twentieth century. On the other hand, there is concern that ever-increasing numbers of older, feeble, and chronically ill people will impose intolerable burdens on society. According to the latter scenario, the longevity revolution amounts to little more than a *longevity epidemic* of long-lived sick people. Thus, given that more people in general are now reaching old age, the key question is whether those additional years of life are being enjoyed in good health or are they being endured in varying states of disability? Much of the debate about that question has crystallized around three main competing hypotheses: *expansion of morbidity*, *compression of morbidity*, and *dynamic equilibrium*, represented schematically in Figure 3.3.

### 3.6.1  Predicting the Level of Morbidity Associated with Gains in Mortality

The *expansion of morbidity* hypothesis predicts that extra years of life due to postponement of mortality contribute to an extension of the period of age-related morbidity (Olshansky et al., 1991). In the context of biomedical healthcare, this pattern has been referred to as the "failure of success" in reference to medical interventions that improve survival rates without curing underlying pathology and associated disability (Gruenberg, 1977). Conversely, the *compression of morbidity* hypothesis predicts the opposite pattern, wherein reduction in major noncommunicable disease **risk factors** is believed to contribute not only to postponement of mortality but also to postponement of the onset of common nonfatal chronic disabling conditions such as arthritis, osteoporosis, and cognitive decline (Fries, 1980; Fries et al., 2011). An intermediate position, the *dynamic equilibrium* hypothesis, predicts no change in overall disability, either because postponement of mortality is accompanied by comparable postponement of morbidity or because prevention-related decreases in severe disability are offset by age-related increases in mild and moderate disability (Manton, 1982).

From the foregoing, it may be inferred that the predicted level of morbidity associated with mortality gains can be expressed as a simple quantitative function. Specifically, it can be seen from Figure 3.3 that all three panels show equal postponement of *mortality* but varying levels of accompanying morbidity, the cumulative amount of which is indicated by the area under the survival curve. If the period of postponement of mortality is assumed to be fixed at the value of

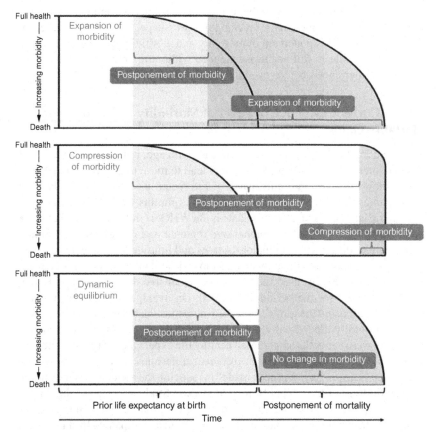

**FIGURE 3.3**    Schematic representation of survival curves for the three main competing hypotheses that seek to explain current and future relative rates of population mortality and morbidity. *Note*: Survival curves showing amount of morbidity (shaded areas) associated with life expectancy at birth before and after postponement of mortality that is (a) greater than postponement of morbidity leading to *expansion of morbidity*; (b) less than postponement of morbidity leading to *compression of morbidity*; and (c) equal to postponement of morbidity leading to *dynamic equilibrium* (i.e., no change in overall morbidity).

1.0, the quantum of *morbidity* predicted by each of the three main hypotheses can be defined as either increased, greater than 1.0 (expansion of morbidity); decreased, less than 1.0 (compression of morbidity); or unchanged, equal to 1.0 (dynamic equilibrium).

Thus, in the upper panel of Figure 3.3, postponement of morbidity is less than postponement of mortality, and results in expansion of morbidity wherein the period of extended life in full health is exceeded by extension of life accompanied by disability. The opposite relationship between life spent in full health and with disability, respectively, is shown in the middle panel representing compression of morbidity. Here, postponement of morbidity is greater than postponement of mortality with the result that morbidity is compressed and the *survival curve* becomes more rectangular in shape. The lower panel of Figure 3.3 shows postponement of mortality and morbidity in dynamic equilibrium such that life in full health is extended without affecting the quantum of disability experienced before eventual death.

### 3.6.2   Current Evidence of Level of Morbidity Associated with Mortality Gains

Because nonfatal disabilities tend to increase with age, it is natural to think that postponement of mortality will inevitably lead to more morbidity. Indeed, evidence shows that the mortality gains of the recent past have generally been accompanied by expansion of morbidity (Crimmins and Beltrán-Sánchez, 2012). The largest relevant study to date is the WHO's Global Burden of Disease Study, which has yielded comprehensive regional and global data concerning mortality and disability from major diseases and injuries making extensive use of summary measures of health such as those described earlier in this chapter. Salomon et al. (2012b) used those uniquely extensive data to examine HALE in 187 countries for the period 1990-2010. On average, life expectancy and HALE increased, but the former increased more than the latter, indicating expansion of morbidity. Specifically, for each 1 year of increased life expectancy at birth there were 8 months of increased healthy life expectancy. Thus, despite substantial postponement of mortality over recent decades, it was concluded that "relatively little progress has been made" in reducing the incidence of morbidity.

The findings of the Global Burden of Disease Study are encouraging in that global postponement of mortality has delivered substantial gains in life expectancy, but disappointing in that longer life has been accompanied by an increase in the average number of years of life with disability. Notwithstanding those results, it should not be assumed that the nexus between population mortality and morbidity is immutable. Rather, it is crucial to consider the main causes of recent mortality gains in order to discern the prospects for retaining past gains in longevity (and making further gains) while simultaneously postponing the onset of morbidity. In that context, it is revealing to consider the mortality gains of recent decades in high-income countries from geographically diverse regions, including Europe, North America, and the Pacific (Björck et al., 2009; Capewell et al., 2000, 2009; Laatikainen et al., 2005; Palmieri et al., 2010; O'Flaherty et al., 2012; Ünal et al., 2005a,b). Extensive study of those gains has shown the immense contribution of leading behavioral risk factors to incidence of noncommunicable diseases.

For example, heart disease mortality in England and Wales for the period 1981-2000 was reported to have decreased by 54%, representing almost 70,000 deaths prevented or postponed, which together contributed approximately 1 million extra years of life (Ünal et al., 2005a,b). When outcomes were partitioned according to cause, behavioral risk factor reduction was found to have had four-times greater impact on life-years gained than medical interventions delivered during the same period. The most important risk factor improvements were reduced rate of smoking and improved diet. However, not all risk-factor change for the period was positive. Despite general improvements in diet (e.g., increased consumption of fruit, fiber, and unsaturated fat, and reduced consumption of saturated fat and salt) being responsible for lowering population levels of blood pressure and total cholesterol, negative trends were observed for physical activity, diabetes, and obesity.

Taking account of reported improvements in particular aspects of diet, it is likely that consistently observed negative trends for diabetes and obesity were due to negative trends in other aspects of diet and probably other risk factors as well, of which increased physical inactivity is a prime candidate. Consequently, if, as evidence suggests, the global burden of morbidity is currently expanding relative to gains in longevity, it is reasonable to ask whether the negative trends in morbidity can be reversed to achieve global compression of morbidity in which people are generally healthy for longer as well as being longer lived. That is a question to which we will return in Part 3. In the meantime, Chapter 4 discusses current common causes of mortality and morbidity with particular reference to the respective roles of biomedical healthcare and risk factor reduction.

## REFERENCES

Barrett, R., Kuzawa, C.W., McDade, T., Armelagos, G.J., 1998. Emerging and re-emerging infectious diseases: the third epidemiologic transition. Annu. Rev. Anthropol. 27, 247–271.

Beaglehole, R., Bonita, R., 2015. Development with values: lessons from Bhutan. Lancet 385, 848–849.

Björck, L., Rosengren, A., Bennett, K., et al., 2009. Modelling the decreasing coronary heart disease mortality in Sweden between 1986 and 2002. Eur. Heart J. 30, 1046–1056.

Bjorklund, D.F., Yunger, J.L., Pellegrini, A.D., 2002. The evolution of parenting and evolutionary approaches to childrearing. In: Bornstein, M.H. (Ed.), Handbook of Parenting: Biology and Ecology of Parenting, vol. 2. Lawrence Erlbaum, Mahwah, NJ.

Bloom, D.E., Chatterji, S., Kowal, P., et al., 2014. Macroeconomic implications of population ageing and selected policy responses. Lancet 384, 649–657. http://dx.doi.org/10.1016/S0140-6736(14)61464-1.

Braun, A.A., 2009. Gross National Happiness in Bhutan: A Living Example of an Alternative Approach to Progress. University of Pennsylvania, Philadelphia, PA. Retrieved 18 October 2013 from, http://repository.upenn.edu/sire/1/.

Brooks, J.S., 2013. Avoiding the limits to growth: gross national happiness in Bhutan as a model for sustainable development. Sustainability 5, 3640–3664. http://dx.doi.org/10.3390/su5093640.

Capewell, S., Beaglehole, R., Seddon, M., McMurray, J.J., 2000. Explaining the decline in coronary heart disease mortality in Auckland, New Zealand between 1982 and 1993. Circulation 102, 1511–1516.

Capewell, S., Hayes, D.K., Ford, E.S., et al., 2009. Life-years gained among US adults from modern treatments and changes in the prevalence of 6 coronary heart disease risk factors between 1980 and 2000. Am. J. Epidemiol. 170, 229–236.

Carnes, B.A., 2011. What is lifespan regulation and why does it exist? Biogerontology 12, 367–374.

Carnes, B.A., Staats, D.O., Vaughan, M.B., Witten, T.M., 2010. An organismal view of cellular aging. Med. Longevite 2, 141–150.

Christensen, K., Doblhammer, G., Rau, R., Vaupel, J.W., 2009. Ageing populations: the challenges ahead. Lancet 374, 1196–1208.

Coall, D.A., Hertwig, R., 2011. Grandparental investment: a relic of the past or a resource for the future? Curr. Dir. Psychol. Sci. 20, 93–98.

Crimmins, E.M., Beltrán-Sánchez, H., 2012. Mortality and morbidity trends: is there compression of morbidity? J. Gerontol. 66B, 75–86.

Diener, E., Seligman, M.E.P., 2004. Beyond money: toward an economy of well-being. Psychol. Sci. Public Interest 5, 1–31, documents/NOTEONHAPPINESSFINALCLEAN.pdf.

Donev, D., Zaletel-Kragelj, L., Bjegovic, V., Burazeri, G., 2010. Measuring the burden of disease: disability adjusted life year (DALY). Retrieved 28 December 2012, http://www.mf.uni-lj.si/dokumenti/6b695fc9385e3e2ab8fb41ec7d34660d.pdf.

Fries, J.F., 1980. Aging, natural death, and the compression of morbidity. N. Engl. J. Med. 303, 130–136.

Fries, J.F., Bruce, B., Chakravarty, E., 2011. Compression of morbidity 1980–2011: a focused review of paradigms and progress. J. Aging Res. 1–10. http://dx.doi.org/10.4061/2011/261702.

Gold, M.R., Stevenson, D., Fryback, D.G., 2002. HALYS and QALYS and DALYS, oh my: similarities and differences in summary measures of population health. Annu. Rev. Public Health 23, 115–134.

Gruenberg, E.M., 1977. The failures of success. Milbank Mem. Fund Q. Health Soc. 55, 3–24.

Harper, K., Armelagos, G., 2010. The changing disease-scape in the third epidemiological transition. Int. J. Environ. Res. Public Health 7, 675–697.

Haub, C., 2011. How Many People Have Ever Lived on Earth? Population Reference Bureau, Washington, DC. Retrieved 3 January 2013 from, http://www.prb.org/Articles/2002/HowManyPeopleHaveEverLivedonEarth.aspx.

Holliday, R., 2009. The extreme arrogance of anti-aging medicine. Biogerontology 10 (2), 223–228.

Ikeda, N., Saito, E., Kondo, N., et al., 2011. What has made the population of Japan healthy? Lancet 378, 1094–1105.

Information Please Database, Pearson Education, 2014. United States statistics, mortality: life expectancy by age 1850-2011. Retrieved 16 February 2014 from, http://www.infoplease.com/ipa/A0005140.html.

Kaneko, R., 2009. The society created by the longevity revolution: historical development and associated issues. Jpn. J. Popul. 9, 135–154.

Laatikainen, T., Critchley, J., Vartiainen, E., et al., 2005. Explaining the decline in coronary heart disease mortality in Finland between 1982 and 1997. Am. J. Epidemiol. 162, 764–773.

Lee, R., 2003. The demographic transition: three centuries of fundamental change. J. Econ. Perspect. 17, 167–190.

Malthus, T., 1798. An essay on the principle of population: An essay on the principle of population, as it affects the future improvement of society with remarks on the speculations of Mr. Godwin, M. Condorcet, and other writers. Electronic Scholarly Publishing Project, 1998, http://www.esp.org.

Manton, K.G., 1982. Changing concepts of morbidity and mortality in the elderly population. Milbank Mem. Fund Q. Health Soc. 60, 183–244.

Mathers, C.D., Stevens, G.A., Boerma, T., et al., 2014. Causes of international increases in older age life expectancy. Lancet 384, 540–548. http://dx.doi.org/10.1016/S0140-6736(14)60569-9.

McGinnis, J.M., Foege, W.H., 2004. The immediate vs the important. J. Am. Med. Assoc. 291, 1263–1264.

McKeown, T., 1979. The role of Medicine: Dream, Mirage or Nemesis? Basil Blackwell, Oxford, UK.

Molla, M.T., Madans, J.H., Wagener, D.K., Crimmins, E.M., 2003. Summary Measures of Population Health: Report of Findings on Methodologic and Data Issues. National Center for Health Statistics, Hyattsville, MD.

Oeppen, J., Vaupel, J.W., 2002. Broken limits to life expectancy. Science 296, 1029–1031.

O'Flaherty, M., Allender, S., Taylor, R., Stevenson, C., Peeters, A., Capewell, S., 2012. The decline in coronary heart disease mortality is slowing in young adults (Australia 1976–2006): a time trend analysis. Int. J. Cardiol. 158, 193–198.

Olshansky, S.J., Ault, A.B., 1986. The fourth stage of the epidemiologic transition: the age of delayed degenerative diseases. Milbank Q. 64, 355–391.

Olshansky, S.J., Carnes, B.A., 2012. Zeno's paradox of immortality. Gerontology 59, 85–92.

Olshansky, S.J., Carnes, B.A., Cassel, C., 1990. In search of Methuselah: estimating the upper limits to human longevity. Science 250, 634–640.

Olshansky, S.J., Rudberg, M.A., Carnes, B.A., et al., 1991. Trading off longer life for worsening health: the expansion of morbidity hypothesis. Aging Health 3, 194–216.

Olshansky, S.J., Carnes, B.A., Désesquelles, A., 2001. Demography: prospects for human longevity. Science 291, 1491–1492.

Olshansky, S.J., Passaro, D.J., Hershow, R.C., et al., 2005. A potential decline in life expectancy in the United States in the 21st century. N. Engl. J. Med. 352, 1138–1145.

Omran, A.R., 1971. The epidemiologic transition: a theory of the epidemiology of population change. Milbank Mem. Fund Q. 49, 509–538.

Palmieri, L., DrStat, K.B., Giampaoli, S., Capewell, S., 2010. Explaining the decrease in coronary heart disease mortality in Italy between 1980 and 2000. Am. J. Public Health 100, 684–692.

Pinheiro, P., Plaß, D., Krämer, A., 2011. The burden of disease approach for measuring population health. In: Krämer, A., Khan, M.H., Kraas, F. (Eds.), Health in Megacities and Urban Areas. Springer-Verlag, Heidelberg, Germany.

Ramos, R., Kennedy, K.I., Visness, C.M., 1996. Effectiveness of lactational amenorrhoea in prevention of pregnancy in Manila, the Philippines: non-comparative prospective trial. Br. Med. J. 313, 909–912.

Salomon, J.A., et al., 2012a. Common values in assessing health outcomes from disease and injury: disability weights measurement study for the global burden of disease study 2010. Lancet 380, 2129–2143.

Salomon, J.A., et al., 2012b. Healthy life expectancy for 187 countries, 1990–2010: a systematic analysis for the global burden disease study 2010. Lancet 380, 2144–2162.

Saramago, J., 2008. Death by Intervals. Vintage Books, London.

Stiglitz, J.E., 2009. GDP fetishism. Economists' Voice. 6 (8). Article 5, http://www.bepress.com/ev/vol6/iss8/art5.

Ünal, B., Critchley, J.A., Capewell, S., 2005a. Modelling the decline in coronary heart disease deaths in England and Wales, 1981-2000: comparing contributions from primary prevention and secondary prevention. Br. Med. J. 331, 1–6. http://dx.doi.org/10.1136/bmj.38561.633345.8F.

Ünal, B., Critchley, J.A., Fidan, D., Capewell, S., 2005b. Life-years gained from modern cardiological treatments and population risk factor changes in England and Wales, 1981-2000. Am. J. Public Health 95, 103–108.

United Nations General Assembly, 2012. Happiness: towards a holistic approach to development. Retrieved 18 October 2013 from, http://www.un.org/esa/socdev/ageing/.

Van de Kaa, D.J., 1987. Europe's second demographic transition. Popul. Bull. 42 (1), 1–59.

Vaupel, J.W., 2010. Biodemography of human ageing. Nature 464, 536–542.

Vaupel, J.W., Rau, R., 2012. Research versus Rhetoric. Gerontology 59, 95–96.

Weon, B.M., Je, J.H., 2009. Theoretical estimation of maximum human lifespan. Biogerontology 10, 65–71.

Zealley, B., de Grey, A.D., 2012. Strategies for engineered negligible senescence. Gerontology 59, 183–189.

Chapter 4

# Biomedicine and Common Causes of Mortality and Morbidity

*In chronic diseases the clinician's first contact with the patient comes late in the natural history of the disease, usually ... when there is already irreversible pathological change.*

(Rose, 1981, p. 1850)

## Contents

In Chapter 1, the WHO definition of health as physical, mental, and social wellbeing was contrasted with the central idea in biomedicine that health is the absence of disease. Because health problems traditionally have been conceptualized as disturbances in physiology, biomedical healthcare has developed to become little more than applied biology aimed at managing those

The Health of Populations. http://dx.doi.org/10.1016/B978-0-12-802812-4.00004-7

disturbances. The dominant medical conceptualization of health and the style of healthcare that it promotes are sometimes referred to as the *medical model*. The first use of that term is credited to the psychiatrist Thomas Szasz (1961) in *The Myth of Mental Illness*. In part, Szasz argued that psychiatry is inherently flawed because, as a branch of medicine, it seeks to apply concepts derived from physiology to diagnose and treat problems that are essentially *problems of living* typified by behavioral, interpersonal, emotional, and economic sequelae.

The essence of Szasz's critique has come to be widely accepted. Yet, due largely to immense influence from the pharmaceutical industry, the biomedical model of *mental health*, as discussed in Chapter 15, has become progressively more rather than less dominant over recent decades. Reflecting the generality of aspects of Szasz's critique, biomedical dominance of *somatic healthcare* has also elicited extensive disquiet. This includes calls for a "new medical model," the *biopsychosocial model*, to take account not merely of biological but also behavioral, psychological, and social influences (Engel, 1977). Nevertheless, diagnosis and intervention in somatic healthcare remain preoccupied with biological processes, wherein psychosocial considerations rarely receive more than occasional perfunctory attention.

With its roots in physiology and other biological and medical sciences, including anatomy, biochemistry, genetics, microbiology, pathology, molecular biology, and pharmacology, the medical model holds that illness, whether somatic or mental, is inherited or acquired. Inherited disorder is caused by an aberrant gene or chromosome present at birth, although expression of the disorder may not become apparent until sometime later. Acquired illness, on the other hand, is held to be due to infection, injury, developmental dysfunction, or processes of degeneration which may be abnormal or not (e.g., normal aging). In all instances, the medical model is explicit: Disease is located *in* the body. Whether inherited or acquired, disease, according to the medical model, is in the organs, tissues, and cells of the body of individuals. Thus, the primary objectives of biomedical healthcare are the diagnosis of physiological disturbance in tissues and organs in individuals and the treatment of targeted sites using biological interventions.

Whereas major limitations of the biomedical model may be relatively easy to comprehend when applied to mental health, those same intrinsic limitations may not be so apparent when considering biomedical healthcare for somatic problems such as the leading causes of death summarized in Table 2.1. Medicine is not merely the main response to those problems, but in the minds of many it offers essentially the *only* effective response. Conversely, throughout these pages, both the causes and outcomes of health are discussed as residing not for the most part in individuals but in *ways of living*, evidenced by individual and collective habits and habitats. This runs counter to the thinking of those to whom it seems natural for healthcare to be directed at biological targets in sick individuals. That thinking, however, is not universal (see Box 4.1), and the current predominance of such thinking may merely reflect the degree to which we have become accustomed to biomedicine's blinkered approach to health.

## BOX 4.1 Hippocrates on Habits and Habitats

Hippocrates (c. 450-380 BCE) *(https://www.google.is/search?q=hippocrates&tbm.)*

The conceptualization of health as a property of habits and habitats encourages consideration of health determinants and outcomes largely ignored in contemporary healthcare. The writings of authors such as Dubos (1959) and McKeown (1955, 1979) show that there was an upsurge of interest in the social, as distinct from biological, determinants of health during the latter half of the twentieth century. However, as described in Chapter 1, the emergence of diverse reform movements in the nineteenth century shows that people had long speculated about the health implications of behavior, social conditions, and the wider environment. To that extent, we can say that appreciation of the importance of habits and habitats for health has existed for at least a couple of 100 years. Even that, however, would be gross underestimation. Thinking about health in relation to habits and habitats predates nineteenth century social reformers by more than 2000 years.

The ancient Greek physician Hippocrates is considered to be the father of Western medicine, which is ironic because Hippocrates' interest in the determinants of health ranged far beyond the biological focus that characterizes contemporary biomedical healthcare. Commenting on variations in the health of populations from different regions, Hippocrates (c. 400 BCE) wrote:

> *"Whoever wishes to investigate medicine properly, should [consider] the mode in which the inhabitants live, and what are their pursuits, whether they are fond of drinking and eating to excess, and given to indolence, or are fond of exercise and labor, and not given to excess in eating and drinking."*

The views expressed by Hippocrates are closely aligned with the collective opinion of the World Health Organization, which, as discussed in Part 3, has launched an
*Continued*

**BOX 4.1 Hippocrates on Habits and Habitats—cont'd**

ambitious plan to try to reduce global noncommunicable disease by 25% by 2025 (WHO, 2012a). The plan identifies four main modifiable risk factors as essential targets for achieving improvements in global personal and population health. In the brief quote above, Hippocrates foresaw three of those four targets: unhealthy diet, physical inactivity, and harmful consumption of alcohol. Only one of the four main targets identified by WHO (2012c, 2014b) was not foreseen by Hippocrates. Cigarette smoking was unknown in ancient Greece. Tobacco smoking as a global health issue did not emerge until after European colonization of the New World in the sixteenth century. Thus, taking account of Hippocrates' evident appreciation of the role of habits and habitats, we might imagine that he would have preferred to be remembered not as the father of biomedicine but as the progenitor of the biopsychosocial understanding of the foundations of health.

In reality, the targeting of biological processes in sick individuals is only one of a host of challenges that must be addressed to achieve optimal personal and population health, and the level of confidence in biomedical healthcare even with regard to that limited objective is disproportionate to the evidence. In that vein, Part 1 concludes with an examination of current and future prospects for biomedical healthcare from the perspective of noncommunicable "diseases," with particular reference to road traffic injury and cardiovascular diseases. This examination serves as an introduction to a more detailed discussion of related concerns in Part 3.

## 4.1 ROAD TRAFFIC INJURY AND BIOMEDICINE: THE RELEVANCE OF POLICY

It can be seen from Table 2.1 and Figure 2.1, based on recent WHO (2014a) estimates for annual global mortality, that there are more than 5 million deaths due to injuries, representing nearly 1 in 10 of all deaths worldwide. Figure 4.1 provides a summary of the main causes of death from unintentional and intentional injuries. About one-quarter are due to road traffic accidents, which are the most common cause of injury-related deaths worldwide, and *the* leading cause of death from *all causes* for those aged between 15 and 29 years. To add further perspective, road traffic deaths substantially exceed death due to self-harm (suicide), which, in turn, substantially exceeds death due to interpersonal violence (homicide). Whereas war and mass civil conflict are ever present in the public media, road traffic accidents go largely unreported except in official statistics, even though the former cause a fraction of the mortality and morbidity of the latter. Indeed, the annual number of deaths from suicide, violence, and wars combined only modestly exceeds that for the total number of road traffic deaths.

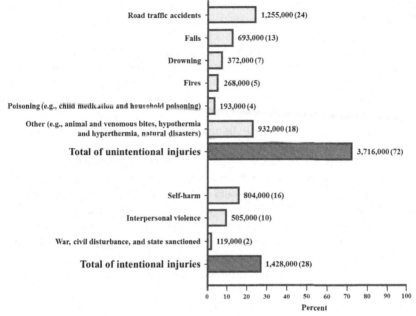

FIGURE 4.1    Number (percent) of the total of 5,144,000 deaths worldwide in 2013 due to different unintentional and intentional injuries. *(Derived from WHO (2014a).)*

## 4.1.1    Reducing the Level of Road Traffic Trauma

Biomedical intervention is at its most effective when treating acute physical trauma, and anyone sustaining serious injury should consider themselves fortunate if they receive emergency medical care with a minimum of delay. Therefore, it may seem logical to conclude that high priority should be given to the widest possible deployment of emergency medical personnel and facilities to address the high global rate of road traffic trauma. This, at least, might appear logical if we accept the premise that biomedical healthcare is the best means available for redressing physical trauma due to injury. However, what may appear to be logical reasoning is flawed due the self-evident reality that traffic-related injuries have primarily nonmedical causes and are largely *preventable*. Traffic injuries are primarily caused by road-user behavior, including inattention, excessive speed, driving while intoxicated, non-wearing of seatbelts, and driver fatigue, in combination with myriad other nonmedical factors such as traffic density, weather conditions, and roadway design. Consequently, exhaustive deployment of biomedical healthcare can never be the most effective way of limiting the toll of road-traffic-related injury nor the most sensible use of limited resources.

Nonfatal injuries exceed deaths by roughly 30-to-one (WHO, 2013), and add an immense burden of harm not visible in statistics pertaining to fatalities. The economic cost of road traffic mortality and morbidity combined is estimated to be 1-2% of the gross national product of most countries, rising to 5% in some, with a cumulative global cost in excess of US$500 billion annually (WHO, 2010). Costs arise from the delivery of emergency medical care, including ambulance and paramedic services, incident investigation, rehabilitation services of indefinite duration for the non-fatally injured, and lost or reduced productivity for the fatally and non-fatally injured.

Taking full account of the benefits of biomedical intervention in addressing road traffic trauma (and ignoring limitations discussed in Part 2), the proven efficacy of behavioral and social policies for reducing road traffic mortality and morbidity is far in excess of anything achievable through biomedical healthcare. The urgent need for those nonmedical interventions to be disseminated more widely grows daily in hand with growth in the road traffic toll, currently standing at about 3500 deaths per day worldwide. Despite the rate of global road traffic injury increasing apace over recent decades, the trend has not been universal. Beginning in the 1970s and 1980s in some high-income countries, previously increasing road accident rates were halted and reversed, thereby sparing millions of lives, billions of dollars, and incalculable human suffering.

### 4.1.1.1   Successful Road Traffic Policy

Several countries with relatively high rates of road traffic fatalities in the 1970s succeeded in improving those rates in the following decades. Australia and France, for example, both had fatality rates in excess of 30 deaths per hundred thousand of population, but by the late 2000s rates were reduced to less than a quarter of where they had been (WHO, 2009). Japan, with a moderately high rate of 20 road traffic fatalities per hundred thousand of population in the early 1970s also succeeded in reducing rates to almost a quarter of where they had been. The fact that reductions in road deaths of approximate equal proportions have been achieved by countries that had different initial rates raises an important question: What is the limit, if any, to the lowering of road trauma? Addressing that question, the Swedish Road Traffic Safety Bill of 1997 legislated for the reduction of fatalities and serious injuries to *zero* by 2020 (Whitelegg and Haq, 2006). The Vision Zero program was expressly founded on ethical and humanitarian considerations, including the principle that *life and health can never be exchanged for other benefits within the society* (Tingvall and Haworth, 2000).

Vision Zero brought together transport, justice, environment, health, and education sectors, and implemented stricter police enforcement of road safety rules. The program, which is currently being emulated in other countries (International Transport Forum, 2013), led to a fall in the number of fatal road crashes in Sweden from the already comparatively low level of about 9 deaths per 100,000 in 1990 to below 3 deaths per 100,000 in 2010. Moreover, that improvement was made against a background of steady increases in traffic volumes throughout the period (Whitelegg and Haq, 2006). The WHO (2010) lists the policies that evidence has shown to be effective, as follows:

- Setting and enforcing appropriate speed limits.
- Setting and enforcing drink-driving legislation.
- Setting and enforcing the wearing of seat-belts by all motor vehicle occupants, use of child restraints, and the wearing of helmets by motorcyclists and bicyclists.
- Developing safer roadway infrastructure, especially separating different types of road users (e.g., pedestrian footpaths and bicycle lanes).
- Traffic calming to reduce speeds in urban areas.
- Implementing vehicle and safety equipment standards.
- Setting and enforcing laws on daytime running of lights for motorcycles.
- Implementing a graduated driver licensing system for novice drivers.

Although policies such as the Vision Zero program have yet to show how close to zero it is possible to bring the annual road fatality rate, the rate of 3 deaths per 100,000 of population that has been achieved in Sweden is less than one-sixth the global rate (International Transport Forum, 2013; WHO, 2013). If all countries achieved the demonstrably achievable rate of 3 deaths per 100,000, more than one million deaths and roughly 30 million non-fatal injuries would be avoided each year. The effective policies are themselves fairly unremarkable and almost banal alongside the sometimes dazzling medical "miracles" used to save life and repair injury. Indeed, if a new prescription drug or medical device were developed that proved capable of delivering comparable benefits to those achievable through public policies to prevent road trauma, the said medical interventions would be demanded as a necessary healthcare service by communities everywhere. It is folly that highly effective policy initiatives do not have the same appeal as innovations in technology. Considering the immense life-saving and injury-preventing benefit that is *not* being achieved due to failure to implement traffic-related public policies that are known to work, road traffic trauma is possibly the most neglected of all major preventable health problems.

### 4.1.1.2 Trends in Road Traffic Trauma

Five or fewer traffic-related deaths per hundred thousand of population were reported in 2009 in only 10 of 178 countries. By comparison, fatality rates were twice or more than that rate (10 or more deaths per hundred thousand) in more than 80% of countries (146 of 178). In more than 30% of countries (55 of 178), the fatality rate was 25 or more deaths per hundred thousand (five or more times the rate in the best performing countries). Commenting on the improved situation in some countries, the WHO (2010) observed that the "downward trend in road traffic fatalities that began in the 1970s and 1980s has started to plateau, suggesting that extra steps are now needed to reduce these rates further" (p. 13). At present, there is no evidence that plateauing is inevitable other than as a consequence of neglect.

Notwithstanding the absence of an obvious plateau preventing further gains from Vision Zero and similar policies, it is nevertheless reasonable to assume that some level of risk of injury (albeit much less than currently exists in most countries) is intrinsic to road transport. As such, prevention strategies, irrespective of effectiveness, will not render biomedical healthcare obsolete as a response to road traffic trauma. Rather, the success of nonmedical strategies in preventing road traffic injury as a leading cause of mortality and morbidity points to the need for a balance between prevention and "cure," with the need for greater emphasis on the former and better appreciation of the limitations of the latter. However, almost everywhere, medicine is reckoned to be the cornerstone of healthcare. The annual toll of human tragedy wrought by road traffic accidents shows how profoundly mistaken such reckoning can be.

With the possible exception of roadway and traffic calming infrastructure, none of the WHO recommended policy steps outlined above involves the commitment of inordinately large public expenditure, especially when compared to current expenditure on biomedical healthcare. Thus, the behavioral, social, legislative, and political initiatives needed to achieve most of what works to lower road traffic injury is in the gift of stable governments everywhere. Failing to exercise that gift not only condemns citizens of those counties to untold human suffering but also incurs the immense cost of clinical invention that produces outcomes which, even when successful, are by definition less satisfactory than preventing the need for clinical invention in the first place.

### 4.1.1.3   All Injuries

Similar conclusions pertain to injury-related death in general. Table 4.1 lists policies recommended by WHO (2010) for preventing deaths due to falls, drowning, poisoning, suicide, and homicide. For those causes of death, as with road trauma, biomedical intervention offers, at best, an adjunctive strategy of last resort. Moreover, the same general conclusion applies to all the major noncommunicable causes of death, including cardiovascular diseases, cancers, chronic respiratory diseases, dementia, and diabetes. All are poorly served by healthcare dominated by biomedical preoccupations. The following section discusses cardiovascular disease, the leading cause of mortality and morbidity worldwide.

## 4.2   CARDIOVASCULAR DISEASE: THE LEADING GLOBAL CAUSE OF MORTALITY AND MORBIDITY

Currently, cardiovascular diseases account for nearly 1-in-3 deaths worldwide (Table 2.1). Despite a declining incidence of cardiovascular deaths in some

**TABLE 4.1** Policy steps to reduce mortality and morbidity due to injury other than road traffic trauma (policy initiatives for the latter are discussed in the text)

**Falls**

Setting and enforcing window guard laws for tall buildings

Redesigning furniture and other products

Establishing standards for playground equipment

**Drowning**

Removing or covering water hazards

Requiring isolation fencing (four-sided) around swimming pools

Wearing of personal flotation devices

Ensuring immediate resuscitation

**Poisoning**

Setting and enforcing laws for child resistant packaging of medicines and poisons

Removing the toxic product

Packaging drugs in nonlethal quantities

Establishing poison-control centers

**Burns**

Setting and enforcing laws on smoke alarms

Setting and enforcing laws on hot tap water temperatures

Developing and implementing a standard for child-resistant lighters

Treating burns patients in a dedicated burns center

**Self-harm**

Ensuring early detection and effective psychosocial treatment of mood disorders

Behavioral therapy for people experiencing suicidal thoughts and behavior

Restricting access to means (e.g. pesticides, guns, unprotected heights)

**Violence**

Developing safe, stable, and nurturing relationships between children and their parents or caregivers

Developing life skills in children and adolescents

Reducing the availability and harmful consumption of alcohol

Reducing access to guns and knives

Promoting gender equality to prevent violence against women

Changing cultural and social norms that support violence

Reducing violence through victim identification, care, and support programs

Adapted from WHO (2010).

countries during the last decades of the twentieth century (discussed below), the global burden of cardiovascular disease has been growing steadily for decades and that trend is expected to continue for decades to come. Forecasting the future for the United States, the American Heart Association (Heidenreich et al., 2011) estimated that, by 2030, 40% of the United States population will have some form of cardiovascular disease. Direct medical costs in the United States (in 2008 dollars) for all cardiovascular diseases, excluding intervention for hypertension (to avoid double counting), is expected to treble from $272 billion in 2010 to $818 billion by 2030. Additionally, indirect costs attributable to lost productivity due to cardiovascular disease are expected to increase from $172 billion to $276 billion for the same period. That is, the total cost of cardiovascular disease in the United States alone is expected to exceed USD1 trillion a year by 2030 (Heidenreich et al., 2011). For high-income countries generally, coronary heart disease has been described as "the most important chronic disease epidemic" of the twentieth century (Taylor et al., 2006, p. 760), an epidemic that is growing ever larger in the twenty-first century.

The popular belief that noncommunicable diseases such as cardiovascular disease mostly afflict high-income populations is misguided. The WHO (2012a) has estimated that the annual number of cardiovascular disease deaths worldwide will increase from about 17 million today to 25 million by 2030. The projected increase partly reflects the epidemiologic transition currently under way in low- and middle-income countries, which are currently transitioning from higher to lower rates of death from infectious diseases combined with rapidly increasing rates of mortality from noncommunicable diseases. Table 2.1 shows that 75% of cardiovascular disease currently occurs in developing countries, and that proportion is expected to grow as more developing countries experience both epidemiologic transition and population aging. Consequently, many developing countries currently suffer double jeopardy imposed by continuing high (albeit decreasing) rates of communicable disease and increasing rates of noncommunicable disease.

The current and anticipated future trajectory for cardiovascular disease incidence means that it is the leading global health problem of today and will remain so over the coming decades. Figure 4.2 lists the main cardiovascular diseases in order of prevalence, with 80% being accounted for by *coronary heart disease* (also referred to as *myocardial ischemia, coronary artery disease*, and *ischemic heart disease*) and *stroke* (*cerebrovascular accident*). Although rates of cardiovascular disease are increasing in most countries, some countries, as mentioned above, experienced decreased rates for both coronary heart disease (Mirzaei et al., 2009) and stroke (Mirzaei et al., 2012) during the last decades of the twentieth century. Notably, the negative forecasts about future rates take account of past improvements in those countries. The United States, for example, experienced substantial gains in cardiovascular health in recent decades, but because of evidence that gains have stalled and appear to be reversing the American Heart Association has issued the dire forecasts outlined above.

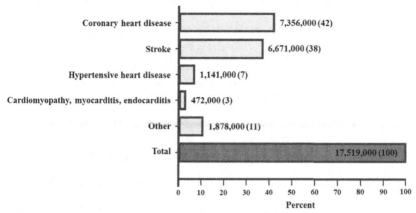

**FIGURE 4.2**   Number (percent) of deaths worldwide in 2013 from major cardiovascular diseases. *(Derived from WHO (2014a).)*

Considering the overall current impact of cardiovascular disease and the gloomy forecasts, it is reasonable to ask about biomedicine's record as well as its likely future effectiveness in containing this most common of noncommunicable diseases. Examining differences in cardiovascular disease prevalence between countries and changes in prevalence over time within countries is important for identifying contributing factors and elucidating effective interventions. Due to extensive study of such patterns, there is now a large body of scientific evidence to illuminate biomedicine's relative (in)effectiveness compared to other factors.

### 4.2.1   What Caused Recent Falls in Rates of Cardiovascular Deaths?

Several high-income countries in Western Europe, North America, and the Western Pacific region experienced pronounced decreases in rates of heart disease mortality during the last decades of the twentieth century (e.g., WHO, 2012a). The trends for six countries of Western Europe are shown in Figure 4.3 (WHO, 2012b). For all six countries, it can be seen that rates of death fell on average by more than 40% during the three decades from 1970. Interestingly, irrespective of absolute rates, the trends were broadly similar between countries. For example, Finland had the highest rate of coronary heart disease death in 1970, and by 2000 the rate had fallen by 43%. On the other hand, whereas the rate in the Netherlands as early as 1970 was only slightly higher than that for Finland in 2000, the Dutch rate of coronary heart disease death had fallen by 48% by 2000.

Impressive as they were, the successes achieved in some European countries evidently do not represent the limits of what is achievable. For the whole of the period that gains were being made in Europe, Japan maintained a rate of

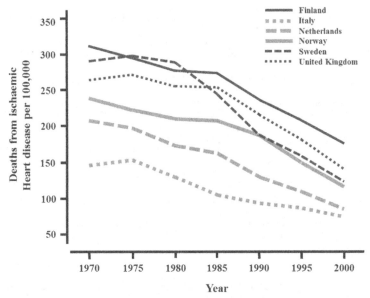

**FIGURE 4.3**   Deaths from coronary heart disease in six countries of Europe from 1970 to 2000. *(Derived from WHO (2012b).)*

coronary heart disease death considerably below that of any European country. By 2000, Japan's rate of death from coronary heart disease was less than 50 per 100,000 of population, well below Europe's best performer, Italy, which had a rate of 75 per 100,000 of population in the same year. Importantly, rates of coronary heart disease, and especially the decreases in rate of death evident in Figure 4.3, were reflected in increased population life expectancy. As shown in Figure 4.4, by 2000, citizens of the same six European countries listed in Figure 4.3 lived 6 years longer than 30 years earlier. A similar proportional benefit was also evident in Japan. With its low rate of ischemic heart disease, Japan also had the longest average life expectancy at birth, 81.2 years, of any country in 2000.

   Differences between countries in rate of disease and life expectancy, such as are evident in Figures 4.3 and 4.4, are not likely to be explained by variation in medical innovation and practice. All of the countries illustrated are high-income and the medical healthcare available within each is approximately uniform. Substantial lack of uniformity would be expected if biomedical healthcare were the main factor influencing disease rates. Additionally, genetic differences are unlikely to explain differences between countries or changes in trend over time. To illustrate, studies of Japanese migration to the United States have shown that rates of coronary heart disease among immigrants come to approximate those of the host country in proportion to the rate of adoption of local ways of living (e.g., diet and other lifestyle factors; Robertson et al., 1977). Because American Japanese share the same genes as their kin who remained in Japan, their poorer

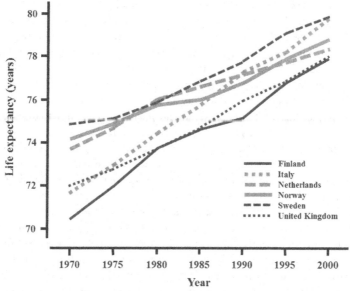

**FIGURE 4.4**   Life expectancy at birth in six countries of Europe from 1970 to 2000. *(Derived from WHO (2012b).)*

cardiovascular health (i.e., poor compared to Japanese living in Japan but similar to average Americans) cannot be attributed to their genes. Rather their changed health status is illustrative of a general principle, namely, that differences in disease burden between countries and over time are primarily attributable to differences in health-related habits and habitats, typically indexed by measuring health-related *risk factors*.

## 4.2.2   Health-Related Risk Factors and Cardiovascular Disease

With respect to human health, a risk factor is any attribute of the person or the environment that increases individual susceptibility to disease or injury. The term, however, should be used with care. The nature of the evidence implicating risk factors is often correlational, and correlation alone does not establish causation. To illustrate by reference to an oft-cited example, being young is associated with increased risk of contracting measles. We might, therefore, conclude (falsely) that young age is a cause of measles. Because children are less likely than adults to have been exposed to the measles virus, immunity will not have developed – leaving children vulnerable to contracting the disease when they do encounter it. Thus, although age and measles are highly correlated, young age does not cause measles. The relevant risk factor in this instance is *exposure* to the measles virus, wherein risk of contracting the disease is high when the virus is encountered for the first time, irrespective of age.

The terms *risk marker, possible risk,* or *association* are sometimes used to refer to variables that are correlated with disease incidence but are either considered not to be causes or their causal status is unknown. Therefore, in the example just given, young age is a risk marker and not a risk factor. When a variable is identified as being associated with disease, any causal contribution it may have is rarely established in the first instance. Hence, the newly identified variable may initially be referred to as a risk marker and only later, when sufficient evidence has accumulated to indicate a causal connection to disease, will the variable be described as a risk factor. Thus, risk factor refers to any variable (a biological indicator, behavior, or environmental factor) that in light of scientific evidence is generally accepted as being a cause of particular disease outcomes. Importantly, most diseases, rather than being caused by a single factor, are the result of *multifactorial* causation.

Emphasis on a particular cause or type of cause to the neglect of other relevant factors can create distorted understanding of disease processes and undermine intervention effectiveness. Indeed, precisely that point is a recurring theme of the present book. Longstanding preoccupation with biological causes of disease has resulted in a legacy of healthcare dominated by biologically based interventions (e.g., pharmacotherapy and surgery) to the neglect of behavioral and environmental targets for intervention that typically account for a larger proportion of the distribution of health and disease than is explained by biological factors.

### 4.2.2.1 Causes of Cardiovascular Disease

Extensive evidence shows that the main causes of cardiovascular disease are cigarette smoking, dietary behavior, and physical inactivity (WHO, 2014b). These long-term *distal* behavioral causes of cardiovascular disease contribute to the development of more *proximal* causes, including high blood pressure, elevated blood cholesterol, obesity, and diabetes. In turn, this complex of interrelated behavioral and biological causes and outcomes occur in the context of economic and sociopolitical influences that determine patterns of health-related behavior from a young age and across the life span. Thus, aspects of the social milieu, including lower socioeconomic status, peers who smoke, exposure to advertising that encourages consumption of non-nutritious foods, and patterns of family life (*domestic culture*) that favor participation in passive entertainments, to name a few, form part of the multiple chains of biopsychosocial causes and effects that determine the pattern and extent of personal and population health and disease. The behavior of smoking, for example, has multiple effects on biology (e.g., physical dependence on nicotine and damage to blood vessels) that contribute to disease outcomes. Similarly, unhealthy dietary behavior (e.g., consumption of processed foods with a high content of fat, salt, or sugar) contributes to increased blood pressure and the atherosclerotic processes that presage disease. The causal chains of events that characterize all of the leading

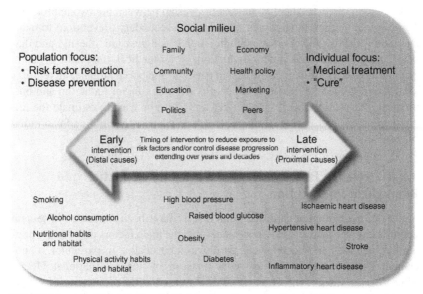

FIGURE 4.5    A 'habits and habitats' overview of cardiovascular disease risk factors and outcomes.

cardiovascular diseases are typically chronic in nature and contribute to disease progression that extends over long time periods.

Figure 4.5 provides a simplified overview of the chronology of events that may broadly be seen as comprising a habits and habitats perspective of cardiovascular disease. Social milieu (e.g., peer influence, family patterns, education, commercial marketing, community services and facilities, public policy, the economy, and the political environment) is the primary habitat within which diverse behaviors (habits) such as avoidance or not of smoking and harmful consumption of alcohol, dietary practices, and level of physical activity protect against or contribute to key intermediate biological outcomes, such as high blood pressure, raised blood sugar, obesity, and diabetes, that may in turn lead to late- and end-stage disease outcomes including heart disease and stroke. The timescale of years and decades over which these events occur provides wide scope for choosing how and when to intervene to reduce the incidence of disease. Figure 4.5 summarizes some of the main intervention options, which approximate a dichotomy of choice. Population-focussed strategies are timed to occur at an early stage of disease progression to prevent disease by reducing exposure to distal risk factors that cause disease. Conversely, individually-focussed biomedical interventions for cardiovascular (and most noncommunicable) diseases typically address proximal causes during late-stage disease to slow disease progression but without delivering a cure.

Although very different in focus and content, risk factor reduction and biomedical intervention are not mutually exclusive strategies for promoting health and reducing disease. There is an area of overlap between the two approaches wherein intervention (sometimes referred to as secondary prevention) to manage biological risk factors, for example, high blood pressure, obesity, and diabetes, may involve risk factor reduction or biomedical treatment or both. Accordingly, a coordinated approach to personal and population health requires knowledge of the likely costs and benefits from risk factor reduction and biomedical intervention. Such knowledge can in turn be used to estimate the relative emphasis and timing of alternative healthcare interventions for optimal overall benefit.

### 4.2.3    Relative Effectiveness of Risk Factor Reduction and Biomedical Intervention

Studies of the burden of cardiovascular disease have been conducted in several countries, with consistent findings concerning the main factors influencing population increases and decreases in cardiovascular deaths. In particular, a substantial body of evidence has been amassed by the International Health Impact Assessment Consortium (IMPACT) based at the University of Liverpool in the United Kingdom (Capewell et al., 2009). The IMPACT model combines data from multiple repositories that describe:

- the number of patients in different disease subcategories of coronary heart disease,
- the percentage of patients who received specific biomedical interventions,
- the effectiveness of those interventions in preventing or postponing coronary heart disease deaths,
- population trends in the prevalence of the leading cardiovascular disease risk factors, and
- the effects of change in *secular* risk factors on coronary heart disease deaths.

In one IMPACT study, referred to in Chapter 3, mortality due to coronary heart disease was comprehensively examined for England and Wales for the period 1981-2000 (Ünal et al., 2004, 2005a,b). As mentioned in Chapter 3, the rate of death due to coronary heart disease was found to have decreased by 54%, resulting in almost 70,000 fewer deaths in 2000 than would have occurred if the rate had remained unchanged since 1981. Apportioning lives saved to two main causes, Figure 4.6 summarizes the number of deaths avoided (prevented or postponed) due to biomedical intervention and changes in level of exposure to known risk factors, respectively, in persons aged 25-84 years in the patient and secular populations, respectively. The patient population was defined as people with diagnosed coronary heart disease, including myocardial infarction, angina, and heart failure, and the secular population was comprised of the general population for the reference age group. Medical intervention included cardiopulmonary resuscitation, coronary artery bypass surgery (coronary artery bypass graft, CABG), angioplasty, thrombolysis, and medication

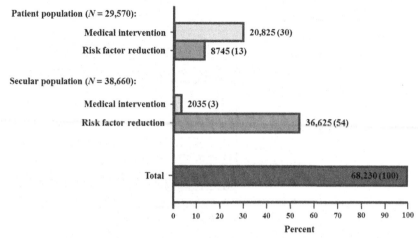

**FIGURE 4.6**   Contribution of medical intervention and risk factor reduction to number (percent) of deaths avoided from coronary heart disease in the patient and secular populations, respectively, in England and Wales between 1981 and 2000. *(Derived from Ünal et al. (2005a).)*

(e.g., β-blockers, ACE inhibitors, warfarin, and statins). The risk factors considered were cigarette smoking, blood pressure level, blood cholesterol concentration, level of socioeconomic deprivation, type 2 diabetes, physical activity, and obesity. Improvements were observed for all except the last three.

Although no direct attempt was made by Ünal et al. (2004, 2005a, b) to identify the causes of secular changes in risk factors, improvements were thought to have been due largely to publicly-sponsored health promotion campaigns and the ongoing dissemination through public media of information about healthy living. These actions appear to have contributed to positive shifts in norms of behavior evident in reduced population rates of smoking and modest positive trends in dietary patterns that contributed to reduced blood pressure (largely attributed to reduced intake of saturated fat and salt) and lower cholesterol levels (attributed to increased intake of fruit, fiber, and unsaturated fats).

As well as estimating the level of biomedical intervention to treat diagnosed cases of heart disease and the extent of secular changes in relevant risk factors, the IMPACT model takes account of the fact that patients receiving medical treatment may also derive benefit from change in risk factors (e.g., quitting smoking) and persons free of heart disease may receive medical treatment (secondary prevention) for relevant but asymptomatic conditions (e.g., drug treatment to lower blood pressure or cholesterol levels). Thus, alternative representations of the results summarized in Figure 4.6 are provided in Figure 4.7, which shows the respective contribution of medical intervention and risk factor reduction in the patient and secular populations combined (Panel a)), and the combined benefit of medical intervention and risk factor reduction for the patient and secular populations, respectively (Panel b)). To summarize, the IMPACT model assesses

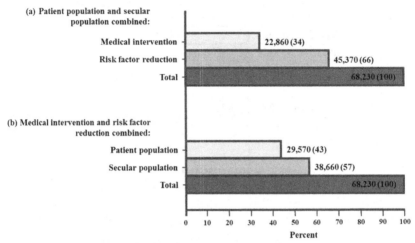

**FIGURE 4.7**  Number (percent) of deaths avoided from coronary heart disease in England and Wales between 1981 and 2000. (a) Respective contribution of medical intervention and risk factor reduction in the patient and secular populations combined and (b) the combined benefit of medical intervention and risk factor reduction for the patient and secular populations, respectively. *(Derived from Ünal et al. (2005a).)*

the overall relative contribution of biomedical intervention and risk factor reduction to avoidance of death from heart disease that occurred during a given period in both the patient and secular populations.

Referring to Figure 4.7, Panel (a), it can be seen that of the 68,230 deaths avoided in England and Wales between 1981 and 2000, 45,370 (66%) were the result of change in risk factors compared to 22,860 (34%) due to medical intervention. That is, risk factor reduction was found to have had an approximate twofold greater life-saving effect than biomedical intervention. In fact, risk factor reduction was even responsible for 30% of deaths avoided among patients receiving medical care (8745 of 29,570; Figure 4.6). Conversely, medical intervention in the secular population (secondary prevention) aimed at reducing the risk factors of elevated blood pressure and cholesterol had comparatively little life-saving effect (about 5%), comprised of only 2035 deaths avoided of a total of 38,660 (Figure 4.6).

Notably, Figure 4.7, Panel (b), shows that the greater portion of the total population benefit from medical intervention and risk factor reduction combined occurred within the secular population (57%) and not the patient population (43%). This is striking given that healthcare is dominated by biomedical interventions targeted at patients, a fact confirmed by healthcare expenditure patterns. Typically, 95% or more of a nation's healthcare budget is spent on biomedical interventions, with the remaining 5% or less being allocated to population-wide health promotion and disease prevention (Marmot and

Bell, 2012; McGinnis et al., 2002). Yet, these findings show that risk-factor reduction for healthy people contributes substantially more to population health and for much less cost than biomedical intervention targeted at patients. Consequently, investing more to improve population risk factors is likely to prove considerably more effective and cost-effective for the health of populations than continuing to invest ever-larger amounts in biomedical intervention.

The pattern of results for England and Wales has been confirmed time-and-again in other countries. Figure 4.8 provides a summary of similar findings from 11 studies conducted in geographically diverse countries. The figure shows that biomedical intervention, though beneficial, has been consistently found to be less effective than risk factor reduction for reducing population levels of death from heart disease. When the results for all 11 studies are combined, the median percentage contribution to the overall reduction in incidence of cardiovascular disease is found to be 40% for biomedical intervention, 55% for secular risk factor reduction, and 5% unexplained. Even then, though undoubtedly impressive, Figure 4.8 underestimates the overall benefit from risk factor reduction. Because health-enhancement from lower exposure to risk factors usually occurs earlier in life than the age at which biomedical intervention occurs, the overall

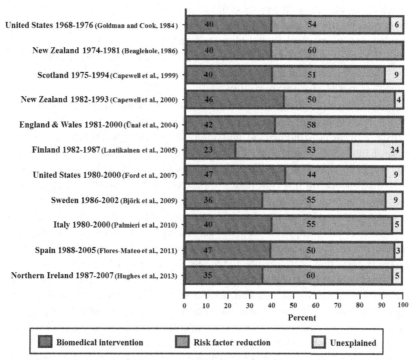

**FIGURE 4.8** Estimated percentage decrease in coronary heart disease deaths attributed to biomedical intervention, risk factor reduction, and unexplained in diverse populations worldwide.

benefit of risk factor reduction is substantially larger when measured in terms of life-years gained than when measured simply as number of deaths avoided or postponed.

## 4.2.4   Beyond Deaths Avoided to Life-years Gained

Analyses show that the number of added years of life gained varies depending on how death was avoided (e.g., Grover et al., 1998), whether by medical intervention or by risk factor reduction. The almost 70,000 deaths that were prevented or postponed in England and Wales between 1981 and 2000 equate to a total of approximately 1 million extra years of life for the people who benefited (Ünal et al., 2005b). When these extra years of life were apportioned by cause of death, it was found that death avoided due to medical intervention added an average 7.5 years of life, whereas death avoided due to risk factor reduction added an average of 20 years of life. This difference is explained in part by the fact that secular improvements in risk factors tend on average to precede the appearance of manifest disease, and thereby confer a greater health benefit by halting or postponing processes responsible for disease. Conversely, as mentioned earlier in this chapter, clinical intervention generally comes relatively late in the process of disease progression, producing less benefit compared to risk factor reduction. Taking account of the number of deaths avoided due to medical intervention and risk factor reduction, respectively, Table 4.2 shows that the former contributed 21% of the total of 1 million extra life-years whereas the latter "accounted for a massive 79% of the total life-years gained" (Ünal et al., 2005b, p. 106).

It is worth emphasizing that, on average, each death avoided due to risk factor reduction is about three times more beneficial in terms of years of additional life than the benefit due to biomedical intervention. Generally speaking, people receive medical intervention when disease has developed, but even the most effective intervention at that time is unlikely to cure disease. On the contrary, intervention is deemed successful if it contributes to postponement of death (by 7.5 years on average in the present instance of coronary heart disease), with patients continuing in varying states of disability for the remainder of life. By contrast, risk factor reduction not only postpones death (20 years on average for coronary heart disease) but also postpones disease onset. In other words, risk factor reduction confers a double benefit, that of longer life and longer disease-free life.

Just as secular improvements in risk factors can deliver great benefits for the health of populations, so too can deterioration in risk factors produce great harm. Referring to Table 4.2, in addition to summarizing benefits from risk factor reduction, the table also summarizes the cost in life-years of negative secular trends in three risk factors: type 2 diabetes, physical inactivity, and obesity. Measured as life-years lost due to coronary heart disease, Table 4.2 shows that the combined negative effect of those three largely *preventable* risk factors was 10%, equivalent to about half of the total benefit from biomedical intervention.

**TABLE 4.2** Contribution of biomedical intervention and secular risk factor reduction to life-years gained due to avoidance of death from coronary heart disease in England and Wales between 1981 and 2000.

Contribution (%) to life-years gained

| Intervention | Medical intervention and risk factor reduction | |
|---|---|---|
| | Separately[a] | Combined[b] |
| **Medical intervention (194,145 life-years):** | | |
| Acute myocardial infarction (MI) | 20 | 4 |
| Post-MI angina, hypertension, elevated cholesterol, coronary surgery, angioplasty, medication, etc. | 80 | 17 |
| Subtotal | 100 | 21 |
| **Risk factor reduction (731,270 life-years):** | | |
| Smoking | 54 | 43 |
| Blood pressure | 28 | 22 |
| Cholesterol | 23 | 18 |
| Socioeconomic deprivation | 7 | 6 |
| Type 2 diabetes[c] | −6 | −5 |
| Physical inactivity[c] | −5 | −4 |
| Obesity[c] | −1 | −1 |
| Subtotal | 100 | 79 |
| Total (925,415 life-years) | | 100 |

[a]Percentage benefit attributed to specific medical interventions and risk factors considered separately.
[b]As a percentage of the total of 925,415 life-years gained due to deaths prevented or postponed from medical intervention (194,145) and risk factor reduction (731,270), respectively.
[c]Negatively-trending risk factors that have a harmful impact on life-years.
Derived from Ünal et al. (2005b).

Moreover, whereas some dietary trends were positive (e.g., reduced salt intake) other more-or-less concurrent trends (e.g., increased consumption of refined sugar) were negative.

### 4.2.4.1   How Effective Are Specific Interventions?

By summarizing the contribution of individual interventions for reducing coronary heart disease, Table 4.2 illustrates the immense importance of risk factor reduction overall as well as the relative contribution of different risk factors.

Reduced prevalence of cigarette smoking was most important, being responsible for 43% of total life-years gained, more than twice the total benefit from all medical intervention combined. Moreover, the benefit from reduced smoking for heart disease takes no account of other substantial benefits, including reduced lung and other cancers among smokers and deaths avoided due to reduced prevalence of *passive smoking*. Smoking was followed in importance by secular reductions in blood pressure and blood cholesterol levels. Each of these individually was approximately equivalent to total benefit achieved by medical intervention. Broadly consistent findings across many studies, such as those summarized in Figure 4.8, leave no doubt about the power of prevention in general and risk factor reduction in particular. These topics are revisited in detail in Part 3.

## REFERENCES

Beaglehole, R., 1986. Medical management and the decline in mortality from coronary heart disease. Br. Med. J. 292, 33–35.

Björck, L., Rosengren, A., Bennett, K., et al., 2009. Modelling the decreasing coronary heart disease mortality in Sweden between 1986 and 2002. Eur. Heart J. 30, 1046–1056.

Capewell, S., Morrison, C.E., McMurray, J.J., 1999. Contribution of modern cardiovascular treatment and risk factor changes to the decline in coronary heart disease mortality in Scotland between 1975 and 1994. Heart 81, 380–386.

Capewell, S., Beaglehole, R., Seddon, M., McMurray, J.J., 2000. Explaining the decline in coronary heart disease mortality in Auckland, New Zealand between 1982 and 1993. Circulation 102, 1511–1516.

Capewell, S., Hayes, D.K., Ford, E.S., et al., 2009. Life-years gained among US adults from modern treatments and changes in the prevalence of 6 coronary heart disease risk factors between 1980 and 2000. Am. J. Epidemiol. 170, 229–236.

Dubos, R.J., 1959. Mirage of Health: Utopias, Progress, and Biological Change. Harper and Brothers, New York.

Engel, G.L., 1977. The need for a new medical model: a challenge for biomedicine. Science 196, 129–136.

Flores-Mateo, G., Grau, M., O'Flaherty, M., 2011. Analyzing the coronary heart disease mortality decline in a mediterranean population: Spain 1988–2005. Rev. Esp. Cardiol. 64, 988–996.

Ford, E.S., Ajani, U.A., Croft, J.B., et al., 2007. Explaining the decrease in U.S. deaths from coronary disease, 1980–2000. N. Engl. J. Med. 356, 2388–2398.

Goldman, L., Cook, E.F., 1984. The decline in ischemic heart disease mortality rates: an analysis of the comparative effects of medical interventions and changes in lifestyle. Ann. Intern. Med. 101, 825–836.

Grover, S.A., Paquet, S., Levinton, C., et al., 1998. Estimating the benefits of modifying risk factors of cardiovascular disease: a comparison of primary vs secondary prevention. Arch. Intern. Med. 158, 655–662.

Heidenreich, P.A., Trogdon, J.G., Khavjou, O.A., et al., 2011. From the American heart association forecasting the future of cardiovascular disease in the United States: a policy statement. Circulation 123, 933–944.

Hippocrates (c. 400 BCE). *On Airs, Water, and Places, Part 1*. Translated by Francis Adams. Retrieved on 9 January 2012 from http://classics.mit.edu/Hippocrates/airwatpl.1.1.html.

Hughes, J., Kee, F., O'Flaherty, M., Critchley, J., et al., 2013. Modelling coronary heart disease mortality in Northern Ireland between 1987 and 2007: broader lessons for prevention. Eur. J. Prev. Cardiol. 20, 310–321.

International Transport Forum, 2013. Road Safety Annual Report 2013. International Transport Forum, Paris. Retrieved 4 April 2014 from, http://www.internationaltransportforum.org/Pub/pdf/13IrtadReport.pdf.

Laatikainen, T., Critchley, J., Vartiainen, E., et al., 2005. Explaining the decline in coronary heart disease mortality in Finland between 1982 and 1997. Am. J. Epidemiol. 162, 764–773.

Marmot, M., Bell, R., 2012. Fair society, healthy lives. Public Health 126, S4–S10.

McGinnis, J.M., Williams-Russo, P., Knickman, J.R., 2002. The case for more active policy attention to health promotion. Health Aff. 21, 78–93.

McKeown, T., 1979. The Role of Medicine: Dream, Mirage or Nemesis?. Basil Blackwell, Oxford, UK.

McKeown, T., Brown, R.G., 1955. Medical evidence related to English population changes in the eighteenth century. Popul. Stud. 9, 119–141.

Mirzaei, M., Truswell, A.S., Taylor, R., Leeder, S.R., 2009. Coronary heart disease epidemics: not all the same. Heart 95, 740–746.

Mirzaei, M., Truswell, A.S., Arnett, K., et al., 2012. Cerebrovascular disease in 48 countries: secular trends in mortality 1950–2005. J. Neurosurg. Psychiatry 83, 138–145.

Palmieri, L., DrStat, K.B., Giampaoli, S., Capewell, S., 2010. Explaining the decrease in coronary heart disease mortality in Italy between 1980 and 2000. Am. J. Public Health 100, 684–692.

Robertson, T.L., Kato, H., Rhoads, G.G., et al., 1977. Epidemiologic studies of coronary heart disease and stroke in Japanese men living in Japan, Hawaii and California: incidence of myocardial infarction and coronary heart disease. Am. J. Cardiol. 39, 239–243.

Rose, G., 1981. Strategy of prevention: lessons from cardiovascular disease. Br. Med. J. 282, 1847–1851.

Szasz, T.S., 1961. The Myth of Mental Illness. Paul Hocher, New York.

Taylor, R., Dobson, A., Mirzaei, M., 2006. Contribution of changes in risk factors to the decline of coronary heart disease mortality in Australia. Eur. J. Cardiovasc. Prev. Rehabil. 13, 760–768.

Tingvall, C., Haworth, N., 2000. Vision zero: an ethical approach to safety and mobility. In: Paper Presented at the 6th ITE International Conference Road Safety & Traffic Enforcement: Beyond 2000, Melbourne, 6–7 September 1999. Retrieved 10 October 2014 from, http://www.avr.lu/web/resources/Microsoft_Word___Vision_Zero_706.pdf.

Ünal, B., Critchley, J.A., Capewell, S., 2004. Explaining the decline in coronary heart disease mortality in England and Wales between 1981 and 2000. Circulation 109, 1101–1107.

Ünal, B., Critchley, J.A., Capewell, S., 2005a. Modelling the decline in coronary heart disease deaths in England and Wales, 1981–2000: comparing contributions from primary prevention and secondary prevention. Br. Med. J. 1–6. http://dx.doi.org/10.1136/bmj.38561.633345.8F.

Ünal, B., Critchley, J.A., Fidan, D., Capewell, S., 2005b. Life-years gained from modern cardiological treatments and population risk factor changes in England and Wales, 1981–2000. Am. J. Public Health 95, 103–108.

Whitelegg, J., Haq, G., 2006. Vision zero: Adopting a Target of Zero for Road Traffic Fatalities and Serious Injuries. Stockholm Environment Institute, Stockholm, Sweden. Retrieved 10 November 2014 from, http://www.sei-international.org/mediamanager/documents/Publications/Future/vision_zero_FinalReportMarch06.pdf.

WHO, 2009. Global health risks: mortality and burden of disease attributable to selected major risks. World Health Organization, Geneva. Retrieved 18 October 2012 from, http://www.who.int/healthinfo/global_burden_disease/GlobalHealthRisks_report_full.pdf.

WHO, 2010. Injuries and Violence: The Facts. World Health Organization, Geneva. Retrieved 31 October 2012 from, http://www.who.int/violence_injury_prevention/key_facts/VIP_key_facts.pdf.

WHO, 2012a. World Health Statistics 2012. World Health Organization, Geneva Retrieved 18 October 2012 from, www.who.int.

WHO, 2012b. European Health for All Database (HFA-DB). World Health Organization, Geneva. Retrieved 23 October 2012 from, http://data.euro.who.int/hfadb/.

WHO, 2012c. A Comprehensive Global Monitoring Framework, Including Indicators, and A Set of Voluntary Global Targets for the Prevention and Control of Noncommunicable Diseases. World Health Organization, Geneva. Retrieved 25 May 2013 from, http://www.who.int/nmh/events/2012/discussion_paper3.pdf.

WHO, 2013. Global Status Report on Road Safety 2013: Supporting a Decade of Action. World Health Organization, Geneva, Switzerland. Retrieved 22 November 2013 from, http://www.who.int/violence_injury_prevention/road_safety_status/2013/en/.

WHO, 2014a. Global Health Estimates 2014 Summary Tables: Deaths by Cause, Age and Sex, by World Bank Region, 2000–2012. World Health Organization, Geneva, Switzerland. Retrieved 29 August 2014 from, http://www.who.int/healthinfo/global_burden_disease/en/.

WHO, 2014b. Global Status Report on Noncommunicable Diseases 2014. World Health Organization, Geneva. Retrieved 13 February 2015 from, http://www.who.int/nmh/publications/ncd-status-report-2014/en/.

# Part 2

# The Harm of Medicine

---

*I observe the physician with the same diligence as he the disease.*

(John Donne, 1572-1631, English poet)

The main threats to human health are from exposure to modifiable disease risk factors associated with human habits and habitats. The evidence discussed in Part 1 shows that the health of populations past and present is only modestly attributable to advances in biomedical healthcare. Several of the themes discussed in Part 1 are revisited in Part 3. Those recurring themes relate primarily to the role of risk factor reduction, the power of disease prevention, and evidence of the urgent need for current healthcare priorities to be radically reordered. In Part 1, infectious diseases in developed countries were described as having been replaced by noncommunicable diseases as the contemporary leading causes of death. The same broad pattern of change continues to spread worldwide. Changing patterns of disease have prompted diverse innovations in clinical medicine, and medicine's dominance in healthcare has grown inexorably. However, dominance should not be confused with effectiveness. Extensive evidence shows that biomedicine's growing dominance is not the basis for solving the global burden of disease but an impediment.

It is evident from popular discourse that clinical medicine is widely believed to be in short supply and that increased healthcare expenditure is the only solution. In important respects, however, the reverse is true, wherein much harm to health is directly attributable to overutilization of biomedical healthcare. Increasingly, observers inside and outside the profession of medicine identify, sometimes despairingly, medicine's greatest challenge as the need to reverse the trend that has led to modern biomedicine becoming a leading global threat to health. No aspect of healthcare is more in need of radical change than the harm caused by biomedical intervention. Accordingly, Part 2 examines medicine's dominance in healthcare from the perspective of the extensive harm it causes.

## WHAT ABOUT THE GOOD OF MEDICINE?

By revealing a scale of harm from biomedical healthcare that exceeds most expectations, Part 2 could provoke questions of "balance." What about the good of medicine? Recently, leading infectious disease experts issued an appeal to colleagues in their field "to find a proper balance between effectiveness and harm" (Fätkenheuer et al., 2015, p. 1148). A similar appeal could be made of every major field of medical practice. However, that exercise should not be taken to imply that what is required is the listing of pros and cons as if constructing a "balance sheet," wherein all the good of medicine is quantified, all the bad is similarly quantified, and the one is subtracted from the other to arrive at an estimate, positive or negative, of the total impact of medicine. A positive balance might then be taken as indicating that medicine is "good," whereas a negative balance might be taken as indicating that medicine is "bad." Such dichotomous thinking overlooks what is the one true goal, namely, to preserve that which is effective in healthcare and to discard that which is harmful.

Moreover, while acknowledging good in contemporary healthcare, it is vital not to ignore the question of *opportunity cost*, which refers to the loss of potential gain from alternative courses of action when a particular action is chosen from among mutually exclusive alternatives. Biomedical dominance in healthcare reflects policy *choices* that produce certain benefits and harms. The tally of benefits should not ignore the possibility that other choices may produce similar benefits and more, and possibly with fewer harms. By focusing on the immense scale of medical harm, Part 2 provides the basis for understanding the great and urgent need for a reordering of priorities in contemporary healthcare policy and practice. Whereas Part 2 focusses on the harm of biomedicine, that which is good in healthcare is discussed in Part 3 in the context of choices that can deliver levels of personal and population health unmatched by biomedical healthcare as currently practiced.

## REFERENCE

Fätkenheuer, G., Hirschel, B., Harbarth, S., 2015. Screening and isolation to control meticillin-resistant Staphylococcus aureus: sense, nonsense, and evidence. Lancet 385, 1146–1149.

Chapter 5

# Medical Harm: What Is It and What Is the Extent?

*Doctors...have more lives to answer for...than ever we generals.*

(Napoleon Bonaparte, 1769-1821)

## Contents

"First, do no harm" is a familiar aphorism which stands as the universal ethical cornerstone of all healing practice. It is taught in medical schools, and espoused by healthcare services throughout the world. Its purpose is to remind practitioners to always be aware that harm may be inflicted when administering care, no matter how well-intentioned. Despite these strictures, harm is endemic to biomedical healthcare, and concern over the potential for medicine to harm

The Health of Populations. http://dx.doi.org/10.1016/B978-0-12-802812-4.00005-9

is as old as medicine itself. The aforementioned aphorism derives from Hippocrates (c. 400 BCE):

*The physician must...have two special objects in view with regard to disease, namely, to do good or to do no harm.*

Although medical harm has been long acknowledged, attempts to understand the problem have been slow to emerge. In the mid-1970s, Austrian philosopher and social critic Ivan Illich (1976, p. 3) wrote that the "medical establishment has become a major threat to health [that] has reached the proportions of an epidemic'." In the decades since, similar laments have been repeated in extensive scholarly commentary. In a brief opinion piece in a recent issue of the *British Medical Journal*, Scottish primary care physician Des Spence (2012) asserted that the main contemporary cause of disease needing the attention of the medical profession is medicine itself. According to Spence, "medicine's challenge" for the twenty-first century is to "fight the pandemic" of medical harm.

Large-scale systematic analyses of medical harm, of which there have been many in recent years, have generally been accompanied by expressions of surprise and concern, a growing sense of urgency, and at times, despair. In the late 1990s, the American Institute of Medicine (IOM), which has been at the forefront of attempts to reveal the scale of medical harm, stated:

*Problems in health care quality are serious and extensive; they occur in all delivery systems [and inflict] a great burden of harm...that is measured in lost lives, reduced functioning, and wasted resources...these problems call for urgent action... The burden of harm...is staggering.*

(Chassin and Galvin, 1998, pp. 1001–1004)

Just how "staggering" were those problems was revealed in a later IOM report (Kohn et al., 2000) discussed below. Those revelations, however, were quickly eclipsed when further refinement of monitoring procedures revealed a level of harm that few would ever have believed possible.

## 5.1 MEDICAL HARM: WHAT IS IT?

There is no universally agreed definition of medical harm, and many different terms are used to refer to it. *Iatrogenic disease*, from the Greek meaning "brought forth" by the physician or healer, is one such term, although it is declining in usage. Perhaps the most general term is *undesirable healthcare event* (see Figure 5.1), used in reference to harm or threat of harm from any category of healthcare, including all branches of medical practice, nursing, physiotherapy, occupational therapy, psychology, speech pathology, and social work. Figure 5.1 tries to capture graphically the approximate relative meaning of the most frequently used terms. Because medical intervention is the main cause of undesirable healthcare events, the term *medical harm* is used here unless other specified healthcare-related causes are involved.

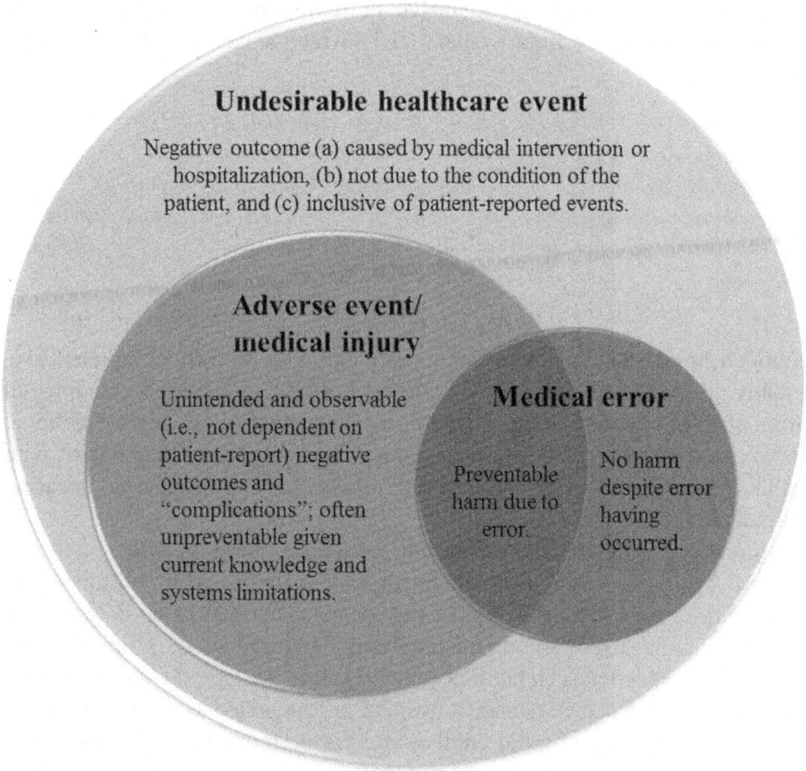

**Undesirable healthcare event**

Negative outcome (a) caused by medical intervention or hospitalization, (b) not due to the condition of the patient, and (c) inclusive of patient-reported events.

**Adverse event/ medical injury**

Unintended and observable (i.e., not dependent on patient-report) negative outcomes and "complications"; often unpreventable given current knowledge and systems limitations.

**Medical error**

Preventable harm due to error.

No harm despite error having occurred.

**FIGURE 5.1**  Some of the related terms used to describe medical harm. *Note.* A shared meaning of all specific terms to describe medical harm is that the harm caused was *not* due to the underlying condition of the patient. The diagram is illustrative and the circles are not to scale (i.e., they are not proportional to the amount of harm recorded in healthcare settings).

In the broadest sense, medical harm may be considered to have occurred when, due to a medical encounter, a patient experiences an undesirable outcome not attributable to the presenting medical condition. Most **operational definitions**, however, are not quite so inclusive. As implied in Figure 5.1, some consider harm has occurred only if it is deemed to be due to error and, therefore, avoidable by definition. Alternatively, others consider harm has occurred only if it is avoidable, whether due to error or not. However, linking harm to error or preventability necessarily underestimates the scale of actual harm. In practice, considerable harm occurs when medicine is applied not in error but in strict accordance with accepted recommendations and guidelines (e.g., side effects of prescribed medications). In some such cases, the harm caused was essentially unavoidable because its occurrence could only have been observed after the intervention that caused it was applied. "However, harm" should not be ignored merely because it was unexpected, and the action that caused it was well-intentioned and expertly administered.

One relatively broad and inclusive definition is that of the American Institute for Healthcare Improvement (IHI), which states in part that medical harm is

> *Unintended physical injury resulting from or contributed to by medical care (including the absence of indicated medical treatment), that requires additional monitoring, treatment or hospitalization, or that results in death. Such injury is considered medical harm whether or not it is considered preventable, whether or not it resulted from a medical error, and whether or not it occurred within a hospital.*

(McCannon et al., 2007, p. 479)

Although broad, the IHI definition cannot claim to be fully comprehensive. Notably, it refers specifically to "physical injury"—thereby excluding all forms of psychological harm, including anxiety, depression, other emotional distress, and distress of an interpersonal nature arising from interactions with healthcare personnel when, for example, these are experienced as uncaring or threatening.

Moreover, while the IHI definition recognizes harm "whether or not it occurred within a hospital," a distinction is made between hospital- and community-caused medical harm in the longer statement of definition from which the quote above was extracted. Harm incurred while hospitalized is recognized even when such harm does not result in "additional hospital days," whereas harm that occurs outside a hospital is recognized as such only "when that harm results in a hospital admission." Furthermore, while committed to patient safety and claiming to adopt a patient-centered perspective, the IHI definition focusses on harm identification by healthcare personnel and makes no explicit acknowledgement of harm reported by patients or families.

*Complications.* Sometimes, an undesirable healthcare event is referred to as a medical "complication," although that term can have different meanings in different contexts. For example, exacerbation of symptoms in the natural course of a disease is sometimes referred to as a complication, but does not constitute medical harm because it is due to the underlying condition of the patient. Used in the context of medical harm, however, the term "complication" is essentially euphemistic; a veiled or polite expression of something more ominous. So-called complications of that kind include undesirable events that may have been anticipated as a possible risk of treatment. Such harm typically refers either to unintended outcomes of intervention that have a known *a priori* risk of occurrence (e.g., a previously reported low-incidence side effect of medication) or are unexpected (e.g., unanticipated life-threatening blood loss during surgery).

*Near misses.* One source of contention regarding medical harm centers on whether the occurrence of actual harm should or should not be a defining feature (Dovey and Phillips, 2004). Intuitively, it might be tempting to ignore events where no actual harm has been done. On the other hand, it would be unwise to ignore events that might have, but fortunately did not, cause harm on a particular occasion. Much can be learned from such "near misses" and that knowledge can be incorporated into harm reduction strategies. Thus, some systems intended to monitor medical harm require only that instances of actual harm are recorded, whereas others aim also to include instances of potential harm.

*Therapeutic failure.* Of course, not all medical intervention is successful, and it is an everyday occurrence that patients who would have died or continued to suffer illness if left untreated die or remain ill *despite* receiving appropriate medical intervention. Such instances, though obviously undesirable, do not constitute medical harm. A patient who does not show improvement after receiving appropriate treatment that is competently delivered has experienced a *therapeutic failure*, not medical harm. Therapeutic failure is the result of current limitations of therapy to deal adequately with the patient's underlying condition, and to that extent is unavoidable. Conversely, medical harm *does* include instances where an intervention is given not because it is believed to be effective but because healthcare personnel feel compelled to do something, possibly under the misapprehension that doing something is better than "standing idly by." If harm occurs that would not have occurred at that particular time, or to the same extent, or at all, had intervention not been administered, the harm caused is attributable to the intervention and qualifies as medical harm. Harm caused while acting with the best of intentions is not exonerated by those good intentions.

## 5.2 SERIAL HARM: WAYWARD DOCTORS AND MALFUNCTIONING SYSTEMS

The absence of a universally agreed definition of medical harm poses an obstacle for attempts aimed at assessing its prevalence and severity. Nevertheless, despite differences in definition and varied nuances of meaning, extensive evidence confirms that even when measured approximately, the scale of medical harm is of shocking proportions. However, before discussing the systematic evidence, it is appropriate to acknowledge the occurrence of isolated instances of harm from causes other than routine clinical intervention. This includes appalling harm perpetrated by aberrant individuals working in healthcare and by groups of healthcare personnel working within dysfunctional systems.

### 5.2.1 Serial Murder

British general practitioner Harold Shipman may be the most notorious of doctors who have murdered. With reference to the number of confirmed victims, Shipman is considered by some to be the most prolific serial killer in history. Shipman murdered women and men of varying ages, but he mostly targeted older women, many in good health, whom he usually killed by lethal injection. An official inquiry found that over a period spanning more than two decades

Shipman killed at least 215 of his patients, although the exact number cannot be established and is possibly substantially higher (Smith, 2002).

Shipman was arrested in 1998, and in 2000 was sentenced to life imprisonment, with a recommendation that he never be released. The presiding judge commented:

> None of your victims realized that yours was not a healing touch. None of them knew that in truth you had brought her death, death which was disguised as the caring attention of a good doctor.

In 2004, while in prison, Shipman committed suicide by hanging. Although the case of Harold Shipman is unusual, he is not alone in being a doctor who has murdered patients. One authority has suggested that the opportunities afforded by medical practice for people "with a pathological interest in the power of life and death" could be responsible for medicine having "thrown up more serial killers than all the other professions put together" (Kinnell, 2000, p. 1594).

Certainly, the healthcare environment offers unusual opportunity for serial murder, as is illustrated by the case of Charles Edmund Cullen, a former nurse in the State of New Jersey, whose record of serial murder possibly exceeds that of Harold Shipman. Cullen confessed to killing up to 40 patients, but some observers believe that he may have been responsible for 10 times that number. Most of his victims died due to injection of lethal overdose of medication and others due to contamination of fluids used in intravenous therapy. During his career, he repeatedly left his employment or was dismissed only to be reemployed shortly thereafter at another hospital. Although often suspected of serious misconduct in the performance of his duties and of possibly causing patient deaths, the murders continued for the whole of his 16-year career. When apprehended, Cullen attempted to explain his actions by claiming that he killed patients so as to end their suffering. This, however, is contradicted by the fact that many of those murdered were not terminal and were scheduled to be released from hospital. For many, his actions are believed to have caused considerable suffering prior to death. Cullen is currently serving life imprisonment without parole.

## 5.2.2    Nonfatal Serial Harm

Besides serial murder, healthcare personnel have also been responsible for intentionally causing repeated and systematic nonlethal harm to patients: essentially, serial physical abuse. For example, Irish consultant obstetrician Michael Neary routinely performed *peripartum hysterectomy* (birth-related removal of the womb). An official inquiry conducted decades after Neary began to practice found that most of the operations were unnecessary (Clarke, 2006). Peripartum hysterectomy carries a high risk of harm, and is performed only after other less radical procedures have been tried and failed. It is typically performed only as emergency surgery to save the life of a woman suffering uncontrolled uterine bleeding. Neary was confirmed to have performed 129 such operations during the span of 25 years. He was responsible for more such operations in a typical year of his practice than would be expected necessary during an entire career for someone in his position (Bateman et al., 2012). Approximately 40% of the hysterectomies involved women having their first or second baby, who were thereby prevented from ever again becoming pregnant. After initially being exonerated by a panel of his professional peers, despite strong evidence of malpractice, Neary was eventually barred from practicing medicine. Like Shipman, he consistently professed innocence, and no motive has ever been established for his actions.

### 5.2.3  Malfunctioning Medical Systems

While Shipman, Cullen, and Neary were the principals in the harm they perpetrated, serial harm has also been performed by groups working in concert. One such example relates to events that occurred at the Bristol Royal Infirmary in the United Kingdom. The hospital had a history of questionable clinical practices and poor outcomes in pediatric cardiac surgery, and this was well known within the hospital and known extensively within the medical community outside the hospital. Despite a poor record, cardiac surgeons within the hospital, apparently impervious to criticism and seeing themselves to be beyond reproach, continued to operate on newborns (Walshe and Shortell, 2004). A subsequent public inquiry concluded that in the years from 1990 to 1995 the actions of the doctors caused the deaths of about 35 infants (Kennedy, 2001). Three doctors were disciplined, two of whom lost their licenses to practice medicine.

Persistent inaction, denial, and occasional obstruction by medical authorities unable or unwilling to intervene on behalf of patients despite complaints and suspicions are common features of instances of harm caused by wayward doctors and malfunctioning healthcare systems (Walshe and Shortell, 2004). History shows that after even the most serious failures, patients bear almost all of the costs while practitioners and healthcare organizations frequently suffer little inconvenience or disruption. As long as the distribution of consequences continues as in the past to be grossly one-sided, genuine change is unlikely. Rather, isolated cases of wayward doctors continuing to practice with relative impunity within malfunctioning systems seem certain to reoccur. Holding healthcare personnel strictly accountable for their actions seems an obvious moral imperative and it is often cited as the solution to eradicate serial medical harm. Unfortunately, the means for achieving the required levels of accountability are far from obvious, as discussed in Box 5.1.

Patients are harmed whenever practitioner self-interest and organizational protectionism take precedence over patient welfare. Not uncommonly, pervasive secrecy is upheld, despite widespread organizational knowledge about the harm being perpetrated. Official reports speak of unsuspecting patients and families being sacrificed ahead of professional and organizational interests. Despite high-profile public revelations and official inquiries, professional and

**BOX 5.1 The Stafford Hospital Scandal**

Robert Francis, QC, chairperson of the most recent public enquiry into the Stafford Hospital Scandal. *(Source: http://www.thelancet.com/pdfs/journals/lancet/PIIS0140-6736(13)60264-0.pdf.)*

During the writing of this book, the scandal of Stafford Hospital near Birmingham, United Kingdom, received extensive, if brief, attention in the international news media. The scandal came to light during the late 2000s amid growing concern over apparently high mortality rates among patients of the hospital under the control of the Mid-Staffordshire National Health Service (NHS) Foundation Trust (Holmes, 2013). Press reports suggested that substandard care had contributed to the deaths of hundreds of patients with countless more having suffered gross violations of their dignity. Revelations in the public media referred to widespread breaches such as patients being neglected to the extent of having to lie in their own urine and being forced by thirst to drink water from flower pots. Several enquiries ensued, culminating in 2010 with a full public inquiry chaired by Robert Francis, QC.

The Francis Inquiry cost EURO13 million (USD 20 million plus) and led to the issuing of a report containing 1782 pages of analysis and recommendations. Reminiscent of earlier inquiries about previous scandals, the report stated that "In the end, the truth was uncovered…mainly because of the persistent complaints made by a very determined group of patients and those close to them" (p. 13). A recurrent theme of the report is the paramount need to keep patient welfare as the main focus of healthcare, with Recommendation 5 in part stating that "Staff [should] put patients before themselves" (p. 1676). Many of the recommendations concern the need for tighter regulatory control, the general tenure of which is conveyed in Recommendation 28, which recommends:

**BOX 5.1 The Stafford Hospital Scandal—cont'd**

*Zero tolerance: A service incapable of meeting fundamental standards should not be permitted to continue. Breach should result in regulatory consequences attributable to an organization in the case of a system failure and to individual accountability where individual professionals are responsible. Where serious harm or death has resulted to a patient as a result of a breach of the fundamental standards, criminal liability should follow and failure to disclose breaches of these standards to the affected patient (or concerned relative) and a regulator should also attract regulatory consequences (p. 1678).*

Notwithstanding the air of indignant outrage that pervades the report, the report itself can be criticized for not being wholly sensitive to the needs of patients and families. While attempting to be thorough and comprehensive, the report omits to provide a succinct account to facilitate comprehension by the patients and families directly affected, and the general public comprising the millions of current and prospective patients and families within the NHS system. The *executive summary* alone is a formidable 125 pages, and the three-volume report contains 290 recommendations. Regarding the recommendations, they are full of "shoulds" and "musts" relating to "fundamental standards," "transparency and candor," and "accountability," which given the history of such reports cannot but leave the reader wondering over the likely fate of this recent addition. The report is particularly vulnerable for failing to counter concerns about serious unintended consequences that could occur as a result of the heightened level of regulatory control and audit (microregulation) that it vociferously recommends.

   Standards in healthcare, traditionally and to a large extent still, depend on practitioners exercising self-regulation to preserve the welfare of the patients they serve. Breaches such as those at Stafford Hospital illustrate gross failures in self-regulation. Whereas self-regulation tends to be ethics based, microregulation is rule based (Fischer and Ferlie, 2013). The Francis report rightly deemed unacceptable breaches such as those that occurred at Stafford Hospital, and proposed rule-based systems of regulatory control involving internal and external scrutiny of individuals and organizations. Though obviously well intentioned, such measures are not free from unintended consequences capable of causing harm to patients. It is misguided to believe that microregulation necessarily leads to improved standards of practice. Such beliefs ignore the fact that microregulation can have reactive effects that undermine existing (if imperfect) systems of self-regulation, thereby potentially increasing rather than decreasing patient risk.

   *Defensive practice* is the term that has been given to one class of reactive effects whereby physician focus is redirected from the patient to self-preservation. Such shifts in focus are exemplified in recent heartfelt advice from one physician to others under her supervision: "... remember there are two people in the consultation. There is you and the patient; safeguard yourself first" (McGivern and Fischer, 2012, p. 292). At the institutional level, defensive practice can lead to the development of protocols that encourage doctors to discuss concerns confidentially with supportive colleagues in the knowledge that what is discussed remains within the group. It is believed that allowing breaches of standards to be revealed in confidence facilitates appropriate corrective action being taken without the fear of reputations, both individual and institutional, being damaged (McGivern and Fischer, 2012). On one

*Continued*

**BOX 5.1 The Stafford Hospital Scandal—cont'd**

hand, such strategies can appear reasonable and constructive. On the other hand, it was precisely within that kind of collegial and organizational framework that the Shipman, Neary, and Bristol abuses (discussed in the main text) as well as those at Stafford Hospital, occurred.

Unquestionably, among the ranks of healthcare personnel there are conscientious, caring, and dedicated practitioners, highly committed to the welfare of patients. Possibly, the majority fits that description, but obviously not everyone does, otherwise tragedies such as happened at Stafford Hospital would not occur. While self-regulation may offer protection for patients when it is being exercised by those who are caring and ethically informed, self-regulation offers no safeguard when left in the hands of the uncaring. Conversely, microregulation as a strategy to safeguard against uncaring practice can produce serious unintended harmful effects due to the individual and institutional defensiveness that external regulation can provoke, even in caring and competent practitioners and institutions. If the ideal blend of self-regulation and microregulation capable of maximizing patient safety exists, it has yet to be discovered. In the meantime, since neither alternative of self-regulation nor microregulation can guarantee patient safety, there is no alternative but to conclude that biomedical healthcare is intrinsically hazardous, whatever form of regulation is applied. Because avoidance of healthcare avoids the dangers therein, it seems that anyone wishing to minimize their own exposure to the dangers of medical harm should strive to minimize their need for biomedical healthcare by remaining as healthy as possible for as long as possible.

organizational self-preservation continue to be deeply embedded in the culture of medical practice. Were it not for the persistence of individuals not part of the inner circle of perpetrators, the secrecy that surrounds tragedies such as those outlined above might never happen. This could suggest that the tragedies that have come to light may be a mere portion of the serial medical harm that actually occurs (Walshe and Shortell, 2004).

Appalling harm, perpetrated by individuals or groups who succeed in finding protection within the ranks of their profession while simultaneously transgressing professional strictures, is a reality of modern medical practice. Yet, despite the gravity of such tragedies, they are but a tiny fraction of the total harm wrought by medicine. Most medical harm is not that perpetrated by aberrant individuals and groups, but that which occurs daily as part of routine clinical practice.

## 5.3   TO ERR IS HUMAN: HARM CAUSED IN THE ROUTINE PRACTICE OF MEDICINE

Despite the universal medical imperative to do no harm, most people will be aware, if not from direct experience then certainly from occasional news reports, of serious harm caused by medical intervention. Many may be of the opinion that such events are relatively infrequent. In reality, serious medical harm is commonplace, although substantially underreported. *To Err is Human* is the title of a an extensive report based on two large studies, one conducted in the State of Utah and the other in New York, published by the American IOM, which concluded that the "horrific cases that make the headlines are just the tip of the iceberg" (Kohn et al., 2000).

### 5.3.1   The Shocking Extent of Medical Harm

The rate of occurrence of medically caused adverse events was estimated by the IOM to be 3-7% of hospitalizations, of which 7-14% were estimated to have caused death. Expressed differently, the IOM found that medical intervention caused the death of between approximately 1-in-100 and 1-in-500 people admitted to hospital (depending on whether the less optimistic estimates of 14% of 7% or the more optimistic estimates of 7% of 3% are used). Extrapolating their findings to the whole of the United States population, the IOM estimated that hospital-based medical harm in the United States was responsible for between 44,000 and 98,000 deaths per year.

To appreciate the magnitude of those numbers, it might help to provide some context. Using the lower range of estimates, the IOM estimated that hospitalization kills more Americans than motor vehicle accidents, breast cancer, or HIV/AIDS individually. Using the higher estimates, hospitalization kills the approximate equivalent of all three of the aforementioned causes of deaths combined. Alternatively, the IOM's estimate of medical harm in American hospitals is equivalent to one death every 5-12 min. To be clear, this estimate refers only to the number of deaths *caused* by medical intervention, and does not include deaths due to therapeutic failures. Of course, many ill patients die *despite* receiving clinical intervention, due to the serious and fatal nature of their underlying medical condition. Those are instances of medical *failure* not *harm*.

Shocking as the statistics are, there is nothing to suggest that the United States is unusual with respect to the scale of medical harm. Broadly similar rates have been reported for Australia (Wilson et al., 1995), Canada (Baker et al., 2004; Wanzel et al., 2000), the 27-State European Union (Conklin et al., 2008), the Netherlands (Zegers et al., 2009), New Zealand (Davis et al., 2002), Spain (Andrés et al., 2005), Sweden (Soop et al., 2009), and the United Kingdom (Vincent et al., 2001). Figure 5.2 summarizes the incidence and nature of medical harm for approximately 75,000 hospital patients representing the five geographically dispersed countries of Australia, Canada, New Zealand, the United Kingdom, and the United States (de Vries et al., 2008). The overall findings were broadly consistent with those of the IOM study, with about 1-in-10 patients being found to have experienced at least one harmful outcome. For slightly more than half of those, the event was minor and "resolved within 1 month." However, of those harmed, approximately equal proportions experienced permanent disability (7%) or death

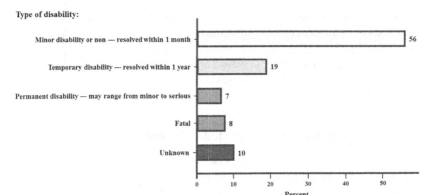

FIGURE 5.2   Percentage distribution by type of approximately 7000 instances of medical harm for a population of approximately 75,000 hospital patients. *(Derived from de Vries et al., 2008.)*

(8%). Even when harm to patients is minor and without permanent disability, the frequency is estimated to be such as to have "massive economic consequences" due to the additional medical intervention and additional hospitalization often required (Vincent et al., 2001).

Whatever beliefs people may have about differences in the quality of healthcare between countries, expert opinion is agreed that the crisis of medical harm is global. Notwithstanding methodological differences that limit the extent to which direct comparisons can be made between countries, a recent study of eight low- and middle-income countries found rates of medical harm, including permanent disability and death, to be generally higher than those for developed countries (Wilson et al., 2012). Notably, although human- and physical-resource limitations are more typical of developing than high-income countries, addressing such limitations offers only a partial solution to high rates of medical harm. Rather, as has been confirmed in studies of high-income countries, medical harm is intrinsic to modern biomedical healthcare.

Possibly few people with a medical condition warranting hospital admission are aware that hospitalization itself exposes them to levels of risk rivaling several known major causes of death. For a substantial number, the potential harm of a medical procedure is greater than the threat to health posed by the underlying condition. Whether harm is sustained in or out of hospital, risk of harm is not distributed randomly. Unsurprisingly, serious and fatal harm is disproportionately higher in patients whose health is already seriously compromised. In other words, biomedical intervention is at its most lethal when it is most needed. It is easy, when citing statistics to lose touch with the human costs of medical harm. Box 5.2 is illustrative, outlining the personal cost, discomfort, and distress associated with serious harm experienced by one individual.

**BOX 5.2 Illustrative Case Profile of Serious Medical Harm**

A 55-year-old man with a history of stroke, multiple-resistant *Staphylococcus aureus* infection,[1] leg ulcers, and heart failure was admitted for the treatment of venous ulceration and cellulitis (bacterial infection of skin) of both legs. The patient was reported to have sustained two harmful outcomes:

1. Failure to manage the leg ulcers aggressively led to the development of osteomyelitis (bone infection). The patient subsequently had below-knee amputation of both legs.

2. Incorrect management of the patient's urinary catheter resulted in necrosis (irreversible tissue death) of the tip of his penis. He then had suprapubic catheterization (to drain the bladder) and developed a further infection at that site.[2]

The patient's hospital stay was extended by 26 days.[3]

[1] *Multiple-resistant* Staphylococcus aureus *(MRSA) infection is most often acquired following contact with healthcare facilities, suggesting an earlier history of medical harm for this patient besides the events described here.*

[2] *It would be reasonable to infer from the description that two separate instances of harm (rather than one as reported by Vincent et al., 2001) occurred at this time, bringing the total for this patient to three (rather than two) sustained during his current hospitalization. Taking account of the information provided of a likely earlier event in the form of hospital-acquired MRSA, the probable total number of serious instances of medical harm for this patient is at least four.*

[3] *Notwithstanding 26 additional days in hospital due to harm sustained during his current hospitalization, the harm caused to this patient included permanent and serious disability likely to require further medical intervention possibly including future rehospitalization.*
*Adapted from Vincent et al., 2001.*

It is notable that a consistent conclusion of several studies of hospital-based medical harm is that about half is preventable (e.g., Conklin et al., 2008; Kohn et al., 2000; Wilson et al., 1995; Vincent et al., 2001). By that estimation, it follows that even under the best of circumstances, wherein all preventable harm is eradicated, about half of the harm currently caused by biomedical healthcare is entirely intractable. While that realization may be slightly unnerving, it should not come as a surprise. Considering the invasive nature of much of routine clinical intervention, involving, for example, ingestion of potentially toxic medications and surgical penetration of the body, occasional misadventure seems inevitable. The precise level of danger for a given individual is difficult to pinpoint, but recent evidence confirms the suspicions of the authors of the IOM Study that their estimates of medical harm were decidedly conservative.

### 5.3.1.1 Community Healthcare Settings

To date, hospital inpatient settings have been the focus of studies of medical harm. Yet, biomedical healthcare is delivered in a plethora of other settings, including outpatient surgical centers, physician offices and clinics, home care, nursing homes, and pharmacies, where prescriptions are filed and substantial patient education is delivered. It is obvious that harm can occur in all of those settings, although knowledge of the scale of any such harm is limited. The IOM report of Kohn et al. (2000) acknowledged that hospital-based intervention represents only a fraction of the biomedicine that is dispensed, and conceded that estimates of rates of death and injury due to hospitalization provide only "a very modest" indication of the total harm caused. In a recent review of evidence that incorporated insurance-industry actuarial data, only about one-third of the total of all medical harm was estimated to be hospital based with the remaining two-thirds estimated to occur in community healthcare settings (Goodman et al., 2011).

### 5.3.2 Harm Exceeds the Upper Limit of Previous Expectations

The acknowledged first step toward dealing with any problem successfully is to measure its dimensions. Methods for measuring medical harm have varied greatly (Kuske et al., 2013). Some rely on practitioners voluntarily recording incidents, whereas others rely on physician, nurse, or automated reviews of patient records. One the many positive effects of the IOM Study (Kohn et al., 2000) was that it stimulated interest in the development of methods for measuring medical harm that are practicable (i.e., not overly cumbersome or demanding of resources) and accurate. One such method is the **Global Trigger Tool**, involving a well-defined protocol for directly and efficiently reviewing inpatient hospital records by small clinical review teams (Griffin and Resar, 2009). Using repeated random sampling of patient records over time, the review team looks for "triggers"; signals of the possible occurrence of adverse events (e.g., specific types of infections, particular diagnoses, laboratory results, prescription orders, and certain events noted in records of surgical operations). When a positive trigger is found, the record is scrutinized to confirm whether or not medical harm occurred, and if so, the harm done is recorded and categorized (e.g., temporary requiring intervention, temporary requiring readmission or extended hospitalization, permanent, intervention required to sustain life, or death). The tool has been shown to possess good inter-observer agreement (Classen et al., 2008), and its use has shown persistent endemic harm to be at levels exceeding those previously deemed to be unacceptable.

In 2011, a large study was undertaken to measure the incidence of medical harm among randomly selected inpatients at three hospitals in the United States using three different methods for measuring medical harm (Classen et al., 2011). The hospitals were chosen because of their high level of commitment to patient safety. All three hospitals offered a full range of medical services, they were part of large integrated healthcare systems, they were teaching hospitals for their respective medical schools, and all had well-established and highly respected patient-safety programs. Of the three methods for measuring medical harm, one method was the Global Trigger Tool outlined above, and another was the Patient Safety Indicators protocol developed for the American

Agency for Healthcare Research and Quality (AHRQ, 2008). Although widely used, both in the United States and increasingly internationally, concerns have been raised about the reliability and validity of the AHRQ tool (Tsang et al., 2008). The third method was the existing voluntary reporting system in each hospital. Such systems are widely used internationally, despite being criticized as likely to miss significant instances of harm (e.g., Aspden et al., 2004). A total of 795 patient records were reviewed from the three hospitals, and the three measurement methods detected a combined total of 393 instances of medical harm.

From results summarized in Table 5.1, the first thing to note is that the incidence of medical harm was found to be substantially higher than rates reported in earlier studies. Specifically, 1-in-3 patients (264 of 795) suffered at least one instance of harm. When the IOM Study (Kohn et al., 2000) reported rates of 3-7%, the figures were regarded as startling. Just 10 years later, Classen et al. (2011) reported a rate of 33%. Although the more recent higher estimate could be an indication of substantial deterioration in safety, results such as these, which came after the seminal IOM studies, are usually interpreted as evidence of the extent to which the earlier IOM results underestimated the true scale of hospital-based medical harm. Because some patients in the Classen et al. study suffered more than one event, the total number of events was 393, which equates to 49 events per hundred hospital admissions. Additionally, clinical intervention was found to have contributed to the death of one in every 100 patients admitted to hospital, confirming the higher levels of the range of rates reported in the IOM study. The Classen et al. (2011) findings have also been substantiated by more recent study (Kennerly et al., 2014).

With reference to the comparative performance of the three measurement methods evaluated by Classen et al. (2011), Figure 5.3 shows that the Global Trigger Tool detected 354 adverse events (90% of the total for all three methods combined), whereas the AHRQ protocol detected 35 events (9% of the total). In their own words, Classen et al. (2011, p. 581):

> *found that the adverse event detection methods commonly used to track patient safety...fared very poorly [and that the] Global Trigger Tool found at least ten times more confirmed, serious events than these other methods.*

To say that the voluntary systems implemented locally within each hospital "fared very poorly" is an understatement. The voluntary systems reported just four events, a mere 1% of the total of 393 events identified.

These findings, confirmed by a more recent Swedish study involving records of 960 patients selected at random over a period of 48 months in a large university hospital, found that only 6% of instances of harm identified using the Global Trigger Tool were reported as part of the voluntary system employed in the hospital (Rutberg et al., 2014). A similar discrepancy between voluntary reporting and cases identified using a pediatric version of the Global Trigger

**TABLE 5.1 Number of instances of medical harm by cause and level of severity sustained by 795 randomly selected patients in three hospitals**

| Cause | Severity level | | | | | Total[a] | Percentage[b] |
|---|---|---|---|---|---|---|---|
| | Temporary requiring intervention | Temporary requiring hospitalization | Permanent harm | Intervention to sustain life | Death | | |
| Medication | 100 | 46 | 2 | 2 | 0 | 150 | 38 |
| Medical procedures (e.g., complication of surgery) | 67 | 26 | 5 | 7 | 4 | 109 | 28 |
| Hospital-acquired (nosocomial) infection | 30 | 37 | 2 | 2 | 1 | 72 | 18 |
| Pulmonary embolism/venous thromboembolism | 8 | 5 | 2 | 0 | 2 | 17 | 4 |
| Pressure ulcers | 10 | 1 | 0 | 0 | 0 | 11 | 3 |
| Device failure | 0 | 6 | 0 | 0 | 0 | 6 | 2 |
| Falls | 2 | 1 | 0 | 0 | 0 | 3 | 1 |
| Other | 10 | 11 | 0 | 3 | 1 | 25 | 6 |
| Total | 227 | 133 | 11 | 14 | 8 | 393 | 100 |

[a]One-third of patients (i.e., 264 of 795) experienced at least one instance of medical harm, and some experienced more than one bringing the overall total to 393. One-in-100 patients (i.e., 8 of 795) died.
[b]Number of instances of harm for each type of cause as a percentage of the overall total of 393.
Adapted from Classen et al. (2011).

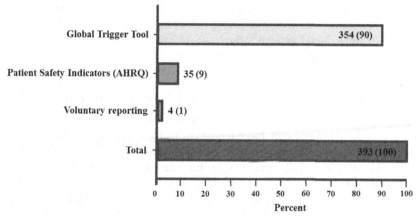

**FIGURE 5.3**   Number (percent) of instances of medical harm detected by each of the three measurement protocols as a percentage of the combined number of events detected by all the three protocols. *(Derived from Classen et al., 2011.)*

Tool was found in a recent Norwegian study (Solevåg and Nakstad, 2014). From these results, it may reasonably be concluded that voluntary reporting by healthcare personnel, a form of self-regulation that relies on the cooperation of healthcare personnel to comply with agreed guidelines, is of little value as a method for monitoring medical harm. In other words, a key aspect of the systems that most hospitals rely upon to secure patient safety is hopelessly unfit for purpose.

A major implication of the Classen et al. (2011) study is that earlier attempts to quantify the incidence of medical harm substantially underestimated the scale of the problem. As mentioned above, the IOM Study (Kohn et al., 2000) showed that occasional news media coverage represented only the "tip of the iceberg" of medical harm. In turn, the Classen et al. findings show that the IOM's estimates themselves missed a previously submerged sizable portion of an even bigger iceberg. Despite that, Classen et al. concluded that what they found "probably represents a minimum number of adverse events actually present" (p. 583). In that context, it is well to remember that the hospitals examined by Classen et al. were selected on the basis of high commitment to patient safety. Based on a more general application of the Global Trigger Tool, the aforementioned American IHI reported that it "consistently observed a rate of 40-50 incidents of harm per 100 admissions" (McCannon et al., 2007, p. 480).

The perspicacity of the authors of the IOM Study (Kohn et al., 2000) deserves praise. When published, their findings received an unprecedented level of attention from governments and policy makers (Leape, 2000). The scale of medical harm reported was shocking to all, not least to those who were disbelieving (e.g., McDonald et al., 2000; Hayward and Hofer, 2001). Later studies, however, confirmed the IOM's own assessment that its conclusions were conservative. Even then, the IOM may not have anticipated how conservative its original estimates would subsequently be shown to have been. In summary, it seems reasonable to conclude that in the best of modern hospitals medical harm occurs in about 1-in-3 patients, about 1-in-100 patients is fatally harmed, and injuries exceed deaths by about 30-fold (Classen et al., 2011).

### 5.3.2.1  Why Does Medical Harm Go Largely Unnoticed?

All things considered, current scientific evidence shows that biomedical health-care, by a very wide margin, has not honored its foundational tenet to: first, do no harm. Contrary to that dictum, the practice of medicine is itself a major cause of disease that easily ranks among the leading causes of mortality and morbidity worldwide. There is little, however, to suggest that the healthcare personnel who daily attend to victims of medical harm have any real appreciation of the scale of the problem or their possible role in it. Perhaps the explanation for medical harm going largely unnoticed is to be found in the nature of disease, the natural course of which is characterized by suffering, disability, and not infrequently, eventual death. Consequently, there may be relatively little in the work of those who daily deal with suffering and death to help them to discriminate between the natural course of disease and the suffering, disability, and death that they themselves cause.

A possible indication of the dynamics at play in these circumstances may be gleaned from a study of adverse drug events described in more detail below (Hazell and Shakir, 2006). In contrast to overall poor rates of reporting of harmful events by healthcare personnel, relatively high reporting rates were found for vaccine-associated harm. Vaccines, it may be noted, are administered to healthy people, whereas medicine is generally administered to patients with preexisting disease. Consequently, vaccine-associated harm occurs against a background of health and is thereby more salient than more typical medical harm which occurs in hospitals containing sick people in whom the inflicted harm is not easily distinguished from preexisting illness. As such, the problem of hospital-based medical harm going largely unnoticed may be partly explained by analogy to figure-ground detection in visual perception and signal-to-noise ratio in auditory perception (and in signaling technologies generally). Thus, vaccine-associated harm may be said to involve a prominent "figure" of illness against a "ground" of healthy people or as providing a good "signal-to-noise ratio." Conversely, medical harm usually occurs in healthcare settings populated by sick patients, representing a not very distinct figure of illness against a ground of illness or a poor signal-to-noise ratio.

Medicine is sometimes portrayed as being at its most effective when it is at its most invasive; namely, in the emergency ward and the critical care unit. The reality, however, does not match that image. Emergency departments worldwide have been described as high-volume environments characterized by a shortage of skilled staff, inadequate resources, and support services; high task variability; little task standardization and consistency; questionable team coordination; and frequent interruptions in which instances of medical harm are common but missed or simply not recorded (Friedman et al., 2008). Many similar challenges are encountered in critical care, which involves the management of life-threatening conditions requiring invasive monitoring and support of vital organs. In terms of patient throughput, critical care consumes a disproportionately high level of resources, both human and material (Finfer and Vincent, 2013). As with emergency care, critical care is fast paced and complex, frequently requires urgent high-risk decision making, often with incomplete information involving personnel with varying levels of relevant training (Rothschild et al., 2005).

Patients enter emergency medical departments and critical care units when their presenting condition ensures that they are at their most vulnerable to further assaults on their health. This is also the point at which they are most likely to be exposed to a high level of risk-laden clinical interventions. Intuitively, we may feel that the high risks attached to intervention are justified. Yet, it remains an irony of the healthcare that most of the world has come to embrace that the moment when we are in greatest need of medical assistance coincides with when we are most harmed by the treatment we receive.

It is a further irony of modern medical practice that a substantial portion of the demand for critical care is created by medical harm. Using the Global Trigger Tool, all unscheduled adult admissions to the intensive care units of five major British hospitals were examined over a continuous 6-week period (Garry et al., 2014). It was found that 27% of patients admitted to critical care had experienced at least one instance of medical harm while in hospital during the preceding week, and that in 80% of those cases medical harm had *caused or contributed* to the critical-care admission. In other words, medical harm caused during a brief stay (up to 7 days) in hospital was *responsible* for more than 1-in-5 of the patient population (80% of 27%) receiving critical care.

### 5.3.3  What Do Patients Say?

An important source of information about medical harm is what patients say. Unfortunately, however, patients are infrequently asked. The few studies of patients that have been done show that they can be a reliable source of information (King et al., 2010). In particular, it has been consistently found that patients report important information about actual and potential harm not picked up by medical personnel and, therefore, not reported in records (Davis et al., 2013). Thus, although largely neglected to date, patients should in future be given priority as a source of healthcare information.

There are many advantages in engaging patients as partners when gathering information about medical harm. Patients are the focus of the intervention process, and are naturally incentivized to be attentive observers of what happens as part of their care. Given a typical scenario, an individual patient is likely to interact with multiple healthcare personnel for varying periods, sometimes fleetingly. In that situation, patients offer a level of observational continuity not easily achieved without their involvement. Not surprisingly, patients who sustain less medical harm tend to be more satisfied and *vice versa* (Agoritsas et al., 2005). Assuming that patient satisfaction is desired, and given that patients are better sources of some important healthcare information than healthcare personnel, it seems obvious that incident reporting systems should be developed for gathering information from patients as a routine feature of healthcare. However, such systems are not widespread and their general introduction will not be without challenges. It is possible, for example, that the eliciting of information from patients may draw their attention to risks they had not previously foreseen, and thereby contribute to harm in the form of anxiety.

At about the time the IOM report was receiving maximum attention in the public media, a survey was conducted in the United States of nationally representative samples of the public ($N = 1207$) and doctors ($N = 777$) in which both groups were asked their opinions about "errors" in medicine (Blendon et al., 2002). About two-thirds of those from both groups who were invited to participate agreed, earning each doctor volunteer USD 100 while volunteers from the public were unpaid. Notably, both groups of participants reported a similarly high incidence of personal experience of medical errors. Specifically, more than one-third of participants in both groups reported incidents in their own care or that of a family member, and about one-fifth of both groups reported that an error had had serious health consequences. However, despite frequent personal experience and widespread media coverage at the time about the high rate of errors in general, when asked to estimate the frequency of errors nationally, both groups gave gross underestimations. On average, estimates were less than 10% of the rates reported by the IOM (Kohn et al., 2000), which in turn (as we have seen) were less than rates reported in more recent studies (e.g., Classen et al., 2011; McCannon et al., 2007). Importantly, neither group named medical errors as being among the main problems in healthcare.

While there was substantial agreement of opinion between the two groups for some survey questions, there were also marked differences of opinion about how instances of harm should be managed. In general, the public favored the imposition of harsher sanctions for instances of harm than did the doctors. Specifically, 50% of the public thought that suspension of the licenses of health professionals would be an effective way to reduce medical errors, whereas only 3% of doctors supported that view. Requirement to report errors to a state agency was endorsed by 71% of the public but only 23% of doctors. A majority of 62% of the public believed that instances of harm should be reported publicly, whereas a substantially larger majority of 86% of doctors thought that reports should be kept confidential.

### 5.3.3.1 Types of Patient-Reported Harm

In a recent study of patients' views following hospitalization in a teaching hospital in the United Kingdom, 80 patients, each with a minimum inpatient stay of 4 days, reported 258 harmful events, more than three events per person

(Davis et al., 2013). Interpersonal problems with healthcare personnel, which usually are not even considered in studies of medical harm, were the most frequently reported (53%) type of event. These included such things as medication side effects not being explained, inadequate provision of information prior to undergoing surgical interventions, poor information about postdischarge recovery, perceived lack of respect from healthcare personnel, and being discouraged by healthcare personnel from asking questions. Interpersonal problems were followed in frequency by harm associated with medical procedures (35%), including inadequate pain control, infection, adverse drug reaction, unexpected transfer to intensive care, and unexpected repeat operation. The next most frequent patient-reported harm concerned problems associated with the healthcare process (12%), which included receiving the wrong diagnosis, being given the wrong drug or the right drug in the wrong amount, being confused with another patient, and errors in diagnostic testing.

An important finding was that whereas the medical records contained very few events (only six in all) that were not reported by patients, patients reported many events that were not in the medical records, thereby confirming patients as a potentially important source of information about medical harm. Interestingly, patients were found to be more willing to report incidents to researchers (for study purposes) than to local or national reporting systems, which tends to suggest a degree of reticence on the part of patients to raise to the level of complaint or disputation matters that may affect their quality of care and safety. Thus, not only is there a need for patient incident-reporting systems, effort is also needed to better prepare patients, and the public generally, about the constructive role patients can have as active participants in incident-reporting processes. In that regard, one way to increase the salience of medical harm for both patients and healthcare personnel and to increase patient participation in incident-reporting may be to couch the issue of patient harm in terms of human rights (Davis, 2004). Specifically, it is the right of everyone to be safe when receiving medical care.

## 5.4  TO ERR IS HUMAN, BUT TO REPEAT THE ERROR IS...INEVITABLE!

"To err is human, but errors can be prevented" proclaimed the authors of the IOM report that precipitated much of the current scientific interest in medical harm (Kohn et al., 2000, p. 5). At the same time, they acknowledged that theirs was not the first report of substantial harm from routine biomedical healthcare (Kohn et al., 2000). In particular, they pointed to the Harvard Medical Practice Study that had produced similar findings almost a decade earlier (Brennan et al., 1991; Leape et al., 1991). Considering the failure of earlier reports to produce change, a main goal of the IOM report was to serve as a catalyst for much needed reform. Indeed, the full title of the IOM report is *To Err is Human: Building a Safer Heath System*. In putting forward their case for change, the authors of the IOM report opined that it would be irresponsible to seek anything less than a 50% reduction in medical harm within 5 years of the report's publication. In fact, the early signs were promising. As mentioned above, the IOM report received wide public and professional attention, provoked much discussion in the professional community, and stimulated extensive scientific study and analysis. New policy initiatives were launched by governments, accreditation bodies, and diverse healthcare providers worldwide. Resolve to address the problem was firm and widespread, and confidence was high.

In 2000, within months of the release of the IOM report, Lucian Leape, and Donald Berwick, members of the Committee on Quality Health Care in America, which participated in the IOM's work, writing in the *British Medical Journal*, asked, "are we up to" implementing the changes necessary to improve safety in healthcare. They addressed their question by referring to the public who "are asking us to promise something reasonable [namely] that they will not be harmed by the care that is supposed to help them" and concluded by proclaiming "We owe them nothing less, and that debt is now due" (Leape and Berwick, 2000, p. 726). Eight years later, also writing in the *British Medical Journal*, and in light of much high-minded rhetoric and considerable effort to improve patient safety, Charles Vincent at the Imperial Centre for Patient Safety and Service Quality, and colleagues, asked "Is health care getting safer?" (Vincent et al., 2008). Their succinct answer was: "Despite numerous initiatives to improve patient safety, we have little idea whether they have worked" (Vincent et al., 2008, p. 1205). Notwithstanding evidence of possible improvement in some areas of practice, the general consensus appears to be that evidence is lacking of long-term trends indicating improvement in overall medical safety (e.g., Amalberti et al., 2011; Goodney et al., 2002; Sari et al., 2007).

## 5.5 IS MEDICINE SAFE? IS IT GETTING SAFER? THE EVIDENCE SAYS "NO" TO BOTH QUESTIONS

In a notably detailed investigation of the extent of improvements arising from the widespread initiatives that followed publication of the IOM report, Landrigan et al. (2010) examined trends in rates of patient harm in a random sample of 10 hospitals in the state of North Carolina. That particular state was chosen for study because it was deemed likely to show substantial improvement based on evidence of a high level of engagement in efforts to improve patient safety and extensive participation by healthcare personnel in safety training programs. The Global Trigger Tool was used to measure medical harm. Among 2341 hospital admissions, a team of reviewers identified 588 incidents; approximately one event for every four admissions. Medical harm was found to have remained common and essentially unchanged from the levels reported previously for the participating hospitals. Of the recorded events, more than 60% were identified by Landrigan et al. (2010) as preventable; though that appears to be a moot point in light of the fact that these so-called preventable events were not prevented despite a high level of engagement by the hospitals with patient safety initiatives and extensive training of healthcare personnel in patient safety.

### 5.5.1   The "Airline Crashes"of Healthcare

In the search for ways to address medical harm, some have suggested that medicine can learn from developments in safety culture and practices in manufacturing and in the construction, mining, and nuclear industries. The IOM report specifically advised that inspiration for addressing medical errors be drawn from experience in the aviation safety industry (Kohn et al., 2000). Indeed, medical errors have been referred to as the "airplane crashes" of the healthcare industry (Walshe and Shortell, 2004). On one hand, the analogy may be helpful in highlighting the need for medical harm to be given the same meticulous attention as that given routinely to airplane mishaps. On the other hand, the analogy overlooks major intrinsic differences between the two "industries," including differences in visibility between medical harm and harm due to airplane crashes. Mortality and morbidity are part of the currency of everyday clinical practice, but when they occur in aviation and other industries they are unmistakable indicators of serious malfunction. Suggestions that medical practice is analogous to other industries in which safety is a priority can be misleading. Worse, such comparisons suggest naive optimism and possible self-deception on the part of those hoping to improve patient safety.

The physical wreck of an airplane crash is impossible to deny. Workplace mishaps are similarly exposed and not easily ignored. In medicine, however, it is commonplace for patients to be ill and to become more so, and for some to die. Unlike other industries, physical evidence of tragedy in medicine is not prima facie evidence of harm caused by medicine. Moreover, the large majority (almost 100%) of instances of medical harm go unrecorded and unacknowledged by those responsible. So invisible are these events that sophisticated tools such as the Global Trigger Tool have had to be invented to expose them. Thus, patients are left to bear the consequences of medical mishaps in relative silence, and there is comparatively little pressure for change despite the high incidence of harm. It should also be acknowledged that the economic consequences of medical harm are profoundly different from harm caused in other industries. Plane crashes are inclined to cause the airline companies involved to go out of business. Medical misadventure, however, often requires additional services to treat the harm that has been caused, thereby generating more, not less, "business." In marketing terms, medical harm could be said to create *repeat business*.

### 5.6   CAN BIOMEDICAL HEALTHCARE BE MADE SAFE?

The IOM report aimed to "break [the] cycle of inaction" surrounding medical harm (Kohn et al., 2000). Considerable credit can be claimed in that regard, because the report did indeed bring forth an era of heightened awareness and concerted efforts to improve patient safety. Unfortunately, evidence indicates little or no actual improvement. Considering the substantial international efforts that have been made to reduce medical harm and the demonstrably limited effectiveness of those efforts to date, it seems reasonable to question whether genuine large-scale improvement in medical safety is achievable.

While it has long been recognized that medicine can be dangerous if not carefully managed, current evidence shows that danger is intrinsic even to practice that is managed carefully. Massive resources, physical, financial, and human, are invested globally to address the world's major diseases. Since medical harm is one such major global "disease," it is appropriate that a proportionate share of available resources be committed to finding "remedies" for medical harm. However, while it is reasonable to assume that some improvement is achievable, evidence accumulated over the past decade suggests that a large core of medical harm is intractable and inevitable. Whatever success (albeit modest) medicine can claim for curing disease, it must also concede the great harm it (inevitably) inflicts.

Confronting the harm of medicine should not be seen as besmirching the genuine conviction, heartfelt compassion, and well-intentioned caring efforts of healthcare personnel. The fact is that much harm occurs *despite* not *because of* lack of caring. The horrible reality is that unlike other industries involving services that are contracted and delivered, clinical practice is alone in requiring elaborate and sophisticated evaluations to discriminate persons who are dead and injured due to "natural" causes from those killed or injured by the service itself. With the possible exception of those engaged in warfare, medical personnel have the unique experience of retiring after a day's work having contributed directly or indirectly, knowingly or unknowingly, to the injury and death of persons in their service, experience those events as routine, and return to work the following day ready to conduct business as usual. Scientific study shows that despite being commonplace, medical harm is substantially unreported, unseen, and unacknowledged. Above all, scientific study shows that substantial harm is intrinsic to biomedical healthcare, that it is largely intractable and therefore inevitable.

## REFERENCES

Agency for Healthcare Research and Quality, 2008. Patient Safety Indicators Technical Specifications Version 3.2 (Revised March 2008). Agency for Healthcare Research and Quality, Rockville, MD.

Agoritsas, T., Bovier, P.A., Perneger, T.V., 2005. Patient reports of undesirable events during hospitalization. J. Gen. Intern. Med. 20, 922–928.

Amalberti, R., Benhamo, D., Auroy, Y., Degos, L., 2011. Adverse events in medicine: easy to count, complicated to understand, and complex to prevent. J. Biomed. Inform. 44, 390–394.

Andrés, J.M.A., Remon, C.A., Burillo, J.V., 2005. National study on hospitalisation-related adverse events. In: ENEAS 2005. Ministry of Health and Consumer Affairs, Madrid, 2006.

Aspden, P., Corrigan, J.M., Wolcott, J.A., Bootman, J.L., Cronenwett, L.R. (Eds.), 2004. Preventing Medication Errors. National Academic Press, Washington, DC.

Baker, G.R., Norton, P.G., Flintoft, V., et al., 2004. The Canadian Adverse Events Study: the incidence of adverse events among hospital patients in Canada. CMAJ 170, 1678–1686.

Bateman, B.T., Mhyre, J.M., Callaghan, W.M., et al., 2012. Peripartum hysterectomy in the United States: nationwide 14 year experience. Am. J. Obstet. Gynecol. 206 (63), e1–e8.

Blendon, R.J., DesRoches, C.M., Brodie, M., et al., 2002. Patient safety: views of practicing physicians and the public on medical errors. N. Engl. J. Med. 347, 1933–1940.

Brennan, T.A., Leape, L.L., Laird, N.M., et al., 1991. Incidence of adverse events and negligence in hospitalized patients: results of the Harvard Medical Practice Study I. N. Engl. J. Med. 324, 370–376.

Chassin, M.R., Galvin, R.W., 1998. The urgent need to improve health care quality: Institute of Medicine National Roundtable on Health Care Quality. JAMA 280, 1000–1005.

Clarke, M.H., 2006. The Lourdes Hospital Inquiry: An Inquiry into Peripartum Hysterectomy at Our Lady of Lourdes Hospital, Drogheda. Stationery Office, Dublin, Ireland.

Classen, D.C., Lloyd, R.C., Provost, L., et al., 2008. Development and evaluation of the Institute for Healthcare Improvement Global Trigger Tool. J. Patient Saf. 4, 169–177.

Classen, D.C., Resar, R., Griffin, F., et al., 2011. 'Global Trigger Tool' shows that adverse events in hospitals may be ten times greater than previously measured. Health Aff. 30, 581–589.

Conklin, A., Vilamovska, A.-M., de Vries, H., Hatziandreu, E., 2008. Improving Patient Safety in the EU: Assessing the Expected Effects of Three Policy Areas for Future Action. Rand, Cambridge, UK.

Davis, P., 2004. Health care as a risk factor. CMAJ 170, 1688–1689.

Davis, P., Lay-Yee, R., Briant, R., et al., 2002. Adverse events in New Zealand public hospitals I: occurrence and impact. N. Z. Med. J. 115, 1167.

Davis, R.E., Sevdalis, N., Neale, G., et al., 2013. Hospital patients' reports of medical errors and undesirable events in their health care. J. Eval. Clin. Pract. 19, 875–881. http://dx.doi.org/10.1111/j.1365-2753.2012.01867.

de Vries, E.N., Ramrattan, M.A., Smorenburg, S.M., et al., 2008. The incidence and nature of in-hospital adverse events: a systematic review. Qual. Saf. Health Care 17, 216–223.

Dovey, S.M., Phillips, R.L., 2004. What should we report to medical error reporting systems? Qual. Saf. Health Care 13, 322–323.

Finfer, S., Vincent, J.-L., 2013. Critical care: an all-encompassing specialty. N. Engl. J. Med. 369, 669–670.

Fischer, M.D., Ferlie, E., 2013. Resisting hybridisation between modes of clinical risk management: contradiction, contest, and the production of intractable conflict. Acc. Organ. Soc. 38, 30–49.

Francis, R., 2013. Report of the Mid Staffordshire NHS Foundation Trust Public Inquiry: Executive Summary, vol. 1-3. The Stationary Office, London, UK. Retrieved 31 March 2014 from: http://www.midstaffspublicinquiry.com/report.

Friedman, S.M., Provan, D., Moore, S., et al., 2008. Errors, near misses and adverse events in the emergency department: what can patients tell us? CJEM 10, 421–427.

Garry, D.A., McKechnie, S.R., Culliford, D.J., 2014. A prospective multicentre observational study of adverse iatrogenic events and substandard care preceding intensive care unit admission (PREVENT). Anaesthesia 69, 137–142.

Goodman, J.C., Villarreal, P., Jones, B., 2011. The social cost of adverse medical events, and what we can do about it. Health Aff. 30, 590–595.

Goodney, P.P., Siewers, A.E., Stukel, T.A., et al., 2002. Is surgery getting safer? National trends in operative mortality. J. Am. Coll. Surg. 195, 219–227.

Griffin, F.A., Resar, R.K., 2009. IHI Global Trigger Tool for Measuring Adverse Events, second ed. Institute for Healthcare Improvement, Cambridge, MA.

Hayward, R.A., Hofer, T.P., 2001. Estimating hospital deaths due to medical errors: preventability is in the eye of the reviewer. JAMA 286, 415–420.

Hazell, L., Shakir, S.A., 2006. Under-reporting of adverse drug reactions: a systematic review. Drug Saf. 29, 385–396.

Hippocrates, c. 400 BCE. Of the Epidemics Book 1 (translated by Francis Adams). Retrieved 1 November 2013 from: http://classics.mit.edu//Hippocrates/epidemics.html.

Holmes, D., 2013. Mid Staffordshire scandal highlights NHS cultural crisis. Lancet 381, 521–522.

Illich, I., 1976. Medical Nemesis: The Expropriation of Health. Random House, New York, NY.

Kennedy, I., 2001. Learning from Bristol: The Report of the Public Inquiry Into Children's Heart Surgery at the Bristol Royal Infirmary 1984–1995. Her Majesty's Stationery Office, London.

Kennerly, D.A., Kudyakov, R., da Graca, B., et al., 2014. Characterization of adverse events detected in a large health care delivery system using an enhanced global trigger tool over a five-year interval. Health Serv. Res. 49. http://dx.doi.org/10.1111/1475-6773.12163.

King, A., Daniels, J., Lim, J., et al., 2010. Time to listen: a review of methods to solicit patient reports of adverse events. Qual. Saf. Health Care 19, 148–157.

Kinnell, H.G., 2000. Serial homicide by doctors: Shipman in perspective. BMJ 32, 1594–1597.

Kohn, L.T., Corrigan, J.M., Donaldson, M.S. (Eds.), 2000. To Err Is Human: Building a Safer Health System. National Academies Press, Washington, DC.

Kuske, S., Maass, C., Weingärter, V., et al., 2013. Patient-safety indicators: a systematic review, criteria-based characterization and prioritization. J. Public Health 21, 201–214. http://dx.doi.org/10.1007/s10389-012-0532-9.

Landrigan, C.P., Parry, G.J., Bones, C.B., et al., 2010. Temporal trends in rates of patient harm resulting from medical care. N. Engl. J. Med. 363, 2124–2134.

Leape, L.L., 2000. Institute of Medicine medical error figures are not exaggerated. JAMA 284, 95–97.

Leape, L.L., Berwick, D.M., 2000. Safe health care: are we up to it? We have to be. BMJ 320, 725.

Leape, L.L., Brennan, T.A., Troyen, A., et al., 1991. The nature of adverse events in hospitalized patients: results of the Harvard Medical Practice Study II. N. Engl. J. Med. 324, 377–384.

McCannon, C.J., Hackbarth, A.D., Griffin, F.A., 2007. Miles to go: an introduction to the 5 million lives campaign. Jt. Comm. J. Qual. Patient Saf. 33, 477–484.

McDonald, C.J., Weiner, M., Hui, S.L., 2000. Deaths due to medical errors are exaggerated in Institute of Medicine report. JAMA 284, 93–95.

McGivern, G., Fischer, M.D., 2012. Reactivity and reactions to regulatory transparency in medicine, psychotherapy and counselling. Soc. Sci. Med. 74, 289–296.

Rothschild, J.M., Landrigan, C.P., Cronin, J.W., et al., 2005. The Critical Care Safety Study: the incidence and nature of adverse events and serious medical errors in intensive care. Crit. Care Med. 33, 1694–1700.

Rutberg, H., Risberg, B.M., Sjödahl, R., et al., 2014. Characterisations of adverse events detected in a university hospital: a 4-year study using the Global Trigger Tool method. BMJ Open 4, 1–7. http://dx.doi.org/10.1136/bmjopen-2014-004879.

Sari, A.B.-A., Sheldon, T.A., Cracknell, A., Turnbull, A., 2007. Sensitivity of routine systems for reporting patient safety incidents in an NHS hospital: retrospective patient case note review. BMJ 334, 79–82.

Smith, J., 2002. First Report of the Shipman Inquiry: Death Disguised. Her Majesty's Stationery Office, London.

Solevåg, A.L., Nakstad, B., 2014. Utility of a Paediatric Trigger Tool in a Norwegian department of paediatric and adolescent medicine. BMJ Open 4, 1–8. http://dx.doi.org/10.1136/bmjopen-2014-005011.

Soop, M., Fryksmark, U., Köster, M., Haglund, A.B., 2009. The incidence of adverse events in Swedish hospitals: a retrospective medical record review study. Int. J. Qual. Health Care 21, 285–291.

Spence, D., 2012. Bad medicine: modern medicine. BMJ 344, 1. http://dx.doi.org/10.1136/bmj.e2346.

Tsang, C., Aylin, P., Palmer, W., 2008. Patient Safety Indicators: A Systematic Review of the Literature. National Institute for Health Research, London.

Vincent, C., Neale, G., Woloshynowych, M., 2001. Adverse events in British hospitals: preliminary retrospective record review. BMJ 322, 517–519.

Vincent, C., Aylin, P., Franklin, B.D., et al., 2008. Is health care getting safer? BMJ 337, 1205–1207.

Walshe, K., Shortell, S.M., 2004. When things go wrong: how health care organizations deal with major failures. Health Aff. 23, 103–111.

Wanzel, K.R., Jamieson, C.G., Bohnen, J.M., 2000. Complications on a general surgery service: incidence and reporting. Can. J. Surg. 43, 113–117.

Wilson, R.M., Runciman, W.B., Gibberd, R.W., et al., 1995. The quality in Australia healthcare study. Med. J. Aust. 163, 458–476.

Wilson, R.M., Michel, P., Olsen, S., et al., 2012. Patient safety in developing countries: retrospective estimation of scale and nature of harm to patients in hospital. BMJ 344, 1–14. http://dx.doi.org/10.1136/bmj.e832.

Zegers, M., de Bruijne, M.C., Wagner, C., et al., 2009. Adverse events and potentially preventable deaths in Dutch hospitals: results of a retrospective patient record review study. Qual. Saf. Health Care 18, 297–302.

Chapter 6

# Sources of Harm: Prescription Drugs, Surgery, and Infections

*He who lives by medical prescriptions lives miserably.*

(Proverb)

## Contents

The Health of Populations. http://dx.doi.org/10.1016/B978-0-12-802812-4.00006-0

The high frequency of medical harm shows that it is largely caused by mainstream clinical practices and not due merely to infrequently used interventions that most of us might hope to avoid. Of the causes summarized in Table 5.1, just three are responsible for more than 80% of all instances of harm: prescription drugs, medical procedures (primarily surgery, but inclusive of other procedures such as diagnostic tests), and hospital-acquired infections. That is, interventions patients are most likely to receive are the ones that cause most harm. Although the full extent of harm from these practices is still emerging, all have in fact been leading causes of harm throughout the history of their use. Chapter 5 discussed the tendency over more than a decade for successive studies of biomedical harm to reveal progressively higher rates than had been reported in earlier studies (e.g., Classen et al., 2011; Kohn et al., 2000; Landrigan et al., 2010). That pattern wherein more detailed analyses have tended to reveal ever greater levels of harm is repeated in the present chapter which considers specific causes of harm. Detailed study of specific causes not merely confirms that biomedical harm is extensive, but that some causes are individually responsible for more death and disability than even relatively recent analyses had suggested were due to all causes collectively.

## 6.1 HARM FROM PRESCRIPTION DRUGS

Harm from prescription drugs is often referred to as an ***adverse drug reaction (ADR)***, which has been defined as a "noxious and unintended" effect of a drug administered in the course of biomedical intervention (WHO, 1969, p. 6). The term ***adverse drug event (ADE)*** is also used, especially in more recent literature, with the two terms sometimes being used interchangeably. It can be useful, however, to distinguish between the two, with ADE having a wider meaning that encompasses two subclasses of events: First, as mentioned, ADR refers to harm caused by prescription drugs. Second, an ***adverse drug interaction (ADI)*** is harm due to a prescribed drug having negative effects when taken simultaneously with one or more other drugs, whether prescribed or not. The distinction between ADR and ADI is warranted, considering the high proportion of patients who take multiple drugs. Even drugs that may not harm when taken individually can interact harmfully when taken simultaneously. ADIs, then, include harm from the interactive effects of multiple prescribed drugs taken by the same patient, as well as single or multiple prescribed drugs taken against a background of other drug use, both over-the-counter and recreational, whether licit or illicit. To reiterate, ADE is inclusive of both ADR and ADI.

Almost 40% of the 393 instances of medical harm reported in Table 5.1 were due to ADEs. Wrong dosage is the most frequent cause, although harm also results from a variety of other actions, including wrong drug prescribed, wrong drug delivered (when the correct drug was prescribed), the prescribing of a drug despite the patient having a known allergy to that drug, and incorrect timing of delivery (too frequent or infrequent dosing). The root causes of such actions are also varied, and include inadequate drug knowledge by the prescribing physician, incomplete patient information, failure to adhere to prescribing recommendations, transcription errors on forms, and lapses in physician judgment. To avoid such errors, guidelines are promulgated, but with limited success. Lack of success cannot be attributed to guideline complexity, since much of the advice is prosaic, including the following requests of doctors to: take account of other drugs the patient may be taking, ask the patient about allergies before beginning the proposed drug treatment, explain potential side effects to patients, ask about side effects after treatment begins, and be aware that new symptom presentation following commencement of medication could be a side effect and not a symptom of the patient's medical condition (Aspden et al., 2007).

Although ADEs are the most frequent adverse medical event summarized in Table 5.1, none resulted in death in the Classen et al. (2011) study. In the wider population, however, fatal ADEs are far from rare. The thalidomide tragedy in the 1960s brought heightened awareness of ADEs, including ADE fatality. In the late 1950s and early 1960s, use of thalidomide as a treatment for morning sickness during pregnancy led to many infant deaths and to many infants being born with major deformities, including radically undeveloped limbs (Smithells and Newman, 1992). Those events prompted large-scale prospective studies of ADEs involving monitoring of defined patient populations to determine ADE incidence.

A **meta-analysis** of 39 studies conducted in the United States over a period of 32 years focussed on two separate patient populations, those admitted to hospital because of an ADE and those who experienced an ADE while in hospital (Lazarou et al., 1998). Extrapolating from the study findings, Lazarou et al. (1998) estimated that, in 1994, 106,000 patients in the United States experienced a fatal ADE, representing approximately 4.6% of fatalities from all causes. By that estimate, death due to the taking of prescribed drugs alone exceeded the upper limit estimated by the Institute of Medicine (IOM; Kohn et al., 2000) for all medically caused deaths. Additionally, taking account of the national mortality statistics for 1994 (Singh et al., 1996), the Lazarou et al. (1998) findings suggest that death from prescribed drugs was the fourth leading cause of mortality in the United States, after heart disease, cancer, and stroke.

The Lazarou et al. (1998) findings are consistent with findings from other studies published before and since. In an early review, Einarson (1993) pooled data from 36 international studies and found that more than 5% of all hospital admissions were the result of ADEs. In a large British study, more than 6% of 18,820 hospital admissions were found to have been medication-related (Pirmohamed et al., 2004). The median bed stay was 8 days, accounting for 4% of the hospital bed capacity. Broadly similar findings have been reported for other countries, including Finland, where 5.0% of 1511 deaths in one large hospital over a period of 1 year were found to be medication-related (Juntti-Patinen and Neuvonen, 2002). In a nationwide Dutch study, almost 2% of all acute hospital admissions during a 1-year period were ADE-related, and of these, 6% were fatal (van der Hooft et al., 2006). In Sweden, fatal ADEs were found to be the country's seventh most common cause of death (Wester et al., 2007).

In the aforementioned Finish study (Juntti-Patinen and Neuvonen, 2002), the drugs most commonly associated with fatalities were cytostatics used in the treatment of cancer, and antithrombotics used to reduce blood clotting. In Sweden, three-quarters of fatal ADEs were found to be due to gastrointestinal and brain hemorrhages, mostly due to the use of antithrombotic drugs (Wester et al., 2007). Gastrointestinal bleeding was also found to be the most common cause of medication-related hospitalization in Britain (Pirmohamed et al., 2004) and the Netherlands (van der Hooft et al., 2006). In the aforementioned Dutch study (van der Hooft et al., 2006), as with similar studies in other countries discussed in more detail below, the frequency of ADE-related hospitalizations in the national central registry grossly underestimated the true rate, with only 1% of all such events being reported to the registry.

The most comprehensive surveillance of medication harm may be that which forms part of the National Poison Data System maintained by Poison Control Centers in the United States. Of over 2 million "exposures" in 2012, including morbidity at all levels of severity and mortality, the most frequent substance class was analgesic medications, which accounted for almost 12% of all cases of poisoning (Mowry et al., 2013). Of a total of approximately 300,000 cases of medication error, the most common were inadvertent double-dosing, ingesting the wrong medication, and incorrect dose.

Systematic studies of harm from prescribed drugs have been ongoing for decades. No one denies that such harm is extensive, often serious, and sometimes fatal. In that context, it would be reasonable to expect that success in containing, if not eradicating, the problem would be a cornerstone of the much-vaunted onward advance of biomedical science and practice. In a large study to address that question, trends in harm from prescribed drugs were analyzed in English hospitals over a 10-year period from 1999 to 2008 (Wu et al., 2010). Over 500,000 admissions were examined showing that the annual number of medication-related admissions increased by almost 80% over the period, and mortality rate following admission increased 10%. That is, despite it being known for a long time that prescribed drugs are a leading cause of patient harm—including death—evidence shows that the problem has not diminished but has worsened.

### 6.1.1 Who Is Harmed by Prescription Drugs?

The overall rates of medication-related hospital admissions and fatalities conceal the fact that risk of harm is not evenly distributed throughout the patient population. Of all patient groups, the elderly are by far at greatest risk, with the prescribing of inappropriate drugs, in particular, consistently being found to be a cause of harm. For example, a Taiwanese study used a comprehensive nationwide database to examine visits by patients aged 65 years and older to community and hospital clinical services over a 3-year period (Lai et al., 2009). Anatomic, therapeutic, and chemical codes were used to identify instances of potentially inappropriate drugs. Almost 20% of nearly 200 million visits were found to have included a prescription for a potentially inappropriate drug, with almost two-thirds of patients in the study being dispensed at least one inappropriate drug each year.

Inappropriate drugs in the Taiwanese study were more likely if the patient was female, and if the prescribing physician was a general practitioner, male, and of older age. Other smaller-scale studies have produced similar findings. In the United States, for example, a study of 389 hospitalized patients aged 75 years or older found that 28% had been administered an inappropriate drug according to standard criteria, and 32% of patients experienced medication-related harm (Page and Ruscin, 2006). In a recent Australian study, 26% of more than 100,000 elderly veterans had at least one medication-related hospitalization within the 5-year period of the study (Kalisch et al., 2012). Similarly, in a recent Irish study of inpatients aged 65 years and older, 26% of 513 patients were identified as having suffered medication-related harm while in hospital (O'Connor et al., 2012).

Notwithstanding high reported rates of harm from prescribed drugs in older patients, there is every possibility that the true rate is higher than that revealed in the studies. Physical instability, psychomotor unsteadiness (e.g., tremor), cognitive decline (e.g., mental confusion), and sleep disturbance—including both drowsiness and insomnia—are symptoms commonly associated with natural aging. However, these are also common side effects of prescribed drugs, especially when multiple drugs are taken simultaneously (Fabian, 2013). It is evident that harm from prescribed drugs in elderly patients often goes unidentified because of confusion with the effects of aging. In a telephone survey of residents aged 65 years and older, the likelihood of having experienced harm from prescribed drugs in the previous 6 months was related to the number of prescribing physicians caring for the patient (Green et al., 2007). The mean number of prescribing physicians was three, and the probability of experiencing medication harm increased by almost one-third for each additional physician in attendance. Similarly, in the Taiwanese study mentioned above, the provision of subsidies for medical expenses appears to have endangered patients by facilitating access to medical services that, in turn, contributed to a higher level of inappropriate drugs being prescribed.

The preceding studies suggest a pattern of cause and effect that is reminiscent of a phenomenon well-known in biomedical science. This is the **dose-response relationship**, which describes the change in effect caused by varying levels of exposure. For example, a small dose of a particular drug may have little effect, a moderate dose may be therapeutic, and a large dose fatally toxic. The principle applies to populations as well as individuals, as for example, when the number of people harmed by environmental air pollution varies proportionately to the density of pollutants in the atmosphere. It is an irony of modern medicine that studies of medical harm suggest that biomedical healthcare itself is similarly fashioned. Broadly speaking, there appears to be a dose-response relationship between harm and level of exposure to clinical medicine (e.g., number of medical consultations, number of drugs prescribed, number of attending physicians, etc.), with harm increasing proportionately to increases in the amount of medical care received.

### 6.1.1.1 Unintentional Self-administered Drug Overdose

In addition to the high incidence of harm from medically supervised use of prescribed drugs, the incidence of unintentional serious harm and death from self-administered overdose of medical drugs, obtained by prescription or illicitly, has increased markedly. The increase has been particularly pronounced in relation to death from **opioid analgesics** prescribed for pain relief. In the United States, for example, the rate of death, usually from respiratory depression, attributable to prescribed opioid overdose more than doubled in the decade after 2000 (Bohnert et al., 2011; Paulozzi et al., 2012a) and currently exceeds fatal overdose from illicit use of heroin and cocaine combined (Paulozzi et al., 2012a, b). As discussed in Chapter 14, sufficient access to pain management involving opioid analgesics is more the exception than the rule worldwide, especially in developing countries (Seya et al., 2011). In contrast, there is a consensus that opioid analgesics are markedly over-prescribed in the United States, an opinion confirmed by the incidence of fatal opioid overdosing in that country.

The situation in the United States appears to be the result of cultural changes in the practice of medicine (Lembke, 2012). Patients, it seems, have come to expect greater compliance from doctors, who in turn are less inclined to refuse patients' requests. Change in medical culture appears to have been abetted by legal mandate in certain circumstances requiring compliance with patients' requests for pain relief. In addition, the convenience of prescribing drugs, and profit from their sale, tend to put patients' requests for pain relief ahead of other considerations, including patients' own ultimate welfare. This is especially evident when drugs such as opioids are prescribed for patients known or suspected of being addicted. Prescribing in those instances is self-evidently not effective as intervention for drug addiction, whereas more appropriate intervention (e.g., education and counseling) is time consuming. According to Lembke (2012, p. 1580):

*for physicians, treating pain pays, whereas treating addiction does not.*

Consequently, there now exists a culture of comparatively unrestrained prescribing of opioid analgesics, and other drugs, especially psychiatric drugs, including benzodiazepines, antidepressants, and antipsychotics, which is responsible for the dramatically increased incidence of death due to prescribed drug overdose (Jones et al., 2013).

## 6.1.1.2   Where Is the Commitment to Solving the Problem of Harm from Prescription Drugs?

Although harm from prescribed drugs has been consistently identified as a leading cause of mortality and morbidity, the precise scale of the problem is obscured by substantial noncompliance with reporting obligations. Many countries have what are known as *spontaneous reporting systems*, which are usually administered by a central or regional regulatory authority, and in some countries reporting is mandatory. Reports of harm from prescribed drugs are received from medical doctors and other health professionals, including pharmacists and nurses. The main function of such systems is to provide a repository of information about the frequency, range, and severity of harm in order to gauge the safety of prescribed drugs currently being used in clinical practice. Such systems are intended to provide a more detailed, representative, and longer-term depiction of drug effects than studies conducted by pharmaceutical companies during premarketing clinical trials of new drugs and postmarketing surveillance once drugs become available for use in clinical practice. Consequently, the quality of the information stored in such repositories is important for public safety, and underreporting poses serious potential threats.

In an assessment of overall levels of reporting, Hazell and Shakir (2006) reviewed 37 separate studies from 12 high-income countries. The studies had used various methods for estimating reporting levels, including comparing spontaneous reports with data from alternative sources that allowed more comprehensive scrutiny—such as hospital admission data, discharge notes, and insurance claims databases. The findings from the 37 studies indicated widespread underreporting. The underreporting rate was estimated for different settings (hospitals and general practice) and for different drugs, and the rates of underreporting were found to be high in all areas of clinical practice. Specifically, the median rate of underreporting was a remarkable 94%.

Generalizing from those results, it appears that medication-related harm is overwhelmingly *not* referred to the reporting agencies established specifically to record such events. It is possible, however, that a single overall rate gives a distorted account of the situation due to a likely bias against reporting common, less serious instances of harm, which comprise the majority in most settings. There was some confirmation of that suspicion in the Hazell and Shakir (2006) study. The median underreporting rate for "serious/severe" harm was indeed found to be lower (i.e., more such events were reported) than the overall rate. However, the difference between the two rates was not great, wherein the median underreporting rate for serious harm was 85%. That is, despite a degree of selective reporting related to severity of events, underreporting remains staggeringly high even for serious harm "including suspected reactions with fatal outcome" (Hazell and Shakir, 2006, p. 391).

Some of the main reasons given by healthcare personnel for not reporting harm from prescribed drugs are lack of time, other priorities, unfamiliarity with the reporting process, and lack of understanding of the purpose of reporting (Hazell and Shakir, 2006). In an attempt to increase reporting levels, many countries have extended existing systems to allow *patients* to report instances of harm they have sustained. One such system is the *Yellow Card Scheme* in the United Kingdom (Fortnum et al., 2012). When introduced in 1964 following the thalidomide tragedy, the scheme was initially available only to doctors, dentists, and coroners, but was later extended to include pharmacists, nurses, and midwives.

Since 2005, electronic, paper, and telephoned incidents from patients have also been accepted, and Fortnum et al. (2012) reported that 18% of notifications to the scheme were from patients. Although that level of participation may appear promising, it should be remembered that underreporting by healthcare personnel consistently exceeds 90% (e.g., Hazell and Shakir, 2006). It follows that—at 18% of all reports in the scheme—notifications from the public represent a small fraction of the total number of adverse events that actually occur. That inference was confirmed by survey results reported by Fortnum et al. (2012) involving a representative sample of more than 2000 of the general population in the United Kingdom. Although about one-quarter of respondents reported having experienced a side effect of medicine, less than 10% had heard of the *Yellow Card Scheme* and just three people in the entire sample of 2000 had ever used it. This amounts to an underreporting rate in excess of 99%.

## 6.2   HARM FROM SURGERY

The highest incidence of medical harm comes from prescribed drugs, but *medical procedures*, especially surgery, possess greater lethality than other causes. Of the comparatively small number of deaths reported in Table 5.1, half were due to medical procedures, a proportion not dissimilar to the estimated 40-45% reported in studies involving larger numbers of deaths (de Vries et al., 2008; Thomas et al., 2000). Prompted by the IOM report of medical fatalities (Kohn et al., 2000), strenuous efforts have been made to reduce rates of surgery-related mortality, and some encouraging trends toward improved safety have been noted in some areas. A recent large national study in the United States found that operative mortality rates for high-risk cancer and cardiovascular surgery declined during a 10-year period for all eight procedures that were studied (Finks et al., 2011). Reduction slightly exceeded one-third for one procedure (abdominal aortic aneurysm repair), but reductions in mortality rate for the remainder were approximately one-fifth or less. Thus, despite encouraging trends, overall progress toward reducing surgery-related mortality has been limited.

### 6.2.1   Never Events

In biomedical healthcare there is a category of avoidable harm referred to as *never events*. Unfortunately, the label signifies an aspiration rather than a reality. Although everyone agrees that these serious events, including *retained foreign objects* and *wrong-site surgery*, should never happen, they are not necessarily any more seriously harmful to patients when they do happen than other categories of medical harm. Rather, what distinguishes never events is not that they should never happen (the invocation of the Hippocratic Oath, *First, do no harm*, informs us that patients should never be harmed), but that they are *conspicuous*. It is their visibility that has led to these events being branded by the medical profession as uniquely unacceptable harms. Objectively, it is impossible to argue that never events are more unacceptable than other forms of serious and fatal medical harm. The distinctive feature of never events is that they embarrass the professionals who are responsible for their occurrence. Other forms of medical harm, including death, that are less conspicuously avoidable, and therefore do not threaten physician accountability, are not regarded as never events by those who cause them.

Of particular concern to physicians is the fact that never events have a comparatively high likelihood of being successfully litigated on a charge of medical negligence. A charge of negligence for failing to remove a surgical object from a patient's body cavity, or of performing surgery on the wrong patient or the wrong body part, is easily proved on the basis of *res ipsa loquitur* ("the thing itself speaks"). This doctrine states that a breach of duty of care under some circumstances is proven without reference to any evidence other than the event itself. Certainly, never events should never happen, but the label has more to do with them being indefensible in medicolegal terms than the scale of harm caused. In other words, never events are deemed unacceptable not because of harm to *patients*, but because of potential reputational and material harm to *physicians*.

## 6.2.1.1   Retained Foreign Objects

The images in Figure 6.1 illustrate the potentially public nature of retained foreign objects. Foreign objects are materials left inside patients upon completion of open surgery, and include surgical sponges, sharps (e.g., syringe, scalpel, and scissors), and other surgical instruments (e.g., clamps). Sponges are the most common, accounting for about one-half to two-thirds of retained objects (Gawande et al., 2003; Lincourt et al., 2007). While some patients with a retained object may remain asymptomatic for extended periods, others present with infection, cramping, obstruction, or other complications that sometimes result in death. *Retained* is a curious expression in this context, because it seems to imply an active effect, even fault, on the part of the unconscious patient; as if the object came to inhabit the patient through some means other than surgeon error.

Counting of sponges, sharps, and other instruments during open surgery is an almost universal practice, conducted several times during each operation (before surgery, before wound closure, and at skin closure). Although intuitively appealing, instrument counting, in the words of one group of researchers

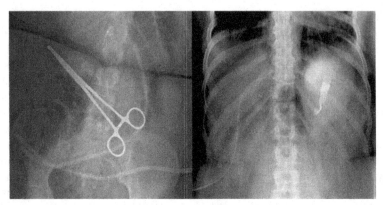

FIGURE 6.1   X-rays showing retained surgical objects of (a) a clamp (Wang et al., 2009) and (b) a laparotomy pad (a type of sponge) (Gibbs, 2011).

has "remained surprisingly primitive" as a way of preventing retained objects (Egorova et al., 2008). Counting is also surprisingly burdensome. In one study of successive patients undergoing general surgery, counting occupied an average of 9 minutes per case (Greenberg et al., 2008). Moreover, counting discrepancies occurred in more than 10% of cases, with each discrepancy requiring an average of 13 minutes to resolve. Delays of that order can have serious and lethal consequences due, for example, to postponement of patient transition to intensive care for monitoring and maintenance of vital functions.

It has been found that in less than 2% of instances of count discrepancies is a retained object actually found (i.e., almost all count discrepancies are *false alarms*) (Egorova et al., 2008). Conversely, in most cases where surgery was later found to have resulted in a retained object, a correct instrument count had been recorded (Cima et al., 2008; Gawande et al., 2003). Considering the delays caused by counting and the possible consequences of such delays, the cost-benefit advantage of counting may not always be obvious to the surgical team. However, the threat of litigation if a retained object is discovered in the absence of instrument counting, guarantees counting's place as an operating-room ritual.

Although universally practiced, instrument counting is known to be incapable of delivering the long-held medical aspiration of making retained foreign objects a never event of surgery. Given that counting, and especially count discrepancies, cause delays in surgery, it may be concluded that instrument counting, which imperfectly addresses the patient harm it is intended to avoid, is itself a cause of harm, although the amount of that harm remains unknown for wont of being comprehensively evaluated. In the meantime, due to the fallibility of manual counting by surgical personnel, work has proceeded on the use of computer-assisted counting of sponges using barcodes and use of gauze sponges tagged with a radiofrequency identification device (Hariharan and Lobo, 2013). Some success has been reported, but elimination of the problem of retained foreign objects is evidently not imminent.

### 6.2.1.2   Wrong-Site Surgery

Another highly visible type of surgery-related harm is that of wrong-site surgery, with all such instances being regarded as serious and reportable never events. Wrong-site surgery subsumes a variety of errors, including operating on the wrong patient, performing the wrong procedure (i.e., not the surgery that was indicated and intended), and operating at the wrong site of the body. The last of these includes performing the intervention on the wrong side of anatomically symmetrical structures (e.g., the left side of the body instead of the right), at the wrong location or body part (e.g., the wrong finger or wrong level on the spine), and wrong location within a structure (e.g., anterior versus posterior). When surveyed, one-in-five hand surgeons reported that they have performed wrong-site surgery (typically wrong-finger surgery) (Meinberg and Stern, 2003) and one-in-two neurosurgeons reported having performed wrong-level lumbar surgery (Mody et al., 2008).

Commensurate with the seriousness of wrong-site surgery, great effort has gone into trying to develop effective protocols to deal with the problem. However, in a comprehensive study, the conclusion of the authors was that wrong-site surgery continues to "occur regularly" (Clarke et al., 2007). The study, which was conducted over a 30-month period in the State of Pennsylvania, recorded 427 events, of which 59% were near misses. Of the remaining 41%, surgery was started, and in half of those it was done to completion. More recent analyses are no more encouraging. A recent review of the evidence concluded that "wrong site surgery [is] a 'never event' [that] continues to occur at an alarming rate" (Cobb, 2012, p. 232).

Casting an optimistic glance in the direction of the airline industry, the same author argued that:

> The checklist has served the airline industry well for many years. However, the … process is distinctly less organized and less strictly enforced in medical institution policy which continues to rely on surgeon and staff memory to avoid medical errors…. Just as pilot compliance with the checklist has successfully minimized errors in flight, so too we as surgeons must accept the process and make the designated crosschecks part of our procedures.
>
> (Cobb, 2012, p. 232)

However, the confidence vested in action checklists may be misplaced. If surgical instrument counting, a form of action checklist, is known to be of limited effectiveness in preventing retained objects (e.g., Egorova et al., 2008), is there reason to expect aviation-style action checklists to be successful in preventing wrong-site surgery?

## 6.2.2   Public Performance Indicators of Safety in Surgery

Considering the relatively high risks associated with surgery and awareness that the risks may not be evenly distributed between hospitals, it has become increasingly common for the performance records of hospitals to be made public. It is assumed that, armed with such information, patients can exercise informed choice, and the resulting "market forces" will create competitive pressure to encourage overall improvements in quality. However, great care is needed in relation to the nature of the information that is made available if it is to be truly helpful. For example, patients considering where to have surgery may be inclined to infer that their chances of survival are highest at hospitals that have a history of zero fatality for the particular operation they require. When this prediction was tested empirically using data from a large national survey of hospitals in the United States, the findings were less than comforting for patients (Dimick and Welch, 2008).

The study focussed on five types of operation having high operative mortality: coronary artery bypass grafting; abdominal aortic aneurysm repair; and resections for colon, lung, and pancreatic cancer. Hospitals where no operative fatalities had been reported for a 3-year period were compared for fatalities with all other hospitals in the subsequent year. The study was designed to see if hospitals that had a continuous 3-year record of no fatalities in specific operative categories (there were almost 3000 such hospitals in the study) maintained their good record in the fourth year when compared to other hospitals (of which there were 12,000) that had experienced one or more fatalities for the same operations during the same 3-year period. This provides a reasonable simulation of what patients might do if supplied with information concerning records of performance for different hospitals.

Contrary to what prospective patients are likely to expect, hospitals with a reported operative mortality of zero did not maintain lower than average mortality in the subsequent year for any of the five types of operation, and actually had a significantly higher mortality rate for pancreatic cancer. That is, the study showed that zero-fatality rate for specific types of operation, in the absence of other information, cannot be interpreted as an indicator of greater surgical safety. One reason for the findings could be the statistical phenomenon of **regression toward the mean**, whereby extreme values or *outliers* that are obtained when a variable is measured first will tend to be closer to the mean when a second measurement is taken.[1] Accordingly, hospitals that posted a mortality-free record may have experienced an aberrant or outlier result that was not maintained when measurements were repeated.

Statistical artifact, however, may only partially account for the seemingly paradoxical results. A fuller explanation is likely to include the fact that the hospitals reporting zero-fatality for specific types of surgery also tended to perform comparatively few of those operations in the 3-year period of initial screening (a tendency that was pronounced for pancreatic cancer resection). All other things being equal, fewer operations equates to fewer opportunities for fatalities to occur. Thus, the apparent good record of the zero-mortality hospitals was

---

1. Less intuitively, the reverse is also true. A value that is close to the mean when a variable is measured first will tend to be further from the mean when a second measurement is taken.

probably partly the result of their relatively low volume of surgeries and con-sequent reduced opportunity for surgery-related fatalities.

That interpretation is supported by other research which shows that high *vol-ume* in relation to particular types of surgery tends to be associated with lower mortality *rate* (Dimick and Welch, 2008) (consistent with the aphorism "practice makes perfect"). Thus, by attending a hospital that has performed comparatively few of a particular type of operation with apparently good results, patients may incur a higher risk than attending a hospital that has performed many such oper-ations despite some of those not being successful. Considered more broadly, the Dimick and Welch (2008) findings show that, despite the undeniable right of public access to information concerning all aspects of healthcare, the provision of overly-simplified information, without relevant contextual detail, may not always be helpful to patients and may sometimes be harmful. Conversely, the provision of more detailed information may also not always be helpful to prospective patients not used to evaluating complex detail.

Notwithstanding doubts about usefulness to patients of public performance indicators intended to communicate the comparative safety of surgery in different hospitals, some healthcare systems have adopted procedures that make public the performance records of individual surgeons. However, if the perfor-mance of fewer operations undermines the utility of hospital-wide information, the alleged safety record of individual surgeons is destined to have even less utility because the frequency of operations performed will be so low for many surgeons as to prevent statistically-meaningful comparison (Walker et al., 2013). Low number of operations performed by individual surgeons means that the reputation of some will benefit unwarrantedly due to a spurious statistical rating of good performance whereas others will suffer reputational damage due to a spurious poor rating.

Moreover, there is variability in the preoperative health status of patients. Therefore, surgeons could be wrongly identified as having a poor record if the patients they treat are at higher risk than the patients of other surgeons per-forming the same operation. Conversely, complacency due to an apparently good record of performance could be encouraged in surgeons whose referral base includes a disproportionately high number of patients at low risk. Such biases could lead to competition between surgeons to treat low-risk patients to the neglect of high-risk patients. Furthermore, reporting the performance of individual surgeons ignores the fact that patient outcomes often depend on the performance of individuals in a surgical team, as well as the performance of separate elements within complex healthcare systems. The quality of periop-erative care, management of patients' vital status during recovery, and follow-up care after discharge can all be important contributors to surgery outcomes, including death. Even when there is patient choice, which often there is not, it remains decidedly unclear whether supplying patients with information about the performance record of individual surgeons benefits patient outcomes or contributes to improved quality of surgery and associated healthcare.

## 6.3   HEALTHCARE-ASSOCIATED INFECTION

A popular image of modern hospitals is that they are (or should be) scrupulously clean: sanctuaries free of the multitude of germs that inhabit the world outside. Certainly, it was not always that way. Figure 6.2 shows a woodcut depicting a scene from a medieval hospital, which evidently would have been far from germ-free. The scene shows patients sharing beds and corpses being stitched into burial shrouds on the floor; a scene that seems remotely distant from what we have today. The true picture of the modern hospital, however, is quite different from popular image, and not as far removed from the hospitals of former times as we might wish. In addition to being places of harm from prescribed drugs and surgical procedures, hospitals are incubators of some of the most intractable and deadly microorganisms to be found anywhere.

Healthcare-associated infection (*HAI* or *HCAI*) refers to infection not present until such time as a patient comes into contact with a healthcare setting. HAIs have been most intensively studied in relation to hospital settings; hence, the essentially synonymous term, *hospital-acquired infection* is widely used, which conveniently is also represented by the acronym, HAI. However, healthcare-associated infection is the preferred term because it better reflects the reality that infection can be acquired as a consequence of medical care delivered in any setting. An equivalent term, which is declining in usage, is *nosocomial infection*; derived from the Greek words, *nosus* (disease) and *komeion* (to take

**FIGURE 6.2**   Triptych showing the Hôtel Dieu in Paris (circa 1500). Note. Comparatively well patients (on the right) were separated from the very ill (on the left). Although hospital routine has evidently changed substantially since the medieval era, problems of hospital safety remain substantially intractable. (*Source: http://commons.wikimedia.org/wiki/File:Hotel_Dieu_in_Paris_about_1500.gif.*)

care of), the term refers to disease (more specifically, infection) contracted while under medical care. As a general guide, infections that become clinically evident after 48 hours of hospitalization are considered to be HAIs. Those that occur after the patient is discharged from hospital are considered healthcare-associated if the infective organism is likely to have been acquired during the hospital stay, especially if the organism is known to inhabit hospitals while being less common in the community. HAIs also include surgical site infections that occur within 30 days after the operative procedure or within 1 year of a surgical implant.

A national survey involving 445 hospitals in the United States in 2002 found that among 37.5 million hospital discharges there were 1.7 million HAIs in a single year, a rate of 4.5% of all hospitalizations (Klevens et al., 2007). The majority of the hospitals in the study were part of a national infections surveillance system, whereby deaths involving patients with an HAI were reviewed and the role of the infection in the death was assessed as having been causal, contributory, not related, or unknown. In cases of multiple HAIs, the assessment was made for each infection separately. By this means, of more than 150,000 deaths among patients with HAI, approximately one-third were attributed to causes unrelated or unknown. Thus, in about 100,000 patients (i.e., almost 6% of all patients with an HAI), the infection was assessed to have caused or contributed to the patient's death. As an annual figure, this number, it might be recalled, is higher than the IOM (Kohn et al., 2000) estimated upper limit for *all* medically caused deaths.

Recently, the Centers for Disease Control, CDC (2015) in the United States published infection data for all states from a sample of more than 14,500 hospitals and healthcare facilities. The study revealed a national rate for healthcare-acquired infection of approximately 1-in-25 patients for 2013. Although that estimate and the earlier one by Klevens et al. (2007) are quite similar, the more recent CDC (2015) result is slightly better (4.0% versus 4.5%) indicating that there may have been some small improvement in recent years. The classification of infection type was similar in the two studies. In the Klevens et al. (2007) study, 80% of infections were of four main types: catheter-associated urinary-tract infections, surgical-site infections, ventilator-associated pneumonia, and catheter-associated bloodstream infections. Table 6.1 shows that urinary-tract infections were the most common, with an incidence almost twice that of surgical-site infections, which in turn were only slightly more common than bloodstream infections and pneumonia. The table also shows that the likelihood of death due to HAI (i.e., the "lethality" of infection) was substantially higher for bloodstream infections and pneumonia than for urinary tract and surgical-site infections.

Despite the high rate of HAIs in the United States, higher rates have often been reported for other countries (see Figure 6.3). In a recent European study of 20,000 patients in 66 hospitals from 23 countries, 7.1% were found to have an HAI (Zarb et al., 2012), which means that about 1-in-14 patients were infected

**TABLE 6.1** National survey of healthcare-associated infections (HAIs) and deaths, inclusive of adults, children, and newborns, involving 445 hospitals in the United States in 2002

| Type of infection | Number of HAIs | Percent of HAIs | Number of deaths | Percent of HAI deaths ("lethality")[a] |
|---|---|---|---|---|
| Urinary tract | 561,667 | 32 | 13,088 | 2 |
| Surgical site | 290,485 | 17 | 8205 | 3 |
| Pneumonia | 250,205 | 15 | 35,967 | 14 |
| Bloodstream | 248,678 | 14 | 30,665 | 12 |
| Other | 386,090 | 22 | 11,062 | 3 |
| Total | 1737,125[b] | 100 | 98,987 | 6 |

[a]Deaths as a percentage of patients with that particular type of infection.
[b]Represents 4.5% of 37.5 million hospital admissions for that year.
Adapted from Klevens et al. (2007).

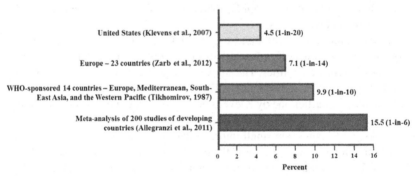

**FIGURE 6.3** Percentage (approximate odds) of HAI among hospital patients in different geographic regions.

compared to the most recent estimate for the United States of 1-in-25. An older, World Health Organization(WHO)-sponsored study conducted in the late 1980s in 14 countries from four regions (Europe, Mediterranean, South-East Asia, and the Western Pacific) found that approximately 1-in-10 patients was infected while in hospital (Tikhomirov, 1987). It is a consistent finding that rates are higher for poorer than for richer countries. A meta-analysis of more than 200 studies of HAI in developing countries reported a pooled prevalence of 15.5% (Allegranzi et al., 2011), more than twice the rate for Europe and almost

**TABLE 6.2** Prevalence of healthcare-associated infection (HAI) by hospital specialty in 66 hospitals from 23 European countries (N = 19,888)

| Medical specialty | Number of patients | Patients with HAI (%) |
|---|---|---|
| General medicine | 7833 | 505 (6.4) |
| Surgery | 6653 | 518 (7.8) |
| Obstetrics and gynecology | 1711 | 32 (1.9) |
| Pediatrics | 1024 | 38 (3.7) |
| Intensive care | 915 | 257 (28.1) |
| Psychiatry | 828 | 2 (0.2) |
| Geriatrics | 502 | 33 (6.6) |
| Other/mixed | 422 | 23 (5.5) |
| Total | 19,888 | 1408 (7.1) |

Adapted from Zarb et al. (2012).

four times that reported for the United States. That is, Figure 6.3 suggests that an average of approximately 1-in-6 inpatients in developing countries acquire an infection as a result of being hospitalized.

Table 6.2 summarizes results for different medical specialties, where it can be seen that the rate of HAI varied considerably from a low of less than 1% for psychiatry to a high of more than 28% for intensive care (Zarb et al., 2012). That is, almost one-third of the most acutely ill and vulnerable patients (those need intensive intervention) acquired an infection while in care. In summary, HAIs are prevalent and represent a serious global endemic source of harm caused by medicine.

The main cause of HAIs is known, and in theory most HAIs are preventable using readily available and relatively simple procedures (Calfree, 2012). However, the reality of the intractability of HAIs again illustrates profound intrinsic limitations of medical practice. It will be recalled from Chapters 1 and 2 that substantial control over infection in the general community was delivered as part of the epidemiologic transition. It is an irony, therefore, that modern health-care practices intended to relieve suffering and spare lives are the cause of a global epidemic of infection-related death. It is a double irony that the main cause of HAI is something as banal as a lack of hygiene by healthcare personnel. Poignant as those ironies are, they are eclipsed by the further realization that decades of routine medical practice have honed new infections that are among the deadliest known.

## 6.3.1 Healthcare Hygiene

While inadequate healthcare hygiene is the main cause of HAIs, the specifics of the pathogenesis of HAIs are nevertheless varied. Infections can be localized to particular bodily sites or they can be systemic (affecting the body generally), and may involve any or all systems of the body. Infectious microorganisms may come from endogenous or exogenous sites. Endogenous sites include organs of the body normally inhabited by microbes, including the nose, mouth, and throat, the gastrointestinal system, and the genitourinary system. Exogenous sites are those other than the patient, including medical devices and equipment, the healthcare environment (e.g., door handles, walls, and other surfaces), visitors, and most importantly healthcare personnel as they move from one location of the hospital to another.

In addition to being located at multiple sites in the healthcare environment, infectious organisms have multiple routes of transmission. Many infections in the nonhospital environment are transmitted by means of a carrier or vector. Vector-borne diseases that remain common in some regions of the world include malaria, dengue fever, hemorrhagic fever, encephalitis, and typhus. Common vectors for such infections include mosquitoes, ticks, flies, fleas, lice, birds, bats, rats, and other vermin. The main vector in the transmission of infections in hospitals is human. While this may include patient-to-patient and visitor-to-patient transmission of disease, by far the most potent infectious-disease vectors in healthcare settings are healthcare personnel—principally doctors and nurses.

Most HAIs (about 80%) occur in association with the use of medical instruments. These include central venous catheters (for delivering fluids and for monitoring body functions), urinary catheters (to drain the bladder; see Box 5.2), endotracheal tubes (to provide a clear airway to and from the lungs), and medical ventilators to assist breathing (Calfree, 2012). The main reason for these devices and procedures causing infection is inadequate hygiene by those tasked with delivering curative care, and the main mechanism by which this happens is failure by doctors and nurses to wash their hands. Although HAI is sometimes portrayed as new, hospitals have always been sites of infection transmission, as may be surmised from the scene depicted in Figure 6.2. Proof that healthcare personnel are vectors of infectious disease and the vital role of hand hygiene by healthcare personnel in preventing infection was demonstrated more than 150 years ago.

### 6.3.1.1 Infection Control

An Austro-Hungarian doctor, Ignas Semmelweis, working in Vienna in the 1840s, is credited with having made the initial breakthrough discoveries showing that doctors could be the vectors of appalling death and disease due to **puerperal fever** in maternity hospitals (see Box 6.1 and Figure 6.4). Puerperal fever, also known as *childbed fever* and the *doctors' plague*, is a bacterial infection contracted by women during childbirth, miscarriage, or abortion, especially

**BOX 6.1  The Father of Infection Control**

Ignas Philipp Semmelweis (1818-1865). *(Source: http://upload.wikimedia.org/wikipedia/commons/f/f8/Ignaz_Semmelweis_1860.jpg)*

During the early period of the medicalization of childbirth, **puerperal fever** was a major cause of death for mothers and newborns. As discussed in the text, physicians were the main cause of what was also known as *childbed fever* and the *doctors' plague*, a highly contagious bacterial infection that can rapidly develop into fatal septicemia (blood poisoning). In the absence of sterile procedures, placental separation during childbirth creates conditions that are highly susceptible to infection, and epidemics of puerperal fever accompanied the expansion of hospitals due to the rich infective environment harbored therein.

Chance circumstances at the University of Vienna General Hospital in the mid-nineteenth century provided near-perfect conditions for a controlled experiment, which revealed that insanitary practice by physicians was a leading cause of death from puerperal fever. Admissions to the obstetrics department, which at that time was Europe's largest, were more-or-less randomly allocated to either of two wards. The wards were identical in almost all respects, including physical facilities, ventilation and temperature control, bed density, patient diet, hygiene of the bed linen, and demographic characteristics of the women. Until 1841, the two wards were staffed by a combination of physicians and midwives, but a change in policy brought a change in staffing arrangements. One ward came to be staffed by physicians and medical students, and the other staffed by midwives and midwifery students (Figure 6.4).

After graduating from medical school at the University of Vienna, Semmelweis trained as an obstetrician at the Vienna General Hospital. While there, he observed that the mortality rate for puerperal fever was substantially higher in the ward staffed by physicians than in the ward staffed by midwives. He surmised that the difference in mortality rate was probably related to the different duties performed by the two groups. He noted, in particular, that the physicians and their students routinely performed autopsy dissections, whereas the midwives did not. Semmelweis hypothesized that puerperal fever was caused by the "conveyance of decomposed animal-organic matter" carried on the hands of physicians and transferred to

*Continued*

**BOX 6.1 The Father of Infection Control—cont'd**

expectant mothers at the time of childbirth. Crucially, he devised an intervention to test his hypothesis. Beginning in 1847, physicians and medical students were required to wash their hands in a solution of bleach prior to assisting with deliveries. Mortality due to puerperal fever in that ward immediately fell to a level approximating that of the ward staffed by midwives.

Figure 6.4 shows that during the period before 1841, when both wards were staffed by a combination of physicians and midwives, the average rate of mortality due to puerperal fever was similar for the two wards. From 1841 to 1846, the mortality rate increased sharply for the ward staffed by physicians and fell for the ward staffed by midwives. The role of physicians in transmitting disease, concealed when both wards were staffed by physicians and midwives, was revealed. That role was confirmed when the new hand-washing regimen mandated by Semmelweis was introduced and the mortality rate in the ward staffed by physicians fell to a level approximating that of the ward staffed by midwives.

**Great Minds**

Semmelweis was not the first to observe that puerperal fever is caused by physical contact between mothers in childbirth and those who attended the births. In 1843, the American physician, poet, and novelist, Oliver Wendell Holmes, published an essay in which he argued that physicians were transmitting the disease between successive patients they visited (Holmes, 1843). He declared that "doctors were instruments of death" by failing to thoroughly clean themselves, their clothing, and their instruments after attending a patient who had the disease and before going onto the next mother-to-be. He even presaged Semmelweis' specific findings with the observation: "A physician holding himself in readiness to attend cases of midwifery, should never take any active part in the post-mortem examination of cases of puerperal fever."

Earlier still, about a half-century before Semmelweis conducted his work, a Scottish physician, Alexander Gordon, asserted that puerperal fever "seized such women, only as were visited, or delivered, by a practitioner, or taken care of by a nurse, who had previously attended patients affected with the disease" (Gordon, 1795). In repudiation of the then popular miasma theory of disease, Gordon also argued that puerperal fever is not due "to a noxious constitution of the atmosphere." Given that Holmes knew of Gordon's work and acknowledged its importance (Lowis, 1993), it is possible that Semmelweis was familiar with both.

Without entering into the longstanding debate over who deserves most credit for discovering the link between insanitary practices and puerperal fever, it is worth mentioning that Semmelweis is distinguished among early pioneers of healthcare-acquired infection by being the first to provide systematic data confirming the efficacy of hand hygiene in limiting the spread of infectious disease. What unites Semmelweis, Holmes, and Gordon is their shared perspicacity and tenacity. They are all likewise united, as are many bearers of uncomfortable ideas before and since, by how they were ignored and sometimes berated by professional colleagues for the views they espoused and the forcefulness with which they argued them. In his day, Semmelweis suffered many rebukes, but today he is applauded as the Father of Infection Control (Best and Neuhauser, 2004).

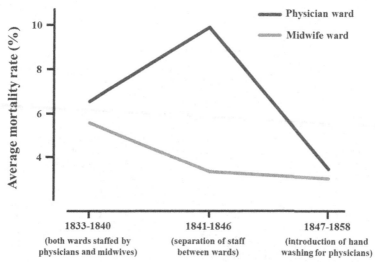

**FIGURE 6.4**    Mid-nineteenth century mortality due to puerperal fever in the University of Vienna General Hospital. Note. The intriguing story of Ignas Semmelweis has entered medical folklore, which may account for discrepancies in detail to be found in the varied literature describing his observations. In particular, some accounts of his work refer to a mortality rate due to puerperal fever of almost 20% when doctors assisted childbirths compared to a rate of 2% for midwife-assisted births. The estimates reproduced here are among the more conservative of those reported. *(Adapted from Funkhouser, 2012, Table 1, p. 5.)*

when these occur in unhygienic surroundings. The term has since been replaced by more specific terminology that identifies the site and severity of infection. Broadly, pathogenic organisms that invade the bloodstream and lymph system may cause potentially fatal septicemia (blood poisoning). When Semmelweis first drew attention to the possibility, there was resistance among physicians to the suggestion that they themselves were a common cause of patient death. However, bacteriological discoveries later in the nineteenth century, including those of Louis Pasteur and Robert Koch, led to acceptance of the *germ theory* of disease over the previously popular *miasma theory*. Confirmation of the germ theory provided convincing evidence that healthcare personnel, practices, and procedures can be a potent cause of disease.

Just as basic hand hygiene has been shown to be effective in limiting the spread of disease in the community (see Box 1.2), the critical role of hand hygiene for limiting the spread of disease in hospitals has been a generally accepted fact for more than a century. Hospital-based protocols for the surveillance, prevention, and control of HAI centered on hand hygiene have been in place since the 1950s (Haidee and Custodio, 2012). In 2005, the WHO launched a global campaign to reduce HAI and identified the promotion of hand hygiene as a priority (Pittet et al., 2009). The WHO guidelines provide healthcare personnel, hospital administrators, and health authorities with specific recommendations for

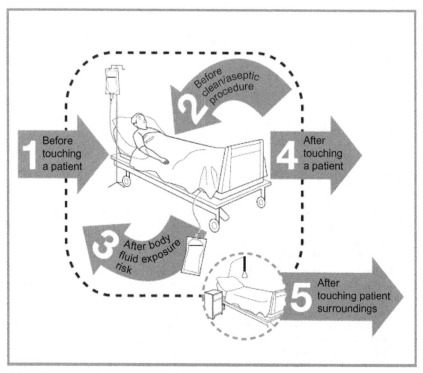

**FIGURE 6.5**   The World Health Organization *My five moments for hand hygiene. (Adapted from Sax et al., 2007 as promulgated by WHO, 2009.)*

improving practices to reduce the transmission of pathogenic microorganisms in healthcare settings (WHO, 2009). Figure 6.5 is illustrative of the standards promulgated by the WHO, wherein healthcare personnel are required to perform hand hygiene between the last hand-to-surface contact with an object outside the "patient zone" and the first contact in the patient zone; before an aseptic task (e.g., giving an injection or performing wound care); immediately after a care task associated with risk of exposing hands to body fluids and before any subsequent hand-to-surface exposure; when leaving the patient zone and before touching an object outside the patient zone; and after hand exposure to any surface in the patient zone but without touching the patient (Sax et al., 2007). For some tasks, the use of gloves is standard procedure, and hand hygiene is required before donning gloves. Use of gloves, however, is not a substitute for hand washing.

Hygiene guidelines and protocols have been shown to reduce HAI incidence (Pittet et al., 2011; WHO, 2009), and for that reason the WHO and health authorities worldwide are justified in making every effort to promote the highest possible level of adherence to such protocols. Despite such efforts, however, observational studies show that both the frequency and the quality of hand hygiene remain persistently suboptimal, thereby ensuring the continuance of

HAI as a common cause of disease and death. Even 90% adherence to hand hygiene protocols is believed to be insufficient to eliminate HAI. Yet, in practice, frequency of hand washing by healthcare personnel routinely falls below (often well below) 50% of that required by accepted guidelines and protocols (Boyce and Pittet, 2002).

In summary, despite concerted effort, the problem of healthcare hygiene has to date remained largely intractable. Realities indicate that the problem is intrinsic to healthcare practice, and therefore healthcare hygiene is certain to continue, and may worsen, as a leading cause of medical harm. The occurrence of infection, due largely to suboptimal healthcare hygiene during routine medical practice, typically requires additional clinical intervention to treat infected patients. The most common interventions for healthcare-acquired infection include the use of antibiotics. However, use of antibiotics contributes to the emergence of antibiotic-resistant infectious microorganisms, a problem that has grown to such proportions as to threaten infection control globally.

## 6.4   MULTIPLE DRUG RESISTANCE

*The art of war is deception; that is deceiving the enemy. But in the war against microbes we have deceived ourselves by misusing, under using and overusing antibiotics.*

(Ghafur, 2010, p. 144)

Antimicrobial drugs (*antibiotics*) are of several types, including antibacterials, as well as antivirals, antifungals, and antiparasitics. Some microorganisms survive exposure to antimicrobial drugs, and are said to be *resistant*. Resistance may be intrinsic or acquired, but the latter is of particular concern because it means that drugs that had previously been used to treat specific infections may no longer be effective. Use of antibiotics creates a selective advantage for the survival of the organisms with genes for resistance, and encourages the spread of antibiotic resistance throughout an ecosystem of bacteria. Due to the intensive use of antibiotics in healthcare settings, hospitals generally, and intensive-care facilities in particular, are major sources of acquired drug resistance. Resistant organisms are not only transmitted within healthcare settings but also into the community by patients when discharged. The problem is especially critical when pathogenic organisms develop resistance to several drugs that had previously been found useful in clinical practice. These multiple drug (or multidrug) resistant (*MDR*) pathogens ("superbugs") pose a growing threat to population health worldwide (WHO, 2012).

Public awareness of multidrug resistance is most strongly associated with various serious infections that have become increasingly endemic in healthcare settings. The more familiar pathogens include the Gram-positive bacteria methicillin-resistant *Staphylococcus aureus* (*MRSA*), vancomycin-resistant enterococci (*VRE*), and *Clostridium difficile (C. difficile)*, and the Gram-negative *Acinetobacter*, *Pseudomonas aeruginosa*, and *Escherichia coli*

## BOX 6.2 Gram-Positive and Gram-Negative Bacteria

Hans Christian Joachim Gram (1853-1938). *(Source: http://www.ncl.ac.uk/dental/ oralbiol/oralenv/tutorials/christian_gram.htm.)*

Hans Christian Joachim Gram was a Danish bacteriologist. While working in the morgue of the city hospital of Berlin, he developed a method, which bears his name, for staining bacteria. The Gram stain is almost always the first step in the identification of a bacterial organism, and it is used to differentiate bacterial species into two main groups. Gram invented the method not for the specific purpose of distinguishing one type of bacterium from another, but for the more general purpose of enabling bacteria to be seen more readily in stained sections under a microscope. As typically employed, a Gram stain is made using an initial stain of crystal violet and a counterstain of safranin. Bacteria that turn blue or purple when stained are called "Gram positive," whereas those that turn pink or red when counterstained are called "Gram negative." Some organisms are Gram-variable, meaning that they may stain either positive or negative, and others are Gram indeterminate. Alternative techniques for identifying bacteria are also available (e.g., genetic markers).

Difference between bacteria in response to the Gram stain technique is due primarily to variability in the chemical structure of the cell wall, which is also associated with differences in susceptibility to antibiotic treatment. For example, vancomycin can kill Gram-positive bacteria, but is ineffective against Gram-negative pathogens. Gram staining provides a method for obtaining a general indication of bacterial type, and its main advantage is the rapidity with which it provides results compared to techniques that depend on the production of bacterial cultures. Thus, patients suspected of bacterial infection have body fluids or biopsy tissues stained by the Gram method, and the results are used to inform initial antibiotic treatment in advance of cultivation of microbiological cultures for purposes of further pathogen identification and treatment selection.

*(E. coli)* (see Box 6.2). Some occurrence of bacterial resistance is possibly an inevitable consequence of antibiotic use, including use that is appropriate and well-regulated. However, the problem of antimicrobial resistance is greatly exacerbated by misuse, which has contributed substantially to the emergence of appalling and fatal infectious bacteria that are resistant to multiple antibiotics.

It is widely accepted that at least half and probably more of the antibiotics used in clinical practice are misused. In one study, 100 consecutive admissions to hospital emergency departments were identified in which a common class of antibiotics *(fluoroquinolones)* was prescribed, and the appropriateness of the intervention was judged according to institutional guidelines (Lautenbach et al., 2003). Of the 100 patients, use of the antibiotic was judged to be inappropriate in 81. Of the 19 patients for whom the antibiotic was appropriately prescribed, only 1 received both the correct dose and duration of therapy.

One of the most common misuses of antibiotics is in treating acute *upper respiratory tract infection*, or the *common cold*, characterized by symptoms of nasal stuffiness and discharge *(rhinitis)*, sneezing, sore throat *(pharyngitis)*, and cough. The frequency of colds tends to decrease with age, with young children having on average 6 to 10 and adults 2 to 5 colds per year (Eccles, 2005; Heikkinen and Järvinen, 2003). Cumulatively, colds impose an immense economic burden in terms of healthcare costs, and absences from work, school, or day care. Many people seek remedies for the discomfort caused by colds, but there is little effective treatment. In general medical practice, antibiotics are widely prescribed for patients who present with a cold, despite the fact that colds are caused by viruses known to be unresponsive to antibiotics (Kenealy and Arroll, 2013; Simasek and Blandino, 2007).

In one study that utilized over 100,000 computerized records covering a 12-month period for 17 general practices in New Zealand, upper respiratory tract infection was noted for almost 1 in 10 consultations (McGregor et al., 1995). Of these, approximately 4 of every 5 were prescribed 1 of 15 different antibiotics. Comparison of clinical outcomes for patients who did and did not receive antibiotic treatment showed no differences in recovery from cold symptoms. In the United Kingdom, consumption of antibiotics increased by 36% in the decade to 2010, and the proportion of patients prescribed antibiotics for coughs and colds rose from 36% in 1995 to 51% in 2011 (Editorial, 2014).

Use of antibiotics for treating colds is not merely ineffective it can induce negative side effects such as diarrhea (Kenealy and Arroll, 2013). Additionally, antibiotic treatment of infection can, paradoxically, expose patients to increased infection risk (Wat, 2004). The normal human gut is colonized by an estimated 100 trillion microbes that assist with vital functions such as digestion, immune defense, and nutrient production (Turner and Thompson, 2013). While increasingly less effective against multiresistant pathogens, antibiotics, especially widely-used broad-spectrum agents, are active against native bacterial species. Thus, by simultaneously undermining the integrity of native flora and being increasingly less effective against pathogenic species, antibiotics have the potential to increase vulnerability to infection.

## 6.4.1    New MDR Threats

While much attention in recent years has been focussed on multiresistant Gram-positive bacteria such as MRSA, VRE, and *C. difficile*, Gram-negative bacteria are currently arousing increased concern (Kumarasamy et al., 2010). The explanation for this shift in attention is that resistance is developing faster in Gram-negative than in Gram-positive bacteria, and fewer antibiotics are either available or in development to provide therapeutic cover against Gram-negative bacteria. These bacteria have not only been responsible for devastating hospital- and community-acquired disease, but their rapid spread has been aided by human travel and migration (Schofield, 2011). One particular form of international travel that has been identified as having markedly enhanced potential to accelerate the rate of bacterial transmission worldwide is *medical tourism*, sometimes called *added value travel*. This involves people traveling between countries in pursuit of lower-cost medical procedures, often for elective surgery, such as cosmetic procedures, but increasingly also for corrective surgery. It is claimed that, in India alone, this trade, which is forecast to increase, currently caters for 450,000 people per year as part of an industry worth about USD2 billion (Walsh and Toleman, 2011).

Particularly dire warnings have been sounded in relation to hospital-acquired New Delhi metallo-β-lactamase-1 *(NDM-1)*,[2] which is not to a single bacterial species but a transmissible genetic element capable of encoding multiple resistance genes (Moellering, 2010). Bacteria containing this genetic element (or variants) acquire resistance to almost all antimicrobial agents. The ability to transmit among usual Gram-negative bacteria confers enormous potential for the NDM-1 gene pool to go largely undetected. It has been estimated that at least 100 million Indian residents carry NDM-1 bacteria as normal gut flora having the potential to cause vast numbers of cases of largely untreatable infection (Walsh and Toleman, 2011). The first confirmed appearance of NDM-1 was in Sweden in 2008 in a patient who returned home with the infection after being hospitalized in India (Yong et al., 2009) and identification of other similar cases quickly followed. Bacteria containing the gene have since been isolated in many countries covering most regions of the world, raising fears about the "uncontrollable spread of pandemic clones for which new and effective antibiotics are currently not available" (Rolain et al., 2010, p. 1700).

It will be recalled that improvements in public sanitation, including sewage disposal and the provision of clean water, in England and Wales during the eighteenth and nineteenth centuries were essential factors in the decline in infectious diseases that define the epidemiologic transition. In that context, it is notable that the reservoir of NDM-1 infection in the Indian subcontinent has been attributed primarily

---

2. An outcry from Indian authorities over the inclusion of "New Delhi" in the labeling of this new discovery has led to calls for a change of name in this instance and for stricter guidelines concerning the naming of scientific discoveries in general (Singh, 2011).

to poor sanitation for hundreds of millions of residents (Walsh and Toleman, 2011). It should also be noted that there is a history in India of easy access to antibiotics with or without a prescription. Indian physician Abdul Ghafur has claimed that "India, is the world leader in antibiotic resistance, in no other country [have] antibiotics been misused to such an extent" (Ghafur, 2010, p. 144). High levels of bacterial pollution of the environment and endemic infection in the population have encouraged antibiotic use in India as a substitute for sanitation. Considering the largely unfettered access to antibiotics, continuing rapid development of antimicrobial resistance in community and healthcare settings is largely guaranteed. International travel, especially medical tourism, threatens to exacerbate the global transmission of increasingly antibiotic resistant pathogens.

## 6.4.2    Animal Husbandry

Notwithstanding the immense quantities of antibiotics consumed by people, that source accounts for less than half of the world's total consumption (WHO, 2012). The remainder goes into animal husbandry and aquaculture, where antibiotics are mass administered to healthy food-producing animals for the purposes of promoting growth and preventing disease. The fact that the quantity of antibiotics that goes to healthy animals exceeds the total amount used to treat disease in humans is universally accepted as gross misuse and a major contributor to the emergence of multidrug resistance. The additional fact that some of the same antibiotics used in medicine are also used in food production further compounds the problem of their overuse in animal husbandry. Such use not only contributes to the emergence and spread of resistant bacteria, but also increases the risk of cross infection from animals to people (Collignon et al., 2009).

Antibiotic use in animals, as in humans, causes increased antibiotic resistance in bacterial pathogens while also contributing to the transfer of resistance genes to other intestinal bacterial species (Khachatourians, 1998). The greater the number of resistant bacteria present in the intestinal flora of animals, the greater is the likelihood of genes encoding resistance being disseminated into the environment. This could include transfer to wild animals, birds, and fish, thereby adding to existing threats from emerging infectious diseases in wildlife populations (Daszak et al., 2000). Additionally, there is increased risk of transfer of resistance genes to humans from the consumption of farmed animals. Consequently, not only can resistant animal pathogens cause human disease, but resistance genes transferred from animals to the normal intestinal flora of humans can also cause disease. Thus, the carriage of resistance genes by pathogenic as well "normal" bacteria greatly increases the threat of infectious diseases being disseminated throughout human populations. Notwithstanding the enormous harm from healthcare-acquired infections, the extent of that harm is thought by some to be dwarfed by the potential threat to health from resistant bacteria emanating from current practices in animal husbandry (van den Bogaard and Stobberingh, 2000).

## 6.5   LIMITING HARM DUE TO ANTIBIOTIC RESISTANCE

A policy package has been proposed by the WHO (2012) for stemming harm from antibiotic resistance. The package includes six actions: surveillance of antimicrobial resistance and use; rational antimicrobial use and regulation; reducing or eliminating antimicrobial use in animal husbandry; infection prevention and control; fostering innovations; and political commitment. The essentials of each strategy are summarized in Box 6.3. Of many strategies proposed to counter antibiotic resistance, particular attention to date has been given to drug innovation, antibiotic stewardship, and isolation practice.

---

**BOX 6.3 Selective Summary of Policy Actions Recommended by the World Health Organization for Tackling the Threat of Antimicrobial Resistance**

| Action | Summary |
|---|---|
| Surveillance to track antimicrobial use and resistance in bacteria | Monitoring of antibiotic use and occurrence of resistant bacteria as a basis for planning strategies, and for mobilizing local, national, and international resources and commitment to action. Currently, surveillance varies between countries and regions, and coordination is needed to provide more extensive and uniform geographic coverage |
| Measures to ensure better use of antibiotics | Much of the antimicrobial resistance problem stems from the misuse of antibiotics, particularly excessive use. If antibiotics were always prescribed appropriately and only when needed, the treatment correctly followed, never used in agriculture or aquaculture, and if substandard and counterfeit products could be abolished, selective pressure on bacteria to become resistant would be reduced. Development of regulations and practical measures is needed, as is political agreement to put into practice the regulations and measures that are required |
| Reducing antimicrobial use in animal husbandry | Antibiotics are used widely and in immense quantities for preventing disease and promoting growth of livestock, poultry, and fish reared for food production. Although some countries have banned the use of antibiotics as growth promoters, the practice remains widespread. Legislation and regulation with enforcement are needed in many countries to control the use of antibiotics in animal husbandry |

**BOX 6.3 Selective Summary of Policy Actions Recommended by the World Health Organization for Tackling the Threat of Antimicrobial Resistance—cont'd**

| | |
|---|---|
| Infection prevention and control in healthcare facilities | As centers for the treatment of serious illnesses, hospitals are unfortunately also where antibiotic-resistant infections are particularly likely to develop and spread. Infections acquired in hospitals and other healthcare facilities caused by resistant bacteria exert a heavy toll in terms of illness and mortality, as well as added direct and indirect costs. The key to limiting the risk lies in the meticulous application of measures for the prevention and control of infection |
| Fostering innovation to combat antimicrobial resistance | Considering the inexorable increase in antimicrobial-resistant infections, a dearth of new antibiotics, and little incentive for industry to invest in research and development in this field, innovative approaches are reputed to be crucial for the development of new products to counter the rise of antimicrobial resistance. Although an enabling environment for innovation depends on support from policy decision-makers, history shows that new antibiotics are also likely to be misused |
| Political commitment to enable options for action | The global health crisis due to antimicrobial resistance concerns us all. It is a question of whether or not there will be effective antibiotics to treat many important life-threatening infections in the future. Antimicrobial resistance can be reduced, and despite knowledge gaps, strategies and practical measures could be applied more widely. Mobilizing the necessary expertise and resources to mount a concerted effort to prevent and control antimicrobial resistance will depend on the cooperation and commitment of policy decision-makers in all countries |

*Adapted from WHO, 2012.*

## 6.5.1 What Innovation? The Pipeline Is Empty

A recurring lament of those working in the field of multidrug resistance is the dual problem of growing demand for antimicrobial agents to combat the relentless spread of resistant organisms and the dwindling supply of suitable agents as existing ones fall prey to increased resistance. One oft-proposed strategy seeks to outstrip resistant strains by developing new antibiotics to replace those for which resistance has emerged. The strategy assumes that in time the new agents will also succumb to the relentless spread of resistance, at which time they must then be replaced by still newer agents. No one has addressed the problem of how many cycles this process might need to be repeated, but in principle it could continue ad infinitum. It seems, however, that this replacement strategy is advocated more often for want of something better to do, rather than genuine belief that it can work. The accepted reality among many commentators is that bacteria "overwhelm us with their superior numbers, they reproduce with remarkable speed, and they develop extremely efficient ways to exchange and promulgate resistance genes" (Moellering, 2010, p. 2379). That the replacement strategy is more aspiration than realizable goal is evidenced by the antibiotic development "pipeline" being largely empty.

Over the past decade there has been a crescendo of claims by industry leaders, health authorities, and policy makers about a putative *innovation crisis* in pharmaceutical research. Reasons for the crisis in drug innovation (in general, not only antibiotics) include overcoming practical and intellectual challenges of new drug discovery, conducting multiple trials to determine safety and efficacy, and navigating the regulatory hurdles involved in obtaining approval for clinical use, with the whole process taking possibly 10-15 years to complete one cycle. Additionally, with respect to antibiotics, there is the likelihood of a relatively short period of effectiveness (and profitability) due to loss of efficacy from resistance, and the need for further replacement with the next new drug. In short, it does not pay the pharmaceutical industry to go to all that trouble and expense to develop drugs that bacterial resistance may render redundant before costs can be recouped and profits reaped.

Light and Lexchin (2012), however, have argued that the innovation crisis, at least, as usually portrayed, is a myth. The real crisis they explain is that current incentives do not reward companies for genuine innovation in drug development. Rather, companies are rewarded for redesigning existing drugs in ways that allow the modified product to be promoted as "new." Typically, the new drugs offer little advantage over earlier versions, and may even be less effective, a topic discussed in the next chapter. The relevant point here is that "telling 'innovation crisis' stories to politicians and the press serves as a ploy" to extract benefits from government such as taxpayer subsidies and protection from competition (Light and Lexchin, 2012). At present, industry strategy is to focus on the easier challenge of developing new drugs of low efficacy that are not needed while ignoring the bigger challenge of developing drugs of high efficacy, including new antibiotics, that everyone believes are needed. Industry's empha-sis on *profitability*, not *need*, is reflected in the global distribution of the research effort dedicated to the development of drugs in general. Moreover, comparing high- and low-income countries, there is an almost fourfold greater number of drugs in the development pipeline for diseases prevalent in high-income countries (where high profits are to be made) than for diseases prevalent in low-income countries (Fisher et al., 2014). In that regard, industry priorities contribute directly to health inequalities globally.

It has been argued that concerted effort by governments and regulatory agencies is required to reorder incentives in favor of genuine innovation without jeopardizing industry's capacity to make profit (Light and Lexchin, 2012). Although such arguments are generally predicated on criticism of past industry priorities and commitments, it may be confidently predicted that industry will support any reordering of incentives designed to safeguard continued profit-making. Intervention of the kind proposed is tantamount to government being urged to use public funds to intervene in the free market for the purpose of underwriting commercial ventures. It is well to remember, however, that profit-based innovation has thus far led to decidedly perverse outcomes in the form of progressive erosion of antibiotic potency and the threat of global pandemic infection from resistant pathogens. Against that background, public guarantee for continued private-sector profit seems an unlikely solution for dealing with the long-term threat from antibiotic resistance.

### 6.5.2  Antibiotic Stewardship

Given the dim prospect that new antimicrobial drugs will achieve any kind of wholesale thwarting of resistant pathogens, a strong movement has emerged over the past decade in support of strengthening policies to slow the pace with which antimicrobial resistance develops (Lim, 2012). The main strategy for achieving that outcome is **antibiotic stewardship**, which refers to coordinated efforts to improve policies, guidelines, education, regulations, and surveillance for the appropriate use of antibiotics, including optimizing doses and duration of treatment and minimizing inappropriate use (e.g., Bartlett, 2011; Charani et al., 2010; Fishman, 2006). Such proposals, however, bear a striking resemblance to proposals intended to address the problem of healthcare hygiene. Both are examples of the evident inability of biomedicine to respond decisively and effectively to longstanding and incontrovertible evidence of the harm it causes.

The need for antibiotic stewardship in policy governing the responsible use of antibiotics has been known for as long as antibiotics have been in use. As such, the sudden strong advocacy of stewardship at this point in history cannot but invite clichéd thoughts along the lines of "too little too late" and "closing the barn door after the horse has bolted." So far, evidence is lacking of any appreciable strengthening of stewardship. Consequently, little can be said about the likely effectiveness of stronger stewardship in averting future pandemics of resistant pathogens which decades of inadequate stewardship have helped to create.

### 6.5.3  Isolation Practice

Considering the general lack of success in curtailing harm caused by antimicrobial resistance, those responsible for caring for infected patients search for practical measures in clinical practice to manage the consequences of dwindling antibiotic efficacy. One such measure is the use of contact isolation, which includes physical separation of patients to prevent person-to-person spread of infection; use of masks, gloves, gowns, and other protection; and special procedures for the handling and disposal of contaminated items, including body fluids and materials used in clinical care. As with instrument counting, discussed earlier in this chapter, isolation has considerable intuitive appeal as a strategy for limiting harm. However, as with instrument counting, studies have found that the efficacy of contact isolation can be disappointing and that it can cause unexpected harm for the patients who are isolated.

Reminiscent of problems discussed above in relation to healthcare hygiene in general, and especially handwashing, a continuing major challenge is to achieve the necessary levels of healthcare personnel compliance with required protocols for contact isolation to be effective as a means for preventing the spread of MDR organisms (Huskins et al., 2011). Additionally, compared to patients not isolated, patients who are isolated have been found to be at increased risk of harm, including delays in treatment and increased incidence of adverse events (Zahar et al., 2013). The fact that harm may be more frequent for isolated patients is salient because of the high levels of surveillance and staffing that accompany isolation, with the staffing ratio in the Zahar et al. (2013) study being one nurse for every two patients. The findings provide further confirmation that irrespective of evident great "care," implied here by the provision of high levels of materials, personnel, and expertise characteristic of intensive isolation care, little happens in clinical medicine without patients being harmed.

As well as increased risk of physical harm, extensive literature shows that contact isolation contributes to psychological harm. A review of evidence found that isolated patients reported less satisfaction and increased levels of anxiety and depression (Morgan et al., 2009), and it appears that hospitalized older patients may be particularly vulnerable to such harm (Tarzi et al., 2001). Fätkenheuer et al. (2014) have proffered an interesting argument based on the notion that isolation, such as imprisonment, is the prototypical punishment that society imposes on wrongdoers. As such, it is unsurprising that patients subjected to isolation may experience increased anxiety, depression, and resentment. Indeed, apart from absence of uniformity concerning suitable isolation *methods*, such as use of gowns, gloves, masks, and isolation rooms, considerable uncertainty exists in hospitals regarding the *circumstances* that warrant the use of contact isolation. Considering the legal and other safeguards that exist in society to prevent unwarranted isolation, such as wrongful imprisonment, Fätkenheuer et al. have questioned the absence of safeguards against "wrongful isolation" in hospitals.

### 6.5.4    Moral Hazard and "Unintended" Consequences

*Moral hazard* refers to the tendency for decision makers to take actions involving higher risks when a greater proportion of benefit and a lesser proportion of cost arising from the action accrue to the decision maker. Such situations are characterized by what economists call *information asymmetry* in which one party in a transaction has more relevant knowledge about what is being transacted than the other party. Medical practice is fraught with moral hazard, wherein greater knowledge about clinical "transactions" resides with practitioners who benefit financially, socially, and professionally, while risks—including negative side effects of treatment—disproportionately accrue to patients. The overuse of antibiotics is illustrative. Antibiotics have provided doctors with a convenient and professionally sanctioned means of responding to patients' requests for care, even when treating self-limiting conditions such as common cold for which antibiotics are ineffective. In that context, it is striking that much of the discourse about harm from antibiotic resistance alludes to "unintended consequences" from well-intentioned actions.

As usually understood, the phrase *unintended consequences* refers to outcomes that were not merely unintended but also were unanticipated (Merton, 1936). In that regard, the emergence of antibiotic resistance might be said to have been *unintended* in that antibiotic therapy is not typically used with the intention of producing MDR organisms. However, the emergence of resistance was always *anticipated*. From the earliest clinical use of antibiotics more than 70 years ago, antibiotic resistance was not merely observed but was well-understood and predicted on the basis of principles of Darwinian selection (Davies and Davies, 2010). Therefore, to regard antibiotic resistance as an unintended consequence is to engage in euphemism and prevarication. The likely reoccurrence of antibiotic resistance following the introduction of each new antibiotic has certainly been anticipated. Global population endangerment from the misuse of antibiotics in the absence of strenuous action to at least impede if not stop the evolution of antibiotic resistance, when its occurrence was anticipated all along, is illustrative of the moral hazard intrinsic to biomedical healthcare and commercial husbandry.

### 6.5.5    Given Past Neglect, What Does the Future Hold?

Besides MDR pathogens, including essentially untreatable strains, truly unintended (i.e., unanticipated) consequences of the dissemination of antibiotics might still occur. Although antibiotic resistance in the natural environment is known to predate the human discovery of antibiotics, most bacterial species are not intrinsically resistant to manufactured antibiotics. Resistance is acquired due to exposure to antibiotics, a process that has been accelerated beyond measure due to the immense quantities of antibiotics manufactured and distributed annually, a large proportion of which is released unchanged into the environment as waste. This creates vast microbial ecosystems containing strong selection pressures that favor antibiotic-resistant strains.

Gillings and Stokes (2012) have argued that the rapid and widespread evolution of antibiotic resistance in bacterial pathogens is evidence of the dramatic influence humans can have on evolutionary processes (much as was discussed in Chapter 2 in the context of niche construction theory), which in this instance has led to change in the *rate* of bacterial evolution. Many antibiotics are designed to have a broad spectrum of activity, which imposes selection pressures on a correspondingly broad range of bacterial species. Gillings and Stokes speculate that an environment saturated with antimicrobial agents has the potential to influence genetic mutation, recombination, and lateral gene transfer in ways capable of dramatically accelerating the natural rate of microbial evolution. In consequence, existing microbial pathogens may become progressively more dangerous, and novel species of virulent and resistant pathogens could emerge. These possibilities underlie much of the concern about the emergence of antibiotic resistance.

### 6.5.5.1 Return of the Pre-antibiotic Era

The oft-repeated claim in the scientific literature as well as in public media that widespread antibiotic resistance signals a possible return to the pre-antibiotic era is partly true while also being untrue in important respects. The part that is true is the possibility of a return to population levels of infection similar to those that existed before the epidemiologic transition. However, implied attribution that the epidemiologic transition was brought about by the advent of antibiotics is mistaken. It is not merely mistaken, it is impossible because the epidemiologic transition preceded the advent of antibiotics.

As described in Part 1, the epidemiologic transition was not due to improvements in clinical medicine, with or without antibiotics. It was the result of far-reaching changes in human habits and habitats precipitated by the Industrial Revolution. Moreover, to the extent that death and morbidity from largely untreatable infections are common in healthcare settings, it could be said that within the confines of the modern hospital the world has already substantially returned to the pre-antibiotic era. The spread of antibiotic-resistant infections from hospitals to communities could well mark wholesale return to epidemic levels of population infection equivalent to, or even greater than, those that existed before the epidemiologic transition. If so, the epidemics will be due not so much to *loss* of antibiotic potency but to *misuse* of antibiotics. In short, the belief that loss of antibiotic efficacy will *by itself* cause a return to high levels of population infection is false. Rather, the real threat lies in population dissemination of new, untreatable, and largely "unnatural" strains of infectious pathogens. Pathogens *caused* by the use of antibiotics and against which humans have little or no natural resistance. Unlike the past, confronted with new pathogens possessing high mortality and efficient transmission, strong immunity buttressed by highly efficient physical barriers such as are provided by modern sanitation and hygiene may no longer be sufficient.

The use of antibiotics in some developing countries as a substitute for the lack of clean water and safe disposal of waste (Editorial, 2014) is reminiscent of industrial-scale animal husbandry wherein animals living in cramped and insanitary conditions are indiscriminately fed antibiotics. The time may come when public infrastructure for the delivery of clean water and disposal of sewage for human populations no longer has the transformative effects on health that it had in the countries that were fortunate enough to adopt policies of public sanitation and hygiene before the advent of antibiotics. New resistant microbes may emerge for which natural defenses are few and technological defenses ineffective, with potentially catastrophic effects for human populations. The advent of antibiotics, sometimes trumpeted as the greatest medical achievement of the twentieth century, has the potential to end in global health disaster from epidemics of novel infectious diseases selectively engineered by the misuse of antibiotics.

## 6.5.6    An Outbreak Anywhere Is a Threat Everywhere

The gravity of the threat to personal and population health from antibiotic resistance may be gauged from the frequency and solemnity of public statements from public health authorities and governments. Reports of new and threatening pathogens such as severe acute respiratory syndrome (*SARS*), Middle East respiratory syndrome coronavirus, and avian influenza A (*H7N9*) appear at regular intervals (Frieden et al., 2014). Additionally, there is the threat of *bioterrorism* involving the intentional release and dissemination of biological agents, including antibiotic-resistant organisms. The World Health Assembly recently passed a resolution requiring the Secretariat of the World Health Organization to urgently draft an action plan on global antimicrobial resistance (WHO, 2014). The European Commission has confirmed the high priority it attaches to measures intended to combat antimicrobial resistance with the awarding of nearly €800 million for transnational collaborative research into antimicrobial resistance in the context of human health, animal health, food supply, and the environment (Geoghegan-Quinn, 2014; WHO, 2014). United States President Barack Obama (2014) recently issued an executive order titled, *Combating Antibiotic-Resistant Bacteria*. The President's announcement declared the threat of antibiotic-resistant bacteria to be a national security priority, and a task force to implement countermeasures has been established under joint oversight of departments of defense, agriculture, and health.

Stories of the threat of new infectious organisms emerging to devastate humanity are ever present in the public media, and they tend to sabotage discussion about more effective ways for promoting personal and population health than can be achieved through provision of evermore biomedical healthcare. In that sense, biomedicine derives succor from stories of disease epidemics, because people are inclined to believe that biomedicine possesses the most (possibly, the only) effective strategies for defeating new diseases. The pharmaceutical industry, too, is much assisted, because such threats can be used to justify evermore taxpayer support for industry "innovation" in drug development.

Discussion about the outbreak of Ebola virus in West Africa has included speculation about the disease having spread from fruit bats to other animals and ultimately to humans. However, perhaps a story having greater prescience than ones often told about novel virulent diseases migrating from remote wilderness areas will have biomedicine itself caste in the role of villain. That story might foretell a future of devastating global disease that has migrated from antibiotic-saturated waste in a densely-populated part of the developing world where antibiotics are used as substitute for sanitation infrastructure. The central plot of that story would be the contribution of profligate biomedical misuse of antibiotics to the emergence of epidemics of multi- and totally-drug-resistant pathogens against which humans may have fewer natural defenses than against any of the past plagues of history. Whereas many look to biomedical healthcare to save us from impending plague, biomedicine may yet prove to be the cause of future plagues of heretofore unknown devastation.

## REFERENCES

Allegranzi, B., Nejad, S.B., Combescure, C., et al., 2011. Burden of endemic health-care-associated infection in developing countries: systematic review and meta-analysis. Lancet 377, 228–241.

Aspden, P., Corrigan, J.M., Wolcott, J., Erickson, S.M. (Eds.), 2007. Patient Safety: Achieving a New Standard for Care. National Academic Press, Washington, DC.

Bartlett, J.G., 2011. A call to arms: the imperative for antimicrobial stewardship. Clin. Infect. Dis. 53 (Suppl. 1), S4–S7.

Best, M., Neuhauser, D., 2004. Ignaz Semmelweis and the birth of infection control. Qual. Saf. Health Care 13, 233–234.

Bohnert, A.S., Valenstein, M., Bair, M.J., et al., 2011. Association between opioid prescribing patterns and opioid overdose-related deaths. JAMA 305, 1315–1321.

Boyce, J.M., Pittet, D., 2002. Guideline for hand hygiene in health-care settings. Am. J. Infect. Control 30, S1–S46.

Calfree, D.P., 2012. Crisis in hospital-acquired healthcare-associated infections. Annu. Rev. Med. 63, 359–371.

CDC, 2015. National and State Healthcare-Associated Infections Progress Report. Centers for Disease Control, Atlanta, GA. Retrieved 24 January 2015 from: http://www.cdc.gov/HAI/pdfs/progress-report/hai-progress-report.pdf.

Charani, E., Cooke, J., Holmes, A., 2010. Antibiotic stewardship programmes: what's missing? J. Antimicrob. Chemother. 65, 2275–2277.

Cima, R.R., Kollengode, A., Garnatz, J., et al., 2008. Incidence and characteristics of potential and actual retained foreign object events in surgical patients. J. Am. Coll. Surg. 207, 80–87.

Clarke, J.R., Johnston, J., Finley, E.D., 2007. Getting surgery right. Ann. Surg. 246, 395–405.

Classen, D.C., Resar, R., Griffin, F., et al., 2011. 'Global Trigger Tool' shows that adverse events in hospitals may be ten times greater than previously measured. Health Aff. 30, 581–589.

Cobb, T.K., 2012. Wrong site surgery—where are we and what is the next step? Hand 7, 229–232.

Collignon, P., Powers, J.H., Chiller, T.M., et al., 2009. World Health Organization ranking of antimicrobials according to their importance in human medicine: a critical step for developing risk management strategies for the use of antimicrobials in food production animals. Clin. Infect. Dis. 49, 132–141.

Daszak, P., Cunningham, A.A., Hyatt, A.D., 2000. Emerging infectious diseases of wildlife: threats to biodiversity and human health. Science 287, 443–449.

Davies, J., Davies, D., 2010. Origins and evolution of antibiotic resistance. Microbiol. Mol. Biol. Rev. 74, 417–433.

de Vries, E.N., Ramrattan, M.A., Smorenburg, S.M., et al., 2008. The incidence and nature of in-hospital adverse events: a systematic review. Qual. Saf. Health Care 17, 216–223.

Dimick, J.B., Welch, H.G., 2008. The zero mortality paradox in surgery. J. Am. Coll. Surg. 206, 13–16.

Eccles, R., 2005. Understanding the symptoms of the common cold and influenza. Lancet Infect. Dis. 5, 718–725.

Editorial, 2014. Prescribing antibiotics: a battle of resistance. Lancet 384, 558.

Egorova, N.N., Moskowitz, A., Gelijns, A., et al., 2008. Managing the prevention of retained surgical instruments: what is the value of counting? Ann. Surg. 247, 13–18.

Einarson, T.R., 1993. Drug-related hospital admissions. Ann. Pharmacother. 27, 832–840.

Fabian, T., 2013. Aging changes and pharmacotherapy. In: Miller, M.D., Solai, L.K. (Eds.), Geriatric Psychiatry. Oxford University Press, Oxford, UK.

Fätkenheuer, G., Hirschel, B., Harbarth, S., 2014. Screening and isolation to control meticillin-resistant Staphylococcus aureus: sense, nonsense, and evidence. Lancet 385, 1146–1149. http://dx.doi.org/10.1016/S0140-6736(14)60660-7.

Finks, J.F., Osborne, N.H., Birkmeyer, J.D., 2011. Trends in hospital volume and operative mortality for high-risk surgery. N. Engl. J. Med. 364, 2128–2137.

Fisher, J.A., Cottingham, M.D., Kalbaugh, C.A., 2014. Peering into the pharmaceutical "pipeline": investigational drugs, clinical trials, and industry priorities. Soc. Sci. Med. 131, 322–330. http://dx.doi.org/10.1016/j.socscimed.2014.08.023.

Fishman, N., 2006. Antimicrobial stewardship. Am. J. Infect. Control 34, S55–S63.

Fortnum, H., Lee, A.J., Rupnik, B., Avery, A., 2012. Survey to assess public awareness of patient reporting of adverse drug reactions in Great Britain. J. Clin. Pharm. Ther. 37, 161–165.

Frieden, T.R., Tappero, J.W., Dowell, S.F., et al., 2014. Safer countries through global health security. Lancet 383, 764–766.

Funkhouser, W., 2012. Semmelweis and puerperal fever. ASIP Pathways 7, 4–5.

Gawande, A.A., Studdert, D.M., Orav, E.J., et al., 2003. Risk factors for retained instruments and sponges after surgery. N. Engl. J. Med. 348, 229–235.

Geoghegan-Quinn, M., 2014. Funding for antimicrobial resistance research in Europe. Lancet 384, 1186.

Ghafur, A.K., 2010. An obituary: on the death of antibiotics!. J. Assoc. Physicians India 58, 143–144.

Gibbs, V.C., 2011. Retained surgical items and minimally invasive surgery. World J. Surg. 35, 1532–1539.

Gillings, M.R., Stokes, H.W., 2012. Are humans increasing bacterial evolvability? Trends Ecol. Evol. 27, 346–352.

Gordon, A., 1795. A Treatise on the Epidemic Puerperal Fever of Aberdeen. GG and J Robinson, London.

Green, J.L., Hawley, J.N., Rask, K.J., 2007. Is the number of prescribing physicians an independent risk factor for adverse drug events in an elderly outpatient population? Am. J. Geriatr. Pharmacother. 5, 31–39.

Greenberg, C.C., Regenbogen, S.E., Lipsitz, S.R., et al., 2008. The frequency and significance of discrepancies in the surgical count. Ann. Surg. 248, 337–341.

Haidee, T., Custodio, M.D., 2012. Hospital-Acquired Infections. Medscape, New York. Retrieved 18 December 2012 from: http://emedicine.medscape.com/article/967022-overview.

Hariharan, D., Lobo, D.N., 2013. Retained surgical sponges, needles and instruments. Ann. R. Coll. Surg. Engl. 95, 87–92.

Hazell, L., Shakir, S.A., 2006. Under-reporting of adverse drug reactions: a systematic review. Drug Saf. 29, 385–396.

Heikkinen, T., Järvinen, A., 2003. The common cold. Lancet 361, 51–59.

Holmes, O.W., 1843. Medical Essays, 1842–1882. Houghton, Mifflin and Co., Boston.

Huskins, W.C., Huckabee, C.M., O'Grady, N.P., et al., 2011. Intervention to reduce transmission of resistant bacteria in intensive care. N. Engl. J. Med. 364, 1407–1418.

Jones, C.M., Mack, K.A., Paulozzi, L.J., 2013. Pharmaceutical overdose deaths, United States, 2010. JAMA 309, 657–659.

Juntti-Patinen, L., Neuvonen, P.J., 2002. Drug-related deaths in a University central hospital. Eur. J. Clin. Pharmacol. 58, 479–482.

Kalisch, L.M., Caughey, G.E., Barratt, J.D., et al., 2012. Int. J. Qual. Health Care 24, 239–249.

Kenealy, T., Arroll, B., 2013. Antibiotics for the common cold and acute purulent rhinitis. Cochrane Database Syst. Rev. 6, 1–58. http://dx.doi.org/10.1002/14651858.CD000247.

Khachatourians, G.G., 1998. Agricultural use of antibiotics and the evolution and transfer of antibiotic-resistant bacteria. Can. Med. Assoc. J. 159, 1129–1136.

Klevens, R.M., Edwards, J.R., Richards, C.L., et al., 2007. Estimating health care-associated infections and deaths in U.S. hospitals. Public Health Rep. 122, 160–166.

Kohn, L.T., Corrigan, J.M., Donaldson, M.S. (Eds.), 2000. To Err Is Human: Building a Safer Health System. National Academies Press, Washington, DC.

Kumarasamy, K.K., Toleman, M.A., Walsh, T.R., et al., 2010. Emergence of a new antibiotic resistance mechanism in India, Pakistan, and the UK: a molecular, biological, and epidemiological study. Lancet Infect. Dis. 10, 597–602.

Lai, H.-J., Hwang, S.-J., Chen, Y.-C., et al., 2009. Prevalence of the prescribing of potentially inappropriate medications at ambulatory care visits by elderly patients covered by the Taiwanese National Health Insurance Program. Clin. Ther. 31, 1859–1870.

Landrigan, C.P., Parry, G.J., Bones, C.B., et al., 2010. Temporal trends in rates of patient harm resulting from medical care. N. Engl. J. Med. 363, 2124–2134.

Lautenbach, E., Larosa, L.A., Kasbekar, N., et al., 2003. Fluoroquinolone utilization in the emergency departments of academic medical centers: prevalence of, and risk factors for, inappropriate use. Arch. Intern. Med. 163, 601–605.

Lazarou, J., Pomeranz, B.H., Corey, P.N., 1998. Incidence of adverse drug reactions in hospitalized patients: a meta-analysis of prospective studies. JAMA 279, 1200–1205.

Lembke, A., 2012. Why doctors prescribe opioids to known opioid abusers. N. Engl. J. Med. 367, 1580–1581.

Light, D., Lexchin, J.R., 2012. Pharmaceutical research and development: what do we get for all that money? BMJ 344, e4348.

Lim, V.K.E., 2012. Antibiotic stewardship. IeJSME 6 (Suppl 1), S75–S79.

Lincourt, A.E., Harrell, A., Cristiano, J., et al., 2007. Retained foreign bodies after surgery. J. Surg. Res. 138, 170–174.

Lowis, G.W., 1993. Epidemiology of puerperal fever: the contributions of Alexander Gordon. Med. Hist. 37, 399–410.

McGregor, A., Dovey, S., Tilyard, M., 1995. Antibiotic use in upper respiratory tract infections in New Zealand. Fam. Pract. 12, 166–170.

Meinberg, E.G., Stern, P.J., 2003. Incidents of wrong-site surgery amoung hand surgeons. J. Bone Joint Surg. 85, 193–197.

Merton, R.K., 1936. The unanticipated consequences of purposive social action. Am. Sociol. Rev. 1, 894–904.

Mody, M., Nourbakhsh, A., Stahl, D.L., et al., 2008. The prevalence of wrong level surgery among spine surgeons. Spine 33, 194–198.

Moellering Jr., R.C., 2010. NDM-1: acause for worldwide concern. N. Engl. J. Med. 363, 2377–2379.

Morgan, D.J., Diekema, D.J., Sepkowitz, K., Perencevich, E.N., 2009. Adverse outcomes associated with contact precautions: a review of the literature. Am. J. Infect. Control 37, 85–93.

Mowry, J.B., Spyker, D.A., Cantilena Jr., L.R., et al., 2013. 2012 Annual report of the American Association of Poison Control Centers' National Poison Data System (NPDS): 30th annual report. Clin. Toxicol. 51, 949–1229.

Obama, B., 2014. Combating Antibiotic-Resistant Bacteria. Executive Order. Office of the Press Secretary, Washington, DC. Retrieved on 26 September 2014 from: http://www.whitehouse.gov/the-press-office/2014/09/18/executive-order-combating-antibiotic-resistant-bacteria.

O'Connor, M.N., Gallagher, P., Byrne, S., O'Mahony, D., 2012. Adverse drug reactions in older patients during hospitalisation: are they predictable? Age Ageing 41, 771–776.

Page, R.L., Ruscin, J.M., 2006. The risk of adverse drug events and hospital-related morbidity and mortality among older adults with potentially inappropriate medication use. Am. J. Geriatr. Pharmacother. 4, 297–305.

Paulozzi, L.J., Kilbourne, E.M., Shah, N.G., et al., 2012a. A history of being prescribed controlled substances and risk of drug overdose death. Pain Med. 13, 87–95.

Paulozzi, L.J., Baldwin, G., Franklin, G., et al., 2012b. CDC grand rounds: prescription drug overdoses—a US epidemic. Morb. Mortal. Wkly. Rep. 61, 10–13.

Pirmohamed, M., James, S., Meakin, S., et al., 2004. Adverse drug reactions as cause of admission to hospital: prospective analysis of 18,820 patients. Br. Med. J. 329, 15–19.

Pittet, D., Allegranzi, B., Boyce, J., 2009. WHO Guideline: The World Health Organization guidelines on hand hygiene in health care and their consensus recommendations. Infect. Control Hosp. Epidemiol. 30, 611–622.

Pittet, D., Panesar, S.S., Wilson, K., et al., 2011. Involving the patient to ask about hospital hand hygiene: a National Patient Safety Agency feasibility study. J. Hosp. Infect. 77, 299–303.

Rolain, J.M., Parola, P., Cornaglia, G., 2010. New Delhi metallo-beta-lactamase (NDM-1): towards a new pandemia? Clin. Microbiol. Infect. 16, 1699–1701.

Sax, H., Allegranzi, B., Uckay, I., et al., 2007. 'My five moments for hand hygiene': a user-centred design approach to understand, train, monitor and report hand hygiene. J. Hosp. Infect. 67, 9–21.

Schofield, C.B., 2011. The anarchy of antibiotic resistance: mechanisms of bacterial resistance. Med. Lab. Obs. 43, 10–16.

Seya, M.J., Gelders, S.F., Achara, O.U., et al., 2011. A first comparison between the consumption of and the need for opioid analgesics at country, regional, and global levels. J. Pain Palliat. Care Pharmacother. 25, 6–18.

Simasek, M., Blandino, D.A., 2007. Treatment of the common cold. Am. Fam. Physician 5, 515–520.

Singh, A.R., 2011. Science, names giving and names calling: change NDM-1 to PCM. Mens Sana Monogr. 9, 294–319.

Singh, G.P., Kochanek, K.D., MacDorman, M.F., 1996. Advance report of final mortality statistics, 1994. Mon. Vital Stat. Rep 45 (Suppl.), 1–78. http://www.nber.org/mortality/1994/docs/mvs45_3s.pdf.

Smithells, R.W., Newman, C.G.H., 1992. Recognition of thalidomide defects. J. Med. Genet. 29, 716–723.

Tarzi, S., Kennedy, P., Stone, S., Evans, M., 2001. Methicillin-resistant *Staphylococcus aureus*: psychological impact of hospitalization and isolation in an older adult population. J. Hosp. Infect. 49, 250–254.

Thomas, E.J., Studdert, D.M., Burstin, H.R., et al., 2000. Incidence and types of adverse events and negligent care in Utah and Colorado. Med. Care 38, 261–271.

Tikhomirov, E., 1987. WHO programme for the control of hospital infections. Chemioterapia 6, 148–151.

Turner, B.L., Thompson, A.L., 2013. Beyond the Paleolithic prescription: incorporating diversity and flexibility in the study of human diet evolution. Nutr. Rev. 71, 501–510.

van den Bogaard, A.E., Stobberingh, E.E., 2000. Epidemiology of resistance to antibiotics: links between animals and humans. Int. J. Antimicrob. Agents 14, 327–335.

van der Hooft, C., Sturkenboom, M.C.J.M., van Grootheest, K., et al., 2006. Adverse drug-reaction related hospitalizations: a nationwide study in The Netherlands. Drug Saf. 29, 161–168.

Walker, K., Neuburger, J., Groene, O., et al., 2013. Public reporting of surgeon outcomes: low numbers of procedures lead to false complacency. Lancet 382, 1674–1677.

Walsh, T.R., Toleman, M.A., 2011. The new medical challenge: why NDM-1? Why Indian? Expert Rev. Anti Infect. Ther. 9, 137–141.

Wang, C.F., Cipolla, J., Seamon, M.J., et al., 2009. Gastro-intestinal complications related to retained surgical foreign bodies (RSFB): a concise review. OPUS 12, 11–18.

Wat, D., 2004. The common cold: a review of the literature. Eur. J. Intern. Med. 15, 79–88.

Wester, K., Jönsson, A.K., Spigset, O., et al., 2007. Incidence of fatal adverse drug reactions: a population based study. Br. J. Clin. Pharmacol. 65, 573–579.

WHO, 1969. International drug monitoring: the role of the hospital, World Health Organisation Technical Report Series No. 425. World Health Organisation, Geneva, Switzerland.

WHO, 2009. WHO Guidelines on Hand Hygiene in Health Care. World Health Organization, Geneva.

WHO, 2012. The Evolving Threat of Antimicrobial Resistance: Options for Action. World Health Organization, Geneva.

WHO, 2014. Draft Global Action Plan on Antimicrobial Resistance, 2014. World Health Organization, Geneva. Retrieved 2 October 2014 from: http://www.who.int/drugresistance/amr_global_action_plan/en/ (accessed 8 August 2014).

Wu, T.-Y., Jen, M.-H., Bottle, A., et al., 2010. Ten-year trends in hospital admissions for adverse drug reactions in England 1999-2009. J. R. Soc. Med. 2010 (103), 239–250.

Yong, D., Toleman, M.A., Giske, C.G., et al., 2009. Characterization of a new metallo-β-lactamase gene, blaNDM-1, and a novel erythromycin esterase gene carried on a unique genetic structure in Klebsiella pneumoniae sequence type 14 from India. Antimicrob. Agents Chemother. 53, 5046–5054.

Zahar, J.R., Garrouste-Orgeas, M., Vesin, A., et al., 2013. Impact of contact isolation for multidrug-resistant organisms on the occurrence of medical errors and adverse events. Intensive Care Med. 39, 2153–2160.

Zarb, P., Coignard, B., Griskeviciene, J., et al., 2012. The European Centre for Disease Prevention and Control (ECDC) pilot point prevalence survey of healthcare-associated infections and antimicrobial use. Euro Surveill. 17. pii: 20316, Retrieved 20 December 2012 from: http://www.eurosurveillance.org/ViewArticle.aspx?ArticleId=20316.

# Chapter 7

# The Commercial Culture of Medicine

*Of all the anti-social vested interests the worst is the vested interest in ill-health.*

(Shaw, 1909)

## Contents

The first 60 years of the twentieth century are sometimes referred to as the "golden age" of medicine (McKinlay and Marceau, 2002). Doctors were generally held in high esteem and they possessed a level of moral authority that few other groups in society enjoyed. Organizations representing doctors were politically influential and effective in protecting privilege for members. By the

The Health of Populations. http://dx.doi.org/10.1016/B978-0-12-802812-4.00007-2

1970s, however, public sentiment soured in response to growing perceptions that doctors had misused the trust bestowed on them for personal gain. While presenting themselves as altruistic healers, their behavior often suggested preoccupation with self-interest and a sense of self-entitlement. Action to restore esteem, trust, and authority have been much discussed within the profession (Stevens, 2002), but once eroded, trust is not easily restored.

Central to medicine's waning fortunes is the fact that clinical practice and research have become profoundly entangled with vested commercial interests. Commercialism in medicine takes many forms and occurs whenever healthcare practice and research are conducted for profit (Walsh, 2006). Profit in healthcare is not intrinsically wrong. However, since the primary concern of healthcare is patient welfare, profit from healthcare delivery can lead to **conflicts of interest** between patient welfare and physician self-interest. The risk of decisions and practices being biased in ways not solely in patients' interests is a key ethical objection to commercialism in biomedicine. Presumption of harm is justified whenever physicians recommend interventions that would not have been recommended if there had been no material or other personal incentive involved. On that basis, it is safe to conclude that harm, potential and real, from conflicts of interest is endemic in medical practice and research.

This and the next chapter, Chapter 8, address the dual concerns of commercialism in clinical medicine and biomedical science, respectively. While the two domains raise common concerns (e.g., *conflict of interest*), the particulars of each are different enough to warrant discussion in separate chapters. However, because it is not possible to discuss the one set of issues without occasional reference to the other, there is some overlap between this chapter and the next. Taken together, the two chapters attempt to provide an integrated discussion of interrelated concerns while keeping repetition to a minimum.

## 7.1 THE MYTH OF SAFE AND EFFECTIVE PRESCRIPTION DRUGS

*The marketing of drugs … is different from that of other types of goods and services, in that the target of … pharmaceuticals is not the final purchaser of the product [but] a select group—physicians—[who control] consumption for millions of others—their patients.*

(Shaughnessy et al., 1994, p. 563)

Entanglement between the pharmaceutical industry and medicine is so extensive that essentially no sphere of medicine remains free of **Big Pharma** influence. Global spending on prescription drugs was almost USD1 trillion in 2011, and is projected to grow by a further 20% by 2016 (IFPMA, 2012). Public perceptions of the pharmaceutical industry are ambivalent. Despite suspicions, a popular image is of a vast array of research and development activities that are the foundations of a dynamic and spectacularly successful industry delivering ever more refined and effective prescription drugs to a needy and grateful public. In that light, large profit for biomedical entrepreneurship can appear just reward for the massive investments required to sustain programs of innovation capable of producing a steady stream of new therapeutic drugs ("biologicals"). The facts, however, tell a different story. Biomedical research and development does indeed consume massive investment, but most of that is sourced from government. Moreover, the consistency with which new biologicals are found to be cost-*in*effective and harmful has led to a crescendo of considered expert opinion that speaks of a crisis in biomedical innovation.

As mentioned in Chapter 6, belief in a drug innovation crisis suits the pharmaceutical industry. Emphasizing the high cost of drug development and the high risk of failure, industry has succeeded in garnering media and public sympathy and using that to mine benefits from government (Light and Lexchin, 2012). Benefits include a range of government protections against free-market competition, including extended patents that restrict competition from **generic drugs**. Compared to branded drugs, *generics* typically contain similar or identical formulations, have the same principal therapeutic effects, are usually brought to the market with little advertising, and are typically lower, often much lower, in price. Competition from generics is something the manufacturers of branded pharmaceuticals do all in their power to limit. Grant-based taxpayer subsidies are an important source of government support for Big Pharma. An analysis of global investment in drug-related basic research found that industry contributes a minor share, only 12%, with 84% coming from the public and the remaining 4% from private nonprofit sources (Light, 2006). The perception that industry is strongly committed to discovering new drug products has helped to bolster public esteem in the pharmaceutical industry, and if not esteem, then at least grudging respect.

If size of investment is an indicator, the pharmaceutical industry is evidently more interested in selling drugs, whether old or new, than in developing better ones. Industry spending on discovery of new biologicals has been estimated to be 1.3% of revenues compared to 25% of revenues spent on promotion, an investment ratio in basic research versus marketing of nearly 1-20 (Light and Lexchin, 2012). In general, investment trends show that as public investment in innovation has grown, industry commitment has retreated. As such, investment risk has transferred to the public sector, allowing the private sector to amass ever larger profits. Expansion of public underwriting of private venture in pharmaceuticals has been argued on the ground that new drugs serve the public good, but that argument amounts to little more than corporate public relations. In reality, evidence is lacking of actual public health benefit from the steady stream of new drugs that come to market.

For reasons of public safety, new therapeutic drugs must comply with regulatory processes governing initial development, subsequent human trials, and final approval before being made available to the public under prescription from certified physicians. A prevalent perception, encouraged by industry, is that a crisis of innovation in pharmaceuticals has been created by overly conservative regulatory processes that obstruct new drug approvals. However, extensive analysis has shown that the annual rate of approval of new drugs has remained essentially unchanged for 60 years (Munos, 2009), indicating that any current crisis of innovation in pharmaceuticals is not due to depressed output. Nor is any loss of innovation attributable to reduced funding for pharmaceutical research and development, since that has grown steadily due largely to investment from the public sector. Rather, absence of innovation is evidenced by poor performance of the majority of new drugs that come onto the market.

The pharmaceutical industry measures innovation on the basis of new molecular entities (*NMEs*), which are active ingredients that have not been previously approved for marketing. However, the absolute *number* of NMEs capable of being brought to market is of secondary importance. What matters is the therapeutic *value* of those new molecules (Light et al., 2013). When measured in terms of therapeutic value, there does indeed appear to be a crisis in innovation, because independent evaluations have consistently shown that the majority of new drugs, despite approval for clinical use, perform poorly. An analysis of Canadian pharmaceutical expenditures, which doubled in the 7 years from 1996 to 2003, found that 80% of the increase in expenditure was due to new higher priced drugs offering no substantial improvements over less expensive alternatives available since before 1990 (Morgan et al., 2005). A review of all new drugs approved in France for the decade from 2002 to 2011 found that more than three-quarters were of "no" or "minimal" added value over existing drugs (Light et al., 2013). Only 8% provided added therapeutic benefit, and twice that number (16%) were more harmful than beneficial.

The image of the pharmaceutical industry as a dynamic global enterprise burgeoning with breakthrough innovations urgently being brought into service for the common good is fantasy (Angell, 2004b). In reality, the gargantuan profits of the pharmaceutical industry derive from a far less alluring business model wherein most new drugs consist of minor variations on existing products (Angell, 2004a,b). In the words of one senior Roche employee, this is the "me-slightly-different-marketed-like-hell" business model (Gagnon, 2013, p. 571). A few new drugs *may* produce genuine clinical advances and become market blockbusters, while heavy promotion and high prices ensure that the remainder also prove lucrative despite being of little clinical value. Even among "blockbusters," many do not actually provide much by way of genuine therapeutic advance when the eventual tally of benefits and harms is known (Gagnon, 2013). For a substantial proportion of new drugs, initial apparent benefits are eclipsed by subsequent realization of harmful side effects.

As discussed in Chapter 6, the supply of new antibiotics has dwindled because the pharmaceutical industry has found it more lucrative to settle for trivial modifications of existing drugs than to pursue genuine innovation. The same free-market forces characterize the current lack of innovation in new therapeutic drugs in general. Contemporary healthcare is shaped not by *efficacy* but by *profit*. The dearth of innovation in drug development is because the pursuit of genuine therapeutic breakthroughs is difficult and returns are uncertain. It is more profitable to produce new drugs of little incremental value, because irrespective of efficacy, it is *marketing* not product that creates most profit and "the newer the drug, the better it sells" (Angell, 2004b, p. 97). Relaxation of regulatory processes, a constant pursuit of industry, further boosts profits by allowing more low-value high-priced drugs to come onto the market.

## 7.1.1    Disease Mongering: There Is Profit in Getting People to Believe They Are Sick

*One of the most widespread diseases is diagnosis.*

(Karl Kraus, 1874-1936, Austrian Writer)

Industry entanglement with medicine is strongly associated with the phenomenon of **disease mongering**; the practice of relaxing the diagnostic boundaries of disease to expand markets for medical interventions (Moynihan et al., 2014). Variously described as *diagnostic creep* and the *selling of sickness*, disease mongering is an aspect of the more general processes of *medicalization* (*biomedicalization*, and *pharmaceuticalization*) of health (Bell and Figert, 2012). Disease mongering is a type of disease awareness-raising process (Moynihan et al., 2002). However, whereas awareness-raising for the common good involves education to inform the public about an underdiagnosed disease, disease mongering is industry-sponsored public "education" to boost sales and generate profit with little or no associated public health benefit and actual likelihood of harm. Disease mongering is possible, in part, because much like *health* (described in Chapter 1), concepts of *disease* are "slippery" (Smith, 2002). Life processes such as birth, aging, sexuality, menopause, unhappiness, and death are examples of non-diseases which, due to the medicalization of health in general, and disease mongering in particular, are increasingly treated as if they were diseases (Moynihan and Smith, 2002).

In principle, disease mongering can occur in association with promotion of any healthcare intervention, but in practice most is attributable to industry marketing of prescription drugs. There are numerous states of health and wellbeing that do not warrant being considered diseases, but are of such imprecise and subjective definition as to be vulnerable to diagnostic creep. In particular, there is a growing tendency for risk indicators of disease, such as elevated cholesterol or low bone mineral density, to be redefined as diseases, thereby legitimizing medical intervention for ever-larger sections of the population. Additionally, while not denying the reality of human distress in all its forms, other diagnoses that are particularly susceptible to medicalization and pharmaceuticalization include mild anxiety and depression, social anxiety and shyness, attention deficit disorder, childhood conduct disorder, restless legs syndrome, erectile dysfunction, female sexual dysfunction, and irritable bowel syndrome.

Shaw (1909) wrote of the public's desire for "a cheap magic charm to prevent, and a cheap pill or potion to cure, all disease." Disease mongering is possible and profitable because it exploits human desire for health and fear of ill-health. Additionally, it exploits human vulnerability to promises of avoidance or postponement of sickness, pain, and death, especially when promises involve interventions that require little individual effort, of which drug-taking is without parallel. At the very least, disease mongering is wasteful of resources. Ever-wider use of prescription drugs for largely inappropriate purposes necessarily depletes resources available for innovation in support of the development of new healthcare interventions genuinely capable of improving health. Moreover, incremental expansion of drug usage to ever-larger populations generally leads to a progressive decline in benefit and a commensurate increase in harm. Details of this relationship, known as the *benefit-to-harm ratio*, are discussed in Chapter 9.

Of the many commercially exploitable human concerns, possibly none is greater than that of personal **genomics**, which offers almost unlimited opportunity for disease mongering. Given that death claims everyone, we all carry within us a complement of excesses and deficits of genetic material that will (purportedly) determine the timing and nature of our ultimate demise[1]. Consequently, irrespective of the soundness or otherwise of our past or present state of health, we are all "diagnosable" as possessing some manner of genetically "programed" disease susceptibility, which is acquired at the moment of conception. Therein may be found an inexhaustible market for pharmaceuticals that is only now beginning to be exploited in the form of the much-heralded innovations of "personalized medicine." Taken to its logical conclusion, genome-based personalized medicine, a topic to which we will return in Chapter 9, can mean only one thing: lifelong medication for all.

## 7.1.2 Pharmaceuticalization of Daily Life

A major growth area for pharmaceuticals is the use of prescription drugs not as cures for underlying conditions but as "lifestyle enhancements" of normal function and of daily living. Increasingly, sales of prescription drugs exploit personal insecurities in the face of life's challenges, aspirations for increased

---

1. In fact, it should be obvious that genetic endowment does not alone determine the timing and nature of our ultimate demise. Rather, death is the result of complex interactions between genes, behavior, and the environment, with chance playing a major role at every level.

physical and social attractiveness, and desire for greater sexual fulfillment. The promise of chemically-assisted enhancement is underpinned by the blurring of distinctions between health and disease. The pharmaceuticalization of daily life is not merely *disease* mongering, but mongering of *non-disease* for the sale of useless and harmful clinical interventions. A notable example is sildenafil citrate (*Viagra*); initially used as an intervention for erectile dysfunction due to diabetes, spinal cord damage, or other medical condition, it has come to be used recreationally to enhance sexual performance in the absence of physical impairment. The immense success of Viagra for non-disease in men encouraged drug companies to search for a "pink Viagra" for women (Hartley, 2006).

The clinical diagnosis of erectile dysfunction long preceded the advent of Viagra, but to guarantee a market for a female equivalent it was necessary to create a relevant medical diagnosis for women where none existed previously. Arising from the work of international groups of researchers extensively entangled with industry, aspects of female sexual experience have come to be medicalized as "female sexual dysfunction" (Moynihan, 2003a). Allegedly afflicting almost half of all women aged under 60 years, this "disease" is diagnosed on the basis of ill-defined indications of unsatisfied sexual desire and arousal. Thus far, industry has failed to discover an effective pharmaceutical for the alleged "symptoms," and such seems unlikely given that the symptoms in question are more plausibly explained on the basis of the emotional life, interpersonal relationships, and work and family stresses experienced by women than any underlying physical pathology.

Nevertheless, in August 2015 in the United States, the Food and Drug Administration (*FDA*) approved the use of flibanserin (Addyi) for *hypoactive sexual desire disorder* in premenopausal women. Flibanserin was originally developed for the treatment of depression, but failed due to absence of benefit. It was then "rescued" (see Chapter 8) for use with female sexual dysfunction. The FDA had twice refused approval for use of the drug as therapy for low sexual desire, finding that it lacked efficacy and posed serious safety concerns (Gellad et al., 2015). However, despite several indicators of sexual performance showing no benefit compared to placebo, advocates claimed efficacy on the basis of women reporting average improvement of one-half of a "satisfying sexual event" per month. Negative side-effects, on the other hand, include adverse cardiovascular effects, sleepiness, fainting, and dizziness. The FDA's eventual decision to approve the drug appears to have been influenced by perceived consumer demand. In particular, an industry-initiated advocacy group called *Even the Score* mounted a campaign demanding "gender equality" in access to drugs for sexual dysfunction.

If the pharmaceutical industry is to be believed, it seems that "female Viagra" may finally have been discovered. However, in reality, there are no approved products specifically intended to address problems of low sexual desire in men, and therefore the equity argument and claims of success for Addyi are largely spurious. Moreover, the main mechanism of action for Viagra involves increased blood flow to the penis, and it is recommended that the drug be taken only prior to sexual activity. Addyi, on the other hand, alters brain chemistry in ways that are poorly understood, and despite being found to interact adversely with alcohol, oral contraceptives, and some prescribed medications, its use requires that it be taken daily. It is a cruel irony that success in obtaining approval for Addyi was obtained on a platform of gender equality for women. Current evidence indicates that the benefit to women is none to minimal on average, whereas the probability of harm from long-term use is not merely high but certain.

The home computer and Internet have been a boon to the pharmaceutical industry, offering unprecedented opportunity for direct marketing to prospective consumers and new distribution channels in the form of online pharmacies (Fox and Ward, 2008). Apart from the as-yet unfulfilled search for a female blockbuster equivalent of Viagra for men, the immense profits generated by Viagra, and subsequent imitators (tadalafil, *Cialis*, and vardenafil, *Levitra*), has provided impetus for "lifestyle marketing" of drugs for a wide range of common and often vaguely-defined conditions, including allergies, acid reflux, and insomnia. Prescription drugs were once limited to patients, a subset of the population under the care of a physician. Today, however, marketeering is a vital aspect of the drug distribution chain (Applbaum, 2006). The new industry objective is to transform entire populations into patients functioning as healthcare consumers exercising "independent" choice as they shop for pharmaceuticals within an unfettered marketplace.

One rapidly emerging source of harm associated with pharmaceutical consumerism comes from the black market rather than the conventional drug industry. A substantial trade has emerged in **designer drugs**, which are synthetic analogs of existing licensed drugs. By operating outside the usual regulatory and commercial framework, manufacturers of analogs avoid payments arising from intellectual property rights. The trade is particularly extensive in substances possessing psychoactive properties, doping agents to enhance sporting performance, and for controlling weight. In specific growth areas such as formulations for erectile dysfunction, illicit sales rival legitimate trade (Venhuis and de Kaste, 2012). Profitability is enhanced because of low production and operating costs, and commensurately poor quality control. Structurally or functionally similar to the parent drug, designer drugs are developed to mimic the effects of the original. However, the precise pharmacology of analogs is typically largely unknown. Consequently, the associated risk of toxic side effects, both short- and long-term, is high.

### 7.1.3   Industry Entanglement with Clinical Practice Guidelines

Risk of harm to consumers from illicit trade in biomedical products underscores the importance of the regulatory frameworks intended to govern legitimate healthcare practices. In reality, however, the frameworks that exist are disappointingly ineffectual. To be maximally effective, regulation must be free of vested interests other than those concerning patient health and safety. Within the evidence-based paradigm of contemporary healthcare, **clinical practice guidelines** are used to promote optimal patient care. Clinical guidelines derive their authority from the quality of the evidence on which the recommendations they contain are based. For that reason, guidelines need to be, and must be seen to be, thoroughly rigorous and dispassionate. Typically, summary consensus statements are prepared that draw upon the best available scientific evidence to determine best practice. Guidelines are usually issued by medical societies or governments, with the aim of standardizing medical care, improving health outcomes, reducing risks, and optimizing cost-effectiveness across all fields of medicine. At its clinical practice guidelines portal (http://www.clinicalguidelines.gov.au/), the Australian National Health and Medical Council lists over 1600 sets of guidelines for physicians to use. The National Guideline Clearinghouse of the United States Agency for Healthcare Research and Quality contains about 2700 guidelines, and the database of the Guidelines International Network representing about 50 countries is said to contain at least 6800 guidelines (Kuehn, 2011). Physicians are obliged to be familiar with guidelines relevant to their practice, using them as decision aids in combination with experience and professional judgment when dispensing medical care to individual patients.

Typically, clinical practice guidelines are prepared by panels of experts using two main types of information: findings from clinical trials and expert opinion, both of which may purport to be independent but rarely are. In practice, much of the available information is sourced from industries whose profits weigh heavily in the balance of guideline recommendations. Regarding clinical trials, as mentioned in previous sections and discussed in more detail in Chapter 8, the pharmaceutical industry conducts most of the trials that assess drug efficacy and safety, while also substantially controlling which results are disclosed or withheld (Angell and Relman, 2002). Regarding expert opinion, it is commonplace for the expert panel members responsible for compiling guidelines to have an extensive profile of past and present entanglement with industry. Panel independence is frequently compromised due to a majority of panel members possessing extensive industry connections, and panels are frequently chaired by a person having close ties to industry. There is overwhelming prima facie evidence for questioning the integrity of most clinical practice guidelines. Additionally, there is extensive **empirical evidence** of guidelines being compromised due to industry influence.

For example, of the nine authors of cholesterol treatment guidelines released in the United States in 2004, eight had financial links to drug companies (Mintzes, 2006). That fact attracted much criticism, which may have had some effect, though apparently not much. In November 2013, the American Heart Association and the American College of Cardiology jointly released new cholesterol guidelines (Stone et al., 2013), which immediately provoked renewed criticism about industry entanglement (e.g., Health Care Renewal, 2013). It emerged that the chairperson and one of two co-chairpersons of the working panel responsible for writing the guidelines had ties to drug companies. As with the previous guidelines of almost a decade earlier, a majority (specifically, 8 of the 15) panelists had ties to industry (Lenzer, 2013). In seeming confirmation of industry influence, specific recommendations in the guidelines advocated wider population use of statins: a group of pharmaceuticals intended to avert cardiovascular disease by reducing blood cholesterol.

Evidence supports statin therapy for selected patients with diagnosed cardiovascular disease, although use is associated with negative side effects, including myopathy, diabetes, muscle pain, fatigue, and others (Golomb and Evans, 2008; Golomb et al., 2012). Wider use would expose many people to risk of harm from side effects but less likelihood of benefit in the form of reduced incidence of cardiovascular disease. Estimates indicate that the wider use of statins recommended in the new guidelines would lead to an approximate 30% increase in the number of adults in the United States eligible for statin therapy, with most of that increase involving people without cardiovascular disease (Pencina et al., 2014). Specifically, almost half of the population aged 40-75 years, including many who do not and never will develop cardiovascular disease, would be eligible for statin therapy under the new guidelines. Among men aged 60-75 years without cardiovascular disease and not currently receiving statin therapy, almost 90% would be candidates for statin intervention. Past experience shows that clinical practice guidelines promulgated by medical authorities in developed countries influence global patterns of marketing and treatment. Worldwide implementation of the new cholesterol guidelines issued for use in the United States would represent disease mongering on a monumental scale, having the potential to create a billion patients globally while threatening "to 'statinize' the planet" (Ioannidis, 2014). In short, there is a strong body of opinion that the new guidelines, if implemented, would be of great commercial benefit to industry and of immense harm to personal and population health.

The progressive lowering of recommended intervention thresholds is a source of continuing concern not merely for statin therapy but for prescription drugs in general (Heath, 2006; Mintzes, 2006; Moynihan et al., 2013). For example, thresholds specified by the European guidelines on cardiovascular disease prevention are such as to indicate that 75% of adult Norwegians aged between 20 and 79 years have an "unfavorable risk profile," despite the population of Norway being among the world's healthiest and longest-lived (Getz et al., 2004). In Germany, an examination of guidelines pertaining to two drugs selected for individual case study revealed clear evidence of manipulation of data and extensive conflict of interest among panel members (Schott et al., 2013). In the case of gabapentin (*Neurontin*), used in the treatment of epilepsy, the guidelines were found to contain recommendations based on industry-manipulated data, which spuriously supported use of the drug for a wide range of conditions, including migraine, **bipolar disorder**, and pain. Additionally, in the case of efalizumab, used in the treatment of psoriasis, comparisons were made between the evaluations of assessors having industry associations and those of others having no evident associations, with the evaluations of the former being found to be substantially more favorable toward the use of efalizumab.

A survey of a random sample of recently-published guideline recommendations assessed whether proposed changes widened or narrowed disease definitions, whether mention was made of potential harms from the changes, and the nature and extent of disclosed ties between panel members and relevant industry (Moynihan et al., 2013). Of 16 proposed changes, 10 involved widening and only 1 involved narrowing of definitions. Less than half of the guideline publications mentioned potential harms of proposed changes to definitions, and none included a rigorous evidence-based account of risks or how they might be mitigated. Among panels disclosing entanglement with industry, almost all chairpersons and a large majority of panel members had financial ties to industry as recipients of funding for research or as consultants, advisers, and speakers. For example, 20 of 24 members of an asthma panel had financial relationships with GlaxoSmithKline, a major producer of pharmaceuticals for asthma, and all 20 were serving as consultants or speakers for the company (Moynihan et al., 2013).

In a rare **quasi-experimental** study of the problem, George et al. (2014) compared two sets of guidelines recommending treatment options for primary immune thrombocytopenia (a group of autoimmune blood diseases). By chance, both sets of guidelines were released in close succession through publication in the same peer-reviewed journal, using a similar body of scientific knowledge, but prepared by two different panels comprised of members having similar professional backgrounds (hematologists). One of the two panels was supported by pharmaceutical companies that sell products used in treating the condition of interest, and three-quarters of the members reported having relevant industry associations. The other panel was comprised of members who reported having no conflicts of interest. Comparisons between the two sets of guidelines showed a "conspicuous" difference wherein stronger recommendations were made by the panel having industry associations for use of treatments manufactured by companies with which panel members were associated.

### 7.1.3.1   Disclosure of Conflicts of Interest

It is widely believed that bias from conflicts of interest is substantially mitigated by requiring potentially conflicted persons to disclose conflicts. However, faith in disclosure is entirely unwarranted, not least because conflicts of interest frequently go unreported, either because individuals omit to disclose potential conflicts or disclosures that are made are not reported in published documents. In a survey of a random sample of clinical practice guidelines archived at the American National Guideline Clearinghouse, information on conflicts of interest was given in fewer than half of the guidelines surveyed (Kung et al., 2012).

When such information was supplied, more than two-thirds of committee chairpersons and 90% of co-chairpersons were found to have conflicts of interest. Overall, the study revealed poor compliance with recommended national standards covering potential conflicts of interest, and showed that there had been little if any improvement in that regard over the previous two decades. Another survey, concerned with guidelines published by national organizations for control of cholesterol and diabetes in Canada and the United States found that more than one-third contained no conflict-of-interest statements (Neuman et al., 2011). Of those that included disclosures, approximately 10% of panelists who formally declared no financial conflicts were found following independent investigation to have one or more such conflicts.

Underreporting of conflicts of interest was found to be even more prevalent in a recent Danish study that examined 45 sets of guidelines from medical organizations representing 14 different specialties (Bindslev et al., 2013). Independent investigation showed that almost all of the guidelines (43 of 45) had authors with conflicts of interest, but these were declared for only *one* set of guidelines. More than half of the conflicts concerned guideline authors being a company employee, or a consultant or advisor for industry. In a separate study of 13 international clinical practice guidelines concerned specifically with glycemic control in type 2 diabetes mellitus, three did not contain conflict-of-interest statements while the percentage of authors with one or more financial disclosures in the remaining 10 guidelines varied from 0% to 94% (Norris et al., 2013). On average, more than half of the drugs recommended in each guideline were from companies with financial links with one or more authors of the guideline.

Although much of the entanglement between guideline panelists and industry represents prima facie evidence of guideline bias rather than proof, entanglement, known or suspected, disclosed or not, causes loss of credibility, which in itself is a form of harm. Doubtful credibility leaves consumers, including physicians and patients alike, in the uncertain position of not knowing whether guideline recommendations can be trusted. Survey research of physician opinion has found that lack of credibility due to industry entanglement and perceived bias on the part of guideline authors are among reasons physicians fail to adhere to guideline recommendations (Cabana et al., 1999). Consequently, irrespective of any actual corruption of guideline recommendations, the fact that conflicts of interest, both real and perceived, are widespread among guideline authors undermines the main objective of clinical practice guidelines, which is to standardize best practice and improve patient health and safety.

Responding to widespread concerns about poor standards and bias, the American Institute of Medicine (Graham et al., 2011) and other authorities (Ransohoff et al., 2013) have issued standards of practice intended to enhance the transparency and objectivity of guideline development. However, recent analyses suggest little overall improvement in the quality of clinical practice guidelines (Ioannidis, 2014; Moynihan et al., 2014), and it is not difficult to discover possible reasons why progress has been slow. An illustration of the problem can be seen in the work of one high-level expert group in the field of gastroenterology supported by several professional societies which was convened to examine options for "an ethics framework" that would enhance the integrity and credibility of clinical practice guidelines (Jones et al., 2012). A main conclusion of the group's deliberations was that "it *may be ethical* (p. 812, emphasis added) in some circumstances to exclude [persons] with *significant* financial conflicts of interest" from involvement in guideline development. The choice of wording seems to reverse usual ethical concerns by implying that it might be *unethical* to exclude persons who have major conflicts of interest. The perversity of that suggestion and the general feebleness of the proposed framework invite the conclusion that the true aim of the expert group was to prevaricate. Notably, a disclosure statement at the end of the article, published in the influential high **impact factor** journal *Annals of Internal Medicine*, mentions that the group's work was supported by "arms-length contributions" from pharmaceutical (Astra-Zeneca and Abbott) and medical supply (Olympus) companies. Whether by coincidence or design, the expert group's main conclusions surely sit comfortably with the business objectives of their so-called "arms-length" corporate sponsors.

Some may have hoped that professional bioethicists would insulate themselves from corporate influence so as to maintain independence, both real and perceived. However, considering the unparalleled profitability of the pharmaceutical industry and its immense financial and political influence (Angell, 2004b), it should have been expected that industry entanglement with the field of bioethics would become a reality much as it has been for decades in biomedicine generally (Elliott, 2004). The fact is that industry has infiltrated the world of bioethics, employing eminent bioethicists to assist corporations to mitigate ethical restraints (especially those emanating from within the profession of bioethics) that could threaten corporate profit.

## 7.2  HAS MEDICINE SOLD OUT TO BIG PHARMA?

*That physicians increasingly function as financially interested middlemen is the result of policies and perceptions that have permitted health care to be subsumed by the health care industry.*

(Jenny-Avital, 2005, p. 733)

Many different types of "entanglement" between doctors and drug companies have been described (Moynihan, 2003b), and the major ones are summarized in Table 7.1. It is noteworthy that entanglement is targeted at every level of healthcare: students, newly qualified physicians, and seasoned practitioners; group practices and clinics; hospital wards and entire hospitals; medical departments and entire medical schools; professional meetings, workshops, and conferences; professional societies and associations; patient-advocacy groups; and scientific journals. Some forms of entanglement, such as payment of consulting fees to actively promote company products, and gifts of lavish travel and entertainment to physicians who frequently prescribe company products, are self-evidently not trivial and would be perceived by most people to be bribes. Other forms of entanglement, such as the pens and pads inscribed with company names and insignia that are ubiquitous in physicians' offices, may appear trivial and of little consequence, but as discussed below, appearances can be deceiving.

Of the many forms of physician-industry entanglement, the mainstay for more than a half-century has been direct person-to-person contact between company representatives and healthcare professionals. The company's representative is the "detail man," or *detailer*, whose job is to establish a personal relationship with physicians wherever they can be found in every healthcare office, clinic, and hospital. The detailer's job is part sales pitch and part "educational," wherein physicians are presented with prescribing information ("details") concerning company products. The first instructional textbook on the art of detailing to physicians was published in 1940. It was written by a

**TABLE 7.1** Types of "entanglement" between doctors and drug companies

Face-to-face visits from drug company representatives

Gifts of office supplies such as pens, note paper, calendars, and appointment pads

Gifts of travel, accommodation, equipment, or software

Sponsored dinners, and social or recreational events

Sponsored educational events, continuing medical education, workshops, or seminars

Sponsored attendance at scientific conferences

Stock or equity holdings

Undertaking paid consultancy work for companies

Sponsored research

Sponsored membership of professional societies and associations

Funding for medical schools, professorial positions, or lecture halls

Sponsored professional societies and associations

Advising a sponsored disease foundation or patients' group

Involvement with or use of sponsored clinical guidelines

Membership of company advisory boards of "thought leaders" or "speakers' bureaux"

Authoring "ghostwritten" scientific articles

Drug-company advertising in medical journals, company-purchased reprints of scientific publications, and sponsored journal supplements

Adapted from Moynihan (2003b).

veteran salesman, Thomas H. Jones, who could be said not merely to have authored the first book but to have had the last word:

> *Detailing is, in reality, sales promotion and every detail man should keep that fact constantly in mind. The primary purpose of the personal call upon the doctor is to promote the sale of the product.*

(Jones, 1940, p. 17)

Jones' credo has not merely stood the test of time—it has also proved to be spectacularly successful. Evidence has consistently shown that physicians who have more contact with pharmaceutical representatives prescribe more drugs and at higher cost (creating more profit for industry) than physicians who have less contact (e.g., Adair and Holmgren, 2005; Caudill et al., 1996; Lexchin et al., 2003; Moskop et al., 2012; Steinman et al., 2007; Zipkin and Steinman, 2005). For example, a survey of over a thousand English general practitioners found that many had contact with drug company detailers as often as once a week or more, and that those who had more frequent contact were more willing to prescribe new drugs (even when this contravened clinical guidelines) and to prescribe more readily to patients who expected a prescription even when one was not clinically indicated (Watkins et al., 2003).

These findings are consistent with results from a meta-analysis of independent studies conducted mostly in the United States but also in Australia, Canada, Holland, and New Zealand (Wazana, 2000). The analysis showed that physician entanglement with industry generally began in medical school and continued thereafter on a weekly basis on average. Despite physicians generally believing that their behavior was not influenced by industry entanglement, there was a *dose-response relationship* between amount of contact and physician behavior, wherein physicians having more contact were more ready to prescribe company products.

The extent of the commercial culture of medicine is revealed by the ubiquity and variety of relationships between physicians and industry. Two national random surveys of physicians in the United States, conducted 5 years apart and involving a wide range of medical specialties (anesthesiology, cardiology, family practice, general surgery, internal medicine, pediatrics, and psychiatry) found that the large majority reported that they received financial benefits from industry (Campbell et al., 2007, 2010). In the interval between the two surveys, physician-industry relationships received extensive attention in the public media, most of it negative and frequently expressing outrage at the variety and extent of secret and illegal industry payments ("kickbacks") to physicians. Specific legislation was introduced and new policies were advocated for medical schools and teaching hospitals to limit, if not eliminate, undue industry influence on healthcare practices. Figure 7.1 summarizes the results of the two surveys. Whereas 9-in-10 physicians reported financial relationships with industry in the first survey, the number was 8-in-10 in the second survey. This could suggest that under pressure of negative publicity and policy initiatives

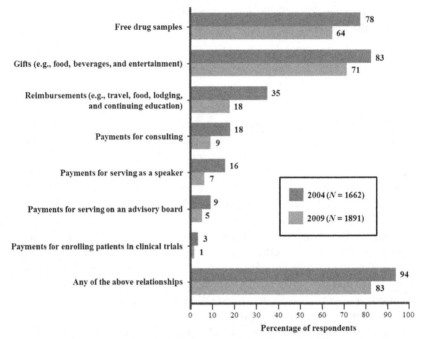

**FIGURE 7.1** Frequency (%) of physician-industry relationships reported in two successive surveys conducted in 2004 and 2009 according to benefit received. *(Derived from Campbell et al., 2007, 2010.)*

physician practices had changed (a little) in the intended direction. Conversely, since the surveys relied on voluntary disclosure, the reduction in reported relationships could just as easily have been due to greater reluctance among physicians to disclose relationships widely perceived to be unacceptable.

## 7.2.1 "Free" Drug Samples

Some of the benefits listed in Figure 7.1 may seem less objectionable than others, and some may even seem entirely acceptable. For example, receipt of free drug samples by physicians who then dispense them to patients may appear to be a benefit to patients rather than doctors. However, dispensing drugs on the basis that they can be supplied to patients without cost may not constitute good medical practice, especially if an alternative drug would have been the intervention of choice were it not for financial considerations (Goodman, 2001). Moreover, winning favor from patients by dispensing free drugs is a "public relations" benefit to the physician that should not be ignored, especially if that benefit is earned at the cost of potential harm to the patient. Furthermore, industry dispenses free samples with the aim of encouraging longer-term adoption of products, particularly drugs that are more expensive than ones of equivalent efficacy they are intended to replace. The future cost of an ongoing course of treatment is likely to exceed savings from initial free samples, leaving patients our-of-pocket for the privilege of having received an initial course for "free." Additionally, by concentrating its free sample supplies on new drugs, industry encourages potentially premature adoption of novel treatments possessing less clear safety and efficacy than existing alternatives of known safety and efficacy, thereby exposing patients to risk of harm.

An additional important consideration is that free samples encourage physicians to administer drugs **off label**, wherein a drug is prescribed for a condition other than that for which it has been officially approved (Kesselheim et al., 2011; Rosenthal et al., 2002). Despite concerns about patient safety and healthcare costs, off-label use of medications is widespread. One study of 160 drugs commonly prescribed in outpatient care estimated that approximately 1-in-5 was off-label despite there being little or no scientific evidence to support the off-label use (Radley et al., 2006). Although it might be thought that greater caution would be exercised when drugs are prescribed to children, the opposite appears to be true with respect to off-label prescription practices. One nationally-representative study in the United States reported that approximately 3-in-5 outpatient pediatric visits involved off-label prescribing (Bazzano et al., 2009). Specialist pediatricians were more likely than general practitioners to prescribe off-label, and the practice was more frequent for younger than older children.

Ultimately, "free" samples are not free at all. They are produced at a cost to the manufacturer, a cost that must be recouped through pricing of products that is passed on, inflating overall healthcare costs that is paid for by patients directly, through insurance premiums, or as a cost to taxpayers in the case of publicly-subsidized care. The same is true of all financial benefits, however small or large, physicians receive from industry. The costs of dinners, trips, "educational" events, and other incentives provided by the pharmaceutical industry for the benefit of physicians are recouped through higher prices for prescription drugs. Physician entanglement with industry representatives is for the material benefit of industry *and* physicians, and is a *tax* on everyone else.

## 7.2.2 The Magic of Gifts Large and Small

Figure 7.1 shows that the most common way in which industry has become entangled with medicine is through the giving of gifts, both large and small. This aspect of physician-industry relationships has been the focus of extensive debate, which has revealed that, among other problems, physician acceptance of gifts from industry representatives is fraught with self-rationalizations, denials, and double standards. In an early study of medical students, Palmisano and Edelstein (1980) observed a "leniency gradient" that extended from "them" to "us" to "me." Whereas a large majority of the students were of the opinion that it is improper for politicians ("them") to accept a gift in the course of their work as a public representative, only a minority of the students found it improper for "us" (medical practitioners) to accept a gift of similar value from a pharmaceutical company.

Responding to a survey about attitudes and behavior toward gifts from industry, physicians generally have been found to report contradictory beliefs (Steinman et al., 2001). In particular, in a confidential survey of newly-qualified physicians, more than 60% said they believed that industry gifts had no effect on their *own* practice, while simultaneously expressing the belief that gifts influenced the prescribing practices of more than 80% of their *peers*. Thus, in common with most people, physicians are subject to **optimism bias** (or *unrealistic optimism*), which entails the belief that one is at less risk than others from a specified hazard or negative event (Sah and Fugh-Berman, 2013).

It is noteworthy that allegations of bias in clinical practice due to influence from industry often elicit indignation from physicians. Negative reactions of that kind stem from the conventional view that such bias is a deliberate act prompted by the prospect of personal gain. Interpreted in that way, the allegation of bias is tantamount to an accusation of impropriety, which not surprisingly is responded to with denial and indignation. That kind of understanding of bias is evident in codes of professional conduct, which typically make a distinction between accepting gifts of small versus large value, with the implication that small gifts do not constitute an ethical breach whereas large gifts do because they create intentional bias. In the aforementioned survey of newly-qualified physicians (Steinman et al., 2001), 100% stated that accepting inexpensive gifts was acceptable and appropriate, whereas a substantially smaller proportion of 60% (still high, and alarmingly so, some might think) felt similarly about expensive items. Notably, even when respondents were of the opinion that particular categories of gifts were inappropriate, many nevertheless reported receiving *and* accepting such gifts.

### 7.2.2.1 The Psychology of the Self-serving Bias and Social Reciprocity

The idea that bias can be eliminated from entanglement with industry through the exercise of deliberate choice is not supported by studies in social psychology. Despite individual resolve founded on good intentions, research findings consistently show that human judgment and decision making are subject to a **self-serving bias** that is unintentional and unconscious (Dana and Loewenstein, 2003). A familiar example is that people tend to attribute individual successes to internal or personal "strengths" while attributing failures to external or situational factors (Campbell and Sedikides, 1999). That is, self-serving bias is an aspect of human nature wherein judgment is reliably skewed in favor of outcomes in which the individual has a stake. Even the perception of fairness, which many people might expect to be relatively stable and immutable, is substantially related to self-interest. Early experiments, for example, found that when given the opportunity to control the sharing of pay for work completed, people consistently perceive fairness in paying themselves more than another person when an equal amount of work has been completed by both parties (Messick and Sentis, 1979). Similarly, when given the opportunity, people generally perceive fairness in paying themselves equally when the other person has completed more work.

Self-serving bias is an element of *social reciprocity*, believed to have evolved as part of the social organization of pre-humans (*Homo heidelbergensis*) and subsequently to have continued as an integral part of human social organization (Layton et al., 2012). People generally feel a strong compulsion or obligation toward reciprocal exchange, including gift giving, which is a distinctive social strategy for encouraging greater cooperation within and between groups. To the extent that gift-giving strategies are subject to self-serving bias and reciprocity, the unconscious nature of their influence runs counter to policies that assume bias is amenable to deliberate choice. Even the provision of education about sources of bias, including specifics about self-serving bias and reciprocity, are generally ineffective. This is because alleged educational counter measures are founded on a faulty model of human behavior that assumes bias in decision making is subject to conscious control (Dana and Loewenstein, 2003). Thus, whatever physicians may honestly intuit, industry gifts, even the ubiquitous trivial trinkets that flood their offices, elicit reciprocity bias that is translated into drug prescribing patterns that boost industry profits.

Orlowski and Wateska (1992) examined the impact on the prescribing patterns of hospital-based physicians who were sponsored by industry to attend expenses-paid symposia at popular vacation sites. The actual educational or scientific components of such trips, then as now, typically occupy only 3 or 4 hours on each of 3 or 4 days, leaving the remainder of the time available for social and recreational activities. In this instance, the impact of the trips was assessed by tracking pharmacy inventory usage patterns for two drugs (an antibiotic and a cardiovascular drug) for 22 months before and 17 months after the symposia. Prescriptions for the drugs increased approximately threefold on average following the symposia, and were substantially higher than at similar facilities nationally. Prescribing increased despite the majority of physicians believing that attendance at the symposia and the drugs they prescribed were unrelated. Sometimes, the intended outcome of industry munificence is less well-concealed, as in the case of one industry-sponsored program in which physicians could earn "frequent-flyer" points toward airline tickets for personal use based on the number of prescriptions they wrote for a particular antihypertensive drug (Graves, 1987). It is more usual, however, for industry (and physicians) to camouflage aims, even if only thinly, as in the case of gifts of expenses-paid holidays disguised as "educational" or "scientific meetings."

Physician lack of concern about being entangled with industry is itself indicative of self-serving bias, wherein individuals in receipt of material benefits judge themselves (though not necessarily their colleagues) to be impervious to industry promotional tactics. In that vein, patients consider pharmaceutical gifts in general to be more influential and less appropriate than do physicians (e.g., Gibbons et al., 1998). Notably, consistent with prediction based on self-serving bias, patients who know that their own doctor has received gifts are less inclined to believe that the gifts were inappropriate. Moreover, both patients and physicians tend to regard small gifts, and especially drug samples, as acceptable and even appropriate. That attitude is reflected in the codes of professional conduct that guide clinical practice. Even when such codes acknowledge industry gift-giving as a strategy to promote sales and as a source of conflict of interest for physicians, they typically condone receipt of small gifts and gifts that might be perceived to be of benefit to patients. In this regard, taking account of the relevant evidence, all parties, physicians, patients, and professional societies, are misguided. The psychological mechanisms responsible for industry's long-standing phenomenal success with gift-giving as a promotional tool do not depend on gift size or reputed understanding of industry intentions.

The usual presumption that large gifts, such as lavish vacations, influence behavior, but low-value items do not, is without support. However improbable it may seem, and unflattering as it is usually interpreted to be, well-controlled studies show that gifts of pens, mugs, and other inconsequential paraphernalia are capable of influencing the clinical decisions of highly skilled and experienced physicians (Katz et al., 2003). Believing otherwise ignores the essential feature of influence in the giving of small gifts, which is the fact that it is a process of *social* rather than material exchange. Irrespective of gift size, the *act* of giving exercises influence over the behavior of the gift recipient. Reciprocity, as discussed above, is a distinctive feature of human social organization, and gift-giving imposes on recipients an obligation to reciprocate in kind; to repay a favor with a favor. Such feelings of obligation do not depend on the size of the gift that is given, but on the act of *acceptance* of the gift by the recipient. Therefore, the only guaranteed defense against influence is not to accept the gift offer in the first place. That few physicians exercise that prerogative is a measure of the moral hazard intrinsic to biomedical healthcare. Discussed in Chapter 6, moral hazard in the present context refers to physician readiness to become entangled with industry, even for relatively small material benefit, when risk of harm to patients is known to be substantial.

The concept of the self-serving bias helps to explain how it is that the large majority of physicians perceive themselves to be unbiased, despite indisputable evidence that physician self-interest is heavily aligned with prescribing patterns, including irrational prescribing such as the use of expensive branded formulations in place of cheaper generic alternatives of equal efficacy (Wazana, 2000). Accepting a gift without reciprocating is socially uncomfortable (Katz et al., 2003). Once evoked, however, the reciprocity rule can be exploited. The initiator in these exchanges is the one who chooses not only the form of the initial favor but also the form of the return favor. The drug company representative is the benefactor and the physician is beneficiary, obligated to reciprocate by supporting the company's products. Because the company's goal is profit, it must ensure that it comes out ahead in the exchange. Thus, a necessary imbalance in outcomes is created, one that is borne not by physicians but by patients in the first instance, in the form of unnecessary and avoidable harm from unneeded or inappropriate interventions, and healthcare in the longer term, in the form of poorer healthcare and higher costs.

As Katz et al. (2003) explained, it is disingenuous to call the small tokens lavishly distributed by industry, "gifts," when in reality they are marketing tools. At the same time, there is little sense in holding pharmaceutical marketeers to account for behaving like marketeers. Industry gift-giving is not a two-way exchange between industry representatives and physicians, but a three-way exchange involving patients. Industry and physicians *both* benefit from the exchange, whereas patients (and ultimately society) pay. Industry does what it is designed to do, it makes profit. The object of healthcare is patient health and safety, which is what patients look for in consultations with physicians. Therefore, the party most to be held to account in this biomedical triad are physicians, who are the point-of-sale agents in the drugs supply chain from manufacturer to consumer, and who benefit materially from a process that harms patients.

A decade ago, Blumenthal (2004) estimated that the pharmaceutical industry in the United States employed about 90,000 drug detailers to service office-based physicians. That number translated to a ratio of approximately one full-time detailer for every five physicians. Therefore, each detailer was on average able to devote the equivalent of almost one full day of every week to each physician client for the sole purpose of encouraging that physician to prescribe company products. This gives an indication of the extent of influence that detailers must have on prescribing practices in order to justify their employment. It would not be possible for industry to support such a large workforce of representatives without immense profits being generated from the influence that those representatives have on physician prescribing patterns. Few people deny that physician-industry entanglement substantially biases the everyday practice of biomedical practice, yet few within medical healthcare appear capable of imagining healthcare without industry largesse to physicians.

Contemporary biomedical healthcare is part of the commercial infrastructure of the pharmaceutical industry. Sadly, even advocates of freedom from industry influence rarely speak of anything more than the evolution of "governing" codes of practice to limit certain excesses, while leaving current arrangements largely intact. However, in a letter to the prestigious *New England Journal of Medicine*, one troubled physician asked:

> *If the honest objective of these relationships is to benefit patients, then why are gifts from pharmaceutical companies necessary for physicians to do the right thing?*

(Lambert, 2005, p. 733)

That question probably resonates with the thinking of many others, especially patients, who are likely to be disappointed to know that the most plausible explanation for physician entanglement with industry, with all its demonstrated harms, is physician material self-interest.

### 7.2.3 Continuing Medical Education

If direct person-to-person contact between company detailers and healthcare professionals is the essential core of physician-industry entanglement, a further influence of potentially equal harm is that of industry involvement in **continuing medical education**. Encouragement of lifelong learning, or **continuing professional development**, is the norm in many professions, and is universally acknowledged within the medical profession as necessary for the maintenance of clinical competence. However, there is a dearth of quantitative evidence concerning the validity, reliability, efficacy, and cost-effectiveness of much of what occurs in the name of continuing medical education (Ahmed et al., 2013). Typically, compliance with mandatory requirements is referenced against "credits" earned by attending educational events rather than against indicators of physician competence or patient outcomes. Uncertainty regarding the effectiveness of continuing medical education is exacerbated by the reality that most of it is sponsored by the pharmaceutical industry.

Estimates suggest that industry sponsors about 75% of continuing medical education in Europe (Ahmed et al., 2013) and about 90% in the United States (Blumenthal, 2004). In fact, industry entanglement in continuing medical education is so extensive as to have led to the creation of a new industry, *medical communication companies* (alternatively, *medical education and communication companies*), which service contracts from the pharmaceutical industry to "educate" doctors. By this means, industry does not need to rely solely on making direct contact with individual physicians, but is able to influence groups of physicians and the professional organizations to which they belong. In addition to providing continuing medical education, medical communication companies also develop campaigns for the launching of new products, and produce digital and print publications (Rothman et al., 2013). Increasingly, continuing medical education is being delivered online, allowing medical communication companies unprecedented opportunities to collect personal data and to create digital profiles of physicians. Typically, such information is collected without explicit consent, wherein participation in a course is taken as providing implicit agreement.

The fact that industry is permitted to participate in the formal continuing professional education that certified practitioners are obliged to pursue gives industry great scope in molding practitioner prescribing behavior. It is possible that physicians may not fully appreciate the extent to which courses are managed by medical communication companies, whose task is to design course content that maximizes sales of industry products. Nevertheless, industry is generally known to be involved in continuing medical education for the purpose of disseminating biased information, and it seems reckless on the part of practitioners to knowingly participate in the charade. The oft-repeated justification given by practitioners and medical educators is that industry entanglement is necessary because it gives doctors access to useful information about new products. Such claims, however, are merely received wisdom; unsubstantiated belief supportive of the status quo. Conversely, there is consistent evidence confirming that industry-sponsored information encourages prescribing habits that are extensively harmful to patients.

### 7.2.4   Patient Advocacy and Charity

Industry's reach is not limited to activities conducted within the formal healthcare system but extends well beyond to include nonprofit consumer and patient advocacy groups and charitable organizations. Such groups, which vary greatly in size from small local assemblies to large international organizations, seek to represent patients with the aim of ensuring patients' needs are understood and are addressed in appropriate ways (Herxheimer, 2003). Public advocacy performs important awareness-raising and public educational functions, and is often effective in lobbying government to commit additional funding for treatment and research for specific medical conditions. While the credibility of patients' organizations depends substantially on being perceived to be independent, information about their structure and funding is often not easily accessed or even available (Colombo et al., 2012). As largely voluntary entities, patient advocacy groups and charities often face major challenges in funding their activities, and many easily fall prey to offers of funding from industry (Rose, 2013). Inevitably, under those circumstances, serious questions arise as to whether independence has been compromised, thereby undermining the group's ability to advocate for patients' best interests.

As an illustration of industry tactics, Herxheimer (2003, p. 1210) at the Cochrane Centre in the United Kingdom has described how, in 2000, the director of the Association of the British Pharmaceutical Industry had privately described the Association's:

> *carefully thought-out campaign ... to employ ground troops in the form of patient support groups, sympathetic medical opinion and healthcare professionals ... which will lead the debate ... This will have the effect of weakening political, ideological and professional defenses ... Then the [Association] will follow through with high level precision strikes on specific regulatory enclaves in both Whitehall and Brussels [referring to the British and European Parliaments, respectively].*

By infiltrating the "consumer advocacy" movement, contemporary pharmaceutical companies are continuing a tradition of marketing practice that has a history of proven efficacy, as discussed in Box 7.1. In addition, companies sometimes circumvent the need to infiltrate existing groups and charities by

## BOX 7.1 The Advent of Public Relations

Edward Bernays (1891-1995). *(Source: http://www.sourcewatch.org/images/c/ca/ Edward_Bernays.jpg.)*

Austrian-American Edward Bernays, often credited with being the *father of public relations*, was a zealous believer in the power of information manipulation to sway public opinion and rouse the masses to action. He considered it the right and responsibility of governing elites to routinely manipulate information in pursuit of perceived national interest. In his book, *Propaganda*, the first chapter, titled *Organizing Chaos*, begins:

*The conscious and intelligent manipulation of the organized habits and opinions of the masses is an important element in democratic society. Those who manipulate this unseen mechanism of society constitute an invisible government which is the true ruling power.*

*(Bernays, 1928, p. 9)*

Bernays distinguished *persuasion* from *authoritarianism*, exulting in the "freedoms of speech, press, petition, and assembly" (not permitted under authoritarian regimes) which guaranteed the "freedom to persuade and suggest." He described the manipulation of public opinion through the use of propaganda as the "engineering of consent," which he considered to be "the very essence of the democratic process" (Bernays, 1947, p. 114).

In his role as public relations consultant, Bernays applied his thinking in the service of industry. He is credited, for example, with having had a pivotal role in the 1920s in reversing the long-established taboo against smoking by women (Amos and Haglund, 2000). A much publicized event of the era was the Easter Sunday parade in New York, where young women were hired to smoke "torches of freedom," *Lucky Strike* cigarettes, as they marched down Fifth Avenue protesting against women's inequality.

> **BOX 7.1 The Advent of Public Relations—cont'd**
>
> The event received widespread coverage in the media and provoked a national debate, all of which contributed to acceptance of smoking by women while helping to make *Lucky Strike* a best-selling brand. Above all, Bernays was a pioneer in the strategy of using *third party* authorities to plead causes on behalf of industry clients. His methods presaged the way today's pharmaceutical industry uses patient advocacy groups and health-related charities, either by infiltrating existing organizations or by creating their own, to pressure government in the pursuit of industry profits while allegedly representing patients' interests.

creating their own "consumer groups" to lobby directly on behalf of industry interests while ostensibly representing patients' interests. The advocacy group, Even the Score, mentioned earlier in this chapter, is an illustrative example.

## 7.3    WHAT DOCTORS DO (AND DON'T DO) TO LIMIT HARMFUL COMMERCIALISM IN CLINICAL PRACTICE

*[Doctors] too easily convince themselves that their professional integrity is immune to seduction by drug companies.*

(Abbasi and Smith, 2003, p. 1155)

Entanglement with the pharmaceutical and device industries is a longstanding source of professional embarrassment and debate in biomedical healthcare. However, it is hard to avoid the conclusion that the persistent rhetoric of concern is substantially disingenuous. The fact that entanglement has continued unabated despite being debated for decades is evidence of the preemptive power of self-interest. Industry could not have succeeded in wresting control over contemporary biomedical healthcare were it not for physicians consenting to economic self-interest taking precedence over patient welfare. Despite benefiting physicians materially, industry's pervasive medical presence has helped to create conditions that physicians frequently complain about most, including regulatory bureaucracy and risk of patient litigation. The price for succumbing to immediate self-interest has been collective loss of public esteem, trust, and professional autonomy. That is, apart from failing as custodians of patients' best interests, physicians may have failed in safeguarding their own ultimate best interests.

### 7.3.1    Unhappy Doctors

There is a perception within the medical profession of pervasive dissatisfaction with the profession and that "unhappy doctors" are the norm (e.g., Edwards et al., 2002; Smith, 2001). However, the current malaise may not be the historical aberration that it is often presumed to be. Rather, it appears that for much of the history of medical practice doctors have been "scrabbling for patients and fees, bitterly defending their turf against healers of other persuasions, and often feeling mortified by the inadequacy of their clinical tools" (Zuger, 2004, p. 73). What distinguishes current practitioner unease is the perception that things have deteriorated, which to a large extent is true. This is because the current era is usually compared to the recent history of the early- and mid-twentieth century "golden age" of medicine, when trust in biomedical healthcare and physician esteem reached their zenith. Even so, physician unhappiness today may not be universal. A survey of Norwegian doctors found that life and job satisfaction had increased rather than decreased over the decade straddling the turn of the century, prompting the authors to acknowledge that this challenges the impression of doctors generally being unhappy (Nylenna et al., 2005).

Although much of the discussion about doctor unhappiness concerns healthcare practice in Western countries, lack of trust and esteem in biomedical healthcare are evidently not unknown in other countries. In China, which has embarked on major healthcare reform, there have been reports of substantial deterioration of doctor-patient relationships (Nanshan, 2014). Part of the problem relates to extreme working conditions wherein a hospital might expect a doctor to see 100 outpatients a day, in contrast to the average number of 15 patients per day seen by primary care physicians in full-time office practice in the United States (Acheson, 2012). However, key aspects of healthcare problems in China are not dissimilar to concerns in Western countries. Chinese media have reported extensively on poor practice by doctors, including self-serving relationships with industry and patient experiences of questionable and expensive interventions perceived to benefit the doctor or hospital more than themselves. Incentives that minimize time with patients and encourage unnecessary and expensive clinical practices are believed to be responsible for recent escalation in incidence of serious physical violence against doctors in China (Nanshan, 2014).

In common with other professions, successful medical practice depends crucially on the trust of those the profession serves. Curiously, when lamenting the loss of patient trust, physicians are inclined to blame patients rather than themselves. Loss of trust, it seems, is due to patients' possessing too high expectations. Yet, high expectations are predictable considering the ceaseless flow of commercially inspired hype about the efficacy and safety of biomedical interventions that circulates in both the professional literature and lay media. Irrespective of high expectations, lack of trust in medical healthcare is inevitable considering the extent and depth of entanglement between biomedicine and industry.

Faced with rising expectations, it has been argued that "new perspectives" in biomedicine are needed to counter patients' beliefs in ever "better treatments and cures [and] another scientific or technological breakthrough just around the corner promising to save even more lives [at] an affordable price" (Braithwaite, 2014, p. 92). However, new perspectives must deal with the central problem of conflict of interest. As a largely self-regulated profession, medicine espouses high integrity for its members and institutions. Considering the anemic state of current biomedical self-regulation, there is great scope for decisive action to be taken to remove the conflicts of interest that are responsible for much of what is wrong with contemporary biomedical practice. Removal of conflicts of interest would encourage realistic patient expectations and increased trust that the true purpose of biomedical healthcare is patient health and safety. Alas, false expectations by patients and practitioners may serve other purposes, such as helping to foster hope and feelings (if not the reality) of safety, and a sense of control (Woolf, 2012). Above all, false expectations support the demand for marketable products and ever more biomedical healthcare. Left in the hands of current vested interests, adoption of mooted new perspectives seems unlikely.

## 7.3.2    Conflict of Interest, Self-regulation, and Professional Integrity

A popular definition of conflict of interest is a

*set of circumstances that creates a risk that professional judgment or actions regarding a primary interest [e.g., patient welfare] will be unduly influenced by a secondary interest [e.g., physician relationship to a company that supplies products relevant to the patient's condition].*

(Roseman et al., 2012, p. 1)

Conflict of interest is not in and of itself wrongdoing, but persons acting in a professional capacity are usually expected to take action to remove the source of any conflicts and if that is not possible to remove themselves from the situation. It is undeniable that entanglement with industry exposes physicians to conflicts of interest. Moreover, there is irrefutable evidence, discussed throughout this chapter and the next, that conflicts of interest influence clinical practice in ways that are harmful to patients. Accordingly, there has been much discussion about what should be done to either prevent conflicts of interest or prevent harm when conflicts occur. To date, the main proposals have included the exercise of individual physician professional integrity with little or no further regulation, public disclosure by physicians of their relationships with industry with or without statutory reporting obligations, and removal of entanglement with industry. The first two proposals are primarily self-regulatory approaches that rely on the exercise of individual practitioner judgment, whereas the third proposal seeks to remove the element of judgment by preventing conflicts from occurring in the first place. Almost all current codes of "best" practice in biomedical healthcare involve aspects of the first two. The third, elimination of entanglement with industry, has hardly advanced beyond tentative theoretical speculation.

With regard to self-regulation in clinical practice, calls abound within the medical literature for physicians to exercise professional integrity in ensuring that patients' interests are not jeopardized as a result of entanglement with industry (e.g., Grande, 2010; Green et al., 2012; Marco et al., 2006; Solomon, 2012). Much of that literature is vague, rarely moving beyond imploring physicians to examine their consciences wherein they will discover how best to protect their patients from harm. However, as discussed elsewhere in this book, self-regulation is known to have a poor record in other areas of clinical practice, notably, the obligation to report instances of medically-caused harm and the endemic misuse of antibiotics. Indeed, the frequent admonition to physicians to exercise integrity and restraint in response to industry entreaties is essentially self-contradictory if not hypocritical. If anyone practicing medicine truly needs to be reminded of the dubious ethicality of entanglement with industry, it rather suggests from the outset that such individuals are not fit to be entrusted to self-regulate patient safety ahead of personal gain.

Appeals to professional integrity are generally predicated on acceptance that industry-physician relationships are a reality and may even be desirable. Regarding desirability, reputed benefit in the form of industry-supplied information and education, as outlined above, are often cited. A particularly strident instance has been forwarded not by a physician but an economist, Paul Rubin. Writing in the *Annals of Emergency Medicine*, Rubin (2012) acknowledged that contact with pharmaceutical representatives, and particularly industry inducements in the form of gifts, influence physician behavior, but Rubin declared these effects to be "trivial to an economist" (p. 99). The gist of Rubin's argument can be distilled to three main propositions, all of which have long been disproven. First, Rubin believes that increased selling of pharmaceuticals is a public good because "pharmaceuticals are good, and newer drugs are better." However, as we have seen, pharmaceuticals are often not good, and certainly not when overused and misused, which is the consensus view about what happens when industry-physician contacts are frequent. Additionally, as we have seen, the failure of newer drugs to be better than existing drugs is so common as to be regarded a major crisis in contemporary biomedical healthcare. Furthermore, increased sales, especially sales of newer and more expensive drugs, is precisely the outcome that has most aroused the concern of authorities who have examined physician-industry entanglement. Although it is arguable that patient safety should be the main concern, it is spiraling medical costs that have provided the greatest impetus for change within regulatory circles. For example, a major objective of the recently enacted United States Physicians Payments Sunshine Act, which came into effect on 1 August 2013, is to "shine a much needed ray of sunlight on a situation that contributes to the exorbitant cost of health care."

Second, Rubin draws a parallel between industry promotion and scholarship, as have other defenders of the pharmaceutical industry (e.g., Hirsch,

2009), arguing that every individual scholar is confronted with a potential conflict of interest between their own arguments and truth, favoring their own arguments when confronted with contrary truths. Rubin (2012) extends the parallel to include peer review of scholarly works. He contends that just as expert peers review a work of scholarship to determine whether the work is worthy of publication, so "it is with physicians." Because "physicians are aware that they are receiving biased information" from industry, they discount the information accordingly. However, the comparison is spurious. Whereas scholars are subject to review by independent others, namely, expert peers, physicians in their interactions with industry are entrusted to review themselves. It is important also to consider the proportionality of alternative sources of potential bias. Bias created by one individual scholar's vested interest (i.e., positing one self-serving "truth" over others) is offset by the countervailing bias of other scholars positing their respective versions of truth, each one under the influence of similar self-serving bias. By comparison, no individual, institution, or organization has the resources to influence the entire process of medical research and practice in the way the pharmaceutical industry can and has. In that respect, as one veteran scholar of industry-physician entanglement has observed, "the industry is in a class of its own" (Lexchin, 2012, p. 258). In short, there is no parallel in proportionality between the dominating influence of industry in biomedicine and the influence of individual scholars competing to be heard above the din of a crowd of equally strongly-opinionated peers.

Third, in common with a few others (Huddle, 2010; Stell, 2009), Rubin (2012) believes that industry-physician contact is essential for good clinical practice, arguing that companies "must spend resources to induce physicians to pay attention to the information" (p. 99) supplied by drug detailers. However, the claim that drug detailing has important educational functions is not supported by evidence. A meta-analysis of studies of physicians' preferences for clinically-relevant sources of information found that physicians rank company-supplied information lowest in value compared to all other sources (Haug, 1997). Indeed, the fact that payments to physicians are necessary to engage physician attention is further testimony to the low educational value of industry-supplied information. Without material inducement, why would physicians give up their time to meet drug detailers and to attend industry-sponsored events which they perceive to be of little value? Of course, the same evidence also supports the view that industry-physician entanglement is economic rather than educational. Finally, even if physicians perceive industry-sponsored information to be useful, it should not be assumed that they are correct in their perceptions. Systematic evidence has consistently shown bias in industry-supplied information, which is usually designed (often successfully) to encourage the use of company products to the point of overuse and misuse, including off-label prescribing (Steinman et al., 2007).

Conflict of interest due to corporate entanglement exists not merely at the level of individual practitioners but extends throughout the organization of professional medical practice to the highest levels. One prominent instance relates to a multimillion dollar corporate arrangement in which the American Medical Association planned to sell its brand name to the Sunbeam Corporation in the form of endorsements for products such as blood pressure monitoring and other medical devices produced by that company (Schlesinger, 2002). The deal was aborted only after extensive negative public reaction. A more recent example concerns acceptance by the American Academy of Family Physicians of substantial corporate sponsorship from the Coca-Cola Corporation for a "patient-education" program on obesity prevention. In response to denouncement of the arrangement by critics, the Academy stood firm. It defended the sponsorship describing it as a "consumer alliance," an evident euphemism for what more clearly is a "corporate alliance" (Brody, 2010). Such arrangements are indicative of the endemic nature of the commercial culture of medicine, wherein the organization of biomedical practice has itself become corporatized. Medical professional associations have become increasingly committed to protecting their own interests as corporate entities, including the pay and privileges of the leadership, even over the interests of their physician members, with lesser regard still to the interests of the patients of those members.

### 7.3.2.1   Codes of Professional Conduct

The inadequacy of self-regulation as a means of controlling undue industry influence in biomedical healthcare is evident in the codes of conduct that are promulgated, sometimes in self-congratulatory tones, by biomedical professional organizations. For example, taking account of relevant scientific evidence, the American College of Emergency Physicians in its policy on *Gifts to Emergency Physicians from Industry* (http://www.acep.org/Content.aspx?id=29482& terms=Gifts%20to%20Physicians%20from%20Industry) concedes that:

> *Gifts create feelings of goodwill and indebtedness that do, in turn, influence choices of therapy (mostly unconsciously) for the wrong reasons [and that] even small favors may create a subliminal sense of gratitude or loyalty that can influence physicians' medical treatment choices.*

Despite such acknowledgements, the policy, which in some respects is progressive compared to many, simultaneously advices physicians that they are "free to interact with industry representatives if they choose," including accepting payment for consultancies, for conducting research, and for speaking in industry-sponsored continuing education programs for other emergency physicians. In the event of such entanglement, the policy states, "emergency physicians should carefully consider the purpose of the gift and the likely consequences of accepting it." Thus, the policy accepts that industry entanglement may cause unconscious bias, and simultaneously advises physicians to exercise (deliberate conscious) care. As such, the policy is profoundly self-contradictory. It is not possible to exercise due care to negate unconscious bias. In short, the code is essentially use-less as a statement of policy intended to protect patients from undue industry influence in biomedical healthcare, although it may be useful as a public relations document in support of professional interests.

Testing credibility of a different kind, the Joint Task Force responsible for preparing recently-published European guidelines on cardiovascular disease pre-vention in clinical practice asserted that "preparation and publication [were] with-out *any* involvement of the pharmaceutical industry" (Perk et al., 2012, p. 1640, emphasis added). What the statement may reasonably be interpreted to mean is that there was no direct or explicit involvement of the pharmaceutical industry. However, any inference that guideline recommendations were free of industry influence lacks credibility. The document acknowledges the involvement of 84 named persons serving as authors (representing nine different professional asso-ciations), "other experts," and "document reviewers," all of whom contributed directly to guideline recommendations, as well as an unreported number of unnamed persons who also contributed expertise to the drafting process. On the certain premise that a substantial proportion, if not a substantial majority, of both named and unnamed persons had ties to industry, the assertion "without any involvement" from industry is indefensible.

### 7.3.3    Professional Ethics: What Are They For?

When discussing physicians' relationships with industry, it is helpful to distin-guish between morals and ethics. Morals concern *the distinction between right and wrong*, and to *conforming to accepted standards of general conduct* (Pearsall and Trumble, 1995). Ethics, on the other hand, refers to the *study* of morals in human conduct, which in the present context refers to *the rules of conduct recognized by professional organizations as appropriate to the prac-tice of biomedicine* (Pearsall and Trumble, 1995). Thus, although the two terms, *morals* and *ethics*, are sometimes used interchangeably, they convey important differences in meaning. In particular, it is possible for actions to be simulta-neously ethical, by complying with codes of professional conduct, and immoral, by breaching generally-accepted standards of behavior. Indeed, contemporary biomedical practice is routinely simultaneously ethical (professionally) and immoral. The policy on gifts from industry adopted by the American College of Emergency Medicine is a case in point. A further example of the apparent contradiction that enables individuals and institutions to be simultaneously eth-ical and immoral is given in Box 7.2.

Given that physicians are daily enmeshed in matters of life and death, the professional bodies that govern biomedical practice might be expected to adopt codes of ethics that adhere to higher standards of moral conduct than the stan-dards that apply in everyday life. In reality, codes of ethics that govern profes-sional conduct in medicine are framed in such a way as to condone behavior that most people, especially patients, would consider to be morally reprehensible

**BOX 7.2 The Irish Medical Council: A Case Study of the Difference Between Morals and Professional Ethics**

Ireland operates a two-tier healthcare system comprised of a public sector, providing healthcare to all residents, and a private sector servicing people who have purchased private health insurance. The quality of care provided in both sectors is officially reputed to be the same, but in reality there are differences such as waiting times which are typically substantially longer for public patients than for private patients. A distinctive feature of the system is that most hospital care, both public and private, is delivered in *public* hospitals staffed by physicians employed on salary to treat public patients but who also treat private patients in the same facilities on a fee-for-service basis. As such, Irish public hospitals have been said to operate a two-tier system within a wider two-tier system of healthcare (Nolan and Nolan, 2005).

In 2006, this author's attention was drawn to a public claim by a senior Irish obstetrician that he and other obstetricians routinely differentiate between their public and private patients by *not* attending normal births involving public patients (this care being devolved to midwives) while *attending* normal births of private fee-paying patients. This admission appeared to confirm the widely-held view of endemic inequality in the Irish healthcare system, wherein more-affluent private patients receive a different level of care than less-affluent public patients. When challenged on that point during a national radio broadcast, the obstetrician in question countered the allegation by asserting that in cases of normal births (which comprise the large majority of all births) there was no need for an obstetrician to be present, and therefore there was no inequality. Despite the difference in attending practices, public and private patients, he asserted, receive an *identical* standard of care.

The claim of equality of care in this instance is implausible and few people believe it. However, by asserting that both public and private patients receive an identical standard of care, the obstetrician in question had revealed a more specific ethical concern. He had admitted routinely attending normal births of his fee-paying patients even though his attendance was not needed, suggesting that his time could be better spent elsewhere. In addition, he admitted that although his attendance at the normal births of his private patients was unnecessary, he nevertheless charged patients a substantial fee for each attendance. Thus, a serious ethical question was raised: If it is true, as claimed, that both public and private patients receive identical care, what justification can there be for charging (private) patients a substantial fee for precisely *no* benefit?

In light of this admission of an apparent ethical breach, I submitted a formal complaint to the Irish Medical Council, the relevant "regulatory body for [Irish] medical doctors ... responsible ... for maintaining professional standards." My hope was that the complaint might either assist the cause of those concerned with equality of care for public and private patients, or help bring an end to the practice of charging fees for a service that delivers no benefit. The matter was referred to the Council's Fitness to Practice Committee, which invited submissions from me and the obstetrician about whom I complained. In due course, the Committee concluded that "having regard to the provisions" of the Council's *Guide to Ethical Conduct and Behavior*, there was no breach. Specifically, the Council referred to provisions that state:

- misconduct is conduct which doctors of experience, competence and good repute, upholding the fundamental aims of the profession, consider disgraceful or dishonorable; and/or

---

**BOX 7.2  The Irish Medical Council: A Case Study of the Difference Between Morals and Professional Ethics—cont'd**

- conduct connected with his or her profession in which the doctor concerned has seriously fallen short by omission or commission of the standards of conduct expected among doctors.

The Medical Council's decision and the relevant provisions contained in the Council's *Guide to Ethical Conduct and Behavior* are revealing on a number of levels, but the point of interest here (assuming I possess a reasonable conception of what most people consider to be moral conduct) is how the episode provides a perfect illustration of the difference between morals and ethics.

My complaint to the Medical Council and the argument presented here are predicated on the assumption that most people consider it immoral to acquire for personal gain a fee for a service that is of no benefit and is identical to that received by others who are not levied. The breach is all the more objectionable in this instance, because the fee is procured under circumstances of heightened vulnerability from soon-to-be-delivered expectant mothers. Perversely, however, the "ethical" principles invoked by the Medical Council make no mention of duty of care but instead relate solely to what doctors "consider" and "expect" from themselves and from one another as members of a privileged élite. Thus, the distinction between morals and ethics is illustrated: the routine conduct of Irish obstetricians is *both* immoral because it breaches generally-accepted standards of conduct (i.e., it is wrong to levy fees for discretionary services that are of no benefit to patients) *and* ethical because it complies with arbitrary standards set by a self-serving and powerful clique. In short, personal aggrandizement by Irish doctors from fees-for-*no*-service does not breach self-appointed and self-serving standards of conduct.

Parenthetically, the emphasis given in the Medical Council's guidelines on what doctors consider suitable for themselves without regard to patient welfare might be noted in light of the case of Irish obstetrician Michael Neary, discussed in Chapter 5. The flimsy moral foundation upon which ethical obligation required under the Medical Council's guidelines may go some way toward explaining how it was that Neary was able to inflict gross physical harm on scores of defenseless patients before finally, after 25 years of unimpeded practice, being judged unfit.

---

(e.g., benefiting materially from entanglement with vested commercial interests knowing that one's professional judgment is likely to be compromised). As described throughout these pages, biomedical healthcare personnel routinely expose themselves to conflicts of interest that are known to influence judgment and decision-making in ways that harm patients. Rather than protect patients, current codes of professional conduct protect wrongdoing, giving practitioners license to endanger patients. Specifically, professional codes of conduct typically condone industry entanglement on the proviso that practitioners *exercise care*, despite it being known that self-serving bias includes unconscious processes not under individual volitional control. After decades of debate and endless posturing, biomedical healthcare continues to operate within a framework of unconscionably low standards of moral conduct.

# REFERENCES

Abbasi, K., Smith, R., 2003. No more free lunches. Br. Med. J. 326, 1155–1156.

Acheson, L.S., 2012. In this issue: through the lens of a clinician. Ann. Fam. Med. 10, 490–491.

Adair, R.F., Holmgren, L.R., 2005. Do drug samples influence resident prescribing behavior? A randomized trial. Am. J. Med. 118, 881–884.

Ahmed, K., Wang, T.T., Ashrafian, H., et al., 2013. The effectiveness of continuing medical education for specialist recertification. Can. Urol. Assoc. J. 7, 266–272.

Amos, A., Haglund, M., 2000. From social taboo to "torch of freedom": the marketing of cigarettes to women. Tob. Control 9, 3–8.

Angell, M., 2004a. Excess in the pharmaceutical industry. Can. Med. Assoc. J. 171, 1451–1453.

Angell, M., 2004b. The Truth About the Drug Companies: How They Deceive Us and What To Do About It. Random House, New York.

Angell, M., Relman, A.S., 2002. Patents, profits & American medicine: conflicts of interest in the testing & marketing of new drugs. Daedalus 131, 102–111.

Applbaum, K., 2006. Pharmaceutical marketing and the invention of the medical consumer. PLoS Med. 3, 445–447. http://dx.doi.org/10.1371/journal.pmed.0030189.g001.

Bazzano, A.T.F., Mangione-Smith, R., Schonlau, M., et al., 2009. Off-label prescribing to children in the United States outpatient setting. Acad. Pediatr. 9, 81–88.

Bell, S.E., Figert, A.E., 2012. Medicalization and pharmaceuticalization at the intersections: looking backward, sideways and forward. Soc. Sci. Med. 75, 775–783.

Bernays, E.L., 1928. Propaganda. IG Publishing, New York, NY.

Bernays, E.L., 1947. The engineering of consent. Ann. Am. Acad. Pol. Soc. Sci. 250, 113–120.

Bindslev, J.B., Schroll, J., Gøtzsche, P.C., Lundh, A., 2013. Underreporting of conflicts of interest in clinical practice guidelines: cross sectional study. BMC Med. Ethics 14, 1–7. http://dx.doi.org/10.1186/1472-6939-14-19.

Blumenthal, D., 2004. Doctors and drug companies. N. Engl. J. Med. 351, 1885–1890.

Braithwaite, J., 2014. The medical miracles delusion. J. R. Soc. Med. 107, 92–93.

Brody, H., 2010. Medicine's ethical responsibility for health care reform—the top five list. N. Engl. J. Med. 362, 283–285.

Cabana, M.D., Rand, C.S., Powe, N.R., et al., 1999. Why don't physicians follow clinical practice guidelines? JAMA 282, 1458–1465.

Campbell, E.G., Gruen, R.L., Mountford, J., et al., 2007. A national survey of physician-industry relationships. N. Engl. J. Med. 356, 1742–1750.

Campbell, E.G., Rao, S.R., DesRoches, C.M., et al., 2010. Physician professionalism and changes in physician-industry relationships from 2004 to 2009. Arch. Intern. Med. 170, 1820–1826.

Campbell, W.K., Sedikides, C., 1999. Self-threat magnifies the self-serving bias: a meta-analytic integration. Rev. Gen. Psychol. 3, 23–43.

Caudill, T.S., Johnson, M.S., Rich, E.C., McKinney, W.P., 1996. Physicians, pharmaceutical sales representatives, and the cost of prescribing. Arch. Fam. Med. 5, 201–206.

Colombo, C., Mosconi, P., Villani, W., Garattini, S., 2012. Patient organizations' funding from pharmaceutical companies: is disclosure clear, complete and accessible to the public? An Italian survey. PLoS One 7 (5), e34974.

Dana, J., Loewenstein, G., 2003. A social science perspective on gifts to physicians from industry. JAMA 290, 252–255.

Edwards, N., Kornacki, M.J., Silversin, J., 2002. Unhappy doctors: what are the causes and what can be done? Br. J. Med. 324, 835–838.

Elliott, C., 2004. Six problems with pharma-funded bioethics. Stud. Hist. Philos. Sci. Part C: Stud. Hist. Philos. Biol. Biomed. Sci. 35, 125–129.

Fox, N.J., Ward, K.J., 2008. Pharma in the bedroom and the kitchen: the pharmaceuticalisation of daily life. Sociol. Health Illness 30, 856–868.

Gagnon, M.A., 2013. The disconnection between health ethics and business models. J. Law Med. Ethics 41, 571–580.

Gellad, W.F., Flynn, K.E., Alexander, G.C., 2015. Evaluation of flibanserin: science and advocacy at the FDA. JAMA 314, http://dx.doi.org/10.1001/jama.2015.8405.

George, J.N., Vesely, S.K., Woolf, S.H., 2014. Conflicts of interest and clinical recommendations comparison of two concurrent clinical practice guidelines for primary immune thrombocytopenia developed by different methods. Am. J. Med. Qual. 29, 53–60.

Getz, L., Luise Kirkengen, A., Hetlevik, I., Romundstad, S., Sigurdsson, J.A., 2004. Ethical dilemmas arising from implementation of the European guidelines on cardiovascular disease prevention in clinical practice: a descriptive epidemiological study. Scand. J. Prim. Health Care 22 (4), 202–208.

Gibbons, R.V., Landry, F.J., Blouch, D.L., 1998. A comparison of physicians' and patients' attitudes toward pharmaceutical industry gifts. J. Gen. Intern. Med. 13, 151–154.

Golomb, B.A., Evans, M.A., 2008. Statin adverse effects. Am. J. Cardiovasc. Drugs 8, 373–418.

Golomb, B.A., Evans, M.A., Dimsdale, J.E., White, H.L., 2012. Effects of statins on energy and fatigue with exertion: results from a randomized controlled trial. Arch. Intern. Med. 172, 1180–1182.

Goodman, B., 2001. Do drug company promotions influence physician behavior? West. J. Med. 174, 232.

Graham, R., Mancher, M., Wolman, D.M., et al., 2011. Clinical Practice Guidelines We Can Trust. National Academies Press, Washington, DC.

Grande, D., 2010. Limiting the influence of pharmaceutical industry gifts on physicians: self-regulation or government intervention? J. Gen. Intern. Med. 25, 79–83.

Graves, J., 1987. Frequent-flyer programs for drug prescribing. N. Engl. J. Med. 317, 252.

Green, M.J., Masters, R., James, B., Simmons, B., Lehman, E., 2012. Do gifts from the pharmaceutical industry affect trust in physicians? Fam. Med. 44, 325–331.

Hartley, H., 2006. The 'pinking' of Viagra culture: drug industry efforts to create and repackage sex drugs for women. Sexualities 9, 363–378.

Haug, J.D., 1997. Physicians' preferences for information sources: a meta-analytic study. Bull. Med. Libr. Assoc. 85, 223–232.

Health Care Renewal, 2013. Confused Thinking About New Cholesterol Guidelines: Were Conflicts of Interest to Blame? Retrieved 20 December 2013 from: http://hcrenewal.blogspot.com/2013/11/confused-thinking-about-new-cholesterol.html.

Heath, I., 2006. Combating disease mongering: daunting but nonetheless essential. PLoS Med. 3, e146.

Herxheimer, A., 2003. Relationships between the pharmaceutical industry and patients' organisations. Br. Med. J. 326, 1208–1210.

Hirsch, L.J., 2009. Conflicts of interest, authorship, and disclosures in industry-related scientific publications: the tort bar and editorial oversight of medical journals. Mayo Clin. Proc. 84, 811–821.

Huddle, T.S., 2010. The pitfalls of deducing ethics from behavioral economics: why the Association of Medical Colleges is wrong about pharmaceutical detailing. Am. J. Bioethics 10, 1–8.

IFPMA, 2012. The Pharmaceutical Industry and Global Health: Facts and Figures 2012. International Federation of Pharmaceutical Manufacturers & Associations, Geneva.

Ioannidis, J.P., 2014. More than a billion people taking statins? Potential implications of the new cardiovascular guidelines. JAMA 311, 463–464.

Jenny-Avital, E.R., 2005. Doctors and drug companies. N. Engl. J. Med. 352, 733.

Jones, T.H., 1940. Detailing the Physician: Sales Promotion by Personal Contact with the Medical and Allied Professions. Romaine Pierson Publishers, New York.

Jones, D.J., Barkun, A.N., Lu, Y., 2012. Conflicts of interest ethics: silencing expertise in the development of international clinical practice guidelines. Ann. Intern. Med. 156, 809–816.

Katz, D., Caplan, A.L., Merz, J.F., 2003. All gifts large and small. Am. J. Bioeth. 3, 39–46.

Kesselheim, A.S., Mello, M.M., Studdert, D.M., 2011. Strategies and practices in off-label marketing of pharmaceuticals: a retrospective analysis of whistleblower complaints. PLoS Med. 8, 1–9. http://dx.doi.org/10.1371/journal.pmed.1000431.

Kuehn, B.M., 2011. IOM sets out "gold standard" practices for creating guidelines, systematic reviews. JAMA 305, 1846–1848.

Kung, J., Miller, R.R., Mackowiak, P.A., 2012. Failure of clinical practice guidelines to meet Institute of Medicine standards: two more decades of little, if any, progress. Arch. Intern. Med. 172, 1628–1633.

Lambert, L.A., 2005. Doctors and drug companies (letter). N. Engl. J. Med. 352, 733.

Layton, R., O'Hara, S., Bilsborough, A., 2012. Antiquity and social functions of multilevel social organization among human hunter-gatherers. Int. J. Primatol. 33, 1215–1245.

Lenzer, J., 2013. Majority of panelists on controversial new cholesterol guideline have current or recent ties to drug manufacturers. Br. Med. J. 347, http://dx.doi.org/10.1136/bmj.f6989.

Lexchin, J., 2012. Those who have the gold make the evidence: how the pharmaceutical industry biases the outcomes of clinical trials of medications. Sci. Eng. Ethics 18, 247–261.

Lexchin, J., Bero, L.A., Djulbegovic, B., Clark, O., 2003. Pharmaceutical industry sponsorship and research outcome and quality: systematic review. Br. Med. J. 326, 1167–1170.

Light, D.W., 2006. Basic research funds to discover important new drugs: who contributes how much? In: Burke, M.A. (Ed.), Monitoring the Financial Flows for Health Research 2005: Behind the Global Numbers. Geneva, Switzerland, Global Forum for Health Research, pp. 27–43.

Light, D.W., Lexchin, J., 2012. Pharmaceutical research and development: what do we get for all that money? Br. Med. J. 344, 1–5. http://dx.doi.org/10.1136/bmj.e4348.

Light, D.W., Lexchin, J., Darrow, J.J., 2013. Institutional corruption of pharmaceuticals and the myth of safe and effective drugs. J. Law Med. Ethics 41, 590–600.

Marco, C.A., Moskop, J.C., Solomon, R.C., et al., 2006. Gifts to physicians from the pharmaceutical industry: an ethical analysis. Ann. Emerg. Med. 48, 513–521.

McKinlay, J.B., Marceau, L.D., 2002. The end of the golden age of doctoring. Int. J. Health Serv. 32, 379–416.

Messick, D., Sentis, K., 1979. Fairness and preference. J. Exp. Soc. Psychol. 15, 418–434.

Mintzes, B., 2006. Disease mongering in drug promotion: do governments have a regulatory role? PLoS Med. 3, e198.

Morgan, S.G., Bassett, K.L., Wright, J.M., et al., 2005. "Breakthrough" drugs and growth in expenditure on prescription drugs in Canada. Br. Med. J. 331, 815–816.

Moskop, J.C., Iserson, K.V., Aswegan, A.L., et al., 2012. Gifts to physicians from industry: the debate evolves. Ann. Emerg. Med. 59, 89–97.

Moynihan, R., 2003a. The making of a disease: female sexual dysfunction. Br. Med. J. 326, 45–47.

Moynihan, R., 2003b. Who pays for the pizza? Redefining the relationships between doctors and drug companies. 1: entanglement. Br. Med. J. 326 (7400), 1189–1192.

Moynihan, R., Smith, R., 2002. Too much medicine? Almost certainly. Br. Med. J. 324, 859.

Moynihan, R., Heath, I., Henry, D., 2002. Selling sickness: the pharmaceutical industry and disease mongering. Br. Med. J. 324, 886–891.

Moynihan, R.N., Cooke, G.P., Doust, J.A., et al., 2013. Expanding disease definitions in guidelines and expert panel ties to industry: a cross-sectional study of common conditions in the United States. PLoS Med. 10, 1–12. http://dx.doi.org/10.1371/journal.pmed.1001500.

Moynihan, R., Henry, D., Moons, K.G., 2014. Using evidence to combat overdiagnosis and over-treatment: evaluating treatments, tests, and disease definitions in the time of too much. PLoS Med. 11, e1001655. http://dx.doi.org/10.1371/journal.pmed.1001655.

Munos, B., 2009. Lessons from 60 years of pharmaceutical innovation. Nat. Rev. Drug Discov. 8, 959–968.

Nanshan, Z., 2014. Violence against doctors: why China? Why now? What next? Lancet 383, 1013.

Neuman, J., Korenstein, D., Ross, J.S., Keyhani, S., 2011. Prevalence of financial conflicts of interest among panel members producing clinical practice guidelines in Canada and United States: cross sectional study. Br. Med. J. 343, d5621.

Nolan, A., Nolan, B., 2005. Ireland's health care system: some issues and challenges. In: Callan, T., Doris, A., McCoy, D. (Eds.), Budget Perspectives 2005. The Economic and Social Research Institute, Dublin, Ireland, pp. 70–89.

Norris, S.L., Holmer, H.K., Ogden, L.A., et al., 2013. Conflicts of interest among authors of clinical practice guidelines for glycemic control in type 2 diabetes mellitus. PLoS One 8, 1–10. http://dx.doi.org/10.1371/journal.pone.0075284.

Nylenna, M., Gulbrandsen, P., Førde, R., Aasland, O.G., 2005. Unhappy doctors? A longitudinal study of life and job satisfaction among Norwegian doctors 1994–2002. BMC Health Serv. Res. 5, 1–8. http://dx.doi.org/10.1186/1472-6963-5-44.

Orlowski, J.P., Wateska, L., 1992. The effects of pharmaceutical firm enticements on physician prescribing patterns: there's no such thing as a free lunch. Chest 102, 210–213.

Palmisano, P., Edelstein, J., 1980. Teaching drug promotion abuses to health profession students. J. Med. Educ. 55, 453–455.

Pearsall, J., Trumble, B. (Eds.), 1995. The Oxford English Reference Dictionary, second ed. Oxford University Press, Oxford, UK.

Pencina, M.J., Navar-Boggan, A.M., D'Agostino Sr., R.B., et al., 2014. Application of new cholesterol guidelines to a population-based sample. N. Engl. J. Med. 370, 1422–1431.

Perk, J., De Backer, G., Gohlke, H., et al., 2012. European guidelines on cardiovascular disease prevention in clinical practice (version 2012). Eur. Heart J. 33, 1635–1701.

Radley, D.C., Finkelstein, S.N., Stafford, R.S., 2006. Off-label prescribing among office-based physicians. Arch. Intern. Med. 166, 1021–1026.

Ransohoff, D.F., Pignone, M., Sox, H.C., 2013. How to decide whether a clinical practice guideline is trustworthy. JAMA 309, 139–140.

Rose, S.L., 2013. Patient advocacy organizations: institutional conflicts of interest, trust, and trustworthiness. J. Law Med. Ethics 41, 680–687.

Roseman, M., Turner, E.H., Lexchin, J., et al., 2012. Reporting of conflicts of interest from drug trials in Cochrane reviews: cross sectional study. Br. Med. J. 345, 1–10. http://dx.doi.org/10.1136/bmj.e5155.

Rosenthal, M.B., Berndt, E.R., Donohue, J.M., et al., 2002. Promotion of prescription drugs to consumers. N. Engl. J. Med. 346, 498–505.

Rothman, S.M., Brudney, K.F., Adair, W., Rothman, D.J., 2013. Medical communication companies and industry grants. JAMA 310, 2554–2558.

Rubin, P.H., 2012. Limiting gifts, harming patients. Ann. Emerg. Med. 59, 99–100.

Sah, S., Fugh-Berman, A., 2013. Physicians under the influence: social psychology and industry marketing strategies. J. Law Med. Ethics 41, 665–672.

Schlesinger, M., 2002. A loss of faith: the sources of reduced political legitimacy for the American medical profession. Milbank Q. 80, 185–235.

Schott, G., Dünnweber, C., Mühlbauer, B., Niebling, W., Pachl, H., Ludwig, W.D., 2013. Does the pharmaceutical industry influence guidelines? Two examples from Germany. Dtsch. Arztebl. Int. 110, 575–583.

Shaughnessy, A.F., Slawson, D.C., Bennet, J.H., 1994. Separating the wheat from the chaff: identifying fallacies in pharmaceutical promotion. J. Gen. Intern. Med. 9, 563–568.

Shaw, G.B., 1909. The Doctor's Dilemma: Preface on Doctors. Retrieved 24 October 2013 from: http://www.online-literature.com/george_bernard_shaw/doctors-dilemma/0/.

Smith, R., 2001. Why are doctors so unhappy? There are probably many causes, some of them deep. Br. Med. J. 322, 1073–1074.

Smith, R., 2002. In search of "non-disease" Br. Med. J. 324 (7342), 883–885.

Solomon, R.C., 2012. Coffers brimming, ethically bankrupt. Ann. Emerg. Med. 59, 101–102.

Steinman, M.A., Shlipak, M.G., McPhee, S.J., 2001. Of principles and pens: attitudes and practices of medicine housestaff toward pharmaceutical industry promotions. Am. J. Med. 110, 551–557.

Steinman, M.A., Harper, G.M., Chren, M.M., et al., 2007. Characteristics and impact of drug detailing for gabapentin. PLoS Med. 4, 1–9. http://dx.doi.org/10.1371/journal.pmed.0040134.

Stell, L.K., 2009. Drug reps off campus! Promoting professional purity by suppressing commercial speech. J. Law Med. Ethics 37, 431–443.

Stevens, R.A., 2002. Themes in the history of medical professionalism. Mt Sinai J. Med. 69, 357–362.

Stone, N.J., Robinson, J., Lichtenstein, A.H., et al., 2013. 2013 ACC/AHA guideline on the treatment of blood cholesterol to reduce atherosclerotic cardiovascular risk in adults. Circulation 128, 1–85. http://dx.doi.org/10.1161/01.cir.0000437738.63853.7a.

Venhuis, B.J., de Kaste, D., 2012. Towards a decade of detecting new analogues of sildenafil, tadalafil and vardenafil in food supplements: a history, analytical aspects and health risks. J. Pharm. Biomed. Anal. 69, 196–208.

Walsh, A.J., 2006. Commercial medicine and the ethics of the profit motive. J. Value Inquiry 40, 341–357.

Watkins, C., Harvey, I., Carthy, P., et al., 2003. The attitudes and behaviour of general practitioners and their prescribing costs: a national cross sectional survey. Qual. Saf. Health Care 12, 29–34.

Wazana, A., 2000. Physicians and the pharmaceutical industry: is a gift ever just a gift? JAMA 283, 373–380.

Woolf, S.H., 2012. The price of false beliefs: unrealistic expectations as a contributor to the health care crisis. Ann. Fam. Med. 10, 491–494.

Zipkin, D.A., Steinman, M.A., 2005. Interactions between pharmaceutical representatives and doctors in training. J. Gen. Intern. Med. 20, 777–786.

Zuger, A., 2004. Dissatisfaction with medical practice. N. Engl. J. Med. 350, 69–75.

Chapter 8

# Big Pharma Entanglement with Biomedical Science

*… the object of the medical profession today is to secure an income for the private doctor; and to this consideration all concern for science and public health must give way when the two come into conflict.*

(Shaw, 1909)

## Contents

Industry entanglement with biomedicine permeates the whole biomedical research infrastructure upon which contemporary healthcare is based. Pharmaceutical corporations have relationships with teaching and research institutes and universities at all levels of organization from individuals and groups to whole institutions. This has led to immense public educational and research resources being brought into service for private gain. Senior administrators and academic leaders within institutions frequently have "personal financial interests in companies whose products and services are related to their

The Health of Populations. http://dx.doi.org/10.1016/B978-0-12-802812-4.00008-4

institutional responsibilities" (Campbell et al., 2004, p. 104). Examples include deans of medical schools and leading researchers owning equity in companies that supply research-related materials to their research groups, senior university academics serving on company boards while also supervising and directing university employees engaged in research work sponsored by the same company, and universities owning equity in companies that conduct research in the university using university resources (Anderson et al., 2014; Campbell et al., 2007; Freshwater and Freshwater, 2011).

Richard Horton, the current editor-in-chief of the preeminent medical journal *The Lancet*, recently stated that academic-industry entanglement has led to universities becoming:

> *instruments of wealth creation, not institutions for knowledge, enlightenment, or cultural engagement.*

(Horton, 2014, p. 117)

With specific reference to "why medicine is killing our universities," he wrote, that universities have succumbed to the:

> *vast sums of money available from medical research funding bodies, together with a political culture that privileges wealth creation above all else [and there] seem to be no leaders to cure this sickness.*

(Horton, 2014, p. 117)

Earlier, Marcia Angell, former editor for two decades of the equally eminent *New England Journal of Medicine*, stated her conclusion, arrived at "slowly and reluctantly," that:

> *It is simply no longer possible to believe much of the clinical research that is published, or to rely on the judgment of trusted physicians or authoritative medical guidelines.*

(Angell, 2009, p. 11)

## 8.1    SCIENCE IN THE SERVICE OF INDUSTRY: THE LOSS OF SCIENTIFIC INTEGRITY IN BIOMEDICAL RESEARCH

A necessary condition of scientific integrity in biomedicine is for research to be conducted independently of vested interests. Entanglement of private and public interests is fraught with conflicts of interest that profoundly undermine the scientific integrity of biomedical research and development (Jorgensen, 2013). It is important to acknowledge that scientific integrity *never* conflicts with public interest, whereas vested interest *may*. Scientific truths inform. Therefore, whether reassuring or menacing, truths can be used to facilitate action in furtherance of the common good. Conversely, revealed facts *may* conflict with private interests, such as when Product A is found to be superior to rival Product B, resulting in a potential loss of profit for the manufacturer of Product B.

Whereas the public interest in such circumstances is always served by knowing the true facts, the owners of Product B have a vested interest in suppressing the facts and are certain to do so if given the opportunity.

Regarding opportunity, behavioral theory predicts that when the cost of a commercially unfavorable outcome from for-profit research exceeds the cost of obtaining a commercially favorable outcome, scientific integrity will be sacrificed for the commercially favorable outcome. Empirical results support behavioral theory by showing that when biomedical science is motivated by commercial profit, research "findings" are reliably predicted on the basis of funding source and independently of the science behind the research. The immense scale of industry entanglement with biomedical research is such that effective ownership of "truth" has been appropriated by the for-profit private sector across a wide spectrum of endeavors of vital public interest, with predictable tragic results for personal and population health.

The pursuit of profit ahead of health can be seen in industry strategies for extracting every possible economic benefit out of each and every investment. For example, when drugs are discontinued as treatments for particular conditions, or partially-developed drugs fail due to poor efficacy or lack of safety as treatments for particular conditions, efforts are made to "rescue" them for possible use with other conditions for which they were never intended (Arrowsmith and Harrison, 2012). Chapter 7 describes one such instance concerning the use of flibanserin (Addyi), which failed in its original purpose as a treatment of depression, but which ultimately succeeded in obtaining FDA approval for use as treatment of so-called female hypoactive sexual desire disorder. Similar tactics include *drug repositioning* or *repurposing*, which is when drugs that have current approval for use with one medical condition are tested for use with other conditions in an attempt to widen their adoption while simultaneously avoiding the expense of developing new drugs for specific purposes. That is, costs are minimized and profits maximized by discarding nothing and using everything for every conceivable purpose regardless of actual therapeutic benefit or none, provided the "new" product can be made to satisfy the lowest regulatory thresholds.

## 8.1.1   Drug Testing

The *randomized controlled trial* (or *randomized clinical trial*) is the scientific method of choice for testing the efficacy and safety of new pharmaceuticals. In its simplest form, a clinical trial compares an experimental intervention (e.g., new prescription drug) to one or more comparator interventions, including an existing intervention (e.g., the "old" prescription drug), a **placebo**, or no intervention. Outcomes of such trials affect the health and longevity of hundreds of millions of patients worldwide. Considering the high stakes, it is imperative that the conclusions of clinical trials not be jeopardized by vested interests. Irrefutably, public and private interests conflict when commercial profit aligns with one outcome of a trial (e.g., the new prescription drug is superior to the old one) rather than another (e.g., the new prescription drug is no better or is worse than the old one). For that reason, it would be reasonable to expect that most drugs coming to market have been tested through an impartial process that is independent of the partisan profit-focussed interests of industry. The reality, however, is the opposite of reasonable expectations. The pharmaceutical industry funds more than 70% of clinical trials (Studdert et al., 2004).

Notably, the pattern of industry funding for clinical trials is the opposite of the funding pattern for *basic research*. It may be recalled from Chapter 7 that industry contributes a relatively small proportion of the funding that goes to basic research, only 12% compared to 84% from public sources (Light, 2006). These proportions are largely reversed for clinical trials, which represent the final step in bringing new pharmaceuticals to market. From an industry perspective on risk management, the distribution of private and public funding for healthcare innovation is rational. Investment risk declines progressively as the research and development chain moves from basic research to the final stages of clinical research, which culminates in human trials to test formulations that are thought to be close to becoming marketable products. Therefore, it is unsurprising that industry is less willing to invest in the higher-risk early stages of research (preferring to leave that task to the public sector) and more willing to invest in later stages when the risk is substantially lower. This is especially likely considering the public's apparent willingness to bear a disproportionate burden of high-risk investment. "Willingness" in this context refers to the vast sums of taxation income that are committed in support of industry ventures by elected representatives, who frequently are intensely lobbied by industry for that purpose. In the words of one commentator:

> The advantages granted to the industry are meant to promote a public good—development of medications to reduce suffering and death. In reality [industry has] learned how to make huge profits with drugs that do not much improve public health and that sometimes are unsafe or are prescribed without need.
>
> (Jorgensen, 2013, p. 562)

However, other than immediate financial gain, there is a deeper strategic reason for industry's focus on the last stage of the research and development process. Basic research, coming as it does at an early stage, aims to discover the most promising molecules among myriad possibilities for potential further development as therapeutic products. This laborious and costly activity is a burden industry willingly consigns to the public account. In contrast, funding later stages of the research and development process not only confers greater likelihood of obtaining outcomes that possess *actual* utility but also cedes control to industry over what may be *claimed* to be useful. In short, by conducting clinical trials, industry takes control of trial results, thereby creating for itself opportunities to bias outcomes in ways that are commercially favorable. Consistent with the aforementioned behavioral theory that predicts such opportunities when they exist will be exploited, there is abundant empirical evidence that

pharmaceutical industry investment in clinical trials is associated with extensive profit-serving bias in reported results (e.g., Als-Nielsen et al., 2003;
Lexchin, 2012; Lexchin et al., 2003; Lundh et al., 2012; Sismondo, 2008a,b).

One study categorized 264 randomized clinical trials on the basis of funding
source as being either (a) non-profit, (b) industry (for-profit), or (c) both (i.e., a
mix of funding from non-profit and industry sources) (Als-Nielsen et al., 2003).
The key outcome that was examined in all trials was whether the experimental
drug, a new formulation undergoing premarket testing, was superior to the comparator (control) intervention. Controls consisted of either placebo or no intervention in about three-quarters of the studies, while the remainder used active
intervention consisting of an alternative drug already in clinical use. Apart from
funding source, the three groups of trials could not be distinguished from one
another, there being on aggregate no discernible methodological or other differences between them. The results of the study are summarized in Figure 8.1,
which shows a strong relationship between funding source and study outcomes.
Compared to non-profit trials, assigned a benchmark **odds ratio** of 1.0, trials
funded by industry had an odds ratio of 5.3, indicating that the likelihood of
a trial finding in favor of industry's commercial interests was more than five
times greater if a trial was funded by industry than if an equivalent trial was
conducted not for profit. The strong inference from these and similar findings
is that trials conducted by industry suffer threats to scientific integrity leading to
study outcomes that are biased in ways that serve industry's commercial
interests.

Further evidence of the biasing influence of industry entanglement with
the trials examined by Als-Nielsen et al. (2003) can be seen in the dose-
response relationship that was observed between funding source and trial
outcomes. Figure 8.1 shows an odds ratio of 2.6 for outcomes from trials that
had mixed industry and non-profit support, which is greater than the odds for

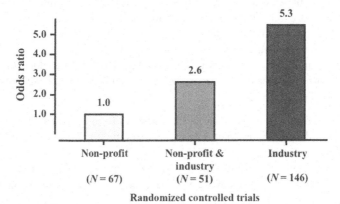

**FIGURE 8.1**   Funding source for clinical drug trials and the odds ratio of conclusions being favorable to industry interests. *(Adapted from Als-Nielsen et al., 2003.)*

non-profit trials but less than the odds ratio for industry trials. In other words, the greater the exposure to industry funding, the more likely study outcomes were biased in favor of industry interests. In addition, using a procedure that concealed funding source, trials were scored on a scale to reflect the degree to which trial conclusions recommended the experimental drug as the treatment of choice without disclaimers or qualifications. New drugs that are recommended without qualification have stronger commercial prospects on entering the market than drugs that come with qualifications and provisos. It is revealing that half of the trials funded by industry recommended the experimental drug without qualification, whereas similarly strong recommendations were made in only 16% of non-profit trials that found in favor of the experimental drug. Again, a dose-response relationship was observed, wherein outcomes for mixed-funded trials recommended the experimental drug without qualification in 35% of trials, a position between that for non-profit and industry-funded trials.

The Als-Nielsen et al. (2003) findings are consistent with studies conducted before and since. An earlier smaller review of 44 reports of oncology (cancer) drugs found that industry-funded studies were nearly eight times less likely to report unfavorable conclusions than non-profit studies (Friedberg et al., 1999). Another study examined the results of meta-analyses of antihypertensive drugs (for lowering blood pressure) and found variability in the level of agreement between results reported in the original studies and the conclusions reported in meta-analyses of those studies (Yank et al., 2007). In meta-analyses funded by industry, agreement between results and conclusions based on those results was found to be poor, with conclusions reported in the meta-analysis being biased in favor of industry interests. Additionally, industry-sponsored meta-analyses of the efficacy of prescription drugs in general have been found to be less transparent in the reporting of relevant details, including reluctance to acknowledge methodological limitations not favorable to industry interests (Jørgensen et al., 2006, 2008). Furthermore, whereas most studies have focussed on clinical drug trials, a recent review extended the search to include trials of medical devices, such as are widely used in medical diagnosis and treatment, and found a similar pattern of bias for both drugs and devices (Lundh et al., 2012).

Of the many ways the biomedical drug and device industries undermine scientific integrity in research, a recent analysis identified five broad sources of "corruption" (Rodwin, 2013):

1. Industry has been allowed to set the priorities for drug research and development, despite the mismatch between its commercial interests and public healthcare needs.
2. Industry conducts most of the clinical trials that test whether drugs are safe and effective, despite its evaluations of products being biased by commercial interests.
3. Industry decides which clinical trial data to disclose and which to withhold, despite the known corrupting effects of selective reporting.

4. The public depends on industry for information about the efficacy and safety of drugs once they are marketed, despite it being known that manufacturers cannot be relied upon to accurately appraise their own products or competitors' products.
5. Physicians depend on industry for product information and for continuing medical education, when the main purpose of those activities is to sell products.

Within these broad categories, industry uses a host of specific strategies, several of which are described later in this chapter.

## 8.2 PARTNERS IN CRIME

When serious misconduct and fraud is discovered, industry frequently seeks refuge in the claim that the wrongdoing was the action of one or a few wayward individuals and not a reflection on the industry as a whole. Recently, however, Peter Gøtzsche at the Nordic Cochrane Centre in Copenhagen, Denmark, used an elegantly simple method to test that claim. On a single day in June 2012, Gøtzsche (2012) conducted one Google search for each of the 10 biggest pharmaceutical companies, combining the name of the company with the word "fraud" (Gøtzsche, 2012). The search yielded between 500,000 and 27 million hits for each company from which Gøtzsche selected one prominent case described in the 10 hits on the first Google page. The resulting 10 cases, summarized in Table 8.1, occurred between 2007 and 2012, and all were discovered in the United States. The most common criminal offences were illegal marketing that recommended drugs for nonapproved (*off-label*) uses, misrepresentation of research results, hiding data about harms, and Medicaid and Medicare fraud. Although the most prominent cases related to the United States, Gøtzsche also found it easy to locate industry crimes in other jurisdictions. Far from being occasional, industry crime is systematic, repetitive, and global.

Despite being repeatedly caught-out and having penalties imposed, even as large as those listed in Table 8.1, there is nothing to suggest any discouragement of industry criminality in drug development and promotion (Braillon, 2012). The explanation for this recidivism may be inferred from the penalties imposed. Although seemingly large in absolute terms, the penalties are paltry relative to profit from illegality. Healthcare crime not only pays, it pays big. In a case involving Pfizer, separate from the one listed in Table 8.1, Gøtzsche (2012) provides a summary of events surrounding *Neurontin* (gabapentin) which— although approved only for use with treatment-resistant epilepsy—was aggressively marketed for off-label use (e.g., anxiety disorders, insomnia, bipolar disorder). Speakers and planted audience members (who asked prearranged questions) were paid to attend meetings to promote use of the drug. High-volume prescribers of the drug were rewarded, and the company paid hundreds of thousands of dollars to influential medical opinion leaders who advocated the drug's wider use. Physicians were recruited for inclusion in ghostwritten

**TABLE 8.1 Summary of recent Big Pharma fraud cases (listed in descending order of settlement size)**

| Company | Settlement (USD) | Summary |
|---|---|---|
| 1. GlaxoSmithKline | 3 billion | The company pleaded guilty to marketing drugs illegally for off-label use, including an antidepressant, an asthma drug, a diabetes drug, and an epilepsy drug. The company paid kickbacks (secret and illegal payments) to doctors, failed to include certain safety data in reports to the regulator, and sponsored programs suggesting cardiovascular benefits in contradiction of warnings about cardiovascular risks from the regulator. The settlement also included allegations of fraud against Federal health agencies. Whistleblowers included senior employees of the company. |
| 2. Pfizer | 2.3 billion | The company pleaded guilty to misbranding drugs "with the intent to defraud or mislead," and was found to have illegally promoted four drugs for uses which had not been approved by regulators: an anti-arthritis drug, an antipsychotic drug, an antibiotic, and an epilepsy drug. Part of the settlement was to resolve allegations that Pfizer paid bribes and offered lavish hospitality to healthcare providers to encourage them to prescribe the four drugs. Six whistleblowers also received settlements from the company |
| 3. Abbott | 1.5 billion | The company settled over illegal marketing of the epilepsy drug Depakote (valproate). Violations included the promotion and sale of the drug for off-label uses; making false and misleading statements about safety, efficacy, and dosing; improper marketing of the product in nursing homes; and payment of kickbacks to induce doctors and others to prescribe or promote the drug |
| 4. Eli Lilly | 1.4 billion | The company settled in relation to a wide-ranging off-label marketing scheme for the drug Zyprexa (olanzapine), which had sales of nearly USD40 billion worldwide between 1996 and 2009. Despite causing serious harm, including heart failure, pneumonia, considerable weight gain, and diabetes, the drug was marketed for numerous off-label uses, including depression and dementia, particularly in children and the elderly. One strategy was to have prepared questions asked by sales people posing as part of the audience during off-label lectures and audio conferences for physicians. The action was triggered by whistleblowers |
| 5. Johnson & Johnson | 1.1 billion | Court proceedings found that the company had lied about the risks of its antipsychotic drug Risperdal (risperidone), potential side effects of which include death, strokes, seizures, weight gain, and diabetes |

| | | |
|---|---|---|
| 6. Merck | 670 million | The company abused government healthcare programs, and paid kickbacks to doctors and hospitals to induce them to prescribe Merck drugs. The company used many different strategies to induce doctors to prescribe its drugs, including payments disguised as fees for "training," "consultation," or "market research." The action was triggered by whistleblowers |
| 7. AstraZeneca | 520 million | The charges were that the company illegally marketed one of its best-selling drugs, intended for treatment of psychosis, to children, the elderly, veterans, and others for nonapproved uses, including anxiety, depression, mood disorder, post-traumatic stress disorder, sleeplessness, dementia, attention-deficit hyperactivity disorder, and anger management. The company targeted its illegal marketing toward doctors who do not typically treat psychotic patients; some of whom were paid kickbacks, and others who received vacations in lavish resorts to encourage them to market and prescribe the drugs for unapproved uses. The settlement included payment to a whistleblower |
| 8. Novartis | 423 million | The penalty was for criminal and civil liability arising from the illegal marketing of an epilepsy drug approved for the treatment of partial seizures, but not for any psychiatric, pain, or other uses. Part of the settlement was to resolve allegations that the company paid kickbacks to healthcare professionals to induce them to prescribe company drugs for a variety of conditions, including for hypertension, irritable bowel syndrome and constipation (later removed from the market because of cardiovascular toxicity), and a drug that mimics a natural hormone. Whistleblowers, all former employees of Novartis, also received payments |
| 9. Sanofi-Aventis | 95 million+ | The settlement was for overcharging of Federal and local health agencies for pharmaceuticals for poor patients |
| 10. Roche | No litigation | Roche marketed Tamiflu (oseltamivir) as an effective intervention against influenza, claiming that it reduces hospital admissions, secondary complications, and lower respiratory tract infections requiring antibiotics. The United States and European governments spent billions of dollars/Euros stockpiling Tamiflu in preparation for what transpired to be the mild influenza epidemic of 2009. Guidance from the WHO repeatedly encouraged member states to stockpile Tamiflu and to gain experience using it (Jefferson et al., 2012). However, Roche based its claims on unpublished trials, and independent analyses do not support claims that Tamiflu prevents influenza complications or reduces the spread of influenza between people. There is evidence that WHO guidance was influenced by persons also working as paid consultants for Roche |

articles purporting to show the drug's effectiveness for a variety of unapproved conditions, and one physician-**whistleblower** testified to having been trained by industry representatives to distort scientific evidence. These activities were highly successful, as about 90% of sales of the drug were for off-label uses. When brought to account, the company paid USD430 million to resolve criminal and civil charges, less than one-sixth of the USD2.7 billion the drug grossed in sales in 1 year alone.

Making profit is one thing, but what about the public opprobrium accompanying litigation? Might that not shame industry into renouncing criminality? After all, companies value reputation which they seek to bolster through self-righteous public statements about commitment to the highest standards of ethical practice. A brief visit to their websites shows that pharmaceutical companies routinely disavow wrongdoing: Eli Lilly, for example, claims that "[f]or more than 136 years [the Company] has shown its commitment to be a responsible global citizen" (https://www.lilly.co.uk/en/responsibility/index.aspx); Pfizer asserts that it "uphold[s] the highest ethical standards in everything from research and development to sales and marketing" (http://www.pfizer.com/responsibility); Abbott Laboratories professes to do "business according to the highest ethical and legal standards" (http://www.abbott.co.uk/about-us/ethics-compliance); Novartis notes that "Responsibility is a core part of our business ... Our patients and customers need to trust us and our products" (http://www.novartis.com/corporate-responsibility/index.shtml); GlaxoSmithKline believes "that being a responsible business ... helps us ... gain the trust of our stakeholders" (http://www.gsk.com.hk/about-worldwide.html); and Sanofi-Aventis states that it is "committed to acting ethically and responsibly at all levels of our activities" (http://www.sanofi.hk/l/hk/en/layout.jsp?scat=5CEBF2C9-F523-41B8-976E-F0CC0060DDEC). Moreover, most of the companies listed in Table 8.1 have corporate integrity agreements with the United States Federal Government, intended to prevent wrongdoing and to detect it when it occurs. Such agreements, however, are evidently largely ineffectual, given that the drug industry remains the biggest defrauder of the United States Government (Braillon, 2012). Rather than serving to encourage ethicality, the record shows that pharmaceutical industry codes of practice and assertions of civic responsibility amount to little more than public relations blather aimed at concealing the selfsame wrongdoing that the codes and assertions disavow.

It is unsurprising that double standards are part of usual pharmaceutical business operations. For example, it will be recalled from Chapter 7 that presenting physicians with conflicts of interest and plying them with gifts are pillars of industry entanglement with biomedical healthcare. It is notable, therefore, that the code of business conduct of Johnson & Johnson (http://www.investor.jnj.com/governance/boardconduct.cfm) states that their own personnel must refrain from accepting gifts, loans, or compensation from suppliers, customers, or competitors. Aware of the biasing influence of gifts and related conflicts of interest with which it entangles physicians, industry is careful to prevent its own employees from being similarly compromised. Taking account of its considerable authority and avowed concern for patient welfare, it is unfortunate that the medical profession does not exercise similar prudence regarding conflicts of interest as that shown by industry.

In the two decades from 1991 to 2012, settlements amounting to more than USD30 billion were agreed between pharmaceutical companies and authorities in the United States, with about 30% being realized in only the last 2 years of that period (Almashat and Wolfe, 2012). As such, industry wrongdoing obviously is not, as is sometimes portrayed, unfortunate one-off or occasional events arising from the clandestine activities of wayward cliques. Rather, wrongdoing entails long-term well-orchestrated operations requiring the complicity of large networks of influential players. Physicians and biomedical researchers are central to such operations. Admittedly, biomedical personnel can be victims as well as perpetrators of pharmaceutical fraud. In a review of antidepressant advertisements in Sweden, a country cited by European politicians and industry as a model of pharmaceutical regulation, numerous breaches were identified (Zetterqvist and Mulinari, 2013). Regulations failed to protect doctors from unreliable information to the extent that one-third of advertisements for antidepressants appearing in the *Swedish Medical Journal* breached regulatory controls. Similar results were obtained a decade earlier in an analysis of advertisements for antihypertensive and lipid-lowering drugs published in Spanish medical journals (Villanueva et al., 2003). Analyses of medical advertisements have consistently shown substantial discrepancies between actual evidence and advertised claims in many countries both high-income and developing (Spielmans et al., 2008; Zetterqvist and Mulinari, 2013).

Although healthcare personnel are undoubtedly victims of fraudulent manipulation of information from pharmaceutical companies, it should nevertheless be acknowledged that within every sphere, including clinics, hospitals, medical schools, medical associations, and research institutes, complicity provides the platform upon which industry delivers a broad program of deceit to patients, the secular public, and to physicians. As described by one whistleblower, "we've got people in the pharmaceutical industry and the healthcare industry all acting in synchrony" (Lenzer, 2005, p. 585). Despite laudable action from diverse quarters, including patients, patient-advocacy groups, whistleblowers, regulatory authorities, and governments, pharmaceutical and medical fraud is relentless. Its continuance is possible only because of self-serving entanglement between healthcare personnel (practitioners and researchers) and industry.

Unscrupulous marketing in healthcare is not new. In an earlier era it was characterized by the merchandising of patent medicines (see Box 8.1). What have changed are the scale and level of coordinated fraud. The pharmaceutical industry is comprised of globalized companies operating in a global market. Straddling multiple jurisdictions, industry is able to maximize

## BOX 8.1 Patent Medicines and the Great Radium Scandal

(Source: Retrieved on 12 December 2013 from: https://www.orau.org/ptp/collection/ quackcures/radithor1.jpg.)

Patent medicines were among the first major products to be marketed using advertising and sales techniques that came to be honed and perfected for use with modern pharmaceuticals. However, the term *patent medicine* is misleading because most such formulations were not patented. In fact, a main difference between many advertised modern medicines and those advertised in the patent-medicine era is that contemporary medicines often are patented whereas so-called patent medicines generally were not. Also, patent medicines were mostly marketed directly to consumers, whereas licensing of contemporary medicines means that most are marketed to physicians who then prescribe them for use by patients. Notably, however, the advent of electronic media has led to a resurgence of direct-to-consumer marketing, which encourages consumers to request specific products from physicians or to order products directly via the Internet. Another main difference is that a large proportion of patent medicines were useless and harmless, whereas many heavily-promoted contemporary medicines of dubious benefit cause harm, including fatalities.

Although many patent medicines were harmless, some were not, containing a variety of potent ingredients such as alcohol, opium, or cocaine, and for a period, some contained radioactive chemicals, including uranium and radium. The most famous of these was *Radithor*, consisting of radium "salts" dissolved in distilled water. Claimed by its inventor, William J.A. Bailey, to be effective against diverse ailments, including headache, constipation, asthma, and diabetes, *Radithor* was marketed directly to the public as well as to physicians, who were offered a rebate on each prescription.

**BOX 8.1  Patent Medicines and the Great Radium Scandal—cont'd**

Bailey became rich from sales of *Radithor*, which had many devotees. These included a wealthy American industrialist, athlete, and socialite, Eben McBurney Byers, who was prescribed *Radithor* in the late 1920s for pain following injury to his arm. Byers continued taking *Radithor*, believing that it improved his health. However, he subsequently developed a "mysterious syndrome … that ravaged his body" (Macklis, 1993). This disease is believed to have been radiation-induced cancers. Shortly after the loss of his jaw due to extensive bone degeneration Byers, aged 52 years, died from "radiation poisoning." He was buried in a lead-lined coffin. Bailey was never tried, and although subjected to official enquiry, his business enterprises flourished until his death in 1949.

It is easy from a contemporary perspective to view the popularity of patent medicines as folly, an illustration of human gullibility during an era when people were less well-informed than today. Conversely, it is instructive to ask how much has really changed. Notwithstanding seemingly tight regulatory controls and reputed benefit, modern prescription drugs, as explained in the main text, are a major cause of mortality and morbidity worldwide.

avoidance of regulatory controls, minimize transparency of its operations, and minimize accountability for its actions. Given the level of corruption and extent of harm caused, it is little wonder that commentators have drawn parallels between the pharmaceutical industry and organized crime (Gøtzsche, 2012). One former Pfizer global vice president of marketing turned whistleblower saw many similarities:

> … *between this industry and the mob. The mob makes obscene amounts of money, as does this industry. The side effects of organized crime are killings and deaths, and the side effects are the same in this industry. The mob bribes politicians and others, and so does the drug industry.*

(Gøtzsche, 2012, pp. 7-8)

Although recognition may have been slow to emerge regarding industry as a leading cause of healthcare harm, it is now widely acknowledged throughout medical education and practice. Perhaps it has simply become too difficult to ignore the death and injury caused, which are of such a scale as to have been referred to as a crime against humanity (Gøtzsche, 2012). Calls for effective action to contain the harm are many and growing (e.g., Lemmens, 2013; Lexchin, 2012; McGauran et al., 2010; Moynihan, 2003). Recommendations include:

- proposed tougher laws, tighter regulation, and better international cooperation
- stricter controls on conflicts of interest
- prison terms for company executives who break the law

- limiting the ability to hide behind industrial and commercial property rights that prevent access to information from industry-funded clinical trials
- government funding of clinical trials to replace industry-funded trials which cannot be believed

Although each proposal has merit, the likelihood of success needs to be considered in the context of the realities of industry entanglement with medicine, an endemic relationship saturated with mutual self-interest.

## 8.2.1   Public-Private Partnership

There is strong evidence that, rather than diminishing, industry influence in (if not control over) healthcare is expanding. To date, sponsorship of healthcare research and practice has been the major strategy industry has employed for generating the enormous profits it currently enjoys. However, in search of ever bigger profits, industry is moving beyond mere sponsorship. One such development involves online social networking, one aim of which is to use specifically selected physicians to influence the mass of other physicians in their daily practice (Landa and Elliott, 2013). Whereas industry contact with physicians has traditionally relied on direct personal visitations by industry detailers, social networking means that direct personal contact can be replaced by virtual visitations involving far greater efficiency and much less expense to industry. Moreover, social networking offers a more compelling form of influence. Due to physician skepticism, industry has always had to go to considerable lengths and expense (e.g., gifts, free samples, and elaborate promotional campaigns disguised as "education") to incentivize personal contact with doctors. Social networking offers new covert ways of overcoming such barriers.

In the virtual world of social media, relatively small numbers of influential physicians, which industry refers to as *thought leaders* or *key opinion leaders*, can be recruited to do the job once done by an army of industry detailers. Social network thought leaders are themselves esteemed physicians whose opinions are respected by their professional colleagues. Whereas the traditional detailer is limited to making contact with one physician at a time, a virtual opinion leader can influence an unlimited number of colleagues simultaneously. Moreover, whereas the traditional detailer is known to be in the pay of industry and therefore committed to promoting an industry line, well-paid social network thought leaders need not disclose any pecuniary relationship with industry and thereby have great scope to influence others without arousing suspicion.

In the United Kingdom, physician social networking is one element of a wider government-initiated campaign of partnership with industry that would allow industry not merely to access physicians but to directly influence government policies concerning overall healthcare delivery. In a program entitled *Moving Beyond Sponsorship: Joint Working Between the NHS [National Health Service] and Pharmaceutical Industry*, the British Department of

Health and the Association of British Pharmaceutical Industries (2010) have partnered in pursuit of "a common agenda to improve patient care outcomes." The partnership effectively outsources public healthcare to private venture. A social networking service will be used to manage healthcare at point-of-delivery by physicians. Additionally, the partnership provides industry with enormous leverage to directly influence *upstream* processes of healthcare, including policy formulation and health services management (Moynihan, 2012). History shows that influence over the direct delivery of healthcare, let alone the policy frameworks within which healthcare is delivered, will contribute greatly to industry profit while simultaneously harming patients. Confronted with those realities, proposed new regulatory strategies such as those cited in the preceding section (e.g., "tougher laws" and "stricter controls"), whether considered individually or collectively, amount to mere tinkering at the margins of the problem and are unconvincing as strategies for containing pharmaceutical industry harm.

*Industry exclusion not partnership is the only viable option for safe and effective healthcare.* The development of public health policy should expressly exclude industry from *all* formative stages. Only after advanced draft policies have been developed, should industry be permitted to proffer opinions for consideration. This should be done primarily for purposes of fact checking and to allow industry to express opinions regarding logistical aspects of policy implementation. With exclusion from early stages of policy development, involvement of industry in later stages should be done in expectation that there will be no change to the underlying policy framework. In particular, industry should not be invited to comment on the science used to develop and support aspects of policy. The science conducted by industry cannot be trusted, nor can industry analysis and interpretation of the science conducted by others be trusted. An invitation to industry to comment on scientific aspects is tantamount to inviting industry to "spin" the evidence in favor of self-interest to the detriment of patients. Any loss of industry profits from the sale of useless or harmful products would be lamented by no one except profiteers.

A top-down process of policy development that does not involve industry during crucial policy-development phases would help to minimize policy distortions, and create a more competitive environment for the production of genuinely effective healthcare products. All things considered, the only certain antidote to harmful influences from industry is the construction of a "firewall" involving:

- complete separation of industry from the processes that shape healthcare policy
- exclusion of industry from the research and development activities that inform new healthcare interventions and practices
- exclusion of industry representatives from contact with the personnel and places responsible for healthcare delivery

Unfortunately, those conditions define a world remotely distant from the one we currently inhabit and possibly farther removed still from the future to which contemporary biomedical healthcare seems inexorably drawn. Despite the absence of evidence of substantive benefit, public-private partnership in healthcare continues to grow (Moodie et al., 2013). Inevitably, this leaves patients vulnerable to evermore voracious predation by industry.

## 8.2.2   Industry and the Biomedical Research Establishment

Mention was made in the previous section of how pharmaceutical companies foster a reputation of social and corporate responsibility. Success in that regard is good for business because enhanced public credibility translates into higher prices and bigger profit margins for products. In banking and finance, the relationship between credibility and pricing of products and services is referred to as *reputational rents*; the greater the credibility the higher the rent (Dinc, 2000). As discussed above, intensive scientific study and numerous lawsuits show how pharmaceutical companies hide, ignore, or misrepresent evidence about new drugs (Light et al., 2013). Knowledge of those activities is now so widespread that it may reasonably be assumed that reputational rents for pharmaceutical companies should be converging to near-zero. How, then, is it possible for pharmaceutical companies to hike prices despite the overall diminishing real value of new pharmaceuticals and low industry credibility?

One reason could be lingering, although declining, public confidence in official regulatory processes. When regulators approve new drugs, the public might reasonably (albeit largely erroneously) infer that the new drugs are safe and probably an improvement on the ones they are intended to replace. A potentially more important influence, however, is the doctor-patient relationship. Patients have a personal relationship with their doctor, a bond founded on confidence and credibility. For industry, success in selling new and more expensive drugs, most of which add little of value to healthcare, depends almost entirely on those formulations being prescribed by doctors. Thus, despite low industry credibility, Big Pharma derives substantial reputational rents from the credibility that doctors individually continue to enjoy in the personal relationships they establish with their patients. Self-serving practitioner entanglement with industry is vital in shielding industry from the negative consequences of its extensive history of corruption and criminality. The friendly, supportive, and caring relationships patients perceive in their doctors is a valuable social resource that is traded for profit and part of the price is patient harm.

Dissemination of information about the efficacy of biomedical practice, including drugs, devices, diagnostics, and intervention, often involves three key parties: the developer or manufacturer who funds clinical trials to evaluate the efficacy of specific practices; researchers who implement the trials, collate data, and interpret results; and editors of medical journals in which the findings of clinical trials are published. *Developers and manufacturers* have an obvious interest in their product or device appearing to be effective, whatever the actual results. Being by far the biggest developers and manufacturers of healthcare-related products, pharmaceutical companies fund most clinical research and thereby position themselves to manipulate much of the "knowl- edge" on which clinical practice is based. As we have seen, the object of the exer- cise is to maximize sales by creating positive perceptions of the efficacy of drugs for treating conditions for which the drugs were developed, while also promoting similar positive perceptions to promote as much use as possible of the same drugs for different medical conditions from those for which the drugs were developed.

*Researchers* who conduct the trials have an important role as thought leaders, and bias from this source in securing product approval from regulatory agencies and shaping clinical practice has been known about for decades (e.g., Melander et al., 2003; Simes, 1986). Product approval usually includes a review of trials, ostensibly all trials, but a common strategy in industry-sponsored research is to systematically cite supportive evidence while selectively ignoring unfavorable findings (Liberati and Magrini, 2003). Additionally, industry may sponsor multiple favorable publications of different specific findings from a single trial, and thereby bias the ratio of favorable-to-unfavorable reports in the overall pool of published studies (Melander et al., 2003). Even when falsi- fication is revealed during regulatory inspection, any falsification that is iden- tified is unlikely to be made public. A recent study examined nearly 60 clinical trials that American Food and Drug Administration inspection had revealed contained significant irregularities, including submission of false information, failure to report adverse patient events, and failure to protect patient safety (Seife, 2015). Only three of nearly 80 (4%) publications in the peer-reviewed scientific literature that reported findings from the trials contained any correc- tions or retractions arising from the issues identified through official inspection.

Another practice is to sponsor published commentaries aimed at discrediting the study design, analysis, or interpretation of unfavorable published findings (Liberati and Magrini, 2003). Whether or not such tactics are directly success- ful, companies can claim that scientific opinion is divided even when there is agreement about findings from independent researchers. Moreover, merely by participating in scientific discussion, companies earn credibility and authority which can be used to influence policy makers and practitioners, especially when double standards are applied with respect to information that is presented to dif- ferent audiences. For example, industry opinion-makers have been shown to acknowledge shortcomings in evidence they present in expert scientific forums, while simultaneously selectively presenting more favorable information when communicating with less-expert approval agencies and practitioners (Liberati and Magrini, 2003).

Publication of the findings of clinical trials depends ultimately on decisions made by *editors of medical journals*. As overseers of the peer-review process, editors have a decisive role as gatekeepers of quality and integrity in science. That role, however, has been drawn into question by the large volume of dubious biomedical science that finds its way into prestigious peer-reviewed medical journals. Richard Horton (2004), Editor of *The Lancet*, expressed the view that medical journals:

> *have devolved into information laundering operations for the pharmaceutical industry.*

Horton described how a pharmaceutical company will sponsor a scientific meeting and pay speakers a "hefty fee" to talk about the company's product. Speakers are chosen on the basis of having views known to be favorable to the product or because they have a reputation for being willing for a fee to adapt their views to industry needs. A medical communication company records the lecture as the basis for an article for subsequent publication, usually as part of a collection of articles arising from the meeting. For payment of substantial fees but little or no peer review, the collection may appear as a special issue or supplement under the imprint of a "reputable" journal.

In a similar vein, Richard Smith, a former editor of the *British Medical Journal*, depicted medical journals as:

> *... an extension of the marketing arm of pharmaceutical companies.*
>
> (Smith, 2005)

Paid advertisements in journals are the most conspicuous but not the only means by which industry promotes its wares in medical journals. Although advertisements are easily seen for what they are, and are likely to be suspected of being misleading, doctors, according to Smith, are probably influenced by advertisements more than they like to believe. Greater harm still, Smith argued, comes from myriad specific tactics (some of which are described in the following section) industry uses to bias the results of clinical trials which are then published as authoritative accounts in medical journals.

## 8.3 MANUFACTURING BIAS IN SCIENTIFIC RESEARCH

Bias in biomedical research is expressly manufactured. Its widespread occurrence is evidenced by the fact that industry-funded clinical trials are markedly more likely to favor the commercial aims of industry sponsors than are similar non-profit trials conducted independently of financial incentive. Nevertheless, there is a high rate of failure even among new drugs trialed by industry. Only one in five new drugs from the 50 largest pharmaceutical companies succeeds when clinically trialed (DiMasi et al., 2010). Thus, most new formulations eventually fail even after having survived the journey from basic research to human clinical trials. As such, companies are under constant pressure to do whatever it takes to minimize failure at the delivery end of what can be a very long pipeline. Reflecting that reality, there are two main speculations to explain why industry-funded research produces a disproportionate number of commercially-favorable outcomes compared to non-profit research.

One speculation is that industry-funded research uses superior scientific method to detect differences between experimental and comparator drugs which reputedly inferior non-profit research fails to detect. The alternative speculation argues the opposite, namely, that industry uses inferior method that produces *spurious* differences between experimental and comparator drugs favoring industry interests. In the language of scientific method, finding spurious differences due to inferior method (possibly characteristic of industry-funded trials) is **Type 1 error**, whereas failing to detect real differences due to inferior method (possibly characteristic of non-profit trials) is **Type 2 error**. There is no evidence that industry consistently employs superior methodology. Comparisons between industry-funded and non-profit trials have shown no consistent differences (e.g., Als-Nielsen et al., 2003). Therefore, the lower rate of "significant" results from non-profit trials is not explained by Type 2 error.

Alternatively, rather than overall quality, there is the suggestion that specific features of method could be superior in industry-funded trials. For example, industry-funded trials have sometimes recruited particularly large numbers of participants, a feature that could contribute to a higher frequency of observed differences in efficacy and safety favoring industry products (Perlis et al., 2005). However, if large numbers of participants are needed to detect differences, the differences observed are likely to be small and may be of questionable clinical importance. Moreover, the opposite claim has also been made that substantial bias in the scientific literature has been caused by the publication of many small industry-funded trials of doubtful reliability that consistently report commercially-favorable findings (Guyatt et al., 2011). Overall, industry-funded and non-profit clinical research employ the same broad type of experimental design, namely, the randomized controlled trial, and examination has failed to reveal any consistent differences in quality between the two.

Assuming no consistent differences in trial methodology, it has been suggested that industry may be particularly skilled at making use of extensive preliminary preclinical data (Fries and Krishnan, 2004). This would include, for example, **translational research** in which laboratory findings are used to ensure that new formulations enter clinical trials only when they are known to have exceptional clinical potential. Strategic exploitation of knowledge acquired before entering the clinical-trial stage is likely to minimize risk of failure, and offers a plausible explanation of industry's comparatively high rate of success in obtaining commercially-favorable outcomes. The evidence, however, does not support this speculation, and what happens when companies compete head-to-head is especially revealing in that regard.

In a study of antipsychotic medicines commonly used in the treatment of patients with schizophrenia, Heres et al. (2006) examined trials in which two drugs from this class were compared head-to-head with the aim of determining if a relationship existed between the sponsor of the trial and trial results. Of 33 industry-sponsored trials, the sponsor's drug was favored 90% of the time, resulting in multiple contradictory conclusions between different studies. That is, whereas sponsors of one drug typically found theirs to be superior to an alternative, sponsors of the alternative drug obtained the opposite result. A similar pattern has been found for statins, the drugs currently in widespread use to lower cholesterol (Bero et al., 2007). In general, the drug that is found to be clinically superior in industry-sponsored head-to-head trials is best predicted by whichever company funded the trial. Drug A is superior to Drug B in trials funded by Company A, whereas Drug B is superior to Drug A in trials funded by Company B.

Thus, on one hand, there is irrefutable evidence that industry-funded trials produce a disproportionate number of results in favor of industry's commercial interests. On the other hand, that high rate of "success" is not explained by use of superior study methodologies. Rather, industry sponsorship of clinical trials is associated with a host of specific tactics, whose purpose is to bias results and conclusions in favor of industry's commercial interests (Lexchin, 2012; Safer, 2002; Smith, 2005). Taken as a whole, so-called drug development is largely an exercise in marketing aimed at selling more drugs at higher prices.

### 8.3.1 Manipulating the Comparator

One biasing tactic relates to the choice and dosage of comparator drugs. In head-to-head trials, for example, industry-funded trials have been found to adhere well overall to standard clinical-trial protocol, except for targeted and systematic manipulation of the dosage regime of the comparator drug (Safer, 2002). This includes administering a low dose of a comparator drug at strategic points in the trial to make the sponsor's own drug seem more effective, and at other times administering a high dose of the comparator to make the sponsor's drug appear to have fewer side effects. Not only do such manipulations produce erroneous conclusions, they also cause harm by depriving patients in the comparator arm of possible benefit while exposing them to possible suffering from avoidable side effects (Lexchin, 2012). Ultimately, of course, such manipulations cause a wider harm when drugs that have been falsely evaluated are disseminated.

### 8.3.2 Salami Slicing, Selective Reporting, and Science Spin

There is widespread industry-sponsored biasing of published findings (Lexchin, 2012; McGauran et al., 2010; Safer, 2002; Schott et al., 2010; Turner et al., 2008). For example, Melander et al. (2003) evaluated evidence from 42 placebo-controlled trials of five antidepressant drugs (all selective serotonin reuptake inhibitors) submitted to the Swedish drug regulatory authority over a 16-year period. The evaluation showed widespread multiple publication wherein the same or similar findings were reported in multiple scientific articles (a practice called *salami slicing*). Spielmans et al. (2010) found extensive evidence of that practice in 43 pooled analyses of the efficacy and safety of duloxetine, a treatment for depression, and almost 90% of the analyses had at least one author who was employed by the drug's manufacturer.

As part of a wide search of medical literature to gain an overview of reporting bias, researchers at the German Institute for Quality and Efficiency in Health Care identified bias in relation to about 50 different interventions, including pharmacological, surgical (e.g., vacuum-assisted closure therapy), diagnostic (e.g., ultrasound), and preventive (e.g., cancer vaccines) treatments, for 40 different medical conditions (McGauran et al., 2010). Regarding pharmacological interventions (which were the most numerous), cases of reporting bias included treatment for the following conditions:

> *acute trauma, Alzheimer's disease, anxiety disorder, atopic dermatitis, attention-deficit hyperactivity disorder, bipolar disorder, cancers (e.g., melanoma and ovarian), cardiovascular disease, depression, diabetes mellitus type 2, gastric ulcers, hypercholesterolemia, infections (e.g. influenza, Hepatitis B, HIV), irritable bowel syndrome, menopausal symptoms, migraine, pain, schizophrenia, thyroid disorders, and urinary incontinence.*

In general, bias was in the form of exaggerated claims for efficacy and safety. The review revealed many instances of data being withheld by companies, including active attempts by companies to suppress publication of findings unfavorable to specific company products.

In the euphemistic language employed by industry, trials that yield primarily negative results are often claimed to be failed *trials* instead of being acknowledged for what they really are, which is demonstrations of failed *pharmaceuticals* (Lexchin, 2012). Extensive analysis of the relevant scientific databases has shown that even when data from industry-funded trials are published without bias, *spin* in the interpretations and conclusions arising from the data often favor industry interests (Golder and Loke, 2008). Industry-sponsored spin includes published meta-analyses of prior clinical trials (Jørgensen et al., 2006; Yank et al., 2007). Industry-sponsored meta-analyses recommended the sponsor's drug twice as often as independent meta-analyses (Jørgensen et al., 2008). Together, these various reporting and publication strategies have created a vast body of extant scientific "evidence" that gives a profoundly biased account of the efficacy and safety of widely used pharmaceuticals and healthcare devices.

### 8.3.3   Cooking Science

Revealing anomalies similar to those mentioned above in relation to regulatory inspection of the conduct of clinical trials (Seife, 2015), several systematic comparisons have shown differences between documentation reporting original results submitted to regulatory agencies, especially the American Food and Drug Administration and the European Medicines Agency, and findings subsequently published in the international scientific literature. Regulatory agencies often act as repositories of large volumes of original clinical-trial data collected by industry, whereas findings published in the scientific literature tend to take the form of concise distillations of original data. Careful comparisons of the two bodies of information have shown that findings published in the scientific literature frequently exaggerate original findings of clinical efficacy while systematically underreporting side effects and other safety concerns (e.g., Gøtzsche and Jørgensen, 2011; Hart et al., 2012; Rising et al., 2008; Schott et al., 2010). Biased reinterpretation of original findings for dissemination in the scientific literature misleads the healthcare community about the likely benefits and possible dangers of products that come to be adopted in clinical practice, with potentially catastrophic consequences for patients.

In one such instance, a class of cardiac arrhythmia drugs that came into widespread use at the beginning of the 1980s is thought to have caused the premature death of about 50,000 Americans each year for most of that decade (Gøtzsche and Jørgensen, 2011). A small early trial suggested that the drugs could be effective against arrhythmia, but that patients had a mortality rate almost 10 times that of patients given placebo (Cowley et al., 1993). However, that study, conducted in 1980, was withheld from publication and it was not until other research conducted at the end of the decade had been published that the high rate of mortality came to light (Echt et al., 1991). Timely publication of the earlier trial would have alerted researchers to the possible high risk of mortality, with the likelihood of many deaths being averted. In Europe, almost 30% of new drugs approved by the European Medicines Agency received safety warnings within a 10-year period of being released onto the market (Light and Lexchin, 2012). Besides cost to patient health, it is estimated that the European Union is paying billions of dollars unnecessarily for new expensive drugs that eventually fail despite earlier approval.

### 8.3.4    "Seeding" Sold as Science

It is known that physicians who participate in clinical trials tend to increase usage of trial drugs in their own practices (Andersen et al., 2006). Taking advantage of that knowledge, the pharmaceutical industry sometimes engages physicians as participants in *seeding* trials, which are essentially bogus clinical trials conducted for marketing purposes rather than to add to scientific knowledge (Hill et al., 2008; Lexchin, 2012). These trials are sometimes conducted during the review phase of a recently-developed drug before the drug has received approval for use in routine clinical practice. As such, seeding trials are designed to encourage early adoption of the drug when the anticipated approval is received. The role of seeding was brought to light when extensive internal company documentation was released into the public domain as part of legal proceedings against Merck over their anti-inflammatory drug rofecoxib (*Vioxx*) (Hill et al., 2008). Although the seeding trial was presented to physicians and patients as a study of the drug's gastrointestinal safety, its true purpose was to create positive perceptions of Vioxx among primary care physicians. In fact, the trial was designed by Merck's marketing division (not its drugs research and development division), which also handled all of the data collection, analysis, and dissemination. That the main purpose of data collection in the Merck trial was marketing, not science, was substantiated by internal Merck documents showing that the then head of Merck's research division did not believe that the trial had any scientific value.

Vioxx was granted approval for clinical use in 1999, but Merck had to withdraw the drug 5 years later after disclosures that it withheld information about risks. Compared to alternative formulations, Vioxx was found to be associated with a substantially increased risk of serious cardiovascular events, including myocardial infarction and stroke (Tanne, 2011). Although Merck claimed that it withdrew the drug as soon as there was "clear evidence" of risk, it has been shown that deaths were occurring more than 3 years prior to withdrawal (Madigan et al., 2012). Intense global marketing led to the drug becoming widely prescribed and highly profitable, which has been interpreted as showing that Merck's interest in the commercial success of Vioxx exceeded its concern about the drug's lethality (Topol, 2004). Use of the drug caused tens of thousands of avoidable deaths worldwide (Ritter et al., 2009).

## 8.3.5    Ghostwriting Medical Research Fictions

*[The] purpose of [published] data is to support ... the marketing of our product.*
(internal pharmaceutical sales document; Moffatt and Elliott, 2007, p. 18)

Industry entanglement with medicine has given rise to the practice of ghostwriting, wherein authors are recruited by industry to write "scientific" articles presenting the results of clinical trials or reviews of the literature for the purpose of promoting the sponsor's products or discrediting competitor products (Bosch et al., 2012). Prominent physicians or academics are recruited to masquerade as the authors without acknowledgment of the involvement of industry-sponsored ghostwriters in the published articles (e.g., Sismondo, 2008a). This brazen form of scientific fraud is systematic and widespread, involves major pharmaceutical corporations, and causes widespread harm to public health (Hart et al., 2012; Lexchin, 2012; Ross et al., 2008; Sismondo, 2008b). In the United States, many cases have been documented and many others remain hidden from the public while under seal by the courts (Bosch et al., 2012). Ghostwriting is part of a wider set of practices euphemistically called publication planning wherein articles and posters produced by drug companies are published in medical journals and presented at scientific meetings to promote key marketing messages (Fugh-Berman, 2010).

Although it is not known how many physicians and academics participate, or how many articles in peer-reviewed medical journals are ghostwritten, the practice has become so common that it may now be regarded as "normal" (Fugh-Berman, 2010). As described in Chapter 7, rather than manage such extensive operations in-house, it is common for drug companies to outsource the work to medical education and communication companies (Moffatt and Elliott, 2007; Sismondo, 2007). The business plan of the communication companies is to portray pseudoscience as science and to hide that fact by enlisting respected biomedical scientists to participate in the fraud for payment. One such company claimed that over a period of 12 years it "planned, created, and/or managed hundreds of advisory boards, a thousand abstracts and posters, 500 clinical papers, over 10,000 speakers' bureau programs, over 200 satellite symposia, 60 international programs, dozens of websites, and a broad array of ancillary printed and electronic materials" (Fugh-Berman, 2010, p. 2).

It is illegal for drug companies to promote their products for non-approved (i.e., off-label) uses. However, scientific articles in reputable journals are not considered promotional vehicles. Consequently, articles in peer-reviewed journals "written" by respected authors offer industry opportunities for bypassing regulatory controls intended to safeguard patients from inappropriate uses of products approved for entirely other uses. One study used information gleaned from whistleblower complaints alleging illegal off-label marketing and found that authors involved in off-label marketing activities rarely adequately disclosed their conflict of interest (Kesselheim et al., 2012). Even after complaints were made, adequate disclosure in subsequent related articles by the same authors was rare, occurring in only 1-in-7 of published articles. That level of non-disclosure indicates not merely author shortcomings but also shortcomings in journal practices. Indeed, editors of several reputable medical journals are directly implicated in fostering ghostwriting through personal entanglement with medical communication companies and the drug companies they represent (Sismondo and Doucet, 2010). The editor of one such journal attending a meeting is quoted as advising representatives of medical communication companies on how

> to avoid practices that are going to slow things up [so that] you can start enjoying the acclaim and the revenue that comes with successful publication in a big journal.
>
> (Sismondo and Doucet, 2010, p. 281)

Documents made public as a result of litigation against Wyeth (now owned by Pfizer) show that the company used all or most of the aforementioned strategies, and especially ghostwritten articles, to promote its bestselling drug for hormone replacement therapy (*HRT*; or menopausal hormone therapy) (Fugh-Berman, 2010, 2013). HRT became popular in recent decades following approval for treatment of symptoms of menopause (e.g., hot flashes), and it was soon promoted as also being effective in the prevention of cardiovascular disease, osteoporosis, Alzheimer's disease, Parkinson's disease, colon cancer, age-related macular degeneration, tooth loss, and wrinkles (Fugh-Berman, 2010). However, research found that HRT is not effective in preventing heart attacks, and actually increases the risk of breast cancer, stroke, dementia, and incontinence. While the toll of harm caused by HRT remains to be quantified, it is evident that ghostwriting contributed to its use by millions of women for whom there was no medical indication of need.

Ghostwriting is one of myriad forms of academic-industry and public-private partnerships that have become commonplace in healthcare and other spheres of life. These arrangements are frequently touted as win-win collaborations by politicians and industry spokespersons. Industry insiders, however, appear to be in no doubt about the true implications of such liaisons. For one former pharmaceutical executive it was obvious that academic entanglement with industry threatens "the academic mission" while being of immense benefit to industry (Fugh-Berman, 2013).

## 8.4    DISCLOSURE OF CONFLICTS OF INTEREST IN BIOMEDICAL RESEARCH

Just as nonreporting of conflicts of interest is common among authors of clinical practice guidelines (discussed in Chapter 7), so it is for authors and sponsors of clinical trials reported in scientific articles. In meta-analyses of pharmaceuticals published in high-impact biomedical journals, information concerning industry sponsorship and author conflicts of interest is rarely reported (Roseman et al., 2011, 2012). As with clinical guidelines, one obvious consequence of the ubiquitous presence of industry in healthcare is that the demand for experts free of industry entanglement cannot be met. In all major fields of contemporary biomedical research, seasoned researchers almost invariably have a history of industry entanglement. Moreover, concerns have been raised that anyone possessing relevant expertise and experience sufficient to warrant serving in an expert capacity is susceptible to a host of potential *secondary interests* related to the particularities of their professional affiliation and research interests, including intellectual investment in a particular research outcome, academic advancement, and competition with others for research funding (Eccles et al., 2012; Norris et al., 2012). As such, expert advice can rarely be said to be free of a host of potential biases, both subtle and not so subtle.

In common with customs surrounding the preparation of clinical guidelines, it is widely presumed that requiring conflicted persons, either voluntarily or by statute, to disclose potential conflicts helps to safeguard against undue influence in the processes of biomedical research. Precisely what corrective effect such disclosures are expected to have has never been properly explained. Given that physicians appear unable to reliably discount biased information about drugs received from company detailers, or even the obviously misleading information contained in advertisements in the medical journals they read, there seems little reason to believe that they are capable of reliably discounting information contained in scientific articles written by authors who disclose industry affiliations. As for patients, they have little choice but to rely on physicians' interpretations of trial results, interpretations which in turn are subject to influence by individual physician entanglement with industry. Mindful of the many flaws in current disclosure procedures, it appears that most commentators believe that concerted efforts should be made to maximize disclosure. Although few would see this as the complete solution to industry-related bias in biomedical research and practice, most appear to believe that maximum disclosure of potential conflicts of interest can only be beneficial.

However, studies show that disclosure may not merely fail in its intended effect of enabling expert opinion to be weighted in proportion to the extent of the source's potential conflicts, but disclosure may actually have "perverse" negative effects (Cain et al., 2005). To begin with, it has been found that people generally do not discount to the extent that they should advice received from a biased source, even when the source discloses conflicts of interest. Moreover, once a conflict of interest has been disclosed, any initial bias the person giving the advice may have is likely to be *exaggerated* following the disclosure (Cain et al., 2005; Loewenstein et al., 2011, 2012). The first of two mechanisms involved is *strategic* exaggeration, which refers to the tendency of the advisor (e.g., a medical expert with industry ties) to provide more biased advice to counteract anticipated discounting by those who receive the advice after being informed that it has come from a conflicted source. Secondly, disclosure may lead to *moral licensing*, which is the feeling, often unconscious, of self-justification to not moderate biased advice because those receiving it have been warned.

Furthermore, due to a mechanism referred to as *insinuation anxiety*, disclosure can lead to *increased* pressure on patients to comply with advice when they would not have done so had there been no disclosure (Loewenstein et al., 2011, 2012). For example, a patient may be invited to participate in the clinical trial of a new drug. In the absence of disclosure, a patient may decline the invitation for a variety of reasons, including being satisfied with the treatment currently being received. However, if in compliance with considered best practice the physician discloses being in receipt of a fee for each successful referral to the trial, the patient might be pressured into accepting due to worry that not to so do could be misinterpreted as disapproval of the doctor's pecuniary interest in the patient's decision. Irrespective of whether or not the patient actually disapproves, the patient may consent despite preferring not to. More generally, this process has been referred to as the *burden of disclosure*, which involves two contradictory effects wherein the patient may have decreased trust in the advice received while simultaneously feeling under increased pressure to comply with the distrusted advice (Sah et al., 2013).

Considering the evidence, although calls for disclosure of conflicts of interest saturate the medical research and publishing worlds, such appeals should be seen either as disingenuous or naïve. On one hand, guidelines concerning disclosure are widely flouted, which merely exacerbates the plight of consumers who may be misled (due to non-disclosure of conflicts) into believing that information received has come from a neutral source. On the other hand, disclosure has little effect on the biasing tendency of conflicted persons other than to encourage *more* bias and is of little help to the consumer who is left not knowing whether information received is actually biased, and if so, to what degree.

None of this implies that disclosure is poor policy. On the contrary, the public has a right to know of potential conflicts of interest that may bias information that physicians receive or that may bias the advice that physicians in turn pass onto patients. Disclosure, however, offers poor, if any, protection, especially for patients as end users of what may be a long chain of information dissemination and advice creation. The solution to such problems is not to forsake disclosure or conversely to tighten disclosure rules, but to eliminate bias at source. Above all, this requires the elimination of self-serving entanglement between biomedical healthcare researchers and industry, but unfortunately there is essentially no prospect of anything of that kind happening soon.

Depending on one's perspective, industry entanglement with biomedical healthcare reads as tragedy or farce, and possibly both. Contemporary healthcare is mired in interests that conflict sharply with the health and safety of the people it is expressly committed to serve. If a way could be found to remove compromising industry influences from biomedical healthcare, the outcome would be a profoundly altered form of healthcare better suited to optimizing personal and population health than that which is currently practiced. Sadly, industry-initiated entanglement finds a receptive audience among biomedical researchers, as well as clinical practitioners and the professional bodies that represent them. It is executed for the mutual benefit of those parties. From industry's perspective, contact with healthcare personnel is designed to incentivize patterns of behavior that boost sales, and benefit to patients, if any, is coincidental. Moreover, any such benefit is more than offset by harm, including harm to patients from being prescribed drugs they do not need and harm to everyone from the mounting costs of healthcare.

## REFERENCES

Almashat, S., Wolfe, S., 2012. Pharmaceutical Industry Criminal and Civil Penalties: An Update. Public Citizen, Washington, DC.

Als-Nielsen, B., Chen, W., Gluud, C., Kjaergard, L.L., 2003. Association of funding and conclusions in randomized drug trials: a reflection of treatment effect or adverse events? JAMA 290, 921–928.

Andersen, M., Kragstrup, J., Søndergaard, J., 2006. How conducting a clinical trial affects physicians' guideline adherence and drug preferences. JAMA 295, 2759–2764.

Anderson, T.S., Dave, S., Good, C.B., Gellad, W.F., 2014. Academic medical center leadership on pharmaceutical company boards of directors. JAMA 311, 1353–1355.

Angell, M., 2009. Drug companies & doctors: a story of corruption. N. Y. Rev. Books 56, 8–13.

Arrowsmith, J., Harrison, R., 2012. Drug repositioning: the business case and current strategies to repurpose shelved candidates and marketed drugs. In: Barratt, M.J., Frail, D.E. (Eds.), Drug Repositioning: Bringing New Life to Shelved Assets and Existing Drugs. John Wiley & Sons, Inc., Hoboken, NJ, pp. 9–32.

Bero, L., Oostvogel, F., Bacchetti, P., Lee, K., 2007. Factors associated with findings of published trials of drug–drug comparisons: why some statins appear more efficacious than others. PLoS Med. 4, 1001–1010.

Bosch, X., Esfandiari, B., McHenry, L., 2012. Challenging medical ghostwriting in US courts. PLoS Med. 9, e1001163.

Braillon, A., 2012. Drug industry is now biggest defrauder of US government. Br. Med. J. 344, 1. http://dx.doi.org/10.1136/bmj.d8219.

Cain, D.M., Loewenstein, G., Moore, D.A., 2005. The dirt on coming clean: perverse effects of disclosing conflicts of interest. J. Leg. Stud. 34, 1–25.

Campbell, E., Moy, B., Feibelmann, S., et al., 2004. Institutional academic industry relationship: results of interviews with university leaders. Account. Res. 11, 103–118.

Campbell, E.G., Weissman, J.S., Ehringhaus, S., et al., 2007. Institutional academic-industry relationships. JAMA 298, 1779–1786.

Cowley, A.J., Skene, A., Stainer, K., Hampton, J.R., 1993. The effect of lorcainide on arrhythmias and survival in patients with acute myocardial infarction: an example of publication bias. Int. J. Cardiol. 40, 161–166.

Department of Health and Association of British Pharmaceutical Industries, 2010. Moving beyond sponsorship: joint working between the NHS and pharmaceutical industry. Accessed 11 November 2014 from: http://webarchive.nationalarchives.gov.uk/20130107105354/http://www.dh.gov.uk/en/Publicationsandstatistics/Publications/PublicationsPolicyAndGuidance/DH_082840.

DiMasi, J.A., Feldman, L., Seckler, A., Wilson, A., 2010. Trends in risks associated with new drug development: success rates for investigational drugs. Clin. Pharmacol. Ther. 87, 272–277.

Dinc, I.S., 2000. Bank reputation, bank commitment, and the effects of competition in credit markets. Rev. Financ. Stud. 13, 781–812.

Eccles, M.P., Grimshaw, J.M., Shekelle, P., et al., 2012. Developing clinical practice guidelines: target audiences, identifying topics for guidelines, guideline group composition and functioning and conflicts of interest. Implement. Sci. 7, 1–8. http://dx.doi.org/10.1186/1748-5908-7-60.

Echt, D.S., Liebson, P.R., Mitchell, L.B., et al., 1991. Mortality and morbidity in patients receiving encainide, flecainide, or placebo. N. Engl. J. Med. 324, 781–788.

Freshwater, D.M., Freshwater, M.F., 2011. Failure by deans of academic medical centers to disclose outside income. Arch. Intern. Med. 171, 586–587.

Friedberg, M., Saffran, B., Stinson, T.J., et al., 1999. Evaluation of conflict of interest in economic analyses of new drugs used in oncology. JAMA 282, 1453–1457.

Fries, J.F., Krishnan, E., 2004. Equipoise, design bias, and randomized controlled trials: the elusive ethics of new drug development. Arthritis Res. Ther. 6, R250–R255.

Fugh-Berman, A.J., 2010. The haunting of medical journals: how ghostwriting sold "HRT" PLoS Med. 7, e1000335.

Fugh-Berman, A., 2013. How basic scientists help the pharmaceutical industry market drugs. PLoS Biol. 11, e1001716.

Golder, S., Loke, Y.K., 2008. Is there evidence for biased reporting of published adverse effects data in pharmaceutical industry-funded studies? Br. J. Clin. Pharmacol. 66, 767–773.

Gøtzsche, P.C., 2012. Corporate crime in the pharmaceutical industry is common, serious and repetitive. Br. Med. J. 345, e8462.

Gøtzsche, P.C., Jørgensen, A.W., 2011. Opening up data at the European Medicines Agency. Br. Med. J. 342, d2686.

Guyatt, G.H., Oxmanc, A.D., Montorid, V., et al., 2011. GRADE guidelines: 5. Rating the quality of evidence—publication bias. J. Clin. Epidemiol. 64, 1277–1282.

Hart, B., Lundh, A., Bero, L., 2012. Effect of reporting bias on meta-analyses of drug trials: reanalysis of meta-analyses. Br. Med. J. 344, d7202.

Heres, S., Davis, J., Maino, K., et al., 2006. Why olanzapine beats risperidone, risperidone beats quetiapine, and quetiapine beats olanzapine: an exploratory analysis of head-to-head comparison studies of second-generation antipsychotics. Am. J. Psychiatr. 163, 185–194.

Hill, K.P., Ross, J.S., Egilman, D.S., Krumholz, H.M., 2008. The ADVANTAGE seeding trial: a review of internal documents. Ann. Intern. Med. 149, 251–258.

Horton, R., 2004. The dawn of McScience. N. Y. Rev. Books 51, March 11, 7–9.

Horton, R., 2014. Why medicine is killing our universities. Lancet 384, 117.

Jefferson, T., Jones, M.A., Doshi, P., et al., 2012. Neuraminidase inhibitors for preventing and treating influenza in healthy adults and children. Cochrane Database Syst. Rev. 1, 1–226. http://dx.doi.org/10.1002/14651858.CD008965.pub3.

Jorgensen, P.D., 2013. Pharmaceuticals, political money, and public policy: a theoretical and empirical agenda. J. Law Med. Ethics 41, 561–570.

Jørgensen, A.W., Hilden, J., Gøtzsche, P.C., 2006. Cochrane reviews compared with industry supported meta-analyses and other meta-analyses of the same drugs: systematic review. Br. Med. J. 333, 782.

Jørgensen, A.W., Maric, K.L., Tendal, B., et al., 2008. Industry-supported meta-analyses compared with meta-analyses with non-profit or no support: differences in methodological quality and conclusions. BMC Med. Res. Methodol. 8, 1–7. http://dx.doi.org/10.1186/1471-2288-8-60.

Kesselheim, A.S., Wang, B., Studdert, D.M., Avorn, J., 2012. Conflict of interest reporting by authors involved in promotion of off-label drug use: an analysis of journal disclosures. PLoS Med. 9, 1–9. http://dx.doi.org/10.1371/journal.pmed.1001280.

Landa, A.S., Elliott, C., 2013. From community to commodity: the ethics of pharma-funded social networking sites for physicians. J. Law Med. Ethics 41, 673–679.

Lemmens, T., 2013. Pharmaceutical knowledge governance: a human rights perspective. J. Law Med. Ethics 41, 163–184.

Lenzer, J., 2005. What can we learn from medical whistleblowers? PLoS Med. 2, 583–586. http://dx.doi.org/10.1371/journal.pmed.0020209.

Lexchin, J., 2012. Those who have the gold make the evidence: how the pharmaceutical industry biases the outcomes of clinical trials of medications. Sci. Eng. Ethics 18, 247–261.

Lexchin, J., Bero, L.A., Djulbegovic, B., Clark, O., 2003. Pharmaceutical industry sponsorship and research outcome and quality: systematic review. Br. Med. J. 326, 1167–1170.

Liberati, A., Magrini, N., 2003. Information from drug companies and opinion leaders. Br. Med. J. 326, 1156–1157.

Light, D.W., 2006. Basic research funds to discover important new drugs: who contributes how much? In: Burke, M.A. (Ed.), Monitoring the Financial Flows for Health Research 2005: Behind the Global Numbers. Global Forum for Health Research, Geneva, Switzerland, pp. 27–43.

Light, D.W., Lexchin, J., 2012. Pharmaceutical research and development: what do we get for all that money? Br. Med. J. 344, e4348.

Light, D.W., Lexchin, J., Darrow, J.J., 2013. Institutional corruption of pharmaceuticals and the myth of safe and effective drugs. J. Law Med. Ethics 41, 590–600.

Loewenstein, G., Cain, D.M., Sah, S., 2011. The limits of transparency: pitfalls and potential of disclosing conflicts of interest. Am. Econ. Rev. 101, 423–428.

Loewenstein, G., Sah, S., Cain, D.M., 2012. The unintended consequences of conflict of interest disclosure. JAMA 307, 669–670.

Lundh, A., Sismondo, S., Lexchin, J., et al., 2012. Industry sponsorship and research outcome. Cochrane Database Syst. Rev. 12, 1–88. http://dx.doi.org/10.1002/14651858.MR000033.pub2.

Macklis, R.M., 1993. The great radium scandal. Sci. Am. 269, 94–99.

Madigan, D., Sigelman, D.W., Mayer, J.W., et al., 2012. Underreporting of cardiovascular events in the rofecoxib Alzheimer disease studies. Am. Heart J. 164, 186–193.

McGauran, N., Wieseler, B., Kreis, J., et al., 2010. Reporting bias in medical research—a narrative. Trials 11, 1–15. http://www.trialsjournal.com/content/11/1/37.

Melander, H., Ahlqvist-Rastad, J., Meijer, G., Beermann, B., 2003. Evidence b(i)ased medicine-selective reporting from studies sponsored by pharmaceutical industry: review of studies in new drug applications. Br. Med. J. 326, 1171–1173.

Moffatt, B., Elliott, C., 2007. Ghost marketing: pharmaceutical companies and ghostwritten journal articles. Perspect. Biol. Med. 50, 18–31.

Moodie, R., Stuckler, D., Monteiro, C., et al., 2013. Profits and pandemics: prevention of harmful effects of tobacco, alcohol, and ultra-processed food and drink industries. Lancet 381, 670–679.

Moynihan, R., 2003. Who pays for the pizza? Redefining the relationships between doctors and drug companies. 2: disentanglement. Br. Med. J. 326 (7400), 1193–1196.

Moynihan, R., 2012. Forget sponsorship and free trips—welcome to Pharmacare. Br. Med. J. 344, d8316.

Norris, S.L., Burda, B.U., Holmer, H.K., et al., 2012. Author's specialty and conflicts of interest contribute to conflicting guidelines for screening mammography. J. Clin. Epidemiol. 65, 725–733.

Perlis, C.S., Harwood, M., Perlis, R.H., 2005. Extent and impact of industry sponsorship conflicts of interest in dermatology research. J. Am. Acad. Dermatol. 52, 967–971.

Rising, K., Bacchetti, P., Bero, L., 2008. Reporting bias in drug trials submitted to the Food and Drug Administration: review of publication and presentation. PLoS Med. 5, 1561–1570.

Ritter, J.M., Harding, I., Warren, J.B., 2009. Precaution, cyclooxygenase inhibition, and cardiovascular risk. Trends Pharmacol. Sci. 30, 503–508.

Rodwin, M.A., 2013. Five un-easy pieces of pharmacological policy reform. J. Law Med. Ethics 41, 581–589.

Roseman, M., Milette, K., Bero, L.A., et al., 2011. Reporting of conflicts of interest in meta-analyses of trials of pharmacological treatments. JAMA 305, 1008–1017.

Roseman, M., Turner, E.H., Lexchin, J., et al., 2012. Reporting of conflicts of interest from drug trials in Cochrane reviews: cross sectional study. Br. Med. J. 345, http://dx.doi.org/10.1136/bmj.e5155.

Ross, J.S., Hill, K.P., Egilman, D.S., Krumholz, H.M., 2008. Guest authorship and ghostwriting in publications related to rofecoxib: a case study of industry documents from rofecoxib litigation. JAMA 299, 1800–1812.

Safer, D.J., 2002. Design and reporting modifications in industry-sponsored comparative psychopharmacology trials. J. Nerv. Ment. Dis. 190, 583–592.

Sah, S., Loewenstein, G., Cain, D.M., 2013. The burden of disclosure: increased compliance with distrusted advice. J. Pers. Soc. Psychol. 104, 289.

Schott, G., Pachl, H., Limbach, U., et al., 2010. The financing of drug trials by pharmaceutical companies and its consequences Part 2: a qualitative, systematic review of the literature on possible influences on authorship, access to trial data, and trial registration and publication. Dtsch. Arztebl. Int. 107, 295–301.

Seife, C., 2015. Research misconduct identified by the US food and drug administration: out of sight, out of mind, out of the peer-reviewed literature. JAMA Intern. Med. 175, 567–577. http://dx.doi.org/10.1001/jamainternmed.2014.7774.

Shaw, G.B., 1909. The Doctor's Dilemma: Preface on Doctors. Retrieved from 24 October 2013 from: http://www.online-literature.com/george_bernard_shaw/doctors-dilemma/0/.

Simes, R.J., 1986. Publication bias: the case for an international registry of clinical trials. J. Clin. Oncol. 4, 1529–1541.

Sismondo, S., 2007. Ghost management: how much of the medical literature is shaped behind the scenes by the pharmaceutical industry? PLoS Med. 4, 1429–1433. http://dx.doi.org/10.1371/journal.pmed.0040286.

Sismondo, S., 2008a. How pharmaceutical industry funding affects trial outcomes: causal structures and responses. Soc. Sci. Med. 66, 1909–1914.

Sismondo, S., 2008b. Pharmaceutical company funding and its consequences: a qualitative systematic review. Contemp. Clin. Trials 29, 109–113.

Sismondo, S., Doucet, M., 2010. Publication ethics and the ghost management of medical publication. Bioethics 24, 273–283.

Smith, R., 2005. Medical journals are an extension of the marketing arm of pharmaceutical companies. PLoS Med. 2, 364–366.

Spielmans, G.I., Thielges, S.A., Dent, A.L., Greenberg, R.P., 2008. The accuracy of psychiatric medication advertisements in medical journals. J. Nerv. Ment. Dis. 196, 267–273.

Spielmans, G.I., Biehn, T.L., Sawrey, D.L., 2010. A case study of salami slicing: pooled analyses of duloxetine for depression. Psychother. Psychosom. 79, 97–106.

Studdert, D.M., Mello, M.M., Brennan, T.A., 2004. Financial conflicts of interest in physicians' relationships with the pharmaceutical industry—self-regulation in the shadow of federal prosecution. N. Engl. J. Med. 351, 1891–1900.

Tanne, J.H., 2011. Merck pays $1 bn penalty in relation to promotion of rofecoxib. Br. Med. J. 343, 1122.

Topol, E.J., 2004. Failing the public health—rofecoxib, Merck, and the FDA. N. Engl. J. Med. 351, 1707–1709.

Turner, E.H., Matthews, A.M., Linardatos, E., et al., 2008. Selective publication of antidepressant trials and its influence on apparent efficacy. N. Engl. J. Med. 358, 252–260.

Villanueva, P., Peiró, S., Librero, J., Pereiró, I., 2003. Accuracy of pharmaceutical advertisements in medical journals. Lancet 361, 27–32.

Yank, V., Rennie, D., Bero, L.A., 2007. Financial ties and concordance between results and conclusions in meta-analyses: retrospective cohort study. Br. Med. J. 335, 1202–1205.

Zetterqvist, A.V., Mulinari, S., 2013. Misleading advertising for antidepressants in Sweden: a failure of pharmaceutical industry self-regulation. PLoS One 8 (5), e62609.

# Chapter 9

# The Charms and Harms of Personalized Medicine[1]

There is something about personal health services that captures the public imagination.

(Blumenthal, 2005, p. 734)

## Contents

---

1. A shorter version of the text of this chapter was published in the *European Journal of Epidemiology* (James, 2014).

The Health of Populations. http://dx.doi.org/10.1016/B978-0-12-802812-4.00009-6

Following completion of the Human Genome Project in 2003 there was considerable optimism about an impending "paradigm shift" in healthcare from conventional to *personalized medicine* (Weston and Hood, 2004). The "revolution" in personal genomics was heralded as the essential catalyst that would transform healthcare from the traditional universalist *one size fits all* approach to personalized healthcare epitomized by the delivery of *the right drug to the right patient at the right time* (Grosse and Khoury, 2006; Khoury et al., 2010; Offit, 2011; Venter, 2010). Though no agreed precise definition has been formulated, personalized medicine, also called *precision medicine*, broadly refers to customized healthcare founded on individualized genomic risk information (biomarkers) referenced against population genomic (biobank) data to prevent, diagnose, and treat disease (National Cancer Institute, 2013).

Even before its completion, the Human Genome Project had created high expectations of transformational gains for health. The belief was widespread that tests and cures informed by personal genomics would soon be available for most diseases (Melzer and Zimmern, 2002). In 2000, the then President of the United States, Bill Clinton, announced that

> *humankind is on the verge of gaining immense, new power to heal. Genome science will have a real impact on all our lives ... It will revolutionize the diagnosis, prevention, and treatment of most, if not all, human diseases.*
>
> (Collins, 2010, p. 674)

Francis Collins, geneticist and director of the National Institutes of Health in the United States from 1993 to 2008, stood beside President Clinton in the East Room of the White House when the President outlined the promising future that lay ahead for genomic medicine. Earlier, Collins (1999) wrote that sequencing of the human genome meant that it would be possible to "understand the literal Book of Life" (p. 36).

Expectations of a rapid revolution, however, faded fairly quickly. Still hopeful a decade later, Collins (2010) expressing a somewhat less expansive view than the one he reported previously, wrote that although the

> *promise of a revolution in human health remains quite real ... The consequences for clinical medicine ... have thus far been modest [and] it is fair to say that the Human Genome Project has not yet directly affected the health care of most individuals (p. 674).*

Similarly, Venter (2010), geneticist, entrepreneur, and prominent advocate of genomic medicine, who had stood with Collins alongside President Clinton 2000, later wrote

> *there is still some way to go before this capability [i.e., sequencing of the human genome] can have a significant effect on medicine and health (p. 676).*

One purported area of special promise for personalized medicine is *pharmacogenomics*, wherein *genomics* and other *omic* technologies (notably, *transcriptomics*, *proteomics*, and *metabolomics*) are used to study the role of inheritance in individual variation in responses to therapeutic drugs with the aim of maximizing efficacy and minimizing adverse reactions. The suffix *-ome* in biology is used to refer to the totality of something, and *-omic* refers to the relevant field of study. Therefore, *genome* refers to the totality of the genetic material of an organism and *genomics* refers to scientific study of the structure, function, and expression of the genome. The *transcriptome* is the total messenger RNA (***mRNA***) in a cell or organism. *mRNA* is the form of ***RNA*** that carries information from ***DNA*** (the "genetic code") to cell sites for protein synthesis in a process called ***translation***. *Proteome* is a portmanteau of *protein* and *genome*, and refers to the full set of expressed proteins, which are responsible for diverse biological functions in a given cell type at a given time. Whereas *proteomic* analysis reveals the set of gene products produced in a cell, *metabolome* refers broadly to the set of metabolic end products of gene ***transcription*** present in a cell, tissue, or organism.

## 9.1  WHY IS GENOMIC MEDICINE SAID TO BE PERSONALIZED?

It is not obvious why the term *personalized* was appropriated on the promise of new advances in genomic-based medicine. From Hippocratic times, at least, medicine has been personal and individual, even if the respective emphasis on medical universalism (the "one size fits all" approach) versus specificity (individually tailored intervention) has fluctuated over the course of history (Tutton, 2012). Indeed, much of the debate concerning evidence-based medicine, discussed in Chapter 10, can be seen as reflecting tensions between those favoring a degree of medical universalism and those defending specificity. Resistance to evidence-based medicine has largely centered on physician perceptions that evidence-based clinical guidelines encourage practices that are overly universal and unaccommodating of clinical judgment intended to take account of individual patient characteristics and needs. It is unclear whether such tensions will be aggravated or relieved by the form of personalized clinical practice that genomic medicine promises. In any event, although medicine has always been personal and individualized, the fashion for individualized healthcare is possibly stronger today than ever before. This appears to be due in part to increased use of technology-intensive medical practices of which genomic medicine is a prime example.

In that context, the designation, *personalized*, to distinguish genomic trends in medicine from conventional medicine, can be attributed largely to *genomic exceptionalism*: the belief that genetic information is uniquely important (Kitsios and Kent, 2012; Melzer and Zimmern, 2002). Indications of exceptionalism can be found in much of the narrative of personalized medicine, especially in relation to the marketing of direct-to-consumer genetic testing, wherein personal genomics is offered as the means for achieving hitherto unattainable understanding of personal health and susceptibility to disease. This presumed privileged understanding has been described as *hegemonic* (Turkheimer, 2012) in the sense that genomic-based claims to truth, at least at the time when enrapture over the Human Genome Project was at its height, assumed such authority as to dominate all competing claims. Exceptionalist appeals form part of what has been referred to as the *rhetoric of empowerment* of personalized medicine, which promises unprecedented control by patients over their individual healthcare needs (Juengst et al., 2012). Some measure of the hyperbole surrounding technology-intensive medical innovation can be gauged from the words of one would-be marketplace leader that "patients of tomorrow will be the CEOs of their own health" (NDRC, 2015).

Expectations of high profit from personalized medicine have encouraged intense commercialization and strenuous competition for a place in the emerging market. Curiously, proponents omit to explain how genomic insights empower patients any more than other diagnostic and clinical information, including the plethora of conventional technologies such as blood-sample analysis and biological imaging. Technology-intense information, whether genomic or conventional, usually requires expert explanation to render it meaningful, and without meaning it cannot be individually empowering. Given that genomic information typically requires relatively high levels of expert explanation for it to be meaningful to laypersons, personalized medicine is likely to leave patients more reliant on authority and less personally empowered than when using conventional medicine.

In fact, personalized medicine, as currently construed, would be more accurately described as *individualized*, and in that regard it may be grouped with most conventional medicine which routinely deals with patient details that are both personal and individual. Thus, it is difficult to escape the conclusion that the appropriation of *personalized* is indicative of beliefs that genomic information is exceptionally and uniquely important. This, in fact, appears to be part of a wider malaise that Horton (2014) wrote about recently in the *Lancet*, where he charged contemporary biomedicine with claiming

> its authority through the ideology of technoevangelism. Health is today defined almost exclusively in individual terms, which has reached its apotheosis in the hubris of personalized medicine.
>
> (Horton, 2014, p. 218)

## 9.1.1 Biological Exceptionalism and "Ultimate" Causes

Discussions about behavioral and environmental influences on health and disease are frequently premised on the assumption that such processes have a biological foundation and therefore biological understanding takes precedence. Indeed, it may be difficult for those steeped in biological science to accept any premise other than that biology is the basis of human agency. In particular, the genome and the brain are widely cited as being preemptively fundamental. Moreover, it is notable that instances of biological exceptionalism are not limited to the discipline of biology. For example, almost every introductory textbook of psychology and many advanced texts contain a chapter titled "The biological basis" of some domain of human psychology such as behavior, personality, or psychopathology. The same books, however, typically do not contain a chapter titled the "The environmental basis" or "The behavioral basis" of biological processes. However, whereas biology is essential to human agency, the inference that biology is *the* fundamental process is false.

Although discussions about "ultimate" causes probably have little meaning, it is important that one set of causal processes (in this instance, biology) is not presumed to have exceptional importance over other key causal processes (e.g., the environment and behavior). In fact, if arguments about ultimate causes have any meaning, they are easily settled by asking one question: What came first? There is but one answer: irrespective of whether the putative basic process relates to genes, the brain, or some other aspect of biology, the *environment* came first. In the beginning, there was the physical universe, which in time included planet earth. From that environment, emerged the biology and behavior of initial primitive life forms and ultimately all life. If precondition is a measure of importance or "basicness," then the *environment*, the ultimate precondition, must be deemed the one true "basis" of life and health. Biology and behavior came later, coevolving in necessary symbiosis.

## 9.1.2   Public Funding for Health Research

The ascendance of technology-intensive biomedicine is reflected in the priorities of public funding for research. Recent examples relate to funding policies of the European Union. A prominent feature of European research policy over the past couple of decades is the emphasis placed on scientific research in the service of *economic growth* founded on *innovation in technology*. For example, *Horizon 2020*, the current EU Framework Program for Research and Innovation, with a publicly funded budget of almost USD100 billion over 7 years (2014-2020), has set as its main priorities "discoveries and world-firsts" for the "market" that "drive economic growth" (http://ec.europa.eu/programmes/horizon2020/en/what-horizon-2020). Economic competitiveness may be a suitable basis for shaping research priorities in some fields of science, including, for example, areas of physical science and engineering. It is far from obvious, however, that the same technology-based economic priorities (e.g., European commission, 2014) should guide health research. Nevertheless, it is commercial, rather than health, priorities that characterize the European research effort, evidenced in official publications describing the level of Europe's financial commitment to private-sector industry to "boost the competitiveness of health-related industries and businesses" (http://ec.europa.eu/research/health/medical-research/index_en.html).

One recently announced initiative, to be undertaken jointly with the pharmaceutical industry, received public funding of more than USD2 billion, approximately one-quarter of the budget in Horizon 2020 that had been allocated for health research (Galsworthy et al., 2014). The priorities of the initiative relate to development of biotechnology with an emphasis on personalized medicine. From an *economic* perspective, that investment may or may not yield dividends, depending on the profitability of the planned exploitation of personalized medicine. From a *health* perspective, however, there is little reason to be optimistic. Whereas proponents of personalized medicine have argued that it possesses unprecedented potential to prevent disease and deliver population-wide gains in health (Hood and Friend, 2011; Khoury et al., 2012; Weston and Hood, 2004), the available evidence suggests the opposite. For diverse reasons discussed below, personalized medicine threatens to do substantially more harm than good for the health of populations.

## 9.2   WHAT IS THE EFFICACY OF PERSONALIZED MEDICINE NOW AND WHAT IS IT LIKELY TO BE?

Omic technologies are employed in much current drug-development work, including efforts aimed at improving diagnostics for selection and dosing in relation to currently available therapeutics (Squassina et al., 2010). An example of comparative success of that kind is the use of the antiviral drug abacavir for human immunodeficiency virus (HIV), which causes a serious hypersensitivity reaction in some patients. Arising from developments in pharmacogenomics, genetic testing has been used to identify patients with the gene variant associated with increased risk of abacavir-induced toxic effects. Offering alternative therapy to those patients has been found to reduce the incidence of hypersensitivity reaction by more than 50%. However, results of testing are not free from error, with some individuals who test positive proving not to be hypersensitive to abacavir (false positives) and others who test negative being hypersensitive (false negatives). Moreover, while genetic testing improves treatment outcomes, hypersensitivity reaction even without testing occurs in only a small minority (less than 10%) of patients treated with abacavir (Squassina et al., 2010).

Regardless of some examples of modest pharmacogenomic success, doubts exist about the extent to which personalized medicine can be truly personalized. While potentially contributing to the precision with which some clinical populations are stratified, there have been few indications of personalized medicine progressing beyond the subgroup level to the delivery of clinical options that are unique to individuals (Manolio et al., 2012). True specificity has been illusive, with risk profiles based on known susceptibility variants in the genome generally having a poor record of predicting individual patient variability in responses to therapy (Pashayan and Pharoah, 2012) or of differentiating between individuals in relation to the progression of common complex diseases such as coronary heart disease, cancer, and type 2 diabetes (Burke et al., 2014). Such are the limitations of the new technologies subsumed under the rubric of personalized medicine that some observers have recommended that genomic information be regarded not as the basis for transformative personalized care but as an additional (albeit imperfect and expensive) resource to be incorporated, along with existing diverse technologies, into conventional individualized medicine (Manolio et al., 2012).

## 9.2.1    Genetic Testing

Developments in genetic testing have contributed to high expectations about genomic personalized medicine. In part, confidence may be attributed to past success in screening for relatively rare *monogenic disorders* caused by modifications in a single gene. There are three main categories of monogenic disorders: recessive, dominant, and X-linked. Given that there are two copies (alleles) of each gene (one copy on each member of a chromosome pair), recessive disorders (e.g., cystic fibrosis, sickle-cell anemia) occur due to damage in both copies. Dominant monogenic disorders (e.g., Huntington's disease) are due to damage to only one gene copy, and X-linked disorders (e.g., fragile X syndrome, hemophilia) are due to damage to genes on the sex (X) chromosome. Success in identifying monogenic disorders encouraged the belief that many, if not most, inherited diseases would follow a similar pattern (Alpert and Chen, 2012).

Today, genetic tests are available for many monogenic disorders, where implementation is relatively straightforward in terms of diagnosis and counseling of patients and relatives (Khoury et al., 2013). However, monogenic disorders account for only a small proportion of mortality and morbidity. In the United States, for example, only about 2% of deaths are attributable to purely genetic diseases (McGinnis et al., 2002). The situation is very different in relation to testing for polygenic diseases, including common complex diseases. These are influenced by genetic factors involving the combined action of alleles of multiple genes. Here, the hereditary pattern is complex, because disease expression depends not merely on the simultaneous presence of several alleles, but also influence from behavioral and environmental factors. Genetic tests exist for some polygenic diseases (e.g., cancers), but their contribution to disease identification is typically small. Breast cancer diagnosis is one area sometimes celebrated as a notable success. However, testing of women, even after stratification for family history of breast cancer, contributes to identifiable genetic inheritance in fewer than 10% of cases (Gage et al., 2012). In those cases, up to 90% are attributable to just one gene variant (the BRCA mutation) which has a strong genetic effect (*high penetrance*). For most women, however, genetic testing for breast cancer is of little predictive value.

In the past decade, genome-wide association studies (*GWAS*) have been popular for examining whether common genetic variants in the population are associated with disease. In contrast to testing one or a few genes, GWAS examine the entire genome for variants in DNA referred to as single-nucleotide polymorphisms (SNPs, pronounced *snips*). From the millions of genetic variants that are revealed, those that are statistically *associated* with disease occurrence are used to identify regions of the human genome that may contribute to risk of disease. Despite GWAS having identified many gene variants associated with common complex diseases, individual variants almost invariably have small statistical effect-size, weak genetic effect (*low penetrance*), and low predictive validity (Bush and Moore, 2012). Consequently, with respect to cardiovascular disease, the main cause of death globally, only a small fraction of the heritability can be reliably explained on the basis of genetic biomarkers (Prins et al., 2012; Zeller et al., 2012).

Similarly, in cancer diagnostics and therapeutics, GWAS have yielded numerous associations generally having small effect-size of limited utility for predicting response to pharmacotherapy, and contributed little to individualized clinical management (Patel et al., 2013). Similarly, GWAS have identified numerous genetic variants associated with risk of type 2 diabetes, but these too have been of limited use in predicting individual risk of the disease even when testing is limited to high-risk individuals (Lyssenko and Laakso, 2013). Furthermore, from a lifecourse perspective, it is notable that human longevity is not compromised by the cumulative effect of alleles shown by GWAS to be associated with increased risk for coronary artery disease, cancer, and type 2 diabetes (Beekman et al., 2010). Even the inclusion in prediction models of as many as 20 independently inherited risk alleles for diabetes was found to produce "only minimal improvement" in the accuracy of predicting diabetes (Talmud et al., 2010). Thus, a recurring conclusion of many empirical studies is that current genetic risk models add little to prediction based on conventional clinical risk factors.

With respect to diabetes, it has been remarked that the added discriminative accuracy contributed by genetic models, even after taking account of all recently identified genetic variants, is slightly better than "flipping a coin" (Lyssenko and Laakso, 2013). One explanation of this poor performance is that known clinical risk factors such as obesity and elevated blood glucose levels are themselves substantially inherited. Consequently, the routine inclusion of these risk factors in conventional risk models limits the scope for further improvement in prediction, despite extensive scrutiny of the genome. While the record to date does not exclude the possibility of important future gene-variant discoveries, there appears to be agreement that major advance in risk prediction, if achievable at all, will require extraordinary levels of investment in human and material resources (Bush and Moore, 2012; Khoury et al., 2013).

In the meantime, considering the disappointingly low predictive validity of current genetic testing for common complex diseases and substantial uncertainty about prospects for major improvements in the future, knowledge of individual genetic risk profile has been touted as being useful, if not for prediction, then as a source of motivation to encourage test-positive individuals to commit more strenuously to relevant disease prevention efforts than they would without such knowledge (McBride et al., 2010). The available evidence, however, does not support that belief. For example, a recent study examined the benefit of genetic risk testing and counseling as part of a diabetes prevention program (Grant et al., 2013). Two groups of overweight participants who tested high and low, respectively, for genetic risk of diabetes received genetic counseling consistent with their test results. A third group of similarly overweight participants was not tested for genetic risk and received no genetic counseling. All three groups participated in a behavioral intervention program for prevention of diabetes, which was successful in producing weight loss in all three groups. However, irrespective of assessed level of genetic risk and whether or not participants received genetic counseling, there were no substantive differences between the groups in adherence to the intervention program or its efficacy.

Rather than encouraging greater personal commitment to disease prevention, it is conceivable that under some circumstances knowledge of individual genetic risk could have the opposite effect due to the operation of *risk compensation*. This refers to a tendency for people to adjust their behavior according to perceived level of risk, wherein higher perceived risk encourages more carefulness and lower perceived risk encourages less carefulness. Risk compensation is a potential, although not inevitable, consequence whenever preventive measures are undertaken to improve safety. There is speculation about its possible occurrence in diverse settings, including road safety (Wilde et al., 2002), HIV prevention (Cassell et al., 2006), human papillomavirus vaccination for cervical cancer (Marlow et al, 2009), and helmet use by skiers and snowboarders (Sulheim et al., 2006) and cyclists (Fyhri et al., 2012; Messiah et al., 2012). Similarly, it is conceivable that anticipated protection from an allegedly "healthy" genome could contribute to lower levels of engagement in readily accessible and genuinely preventive health-promoting habits such as eating nutritious foods and regular physical activity.

## 9.3   TO SCREEN OR NOT TO SCREEN? WHY IS THAT A QUESTION?

*The over-use of drugs is … a near-inevitable consequence of mass screening.*
(Rose, 1992, p. 112)

Screening involves the testing of healthy people to detect those at increased risk of disease. The high intuitive appeal of screening probably derives in part from realization that it is usually advisable for *manifest* disease in *individuals* to be treated earlier rather than later, and generally the earlier the better. As such, it is tempting to conclude that early detection and treatment of *possible* disease in *populations* is also advisable. However, although screening can have benefits, those must be weighed against the fact that screening "always causes harm" (Gøtzsche, 2015). Unfortunately, the compelling intuitive appeal of screening has for decades contributed to its widespread adoption in anticipation of benefit, often with relatively little consideration of harm.

Criteria for identifying when to conduct screening have long been agreed (Wilson and Jungner, 1968), and include:

- the condition screened should be an important health problem;
- there should be an accepted treatment for patients with the screened disease;

- facilities for diagnosis and treatment should be available;
- there should be a recognizable latent or early-symptomatic (predisease) stage;
- there should be a suitable test or examination; and
- there should be good understanding of the natural history of the condition, including development from the latent stage to manifest disease.

Notwithstanding agreement about operational criteria, based on an overview of the utility of mass screening, Harris (2011) suggested that

> the payoff for population health would be greater if we shifted some of the resources we now devote to screening to developing, testing, and implementing alternative approaches to prevent the important threats to population health.

(Harris, 2011, p. 5)

Harris' comments allude to potential inefficient use of resources, and certainly testing large numbers of people to identify a small number of cases can risk being cost-ineffective. This is especially so in developing countries where the infrastructure for testing, not to mention the resources needed to treat identified cases, may be insufficient to provide adequate coverage of the population. However, cost is usually not the main concern, and in high-income countries where most screening takes place, cost is rarely the critical factor in decisions about whether or not to implement mass screening.

## 9.3.1   Some Basics About Screening

Some, but by no means all, of the problems of mass screening would be solved if biomedical tests discriminated perfectly between those who have early signs of disease and those who do not. Alas, no diagnostic test has achieved that level of perfection. One reason for diagnostic imperfection is that disease is usually continuous and not dichotomous (discussed further in Chapter 12) and therefore disease is not necessarily easily categorized as absolutely present or absent. Two commonly-used indicators of the ability of tests to discriminate between disease presence and absence are sensitivity and specificity. Figure 9.1 shows that *sensitivity* (sometimes referred to as *detection rate*) is the proportion of identified genuine cases (*true positives*) among all cases of the disease in the screened population, including cases that are missed (*false negatives*). *Specificity*, on the other hand, is the proportion of genuine noncases (*true negatives*) among all noncases, including those falsely identified as diseased (*false positives*).

High sensitivity minimizes failures to detect disease when it is present, and high specificity minimizes false alarms. Importantly, the relationship between sensitivity and specificity is reciprocal in that precision in one can be boosted by sacrificing precision in the other, and vice versa. That is, relaxing the criteria for inferring disease presence decreases failures to detect disease but increases the number of false detections, whereas imposing more stringent criteria for inferring disease presence has the opposite effect (increases failures to detect

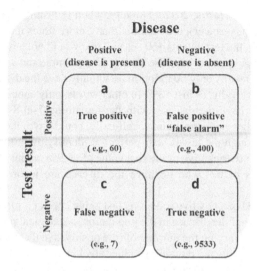

**Disease**

**Positive** (disease is present) | **Negative** (disease is absent)

Test result — Positive / Negative

| | Positive (disease is present) | Negative (disease is absent) |
|---|---|---|
| **Positive** | **a** True positive (e.g., 60) | **b** False positive "false alarm" (e.g., 400) |
| **Negative** | **c** False negative (e.g., 7) | **d** True negative (e.g., 9533) |

Worked example based on 10,000 people who have been screened for disease (note that fractions have been converted to percentages):

$$\text{Sensitivity} = \frac{a}{a+c} = \frac{60}{60+7} = 90\%$$

$$\text{Specificity} = \frac{d}{b+d} = \frac{9533}{400+9533} = 96\%$$

$$\text{Positive predictive value} = \frac{a}{a+b} = \frac{60}{60+400} = 13\%$$

$$\text{Negative predictive value} = \frac{d}{c+d} = \frac{9533}{7+9533} = 99.9\%$$

**FIGURE 9.1**   Computational basis for evaluating the validity of a screening test. *(The example is adapted from Woolf and Harris (2012).)*

but decreases the number of false detections). Achieving the right balance in diagnostic criteria is partly informed by the relative seriousness of the two types of error (false negatives and false positives) shown in the two-by-two table in Figure 9.1. Failing to detect genuine cases (false negatives) could be serious if treatment is available that is safe and effective against the disease when given early. Conversely, detecting disease where none exists (false positives) could be serious if it leads to further testing and unnecessary treatment, either or both of which may physically harm the patient, not to mention the needless worry that may be caused.

Some key principles for assessing the performance of screening tests are illustrated in Figure 9.1, which shows a worked example for a test possessing 90% sensitivity and 96% specificity used to screen for a condition with a prevalence of 60 cases per 10,000 people screened. That level of test accuracy is better than most screening tests, and that level of prevalence is typical of some cancers (Woolf and Harris, 2012). Notably, when returning a positive finding (e.g., the person has cancer), the result is many more times incorrect than correct. Specifically, the test detects 460 "cases" that would not have been detected without screening, but most are false alarms. The example shows that among the 460 positive test results only 60 individuals actually have the disease, yielding a positive predictive value of just 13%. In other words, only about 1-in-8 who test positive actually have the disease, with the remaining 7-in-8 being informed incorrectly that they have (or, more specifically, could have) the disease. Additionally, even though the test is near-perfect at identifying individuals who do not have the disease (negative predictive value of 99.9%) about 10% of genuine cases (i.e., 7 out of 67) who do have the disease are incorrectly given the all clear.

The likelihood of a positive screening test being correct is markedly affected by the prevalence of the disease in the population, a statistical phenomenon that can be verified by varying the number of true positives in the example shown in Figure 9.1 and deriving the remaining values algebraically. In the worked example, prevalence is 0.6% of the population (i.e., 60 true cases per 10,000 of the population). Imagine, however, that we use the same test having 90% sensitivity and 96% specificity to screen for a disease that has a prevalence of 10% (i.e., 1000 cases per 10,000). The positive predictive value would be 74%, meaning that 3-in-4 who test positive do in fact have the disease (compared to about 1-in-8 in the previous example). Thus, when disease prevalence is low, the proportion of false-positive test results inflates dramatically, creating all of the attendant problems of further testing, possible unnecessary treatment, and worry for the proportionately large number who test positive but are healthy. Therefore, even for a comparatively good test, such as the one in the worked example, there is the real prospect that screening will do more harm than good.

Part of the challenge to rational evaluation of screening is that people, both lay and expert, tend to focus on the *presumed* rather than *actual* benefit of early detection of true cases, while also giving insufficient thought to potential harm for large numbers of individuals incorrectly diagnosed (Baum, 2013; Gøtzsche et al., 2012; Woloshin et al., 2012; Woolf and Harris, 2012). A crucial consideration is the extent to which early diagnosis is truly beneficial in terms of the length and quality of life for genuine cases. If existing interventions are only modestly effective in reducing disease morbidity and mortality, early diagnosis from screening may be overshadowed by the long period during which quality of life is reduced due to knowledge of the disease and negative side effects from treatment. Paradoxically, effective interventions under some circumstances can also reduce the potential benefit from screening. Specifically, if interventions exist that are effective in relieving symptoms and controlling disease progression after the disease is well-established, then the importance of detecting disease early is diminished.

Natural history of a disease is an important factor influencing the ratio of benefit to harm. In the event of slow-progressing diseases, individuals may derive no benefit from early diagnosis because the disease may not progress sufficiently to cause serious discomfort during the person's lifetime. Indeed, one important difference between *conventional diagnostic testing of individual patients* and *screening for disease in the general population* is that patients in effect have already "screened" themselves. Patients present with symptoms which function analogously to screening and guide clinical testing to confirm the presence or absence of disorders consistent with the symptom presentation. Symptom pattern provides clues that improve the reliability of testing and diagnosis, and symptom presence guides clinical decisions about suitable interventions. Much as a symptom may lead a patient to self-refer, positive results from population screening are taken as need for follow-up consultation, additional testing, and possible intervention.

However, it must always be kept in mind that people who test positive at screening are typically free of symptoms and may remain so for extended periods without intervention. Therefore, any discomfort (despite possible benefit) from interventions they may be offered for their "disorder" should take account of their current symptom-free state. Accordingly, compared to clinical testing of patients who present with discomfort from existing symptoms, screening of healthy people demands a higher standard of proof of benefit, which is often not met (Gøtzsche, 2015). The proliferation of harm from mass screening has been due substantially to insufficient appreciation of the risks involved for those who have the disease, but especially for those who do not.

### 9.3.2   The Elusive Benefits and Substantial Harms of Screening

Much contemporary screening is concerned with noncommunicable diseases that are common causes of death, and a key question is whether mass screening of healthy adults for these diseases is effective in reducing mortality. Recently, Saquib et al. (2015) conducted an extensive review of meta-analyses and individual trials in order to compare mortality between screened and nonscreened randomized groups. *Disease-specific mortality* was defined as death attributed to the disease that had been screened and *all-cause mortality* was defined as death from any cause. The review focussed on cardiovascular diseases, cancers, chronic obstructive pulmonary disease, and diabetes, and found that among currently available screening tests for those diseases reductions in disease-specific mortality were "uncommon" and reductions in all-cause mortality were "very rare or nonexistent." In other words, the screening programs that for decades have been a prominent feature of contemporary healthcare and that are widely believed to contribute substantially to the saving of life appear to be largely ineffectual in that regard.

Apart from indicating limited benefit for mortality, the Saquib et al. (2015) study also serves to highlight an important methodological issue relevant to discussion below about screening for specific diseases. It has been usual to regard screening as successful if it is followed by reduced mortality from the disease that has been screened, but it is now recognized that disease-specific mortality can be misleading if interpreted without taking account of all-cause mortality. Evaluations of the effectiveness of screening for cancer, for example, have tended to focus on mortality outcome from the specific cancer that was screened, with decreased mortality attributable to that specific cancer taken as evidence of successful screening. However, if all-cause mortality (inclusive of death due to the specific cancer and all other causes) is examined and found not to have decreased, rate of death due to causes other than the specific cancer must have increased. This could occur, for example, if early diagnosis followed by early intervention, including chemotherapy and radiation therapy, decreased cancer deaths while causing an increase in, say, cardiovascular death.

To illustrate with an extreme hypothetical example, imagine that cancer chemotherapy prescribed for patients who tested positive for cancer susceptibility caused cardiovascular death in all cases before the suspected cancer had time to progress to a fatal stage. Measuring disease-specific mortality alone (in this case cancer death) could suggest that screening had been 100% successful. That is, no screened patients died from cancer, even though all had their life shortened due to intervention-induced cardiovascular disease. In general, then, when mortality is the main outcome of concern (which is most often the case), it is crucial to take account of all-cause mortality in order to assess the possibility that early intervention for identified disease susceptibility is not merely reducing disease-specific (e.g., cancer) mortality at the cost of increased mortality from other causes (e.g., cardiovascular death). Taken together, results for cause-specific and all-cause death in this hypothetical instance would indicate that rather than being 100% successful, early detection followed by intervention had failed completely.

### 9.3.2.1 Cervical Cancer

Of the many targets of mass screening, cervical cancer is at present among the least controversial. There is strong evidence that use of the Papanicolaou test or smear (*Pap smear*) with or without human papillomavirus cotesting has contributed substantially to reductions in mortality from cervical cancer (e.g., Dickinson et al., 2014; Peirson et al., 2013; Sasieni et al., 2009). The general consensus is for regular testing of woman aged 30 years and above (to 65 or 70 years). Unfortunately, few studies have reported all-cause mortality, and therefore it is difficult to gauge the total benefit (Saquib et al., 2015). Overall, however, mass screening for cervical cancer appears to be successful.

### 9.3.2.2 Breast Cancer

Of all screening interventions, that for breast cancer has possibly been the most strongly contested (Taylor, 2015). Screening for breast cancer with mammography (X-ray examination) aims to detect lesions before a lump can be felt, with the goal of treating the cancer early when cure is thought to be more likely. Early evaluations, conducted in the 1970s and 1980s, concluded that screening for breast cancer saves lives, and many countries introduced or expanded the practice (Taylor, 2015). Later evaluations, however, including a series undertaken at the Nordic Cochrane Centre have been less positive. An earlier publication in the series concluded that breast cancer screening with mammography is "unjustified" (Gøtzsche and Olsen, 2000) and successive comprehensive reviews have concluded similarly (Gøtzsche and Jørgensen, 2013). The latest review included results from randomized controlled trials involving 600,000 women in the age range 39-74 years. Rate of mortality, including breast cancer mortality after 10 years and all-cause mortality after 13 years, revealed no effect in favor of screening.

Bleyer and Welch (2012) examined trends from 1976 to 2008 in incidence of early-stage and late-stage breast cancer among women aged 40 years and older in the United States. The authors pointed out that for breast cancer screening to have been successful there should have been two major changes in the population pattern of disease incidence over the period: incidence of early-stage cancers should have increased (because presymptomatic disease was being detected that would not have been detected without screening), and there should have been a concomitant decrease in late-stage cancers (because early intervention should have arrested disease at an early stage). The first condition was met, evidenced by a greater than twofold increase in the number of cases of early-stage breast cancer detected each year. However, the second condition was not met in that incidence of late-stage cancer decreased only marginally. Bleyer and Welch estimated that almost one-third of detected breast cancers are attributable to overdiagnosis. During the 30-year period covered by the study, it was estimated that 1.3 million women were diagnosed as having breast cancer who would never have experienced clinical symptoms if their "disorder" had gone undiagnosed.

Many countries invite women to attend for breast mammography screening annually or biennially, and each testing contributes to the cumulative risk of returning a false-positive result. The risk of false positives varies markedly between countries, ranging from 1-in-5 to more than half of women tested having a cumulative risk of receiving a positive test result at least once in 10 screening rounds (Brodersen and Siersma, 2013). Whether accurate or not, breast cancer diagnosis is responsible for a substantial burden of harm. Overdiagnosed women undergo a host of interventions, including surgery, radiation therapy, hormonal therapy, chemotherapy, and often a combination of two or more such interventions for detected abnormalities that otherwise would never have manifested as symptomatic disease (Bleyer and Welch, 2012; Gøtzsche and Jørgensen, 2013). Even women with a positive test result, who, on further testing, are declared free of suspected cancer, and therefore do not undergo treatment, have been found to experience negative psychosocial consequences (Brodersen and Siersma, 2013). Negative effects on wellbeing within the initial 6 months of a false alarm were found to be as pronounced as for women with confirmed breast cancer. Although initial negative psychosocial effects of a false alarm subsided to a degree over time, the women who had had a false-positive result continued to report lower levels of wellbeing even after 3 years compared to women who had never tested positive.

### 9.3.2.3 Prostate Cancer

Some of the same problems concerning breast cancer screening in women have also been present in relation to prostate cancer screening in men. Prostate cancer is frequently *indolent*, meaning that it often progresses slowly and causes little discomfort (Hoffman, 2012). Slightly less than 1% of men worldwide die from prostate cancer, although the rate is higher in high-income countries where it approximates 3% (WHO, 2014). Incidence of high-risk disease is higher in older men, and most deaths occur among older men. Comorbid medical conditions are also more prevalent in later life and can limit both tolerance of and benefit from aggressive treatment for prostate cancer. As for breast cancer, screening has been advocated as a means of detecting prostate cancer in the early stages when it is reputed to be amenable to interventions intended to arrest its progression. However, although findings have been discrepant, there is little compelling evidence of overall benefit from screening for prostate cancer.

Relevant evidence from screening for prostate cancer using serum prostate-specific antigen (*PSA*) was largely unavailable until as recently as 2009 (Andriole et al., 2009), despite PSA screening having been in widespread use throughout the previous 2 decades (Legler et al., 1998). An early randomized trial in the United States evaluated the effectiveness of PSA screening and digital rectal examination on deaths from prostate cancer and from all causes (Andriole et al., 2009). After 7-10 years follow-up, rate of death from prostate cancer and from all causes did not differ between groups of men who had been screened repeatedly and those who had not been screened.

Dubben (2009) similarly found little evidence to support prostate-cancer screening. Whereas any benefit to mortality from correct diagnosis happens only after a long delay, if at all, harmful effects from false-positive diagnosis and overdiagnosis occur almost immediately. Moreover, Dubben estimated that trials need to include more than 200,000 men to meet minimum statistical requirements. Additionally, he estimated that trials need to run for 15 years or more, during which time the relevance of findings may be undermined by changes in population composition, diagnostic practices, and treatment options. For those reasons, Dubben concluded that it is unrealistic to expect to be able to demonstrate benefit from screening for prostate cancer, and that the same might also be true for other diseases with low specific mortality.

Those conclusions are consistent with successive reviews of randomized controlled trials conducted in North America and Europe, which reported little or no reduction in prostate cancer-specific or all-cause mortality following PSA screening with or without digital examination (Chou et al., 2011; Djulbegovic et al., 2010; Ilic et al., 2013). One exception to these otherwise consistent findings is a recent report of findings from a European randomized controlled trial with 13-year follow-up of PSA screening (Schröder et al., 2014). A modest statistically significant reduction in prostate cancer mortality was reported, although the clinical significance of that finding is ambiguous in that no difference was observed for all-cause mortality between screened groups and those not screened. Additionally, it was concluded that further quantification of harms was needed. The consistent conclusion of other reviewers has been that substantial harm from overdiagnosis and overtreatment provides strong grounds against recommending routine screening for prostate cancer (Andriole et al., 2009; Chou et al., 2011; Ilic et al., 2013). As outlined in Chapter 2, even when prostate cancer is present, much harm is caused by treating men who would have remained largely asymptomatic and in whom prostate cancer would not have been detected in their lifetime had it not been for screening (Andriole et al., 2009).

### 9.3.3    Genetic Screening

The preceding discussion about screening provides relevant context for considering the implications of calls for screening services to be expanded to accommodate developments in personal genomics and personalized medicine. Limited effectiveness and harm from conventional screening such as discussed above has led many to look to genomics as a promising new avenue for improving disease prediction and prevention (e.g., Chowdhury et al., 2013; Khoury et al., 2007, 2011a; Schmalfuss and Kolominsky-Rabas, 2013). However, initial high expectations of transformational change, which at one time appeared close at hand, have receded. In the same year that the Human Genome Project was completed, geneticist, Muin J. Khoury, founding director of the Office of Public Health Genomics at the United States Centers for Disease Control, wrote:

> *Over the next decade or two, it seems likely that we will screen entire populations or specific subgroups for genetic information in order to target interventions to individual patients that will improve their health and prevent disease [including] individual susceptibility to common [complex diseases] such as heart disease, diabetes, and cancer.*

(Khoury et al., 2003, p. 50)

Within a decade of that prediction, however, Khoury reversed his position, writing:

> *For common complex diseases ... genomic information ... is generally not useful for diagnosis and individual prediction.*

(Khoury et al., 2013, p. 437)

In general, the prospects for early detection and prevention of disease as part of personalized medicine appear to have receded in proportion to growth in the number of new genetic tests. Routine screening for disease in newborns has been a particularly active area of expansion. The first successful screening for disease in newborns began more than a half-century ago. In a landmark discovery, American microbiologist, Robert Guthrie, developed a simple screening test for the disorder *phenylketonuria*, a rare inherited inability to metabolize the essential amino acid phenylalanine (Guthrie, 1961). Phenylketonuria is effectively treated by dietary restriction and supplementation, but, if untreated, it leads to profound developmental problems, including intellectual disability. Screening involves assaying phenylalanine in heel-prick blood samples (Guthrie test) dried on filter paper (Guthrie cards), which can be mailed to a central laboratory for analysis. Development of such a simple and effective method demonstrated the feasibility of mass screening and marked the beginning of an era of enthusiasm for the discovery of other disorders for screening, a tradition invigorated by the more recent advent of genetic testing.

More than 1800 tests for genetic diseases are available (Khoury et al., 2011b). Although screening policies differ between countries, the tendency has been for testing to expand as new tests are developed. Despite the absence of evidence of benefit, due partly to most of the disorders being rare, expansion of testing has continued (Wilcken and Wiley, 2015). Advances in technology and rapidly falling cost have led to suggestions for whole-genome sequencing of newborns (e.g., Landau et al., 2014). Although possibly of some clinical benefit to a small number of cases, such moves would have far-reaching negative consequences (Burke et al., 2011; Lohn et al., 2014; Miller et al., 2009), including:

- decreased cost-effectiveness due to diminished reliability in detecting disorders of polygenic origin;
- increased overdiagnosis and overtreatment;
- proliferation of incidental findings, including possible genetic susceptibilities to common complex disorders that may not emerge until adulthood, or never;
- diverse findings of unclear significance that create long-term uncertainty and anxiety within families; and
- in later years, long-term uncertainty and anxiety in those who tested positive as infants.

Notwithstanding the formidable challenges and threats associated with genetic screening of newborns, alleged success in that area has prompted calls for expansion of screening for adults (Caskey et al., 2014). Of particular concern is harm likely to accompany applications of genetic screening intended to *prevent* disease in healthy people. Overconfidence in the new technologies, and intense commercialization, threaten to produce excess diagnoses of *predisease* (Moynihan et al., 2013; Viera, 2011) and *nondisease* (Smith, 2002), which are detectable "conditions" that may never progress beyond potentiality. As discussed in Chapter 7, the term *disease mongering* has been used to refer to the marketing of interventions for alleged diseases in healthy people (Moynihan et al., 2012). Ill-founded optimism in genomic personalized medicine threatens to increase the demand for screening and thereby contribute substantially to disease mongering. Specifically, early diagnosis of reputed disease indications is likely to encourage wider use of interventions, including surgical excision of tissue (e.g., prostate or breast) and long-term pharmacotherapy for conditions that may never cause substantial discomfort if left alone.

## 9.4    THE GENOME IS BUT ONE PIECE OF THE HETEROGENEOUS MOSAIC OF LIFE

The vision of personalized medicine making major contributions to the prevention, diagnosis, and treatment of disease faces major challenges from the fact that genes do not act in isolation. Rather, disease and health are the result of complex interactions between networks of genomic and epigenomic processes, behavior, and the environment. Whereas variants in DNA are responsible for genetic effects, *epigenetics* refers to biochemical processes that influence gene function, including gene expression, in the absence of variation in DNA sequence. *Gene expression* is the manifestation of observable characteristics (the *phenotype*) of the individual, and the process can be turned "on" or "off," facilitated or impeded, by epigenetic processes. In turn, both genomic and epigenomic processes are known to be influenced by behavioral patterns and the environment (i.e., habits and habitats) during the lifecourse (Abouheif et al., 2014; Aguilera and García-Muse, 2013; Bizzarri et al., 2013; Braveman and Gottlieb, 2014; Feinberg and Irizarry, 2010; Frank, 2010; Issa, 2014).

Thus, whereas genotype has traditionally been regarded as conferring a more-or-less immutable phenotype, gene expression is now known to be subject to a variety of influences. Such influences, it seems, extend even to biomedical intervention through a process called phenoconversion, evidenced by a mismatch between an individual's genotype and their observed responses (Shah and Shah, 2012). Drug-induced phenoconversion has been reported for individually-prescribed drugs and drug combinations. The taking of one drug, for example, may alter a patient's expected reaction, based on known genotype, to a second drug. Consequently, even with firm data associating a particular genotype with specific clinical responses, the genomic focus of personalized medicine may lead to clinical failure due to phenoconversion (Shah and Smith, 2015). Under those circumstances, rather than realizing the promise of personalized medicine to deliver "the right drug to the right patient at the right time," pharmacotherapy-induced phenoconversion could result in some patients being unresponsive and others being harmed by drugs and drug doses that have been specifically recommended on the basis of individual genomic profile.

Recognition of the various factors that complicate disease heritability and response to clinical intervention has led to suggestions that a *systems biology* approach (discussed in Chapter 12) should be adopted that integrates extensive molecular, behavioral, and environmental information, with the aggregated whole being labeled *P4 medicine*, meaning predictive, personalized, preventive, and participatory (Hood and Friend, 2011; Weston and Hood, 2004). However, even that aggregation has been found wanting, leading to the proposal that the elements of P4 medicine should be infused with a "population perspective" (Khoury et al., 2012). It is speculated that this new conglomeration, *P5 medicine*, will enable healthcare to become simultaneously personalized and integrated into preventive public health practice.

## 9.4.1    Big Data

The main strategy that is recommended for realizing the vision of personalized medicine, especially the reputed preventive capabilities of P5 medicine, is to amass data of gargantuan proportions. Proponents envision a future in which

> *each patient will be surrounded by a "virtual cloud" of billions of data points that will uniquely define their past medical history and current health status [and advanced technologies will be developed] to mine the billions of data points from hundreds of millions of individuals.*

(Hood and Friend, 2011, p. 185)

To obtain the requisite data, it is proposed that multicenter multinational epidemiologic studies of unprecedented scale should be conducted with the assistance of immense public investment.

The collection, storage, and retrieval of the data, not to mention its interpretation, present additional mammoth human and material challenges (Manolio et al., 2012). Undaunted, optimism runs high that these challenges can be overcome, as is reflected in the following:

> *Big data in medicine - massive quantities of health care data accumulating from patients and populations and the advanced analytics that can give those data meaning - hold the prospect of becoming an engine for the knowledge generation that is necessary to address the extensive unmet information needs of patients, clinicians, administrators, researchers, and health policy makers.*

(Krumholz, 2014)

Disconcertingly, much of the fabled promise of personalized medicine is founded on the fragile belief that the problem of individual prediction will be solved by integrating ever larger banks of multilevel information (Hood and Friend, 2011; Khoury et al., 2012). Even if practicable, the mining of patient "data clouds" is likely to prove disappointingly unproductive for both clinical medicine and epidemiology.

The analysis of ever larger populations into ever more-finely layered strata, down to the theoretical limit of the individual, will indeed guarantee discovery of large numbers of associations between indicators of disease and genomic, behavioral, and environmental variables. However, large numbers of associations should not be confused with precision. Most of the anticipated newly discovered associations are destined to have small effect-size and low predictive validity (Turkheimer, 2012). Experience shows that multiple associations at the subgroup level, let alone at the level of the individual, tend not to be additive and do not replicate well between studies. Moreover, observed associations cannot be assumed to be anything other than just that, namely, *associations*. It is unrealistic to expect more than a tiny fraction of the immense number of associations emerging from patient "virtual data clouds" to be causal. Even then, prediction is substantially undermined by background random events.

### 9.4.1.1  That Pesky Problem of Chance

Cause-effect relationships at the individual level exhibit substantial stochastic (i.e., random or chance) variability that undermines individual and even subgroup predictive validity. Such randomness is intrinsic to biological and behavioral systems, and occurs at all levels of organization from the subcellular to the biographical (Abouheif et al., 2014; Bizzarri et al., 2013; Davey Smith, 2011; Qian, 2013; Turkheimer, 2012). The stochastic nature of individual biobehavioral diversity has an evolutionary function in that it confers survival advantage for the population in the face of habitat change (Wagner, 2012). When aggregated at the population level, individual randomness yields to statistical analysis to reveal systematic patterns between broad risks (e.g., diet, physical activity, socioeconomic status) and outcomes for common complex diseases such as cardiovascular disease, cancer, and diabetes. However, while patterns of associations (so-called *emergent properties*) at the population level may be observed with unerring consistency, the stochastic nature of individual-level events renders individually specific prediction incorrigibly uncertain.

For example, there is a consistent population relationship between blood pressure level and cardiovascular disease such that, across the spectrum of blood pressure levels, groups with higher blood pressure consistently show more cardiovascular mortality and morbidity than groups with lower blood pressure (MacMahon, 2000; Lewington et al., 2002). On the other hand, although the statistical likelihood of disease for an individual of given blood pressure can be computed on the basis of population statistics, actual individual outcomes show variation around the statistical "average." Thus, many individuals having higher blood pressure than others escape disease while many with lower blood pressure do not. Therefore, just as biobehavioral events observed for a particular individual cannot be relied upon to reveal generalized patterns applicable to the entire population, consistent associations at the population level cannot be relied upon to predict individual cases.

Empirical evidence across a range of medical conditions indicates that prediction based on individual genomes compares poorly in clinical value to that of clinical risk markers (e.g., age, gender, smoking status, body mass index, blood glucose level) currently employed in conventional medicine (Smolen and Aletaha, 2013). Similarly, evidence does not support claims by some direct-to-consumer purveyors of personalized medicine that the use of individual genomics is helpful in the framing of lifestyle counseling for individuals. Personalized genomic information does not permit any more precise matching of advice about risk factors, including smoking, unhealthy diet, and physical inactivity, than is achieved by general advice about such risk factors drawn from existing health promotion guidelines that are directed at everyone within major population strata irrespective of personal genome (Ioannidis, 2010). Moreover, for the present at least, advice supplied by direct-to-consumer companies has been found to lack consistency (apart from questions of validity even if the advice were consistent). When identical DNA samples were sent to leading commercial suppliers, a comparison of genomic test results found that the information supplied about individual risk profiles was not merely without clinical value but that the level of risk reported differed markedly between the companies (Adams et al., 2013).

The role of chance in disease occurrence appears not to have been adequately appreciated by those who had hoped to develop precision genomic tools for disease prediction. Applying mathematical models, Tomasetti and Vogelstein (2015) found that, of *all* cancers, 65% are due to chance, or as they called it, "bad luck." These chance occurrences are due to random mutations during DNA replication in normal noncancerous stem cells. The lifetime risk of cancer arising at a particular site in the body is strongly correlated with the total number of self-renewing divisions that characterize cells of the tissue type at that site. Of the 35% of cancers not due to chance, 25-30% were found to be due to behavior and the environment, leaving only 5-10% due to genetic inheritance. Accordingly, the envisioned use of genomics in personalized medicine to precisely predict individual occurrences of cancer appears to be a distant prospect. Conversely, it appears that a substantial number of cancers, approximately one-quarter to one-third, may be avoided by minimizing behavioral and environmental risks.

### 9.4.1.2   The Inverse Care Law

Apart from a litany of formidable technical and clinical challenges, the problem of the resource-intensive nature of personalized medicine warrants attention in its own right. More than 40 years ago, British general practitioner, Julian Tudor Hart, wrote that the

> *availability of good medical care tends to vary inversely with the need for it in the population served.*

> (Hart, 1971, p. 405)

This pithy aphorism, which Hart named the *inverse care law*, was conceived not as a law of nature but as a heurist to elucidate aspects of healthcare delivery in the market economy. Considering the immense quantity of human and material investment projected to be necessary to successfully exploit relevant genomic and information technologies, the inverse-care-law heuristic provides insights

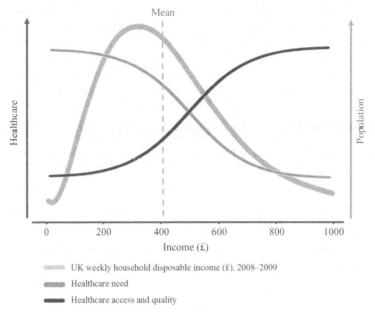

Mean

Healthcare

Population

0        200        400        600        800        1000

Income (£)

▭▭▭ UK weekly household disposable income (£), 2008–2009

▭▭▭ Healthcare need

▬▬▬ Healthcare access and quality

**FIGURE 9.2**   Schematic representation of Hart's (1971) inverse care law.

into possible future population harm from the promulgation of personalized medicine, especially preventive personalized medicine. Figure 9.2 is a schematic representation showing income distribution for the population as background for interpreting predictions of the inverse care law in the context of personalized medicine.

Although the specific income distribution in the figure is for the United Kingdom (Carrera and Beaumont, 2010), it is broadly representative of national incomes globally. That is, national income distribution tends to be skewed toward lower levels, with the majority of people having incomes below the mean and a minority forming a "tail" of higher income earners. Against that background, Figure 9.2 shows a progressively decreasing (*monotonic*) function representing healthcare need, which varies inversely with income wherein those on lowest incomes tend to be most needy. Additionally, the figure shows a progressively increasing function representing access and quality of healthcare. The inverse care law articulates the interaction between the two functions in that there tends to be less healthcare for those who need it most and more healthcare for those who are less needy (Hart, 1971).

To the extent that health expenditures draw upon finite resources, development of the genomic and information technologies that underpin personalized medicine will have the unintended consequence of diverting resources away from other healthcare priorities, exacerbating existing health inequalities. Moreover, the health inequalities predicted by the inverse care law are not likely to be limited to the high-income countries that are leading the development of personalized medicine. Given that income inequality in some low- and medium-income countries may exceed that in high-income countries (United Nations Development Programme, 2010), the contribution to health inequalities caused by the uptake of personalized medicine in poorer countries has the potential to exceed that experienced in richer countries. In general, access to and benefits (if any) from personalized medicine are likely to disproportionately favor the wealthy, who generally have greater access to resource-intensive care. In the event of the real possibility that the affluent who are able to access personalized medicine benefit little from it, then neither poor nor wealthy will have gained and both may be harmed.

### 9.4.1.3  The Inverse Benefit Law

As a logical extension of the inverse care law, Hart (1971) proposed that the predicted health inequalities associated with inequitable distribution of wealth are most evident

> where medical care is most exposed to market forces, and less so where such exposure is reduced.

(Hart, 1971, p. 405)

Four decades after Hart's original proposals, Brody and Light (2011) drew on his work as the conceptual foundation for their *inverse benefit law*, which states that

> the ratio of benefits-to-harms among patients [receiving drug treatment] tends to vary inversely with how extensively the drugs are marketed.

(Brody and Light, 2011, p. 399)

A schematic representation of the inverse benefits law, shown in Figure 9.3, parallels that for the inverse care law shown in Figure 9.2. In Figure 9.3, a

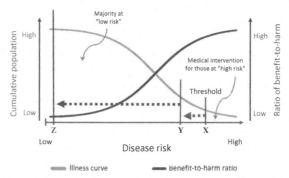

**FIGURE 9.3**    Schematic representation of Brody and Light's (2011) inverse benefit law. *Note*: The vertical lines marked "X," "Y," and "Z" indicate progressively more relaxed treatment thresholds (see text for detail).

disease axis signifying severity of risk factors, symptoms, and pathophysiological markers is intersected by a vertical line at "X," which represents an evidenced-based threshold at and above which treatment is recommended. It can be seen that, in general, and at any given time, a relatively small proportion of the population suffers disease while the majority are more-or-less disease free.

Because intervention always entails some risk of harm, an important question in healthcare concerns where on the disease axis should a given intervention be targeted to achieve most good representing the optimal trade-off between benefits and harms. This is that portion of the illness curve subtending what British epidemiologist Geoffrey Rose called *high-risk* individuals (Rose, 1985). These are the patients targeted by conventional biomedicine; the individuals to the right of the treatment threshold at "X" in Figure 9.3. This is the region of disease progression where the benefit-to-harm ratio is most favorable. Clinical guidelines, such as those pertaining to, for example, elevated blood pressure, blood glucose, and serum cholesterol, incorporate decision rules intended to maximize the ratio of benefits-to-harms. In contrast, *low-risk* individuals, those to the left of the treatment threshold, who are currently disease free but who may be at varying risk of developing disease, are not targeted for conventional biomedical intervention because this would expose them to an unacceptably poor benefit-to-harm ratio wherein harm from negative side effects exceeds benefit.

Although not evidencing manifest disease, low-risk individuals, as the label suggests, are nevertheless at some risk. It is precisely for that reason that preventive intervention exists. By lowering further the already low risk of the healthy undiagnosed population, extensive population benefit is achieved by way of further postponement or avoidance of disease. Indeed, as discussed in Chapter 4 and in more detail in later chapters, intervention to prevent disease in the low-risk (secular) population is responsible for greater gains in personal and population health than intervention in the high-risk (clinical) population. This, then, raises the question of what is the best strategy for optimizing health promotion and disease prevention in the low-risk population. There are two main strategies. On one hand, there are preventive interventions that aim to reduce health-related risk factors such as cigarette smoking, poor diet, physical inactivity, and harmful consumption of alcohol. On the other hand, proponents of preventive personalized medicine (including P4 and P5 medicine) envisage population prevention of disease through the use of personalized pharmacogenomics to arrest disease progression while individuals are still disease-free and at low risk.

Whereas Hart's (1971) inverse care law elucidates harm that exposure to market forces may have on the distribution of healthcare, its generalization in the form of the inverse benefit law (Brody and Light, 2011) illuminates market-related harm of a different kind, namely, the treatment of ever more healthy populations. The predictions of the inverse benefit law are supported by extensive empirical evidence of interventions shown to be efficacious for patients with clear signs of disease that are ineffective or harmful when persons with fewer signs of disease are administered the same interventions (Grady and Redberg, 2010). For cardiovascular disease, examples include diverse and common interventions such as aspirin (Kolber and Korownyk, 2014), statins (Redberg and Katz, 2012), coronary artery bypass graft (Maron and Ting, 2013), and carotid artery stenting (Abbott et al., 2013).

*Shifting thresholds.* In the pharmaceutical marketplace, commercial returns from shifts in treatment thresholds can be particularly alluring because the shape of the illness curve is such that a small relaxation of the threshold captures a disproportionately large number of new patients within the expanded area under the curve. The resulting commercial pressure, it has been claimed, is relentless (Moynihan et al., 2012), with "the market" forever seeking to shift the treatment threshold further to the left of the line marked "X" in Figure 9.3, for example, to "Y." The inverse care law predicts that relaxation of treatment thresholds is accompanied by deterioration in the benefit-to-harm ratio, wherein intervention produces progressively more harm relative to benefit.

By way of illustration, evidence supports the use of antiosteoporotic medications for reducing risk of bone fractures in postmenopausal women with osteoporosis (Eriksen et al., 2014). The usual definition of osteoporosis includes measured bone mineral density 2.5 or greater standard deviations below the population mean for healthy adult women, which for illustrative purposes can be equated with "X" (the treatment threshold) in Figure 9.3. In a move that would affect a large proportion of healthy postmenopausal women globally, pharmaceutical companies have sought to extend treatment to a new "disorder" known as *osteopenia* or *preosteoporosis* (Alonso-Coello et al., 2008), defined as bone mineral density 1.0-2.5 standard deviations below the mean. This relaxation of intervention threshold, represented by a shift from "X" to "Y" in Figure 9.3, would result in many millions of healthy women at low risk of bone fracture being offered unnecessary treatment. For little or no benefit, they would be put at risk of a variety of side effects, including diarrhea, vascular and neurological abnormalities, potentially serious gastrointestinal problems, and rare but catastrophic osteonecrosis (bone death due to poor blood supply) of the jaw (Alonso-Coello et al., 2008).

Even if new technologies associated with personalized medicine result in particular interventions being administered more precisely to selected patients to the right of "X" in Figure 9.3 (an aspiration yet to be achieved, apart from a few comparatively minor exceptions), creeping relaxation of treatment thresholds will create legions of patients who are harmed due to overdiagnosis and overtreatment. Ultimately, since everyone is at risk of developing disease at some time, conceivably everyone could be considered as being *at risk of being at risk* (Alonso-Coello et al., 2008). In that event, everyone to the left of "X" would receive a diagnosis warranting intervention indicated by a shift in treatment threshold to "Z."

## 9.4.2    At Risk of Being at Risk

The idea that entire populations might be encouraged to submit to biomedical interventions for disorders they do not have may appear fanciful, but it is a prospect that has long been seriously contemplated in the biomedical and pharmaceutical communities. It is a prospect that derives obvious impetus from the extraordinary commercial opportunities it offers. The manufacture of pharmaceuticals is among the most profitable of all industries (Angell, 2004). Extending use of "therapeutic" drugs to entire populations offers prospects for profit that are orders of magnitude greater than those currently achieved. Not surprisingly, two of the main contenders for population-wide use, statins and the "polypill," target cardiovascular disease, which, as we have seen in earlier chapters, is the main cause of death worldwide.

### 9.4.2.1    The Great Statin Controversy

Statins are widely prescribed for lowering cholesterol in patients with diagnosed cardiovascular disease. As discussed in Chapter 7, there is evidence of substantial industry involvement in new proposals to relax clinical guidelines that would encourage use of statins by a large proportion of the population who do not have cardiovascular disease but are deemed to be "at risk." The issue has been hotly debated (Abramson et al., 2013; Godlee, 2014a, b; Hawkes, 2014; Malhotra, 2013), with the *British Medical Journal* finding itself the object of criticism for allegedly promoting misinformation which then led to the journal initiating an independent formal inquiry (Heath et al., 2014). Were large profits to industry not at stake, it seems unlikely that the debate would have intensified to the extent that it has. Aside from large financial stakes, the intensity of the debate is out of all proportion to evidence of reputed population benefit from statins.

Notably, the argument for expanded use of pharmacotherapy is exceedingly slight when considered in the context of evidence of benefit from even modest improvements in health-related behavior. For example, in a recent Spanish study of patients at high cardiovascular risk, two versions of a Mediterranean diet (primarily, white meat, seafood, fruit, and vegetables), one supplemented with extra-virgin olive oil and the other supplemented with nuts, were compared to a control comprised of a low-fat diet (Estruch et al., 2013). So effective were the two Mediterranean diets in reducing incidence of cardiovascular events that the trial was stopped so as to no longer deprive participants in the control group of the benefits enjoyed by those in the other two groups. Given the alternatives, it is hard to imagine why health authorities would choose to promote wider use of statins, with all the attendant risks of negative side effects, over encouraging the uptake of a varied and nutritious diet, exemplified by the Mediterranean diet.

Possibly more revealing, a recent study modeled the effect on vascular mortality of prescribing daily statin use for everybody in the United Kingdom over age 50 years (Briggs et al., 2013). The mortality benefit was estimated to be no greater than that which would result from the same people consuming one extra apple a day. Not surprisingly, on all other relevant measures, including negative side effects and cost, apples consistently proved superior to statins.

### 9.4.2.2   The Polypill Puzzle

The polypill combines a variety of fixed-dose drugs, including at least one blood-pressure-lowering drug, a statin, and often aspirin, each of which is supported as being useful in the treatment of cardiovascular disease. In 2003, Nicholas Wald and Malcolm Law, at the Wolfson Institute of Preventive Medicine in London, proposed that rather than giving several drugs to treat a variety of risk factors, such as elevated blood pressure and cholesterol, individual drugs could be combined into a single combination drug, which they named the *polypill*. Although the idea of a polypill is not particularly striking in itself, Wald and Law (2003) made the decidedly radical claim that the polypill would largely eradicate "heart attacks and stroke if taken by everyone aged 55 years and older" irrespective of risk factors. They estimated that coronary events would be reduced by 88% and that one-third of people taking the pill would gain an average of 11 disease-free years of life. They asserted that

> the risk factors are high in us all, so everyone is at risk … there is much to gain and little to lose by the widespread use of these drugs. (p. 1422).

It is impossible to know whether self-interest played any part in this enthusiastic appraisal, but in a footnote, Wald and Law (2003) reported having filed a patent and trademark application on the polypill. A dozen years later, following much deliberation, research, and argumentation, initial claims and predictions about effectiveness, cost-effectiveness, and safety remain unsubstantiated. A recent comprehensive review found no evidence of reduced mortality and that reports of improvements in cardiovascular risk factors (e.g., blood pressure and cholesterol) "should be viewed with caution" (de Cates et al., 2014). Additionally, the polypill was found to be associated with nontrivial negative side effects. Even evidence of improved adherence to the polypill (which everyone tended to take for granted since taking one tablet should encourage better adherence than taking multiple tablets) has been disappointing, with one review reporting that the polypill "may" improve adherence (de Cates et al., 2014) and another reporting that improved adherence remains to be proven (Carey et al., 2012). A third study, however, reported poorer adherence for the polypill than for placebo (Charan et al., 2013).

Although evidence of cost-effectiveness of the polypill in primary prevention is lacking (de Cates et al., 2014), a recent Australian study involving a high-risk population of patients with metabolic syndrome found that the polypill was not a cost-effective intervention for reducing cardiovascular events (Zomer et al., 2013). Furthermore, warning of complications that can arise from long-term medication use, Julian and Pocock (2015) recounted the personal history of one of them (Desmond Julian), a cardiologist. He personally experienced "two potentially fatal events almost certainly due" to long-term use of a beta-blocker of the kind typically included in the polypill. Julian and Pocock warn that drugs that may be effective and safe in the short-term may not remain so in the long-term, especially for elderly patients who may have been prescribed the drugs many years previously and possibly many years before they were needed. This describes precisely what would be the situation if the polypill were used for purposes of primary prevention as envisaged by Wald and Law (2003).

Notwithstanding continuing interest in multidrug formulations for use with patients who have existing cardiovascular disease (e.g., Huffman and Yusuf, 2014), Wald may be alone in persisting with the "polypill concept" as mass primary prevention for the "general population [of] healthy people regardless of their blood pressure or cholesterol levels" (Wald and Wald, 2010). While suggesting that "[d]octors could *prescribe* such a polypill (off-label)," Wald and Wald lament that "off-label commercial *promotion* would be prohibited" (emphasis added). They offer a solution to the regulatory control they hope to avoid in the form of a proposal that pharmacists should be allowed to sell the polypill directly to consumers. Such a move would pave the way for direct-to-consumer marketing of the polypill concept, a strategy for multidrug medication of the masses that would almost certainly do little for health while wreaking substantial harm.

In contrast, consistent with suggestions concerning the advantages of healthier dietary choices over mass consumption of statins (Briggs et al., 2013), alternative advice to the taking of the polypill emerges from an examination of the physiology of physical exercise and the epidemiology of the benefits of regular physical activity. Specifically, it has been suggested that physical activity is the "real polypill," a drug-free intervention of proven effectiveness and safety that is consonant with the evolved human genome (Fiuza-Luces et al., 2013).

### 9.4.2.3    Preventive Personalized Medicine

The proposed use of statins and the polypill for primary prevention of cardiovascular disease in low-risk populations parallels much of the envisioned deployment of personalized medicine for common complex diseases. Accordingly, the challenges and limitations of the former portend a similar outcome for the latter, an outcome that Rose (1981) long ago foresaw when he warned:

> If a preventive measure exposes many people to a small risk, then the harm it does may readily … outweigh the benefits, since these [benefits] are received by relatively few … Consequently we cannot accept long-term mass preventive medication (p. 1850).

In contrast, population-wide intervention that encourages not smoking, for example, has benefits and no negative side effects. Similarly, not eating junk food is safe, promotes health, and prevents disease. Being moderately physically active has pervasive health benefits and minimal negative side effects, and so it is for other health-related behavior change. That is, the ratio of benefit-to harm, irrespective of health-risk status, is almost always positive when promoting health through lifestyle change. Conversely, pharmacological intervention always involves risk of negative side effects that outweigh benefit for the majority low-risk population. Consequently, with reference to Figure 9.3, pharmacological intervention is ill-advised for the majority not at high-risk, because for them the benefit-to-harm ratio from pharmacological intervention is almost always unfavorable.

## 9.5   GENOMIC PERSONALIZED MEDICINE: FUTURE PROSPECTS AND LEGACY

Personalized medicine faces many challenges, not least the massive investment of human and material resources to gather the vast individual and population data arrays reputed to be necessary for its implementation (Hamburg and Collins, 2010). There are many reasons to doubt the likelihood of alleged transformative benefits from personalized medicine, especially in relation to prevention of common complex diseases. Prominent among those reasons is the fact that current proposals for preventive personalized medicine contain no viable solutions for addressing confounding from disease-related individual molecular, behavioral, and environmental interactions, the complexities of which are increased by orders of magnitude due to stochastic (chance) influences at every level (Davey Smith, 2011; Ioannidis, 2010; Tomasetti and Vogelstein, 2015). Meaningful prediction is further undermined due to amplification of uncontrolled variability in the data from the long developmental latencies characteristic of common complex diseases.

Setting aside potential for physical harm due to overdiagnosis and overtreatment, it was recently suggested that concerns about personalized medicine may be exaggerated because of absence of confirmatory evidence from potential ethical, legal, and "social harm" (Caulfield et al., 2013). While it is legitimate to draw attention to the absence of evidence of specific sources of harm, the opposite question posed by Caulfield and colleagues may be more salient, namely: considering the resource-intensive nature of personalized medicine and the absence of evidence of clinical effectiveness, what evidence exists to suggest commensurate social good? Moreover, the claim of absence of evidence by Caulfield et al. is misleading because of the restricted definition of social harm that was employed. Specifically, relative absence of harm was argued in relation to whether gene patents limit genomic research, whether positive genetic-test results create anxiety, and whether those who test positive for genetic disorders experience insurance or ethnic discrimination. These few specific potential harms, each important in its own right, ignore psychosocial harm associated with the diverse physical harm for which considerable empirical evidence does exist. By definition, the sequelae of overdiagnosis and overtreatment, including pain, morbidity, lost productivity, and mortality, have substantial associated psychosocial harms. Therefore, while the precise extent and nature of harms from personalized medicine may be debated, there can be no doubt that psychosocial harm from personalized medicine, if widely implemented, would be considerable.

In summary, the overall potential for harm from genomic personalized medicine, especially that intended to prevent disease, entails no less than triple jeopardy. First, the extreme levels of human and material resources required by the genomic and informational technologies upon which personalized medicine is based threaten to disproportionately disadvantage those with greatest healthcare needs (inverse care law; Hart, 1971). Secondly, ill-founded notions of precision in genomic risk prediction threaten harm to ever larger populations of healthy individuals through overdiagnosis and overtreatment (inverse benefit law; Brody and light, 2011). Finally, there are opportunity costs arising from undue focus on genomic determinants and commensurate lack of attention to behavioral and social determinants.

Due to a mix of excessive confidence in genomic prediction, high expectations of benefits, perceived commercial opportunities, and insufficient attention to harmful consequences, personalized medicine has the potential to "colonize the future" of healthcare (Williams et al., 2011). That is, there is a risk that personalized medicine will continue to capture attention and resources at the expense of strategies that address important behavioral and social pathways capable of contributing substantially to personal and population health. Current evidence, illuminated by the inverse care law and the more recent inverse benefit law, suggests that any large-scale implementation of resource-intensive personalized medicine would in all likelihood do more harm than good to healthcare nationally and globally.

Finally, it should be acknowledged that the advances in knowledge arising from the Human Genome Project are impressive. However, the transformational change in biomedical healthcare envisioned by proponents of personalized medicine has not happened and almost certainly will not. Rather, awareness of the imperfections of genomic prediction has deepened in proportion to growth in knowledge of the genome. It is not the genome alone that determines health, but the dynamics of human habits and habitats comprised of complex interactions between the genome, behavior, and the environment. Arguably, the most important legacy of the Human Genome Project, with its immense investment of human and material resources, is its role in demonstrating the intrinsic limitations of genetic determinism.

## REFERENCES

Abbott, A.L., Adelman, M.A., Alexandrov, A.V., et al., 2013. Why calls for more routine carotid stenting are currently inappropriate: an international, multispecialty, expert review and position statement. Stroke 44, 1186–1190.

Abouheif, E., Favé, M.J., Ibarrarán-Viniegra, A.S., et al., 2014. Eco-evo-devo: the time has come. In: Landry, C.R., Aubin-Horth, N. (Eds.), Ecological Genomics: Ecology and the Evolution of Genes and Genomes. Springer, Dordrecht, Netherlands, pp. 107–125.

Abramson, J.D., Rosenberg, H.G., Jewell, N., Wright, J.M., 2013. Should people at low risk of cardiovascular disease take a statin? Br. Med. J. 347, 1–5. http://dx.doi.org/10.1136/bmj. f6123.

Adams, S.D., Evans, J.P., Aylsworth, A.S., 2013. Direct-to-consumer genomic testing offers little clinical utility but appears to cause minimal harm. N. C. Med. J. 74, 494–498.

Aguilera, A., García-Muse, T., 2013. Causes of genome instability. Annu. Rev. Genet. 47, 1–32.

Alonso-Coello, P., García-Franco, A.L., Guyatt, G., Moynihan, R., 2008. Drugs for pre-osteoporosis: prevention or disease mongering? Br. Med. J. 336, 126.

Alpert, J.S., Chen, Q.M., 2012. Has the genomic revolution failed? Clin. Cardiol. 35, 178–179.

Andriole, G.L., Crawford, E.D., Grubb III, R.L., et al., 2009. Mortality results from a randomized prostate-cancer screening trial. N. J. Engl. Med. 360, 1310–1319.

Angell, M., 2004. Excess in the pharmaceutical industry. Can. Med. Assoc. J. 171, 1451–1453.

Baum, M., 2013. Harms from breast cancer screening outweigh benefits if death caused by treatment is included. Br. Med. J. 346, 1–4. http://dx.doi.org/10.1136/bmj.f385.

Beekman, M., Nederstigt, C., Suchiman, H.E.D., et al., 2010. Genome-wide association study (GWAS)-identified disease risk alleles do not compromise human longevity. Proc. Natl. Acad. Sci. U. S. A. 107, 18046–18049.

Bizzarri, M., Palombo, A., Cucina, A., 2013. Theoretical aspects of systems biology. Prog. Biophys. Mol. Biol. 112, 33–43.

Bleyer, A., Welch, H.G., 2012. Effect of three decades of screening mammography on breast-cancer incidence. N. Engl. J. Med. 367, 1998–2005.

Blumenthal, D., 2005. Doctors and drug companies. N. Engl. J. Med. 352, 733–734.

Braveman, P., Gottlieb, L., 2014. The social determinants of health: it's time to consider the causes of the causes. Public Health Rep. 129 (Suppl. 2), 19–31.

Briggs, A.D., Mizdrak, A., Scarborough, P., 2013. A statin a day keeps the doctor away: comparative proverb assessment modelling study. Br. Med. J. 347, 1–6. http://dx.doi.org/10.1136/bmj.f7267.

Brodersen, J., Siersma, V.D., 2013. Long-term psychosocial consequences of false-positive screening mammography. Ann. Fam. Med. 11, 106–115.

Brody, H., Light, D.W., 2011. The inverse benefit law: how drug marketing undermines patient safety and public health. Am. J. Public Health 101, 399–404.

Burke, W., Tarini, B., Press, N.A., Evans, J.P., 2011. Genetic screening. Epidemiol. Rev. 33, 148–164.

Burke, W., Trinidad, S.B., Press, N.A., 2014. Essential elements of personalized medicine. Urol. Oncol. Semin. Orig. Investig. 32, 193–197.

Bush, W.S., Moore, J.H., 2012. Genome-wide association studies. PLoS Comput. Biol. 8, 1–11. http://dx.doi.org/10.1371/journal.pcbi.1002822.

Carey, K.M., Comee, M.R., Donovan, J.L., Kanaan, A.O., 2012. A polypill for all? Critical review of the polypill literature for primary prevention of cardiovascular disease and stroke. Ann. Pharmacother. 46, 688–695.

Carrera, S., Beaumont, J., 2010. Income and wealth. Social Trends 2010; 41: ISSN 2040–1620, Office for National Statistics, Newport, UK. www.statistics.gov.uk/ (accessed 20.03.13.).

Caskey, C.T., Gonzalez-Garay, M.L., et al., 2014. Adult genetic risk screening. Annu. Rev. Med. 65, 1–17.

Cassell, M.M., Halperin, D.T., Shelton, J.D., Stanton, D., 2006. HIV and risk behaviour: risk compensation: the Achilles' heel of innovations in HIV prevention? Br. Med. J. 332, 605–607.

Caulfield, T., Chandrasekharan, S., Joly, Y., Cook-Deegan, R., 2013. Harm, hype and evidence: ELSI research and policy guidance. Genome Med. 5, 21.

Charan, J., Goyal, J.P., Saxena, D., 2013. Effect of pollypill on cardiovascular parameters: systematic review and meta-analysis. J. Cardiovasc. Dis. Res. 4, 92–97.

Chou, R., Croswell, J.M., Dana, T., et al., 2011. Screening for prostate cancer: a review of the evidence for the US Preventive Services Task Force. Ann. Intern. Med. 155, 762–771.

Chowdhury, S., Dent, T., Pashayan, N., et al., 2013. Incorporating genomics into breast and prostate cancer screening: assessing the implications. Genet. Med. 15, 423–432.

Collins, F.S., 1999. Medical and societal consequences of the human genome project. N. Engl. J. Med. 341, 28–37.

Collins, F., 2010. Has the revolution arrived? Nature 464, 674–675.

Davey Smith, G., 2011. Epidemiology, epigenetics and the "gloomy prospect": embracing randomness in population health research and practice. Int. J. Epidemiol. 40, 537–562.

de Cates, A.N., Farr, M.R.B., Wright, N., Jarvis, M.C., Rees, K., Ebrahim, S., Huffman, M.D., 2014. Fixed-dose combination therapy for the prevention of cardiovascular disease. Cochrane Database Syst. Rev 4, 1–67. http://dx.doi.org/10.1002/14651858.CD009868.pub2.

Dickinson, J.A., Miller, A.B., Popadiuk, C., 2014. When to start cervical screening: epidemiological evidence from Canada. BJOG 121, 255–260.

Djulbegovic, M., Beyth, R.J., Neuberger, M.M., et al., 2010. Screening for prostate cancer: systematic review and meta-analysis of randomised controlled trials. Br. Med. J. 341, 1–9. http://dx.doi.org/10.1136/bmj.c4543.

Dubben, H.H., 2009. Trials of prostate-cancer screening are not worthwhile. Lancet Oncol. 10, 294–298.

Eriksen, E.F., Díez-Pérez, A., Boonen, S., 2014. Update on long-term treatment with bisphosphonates for postmenopausal osteoporosis: a systematic review. Bone 58, 126–135.

Estruch, R., Ros, E., Salas-Salvadó, J., et al., 2013. Primary prevention of cardiovascular disease with a Mediterranean diet. N. Engl. J. Med. 368, 1279–1290.

European Commission, 2014. Horizon 2020 Work Programme 2014–2015: health, demographic change and wellbeing. Retrieved 30 January 2015 from, http://ec.europa.eu/research/participants/data/ref/h2020/wp/2014_2015/main/h2020-wp1415-health_en.pdf.

Feinberg, A.P., Irizarry, R.A., 2010. Stochastic epigenetic variation as a driving force of development, evolutionary adaptation, and disease. Proc. Natl. Acad. Sci. U. S. A. 107 (Suppl. 1), 1757–1764.

Fiuza-Luces, C., Garatachea, N., Berger, N.A., Lucia, A., 2013. Exercise is the real polypill. Physiology 28, 330–358.

Frank, S.A., 2010. Somatic evolutionary genomics: mutations during development cause highly variable genetic mosaicism with risk of cancer and neurodegeneration. Proc. Natl. Acad. Sci. U. S. A. 107 (Suppl. 1), 1725–1730.

Fyhri, A., Bjørnskau, T., Backer-Grøndahl, A., 2012. Bicycle helmets: a case of risk compensation? Transport. Res. F: Traffic Psychol. Behav. 15, 612–624.

Gage, M., Wattendorf, D., Henry, L.R., 2012. Translational advances regarding hereditary breast cancer syndromes. J. Surg. Oncol. 105, 444–451.

Galsworthy, M.J., Palumbo, L., McKee, M., 2014. Has Big Pharma hijacked the European health research budget? Lancet 383, 1210.

Godlee, F., 2014a. Adverse effects of statins. Br. Med. J. 348, 1–2. http://dx.doi.org/10.1136/bmj.g3306.

Godlee, F., 2014b. Statins and the BMJ. Br. Med. J. 349, 1–2. http://dx.doi.org/10.1136/bmj.g5038.

Gøtzsche, P.C., 2015. Screening: a seductive paradigm that has generally failed us. Int. J. Epidemiol. 44, 278–280. http://dx.doi.org/10.1093/ije/dyu267.

Gøtzsche, P.C., Jørgensen, K.J., 2013. Screening for breast cancer with mammography. Cochrane Database Syst. Rev 6, 1–81. http://dx.doi.org/10.1002/14651858.CD001877.pub5.

Gøtzsche, P.C., Olsen, O., 2000. Is screening for breast cancer with mammography justifiable? Lancet 355, 129–134.

Gøtzsche, P.C., Hartling, O.J., Nielsen, N., Brodersen, J., 2012. Screening for Breast Cancer with Mammography. The Nordic Cochrane Centre, Copenhagen, Denmark. Retrieved on 21 January 2015 from, http://www.cochrane.dk/screening/mammography-leaflet.pdf.

Grady, D., Redberg, R.F., 2010. Less is more: how less health care can result in better health. Arch. Intern. Med. 170, 749–750.

Grant, R.W., O'Brien, K.E., Waxler, J.L., et al., 2013. Personalized genetic risk counseling to motivate diabetes prevention: a randomized trial. Diabetes Care 36, 13–19.

Grosse, S.D., Khoury, M.J., 2006. What is the clinical utility of genetic testing? Genet. Med. 8, 448–450.

Guthrie, R., 1961. Blood screening for phenylketonuria. JAMA 178, 863.

Hamburg, M.A., Collins, F.S., 2010. The path to personalized medicine. N. Engl. J. Med. 363, 301–304.

Harris, R., 2011. Overview of screening: where we are and where we may be headed. Epidemiol. Rev. 33, 1–6.

Hart, J.T., 1971. The inverse care law. Lancet 297, 405–412.

Hawkes, N., 2014. Risks in the balance: the statins row. Br. Med. J. 349, 1–3. http://dx.doi.org/10.1136/bmj.g5007.

Heath, I., Evans, S., Furberg, C., et al., 2014. Report of the independent panel considering the retraction of two *BMJ* papers. Retrieved 1 February 2014 from, http://journals.bmj.com/site/bmj/statins/Final%20report%20of%20the%20independent%20panel%20310714.pdf.

Hoffman, K.E., 2012. Management of older men with clinically localized prostate cancer: the significance of advanced age and comorbidity. Semin. Radiat. Oncol. 22, 284–294.

Hood, L., Friend, S.H., 2011. Predictive, personalized, preventive, participatory (P4) cancer medicine. Nat. Rev. Clin. Oncol. 8, 184–187.

Horton, R., 2014. Offline: reimagining the meaning of health. Lancet 384, 218.

Huffman, M.D., Yusuf, S., 2014. Polypills: essential medicines for cardiovascular disease secondary prevention? J. Am. Coll. Cardiol. 63, 1368–1370.

Ilic, D., Neuberger, M.M., Djulbegovic, M., Dahm, P., 2013. Screening for prostate cancer. Cochrane Database Syst. Rev. 1, 1–76. http://dx.doi.org/10.1002/14651858.CD004720.pub3.

Ioannidis, J.P., 2010. Genetics, personalized medicine, and clinical epidemiology: expectations, validity, and reality in omics. J. Clin. Epidemiol. 63, 945–949.

Issa, J.-P., 2014. Aging and epigenetic drift: a vicious cycle. J. Clin. Investig. 124, 24–29.

James, J.E., 2014. Personalised medicine, disease prevention, and the inverse care law: more harm than benefit? Eur. J. Epidemiol. 29, 383–390.

Juengst, E.T., Flatt, M.A., Settersten, R.A., 2012. Personalized genomic medicine and the rhetoric of empowerment. Hastings Cent. Rep. 42, 34–40.

Julian, D.G., Pocock, S.J., 2015. Effects of long-term use of cardiovascular drugs. Lancet 385, 325.

Khoury, M.J., McCabe, L.L., McCabe, E.R., 2003. Population screening in the age of genomic medicine. N. Engl. J. Med. 348, 50–58.

Khoury, M.J., Gwinn, M., Yoon, P.W., et al., 2007. The continuum of translation research in genomic medicine: how can we accelerate the appropriate integration of human genome discoveries into health care and disease prevention? Genet. Med. 9, 665–674.

Khoury, M.J., Evans, J., Burke, W., 2010. A reality check for personalized medicine. Nature 464, 680.

Khoury, M.J., Bowen, M.S., Burke, W., et al., 2011a. Current priorities for public health practice in addressing the role of human genomics in improving population health. Am. J. Prev. Med. 40, 486–493.

Khoury, M.J., Clauser, S.B., Freedman, A.N., et al., 2011b. Population sciences, translational research, and the opportunities and challenges for genomics to reduce the burden of cancer in the 21st century. Cancer Epidemiol. Biomarkers Prev. 20, 2105–2114.

Khoury, M.J., Gwinn, M.L., Glasgow, R.E., Kramer, B.S., 2012. A population approach to precision medicine. Am. J. Prev. Med. 42, 639–645.

Khoury, M.J., Janssens, A.C.J., Ransohoff, D.F., 2013. How can polygenic inheritance be used in population screening for common diseases? Genet. Med. 15, 437–443.

Kitsios, G.D., Kent, D.M., 2012. Personalised medicine: not just in our genes. Br. Med. J. 344, 1–5. http://dx.doi.org/10.1136/bmj.e2161.

Kolber, M.R., Korownyk, C., 2014. An aspirin a day? Aspirin use across a spectrum of risk: cardiovascular disease, cancers and bleeds. Expert Opin. Pharmacother. 15, 153–157.

Krumholz, H.M., 2014. Big data and new knowledge in medicine: the thinking, training, and tools needed for a learning health system. Health Aff. 33, 1163–1170.

Landau, Y.E., Lichter-Konecki, U., Levy, H.L., 2014. Genomics in newborn screening. J. Pediatr. 164, 14–19.

Legler, J.M., Feuer, E.J., Potosky, A.L., et al., 1998. The role of prostate-specific antigen (PSA) testing patterns in the recent prostate cancer incidence decline in the United States. Cancer Causes Control 9, 519–527.

Lewington, S., Clarke, R., Qizilbash, N., et al., 2002. Age-specific relevance of usual blood pressure to vascular mortality: a meta-analysis of individual data for one million adults in 61 prospective studies. Lancet 360, 1903–1913.

Lohn, Z., Adam, S., Birch, P.H., Friedman, J.M., 2014. Incidental findings from clinical genome-wide sequencing: a review. J. Genet. Couns. 23, 463–473.

Lyssenko, V., Laakso, M., 2013. Genetic screening for the risk of type 2 diabetes worthless or valuable? Diabetes Care 36 (Suppl. 2), S120–S126.

MacMahon, S., 2000. Blood pressure and the risk of cardiovascular disease. New England J. Med. 342, 50–52.

Malhotra, A., 2013. Saturated fat is not the major issue. Br. Med. J. 347, 1–2. http://dx.doi.org/10.1136/bmj.f6340.

Manolio, T.A., Weis, B.K., Cowie, C.C., et al., 2012. New models for large prospective studies: is there a better way? Am. J. Epidemiol. 175, 859–866.

Marlow, L.A., Forster, A.S., Wardle, J., Waller, J., 2009. Mothers' and adolescents' beliefs about risk compensation following HPV vaccination. J. Adolesc. Health 44, 446–451.

Maron, D.J., Ting, H.H., 2013. In mildly symptomatic patients, an invasive strategy with catheterization and revascularization should not be routinely undertaken. Circulation 6, 114–121.

McBride, C.M., Bowen, D., Brody, L.C., et al., 2010. Future health applications of genomics: priorities for communication, behavioral, and social sciences research. Am. J. Prev. Med. 38, 556–565.

McGinnis, J.M., Williams-Russo, P., Knickman, J.R., 2002. The case for more active policy attention to health promotion. Health Aff. 21, 78–93.

Melzer, D., Zimmern, R., 2002. Genetics and medicalisation. Br. Med. J. 324, 863–864.

Messiah, A., Constant, A., Contrand, B., et al., 2012. Risk compensation: a male phenomenon? Results from a controlled intervention trial promoting helmet use among cyclists. Am. J. Public Health 102 (Suppl. 2), S204–S206.

Miller, F.A., Hayeems, R.Z., Bombard, Y., et al., 2009. Clinical obligations and public health programmes: healthcare provider reasoning about managing the incidental results of newborn screening. J. Med. Ethics 35, 626–634.

Moynihan, R., Doust, J., Henry, D., 2012. Preventing overdiagnosis: how to stop harming the healthy. Br. Med. J. 344, 1–6. http://dx.doi.org/10.1136/bmj.e3502.

Moynihan, R.N., Cooke, G.P., Doust, J.A., et al., 2013. Expanding disease definitions in guidelines and expert panel ties to industry: a cross-sectional study of common conditions in the United States. PLoS Med. 10, 1–12. http://dx.doi.org/10.1371/journal.pmed.1001500.

National Cancer Institute, 2013. NCI dictionary of cancer terms. Retrieved on 20 March 2013 from http://www.cancer.gov/dictionary/?CdrID=561717.

NDRC, 2015. Future health. Retrieved 14 January 2015 from, http://www.ndrc.ie/projects/futurehealth-3/.

Offit, K., 2011. Personalized medicine: new genomics, old lessons. Hum. Genet. 130, 3–14.

Pashayan, N., Pharoah, P., 2012. Population-based screening in the era of genomics. Pers. Med. 9, 451–455.

Patel, J.N., McLeod, H.L., Innocenti, F., 2013. Implications of genome-wide association studies in cancer therapeutics. Br. J. Clin. Pharmacol. 76, 370–380.

Peirson, L., Fitzpatrick-Lewis, D., Ciliska, D., Warren, R., 2013. Screening for cervical cancer: a systematic review and meta-analysis. Syst. Rev. 2, 1–14.

Prins, B.P., Lagou, V., Asselbergs, F.W., et al., 2012. Genetics of coronary artery disease: genome-wide association studies and beyond. Atherosclerosis 225, 1–10.

Qian, H., 2013. Stochastic physics, complex systems and biology. Quant. Biol. 1, 50–53.

Redberg, R.F., Katz, M.H., 2012. Healthy men should not take statins. JAMA 307, 1491–1492.

Rose, G., 1981. Strategy of prevention: lessons from cardiovascular disease. Br. Med. J. 282, 1847–1851.

Rose, G., 1985. Sick individuals and sick populations. Int. J. Epidemiol. 14, 32–38.

Rose, G., 1992. The Strategy of Preventive Medicine. Oxford University Press, Oxford, UK.

Saquib, N., Saquib, J., Ioannidis, J.P., 2015. Does screening for disease save lives in asymptomatic adults? Systematic review of meta-analyses and randomized trials. Int. J. Epidemiol. 44, 264–277. http://dx.doi.org/10.1093/ije/dyu140.

Sasieni, P., Castanon, A., Cuzick, J., 2009. Effectiveness of cervical screening with age: population based case–control study of prospectively recorded data. Br. Med. J. 339, 1–7. http://dx.doi.org/10.1136/bmj.b2968.

Schmalfuss, F., Kolominsky-Rabas, P.L., 2013. Personalized medicine in screening for malignant disease: a review of methods and applications. Biomarker Insights 8, 9–14.

Schröder, F.H., Hugosson, J., Roobol, M.J., et al., 2014. Screening and prostate cancer mortality: results of the European Randomised Study of Screening for Prostate Cancer (ERSPC) at 13 years of follow-up. Lancet 384, 2027–2035.

Shah, R.R., Shah, D.R., 2012. Personalized medicine: is it a pharmacogenetic mirage? Br. J. Clin. Pharmacol. 74, 698–721.

Shah, R.R., Smith, R.L., 2015. Addressing phenoconversion: the Achilles' heel of personalized medicine. Br. J. Clin. Pharmacol. 79, 222–240.

Smith, R., 2002. In search of "non-disease". Br. Med. J. 324, 883–885.

Smolen, J.S., Aletaha, D., 2013. Forget personalised medicine and focus on abating disease activity. Ann. Rheum. Dis. 72, 3–6.

Squassina, A., Manchia, M., Manolopoulos, V.G., et al., 2010. Realities and expectations of pharmacogenomics and personalized medicine: impact of translating genetic knowledge into clinical practice. Pharmacogenomics 11, 1149–1167.

Sulheim, S., Holme, I., Ekeland, A., Bahr, R., 2006. Helmet use and risk of head injuries in alpine skiers and snowboarders. JAMA 295, 919–924.

Talmud, P.J., Hingorani, A.D., Cooper, J.A., et al., 2010. Utility of genetic and non-genetic risk factors in prediction of type 2 diabetes: Whitehall II prospective cohort study. Br. Med. J. 340, 1–10. http://dx.doi.org/10.1136/bmj.b4838.

Taylor, P., 2015. Tempering expectations of screening: what is the most authoritative advice we can give, given the data that we have? Int. J. Epidemiol 44, 280–282. http://dx.doi.org/10.1093/ije/dyu269.

Tomasetti, C., Vogelstein, B., 2015. Variation in cancer risk among tissues can be explained by the number of stem cell divisions. Science 347, 78–81.

Turkheimer, E., 2012. Genome wide association studies of behavior are social science. Philos. Behav. Biol. 282, 43–64.

Tutton, R., 2012. Personalizing medicine: futures present and past. Soc. Sci. Med. 75, 1721–1728. http://dx.doi.org/10.1016/j.socscimed.2012.07.031.

United Nations Development Programme, 2010. Human Development Report 2010. 20th Anniversary Edition. The real wealth of nations: pathways to human development. UNDP, New York, NY.

Venter, J.C., 2010. Multiple personal genomes await. Nature 464, 676–677.

Viera, A.J., 2011. Predisease: when does it make sense? Epidemiol. Rev. 33, 122–134. http://dx.doi.org/10.1093/epirev/mxr002.

Wagner, A., 2012. The role of randomness in Darwinian evolution. Philos. Sci. 79, 95–119.

Wald, N.J., Law, M.R., 2003. A strategy to reduce cardiovascular disease by more than 80%. Br. Med. J. 326, 1419–1423.

Wald, N.J., Wald, D.S., 2010. The polypill concept. Postgrad. Med. J. 86, 257–260.

Weston, A.D., Hood, L., 2004. Systems biology, proteomics, and the future of health care: toward predictive, preventative, and personalized medicine. J. Proteome Res. 3, 179–196.

WHO, 2014. Global Health Estimates 2014 Summary Tables: Deaths by Cause, Age and Sex, by World Bank region, 2000–2012. World Health Organization, Geneva, Switzerland. Retrieved 29 August 2014 from, http://www.who.int/healthinfo/global_burden_disease/en/.

Wilcken, B., Wiley, V., 2015. Fifty years of newborn screening. J. Paediatr. Child Health 51, 103–107.

Wilde, G.J., Robertson, L., Pless, I.B., 2002. Does risk homoeostasis theory have implications for road safety? Br. Med. J. 324, 1149–1152.

Williams, S.J., Martin, P., Gabe, J., 2011. The pharmaceuticalisation of society? A framework for analysis. Sociol. Health Illn. 33, 710–725.

Wilson, J.M.G., Jungner, G., 1968. Principles and Practice of Screening for Disease. Public Health Papers No. 34 World Health Organization, Geneva.

Woloshin, S., Schwartz, L.M., Black, W.C., Kramer, B.S., 2012. Cancer screening campaigns: getting past uninformative persuasion. New England J. Med. 367, 1677–1679.

Woolf, S.H., Harris, R., 2012. The harms of screening. JAMA 307, 565–566.

Zeller, T., Blankenberg, S., Diemert, P., 2012. Genomewide association studies in cardiovascular disease: an update 2011. Clin. Chem. 58, 92–103.

Zomer, E., Owen, A., Magliano, D.J., et al., 2013. Predicting the impact of polypill use in a metabolic syndrome population: an effectiveness and cost-effectiveness analysis. Am. J. Cardiovasc. Drugs 13, 121–128.

# Part 3

# Achieving Optimal Health Sustainably

Parts 1 and 2 challenge the basis of biomedicine's dominance in healthcare, and Part 3 discusses more successful paths for achieving optimal personal and population health. Several of the themes discussed in Parts 1 and 2 are revisited in Part 3, including the power of prevention, the role of risk factor reduction, and specific intractable weaknesses of biomedical healthcare. Biomedical dominance impedes rather than promotes optimal health because biomedicine's primary focus is disease not health. The cumulative evidence indicates urgent need for radical redirection. The overarching need is for the focus of healthcare to be redirected away from attempts to control the progression of manifest disease in sick individuals. Instead, attention should be directed toward population-wide promotion of health and disease prevention, with biomedicine having an *adjunctive* rather than dominant role.

Limited knowledge and finite resources will always conspire to limit the achievement of optimal health. As such, universal optimal health is an aspiration to be continually strived for rather than a realizable finite objective. Operationally, *optimal personal and population health* is a balance representing the fairest distribution of resources to achieve the fairest distribution of health. Even if biomedical healthcare were well suited to optimizing the health of individuals, it would still represent an inferior means of achieving health for all. Its individual-orientation, disease-focus, and increasingly resource-intensive nature ensure that its dissemination is unequal, with the more socially and economically advantaged within and between countries having greater access to it. In truth, biomedical healthcare falls far short of being well suited to its professed purpose, even for those who have good access. While the health of many is ill-served by too little access, the health of many others is threatened by an excess of biomedical intervention.

There is much pretense, misrepresentation, and deceit in contemporary healthcare. Not merely the ubiquitous deceit perpetrated by healthcare industries and the physicians and biomedical institutions with which they are entangled, but deceit premised on the fantasy that the main contemporary threats to personal and population health are amenable to technological solutions. The evidence is wholly to the contrary. Optimal personal and population health depend primarily on behavioral and social solutions that foster health-enhancing ways of living. Actions that prevent and postpone disease and injury by reducing population exposure to risk factors contribute more to optimizing health, and do so with greater cost-effectiveness than can ever be achieved using the best available biomedicine.

Chapter 10

# Healing Practices and Evidence-Based Medicine

*Formerly, when religion was strong and science weak, men mistook magic for medicine; now, when science is strong and religion weak, men mistake medicine for magic.*

(Szasz, 1974, p. 128)

## Contents

Whereas the origins of health were discussed in Chapter 1, the present chapter is concerned with the origins of healing. It will be recalled that the World Health Organization's definition of health refers to wellbeing (physical, mental, and social), which includes, but is not limited to, the mere absence of disease. From this, we may infer that healing is the *restoration of wellbeing*, which sits comfortably with the dictionary definition of healing as the process of making or becoming *sound or healthy again* (Pearsall and Trumble, 1995). By now, it will come as no surprise to read that understanding the nature of healing requires more than an understanding biology.

The Health of Populations. http://dx.doi.org/10.1016/B978-0-12-802812-4.00010-2

## 10.1 THE NATURE OF HEALING

*The art of medicine consists of amusing the patient while nature cures the disease.*
(Voltaire, 1694-1778, French writer, historian, and philosopher)

That the body has remarkable natural self-healing properties is evident to everyone. We sustain a cut, there is blood, we wipe the cut clean, observe basic hygiene, and within days or weeks the lesion heals, often leaving the merest trace or none of our former injury. If we are unlucky, or careless, and the lesion becomes infected, the odds are hugely in our favor that natural defenses of the body will defeat the infection and return us to full health. Such self-healing properties are not uniquely human, but have been assembled by natural selection in the long course of evolution and inherited from our prehuman ancestors (Humphrey and Skoyles, 2012). In addition to wound healing and infection fighting, natural defenses that encourage healing include:

- inflammation and pain that restrict movement to avoid further injury;
- fatigue to encourage recuperative rest;
- nausea, diarrhea, and vomiting to defend against (by expelling) ingested toxins; and
- fever which may facilitate immune activity and limit the virulence of some pathogens that have strict temperature preferences.

It is important, when considering the efficacy of medical care, not to ignore the role of natural healing. As suggested by the quote from Voltaire above, for some, and possibly many, cures that follow medical intervention it is natural healing rather than the intervention that is primarily responsible. Nevertheless, there are limits to unaided natural healing, and circumstances sometimes call for intervention. In the event of a deep wound, for example, the natural healing processes that are summoned are likely to be assisted if sutures are applied. Even then, the sutures are not *the* cure but merely part of the cure. As such, we may be tempted to ask about the relative contribution of each, natural healing and biomedicine, in the health of individuals and ultimately of populations. Unfortunately, for the present, such questions are not easily answered.

Imagine, for example, that you fall and break your arm. Fortunately, personnel at the emergency care unit of the regional hospital are able to correctly realign the bone, a cast or splint is applied, and removed 6 weeks later. Within a relatively short period thereafter function is restored to your arm. Has medicine cured your broken arm? Yes, but not solely. Natural healing processes were responsible for a share of the work. This is not to take anything away from the medical care you received. Having your arm immobilized and splinted correctly may have been crucial to recovering full function in your arm, and credit is due to those who did that work. However, that work would be of little use in the absence of natural healing processes. Therefore, both medical care and natural healing were crucial in this instance, and it is not easy to see how relative merit could be apportioned. Thus, rather than crediting medical intervention with curing your broken arm (a common attribution), it would be more accurate to credit medicine with having facilitated recovery. Natural healing is never *replaced* by medicine, which at its most helpful only ever *complements* natural processes that maintain and restore health.

The everyday practice of clinical medicine, epitomized by general practice (primary care), is illustrative of natural healing doing much of the work. General practice is crowded with patients presenting with everyday maladies that are more-or-less indifferent to the ministrations of physicians. The common cold and influenza are prime examples. Acute upper respiratory tract viral infections, as discussed in Chapter 6, are among the most common diseases that humans experience (Eccles, 2005). Yet, most treatments for colds and flu are symptomatic and have little effect on the course of the underlying infection (Heikkinen and Järvinen, 2003). As exemplified by colds and flu, the general experience of everyday illness is recovery, sometimes aided by medical intervention and sometimes despite intervention. It is certain that what transpires in the routine practice of biomedical healthcare, whether involving much benefit or little, more is going on than can be explained by the specific actions of interventions physicians dispense. What comprises the *more that is going on* explains not only a large portion of the healing potential of biomedical healthcare but also that of all healing practices, ancient and modern.

## 10.1.1   Healing Practices

Comparative studies show that, despite great diversity between cultures, some aspects of healing practice appear to be universal (Kirmayer, 2004). In particular, almost all healing *practice* is interpersonal and social, wherein an individual (e.g., shaman, soothsayer, or physician) or group (family, clan, or hospital), invested with special knowledge, skills, or powers, ministers to the afflicted person to encourage healing. Parenthetically, it might be noted that the impression that all healing practice is interpersonal may derive from the fact that much cross-cultural study of healing has focussed on *healers*, and in that context healing is by definition interpersonal (the healer ministers to the patient). That does not mean, however, that all healing practice is necessarily interpersonal. The study of healers tends to divert us from taking account of healing practices in which individuals minister to themselves. For example, in a study of cancer patients receiving conventional biomedical healthcare, more than 90% were found to be simultaneously using **complementary and alternative medicine** (*CAM*) of which prayer, relaxation, and exercise were most common (Yates et al., 2005). While each of these can be experienced communally (e.g., with priest, therapist, or trainer), they can also be undertaken solitarily. That is, some healing practice may be personal rather than interpersonal. Nevertheless, although some healthcare in all cultures is possibly self-administered, it appears that all cultures practice social forms of healing in which at least one healer and one patient participate.

Rather than coincidence, the universal practice of socially mediated healing is probably underpinned by an evolutionary history. On that premise, it is reasonable to infer that interpersonal interaction is facilitative of natural healing processes and may be directly healing in its own right. For example, it is widely accepted that despite earlier advances in anatomy and physiology, the corpus of medical practice contained comparatively little specific healing potential until well into the twentieth century. Yet, before that, physician consultation was as popular as it is today. Should we conclude that patients in former times were simply duped en masse into believing falsely that there was benefit to be had from consulting physicians? Should we similarly presume lack of merit for healing practices of all cultures past and present that are outside the recently developed evidenced-based Western medical canon? If so, then the same must also be concluded about much contemporary medical practice. For in reality, although contemporary biomedicine *aspires* to be evidence-based, the evidence for much of what transpires as *intervention* in standard contemporary biomedical healthcare is slender to nonexistent. We will return to questions about the therapeutic value of healing practices that are *unevidenced* later in this chapter, but first it is important to examine the truth of the popular claim that contemporary biomedical practice is evidence-based.

## 10.2 EVIDENCE-BASED MEDICINE

*…it's easy to write prescriptions, but difficult to come to an understanding with people.*

(Kafka, 1919, A Country Doctor)

The ties between medical practice and medical science can be traced to the mid-nineteenth century, at least. However, it was not until the late-twentieth century that the incorporation of scientific evidence into routine clinical practice came to be broadly systematized. The conviction emerged in some quarters that a bridge was needed to span the wide gap that existed between interventions shown by repeatable (i.e., publicly verifiable) research to be effective and physician practices based on individual clinical experience. *Evidence-based medicine* was the proposed solution to the problem, and the term quickly acquired a mantra-like character with respect to all manner of issues concerning medical education, practice, management, and policy. Indeed, the evidence-based approach quickly came to be adopted as a guiding principle within public policy-making in general, and the term is commonly encountered in contexts only tangentially related to medicine or healthcare.

It might be noted that, in the field of human services, incorporation of an explicit model of evidenced-based practice occurred in the field of clinical psychology some decades before similar tenets were adopted in biomedical healthcare. Discussions within professional associations of clinical psychology were ongoing throughout the 1940s about unmet need for psychological services for veterans from the World Wars, and those discussions culminated in 1949 in a model of psychological practice, the *scientist-practitioner model*, being formally ratified by the American Psychological Association at a celebrated conference held in Boulder, Colorado (Davison, 1998). Ever since, the scientist-practitioner model (alternatively, the *Boulder model*) has been influential worldwide in the training and practice of clinical psychology. However, reflecting medicine's hegemony in all matters related to health, contemporary psychologists are more disposed to make reference to *evidence-based practice* when previously the *scientist-practitioner model* would have been invoked.

With regard to biomedical healthcare, it has been claimed "that until post-World War II, modern medicine was essentially the medicine of placebo effects" (Jubb and Bensing, 2013, p. 2710), implying that little of what physicians did produced any specific therapeutic benefit. The long history of uncertainty about biomedical efficacy is epitomized by the advent of evidence-based medicine, predicated on revelations that widely used medical practices lacked scientific foundation. Today, after a mere couple of decades of general dissemination of evidence-based medicine, the kind of public-media coverage medicine receives, with frequent tales of alleged major scientific advances and breakthroughs, is suggestive of widespread confidence in the effectiveness and scientific foundations of current practice.

Optimism in biomedicine may derive in part from the fact that doubts raised about current practices are frequently met with promises of impending transformational breakthroughs, which it seems constantly lie in wait, almost but not quite in full sight, somewhere near the next bend on the highway to discovery. It is curious, however, that each generation appears to have a generally optimistic view of the medicine of its day, while simultaneously spurning the medicine practiced in the preceding generation. In light of that history, perhaps all generations should guard against complacent belief in the superiority of the medicine with which they are familiar. Need for caution is indeed suggested by the record of achievements and limitations of evidence-based medicine; a record which (as described below) offers little support for alleged general effectiveness or scientific soundness of contemporary biomedical practice.

## 10.2.1   Evidence-Based Practice Versus Clinical Autonomy

Unfortunately, the advent of evidence-based medicine, with its promise of uniformly improved effectiveness, has not been as transformative as some had hoped and others continue to believe. By promising to be one thing but being another, evidence-based medicine has elements of sleight of hand. In fact, evidence-based medicine has been as much about epistemology as it has the exploitation of science to elevate practice to higher levels of effectiveness.

*Epistemology*, the branch of philosophy concerned with the theory of knowledge, addresses questions concerning how we know what we know and how should we distinguish between justified belief and opinion. In general, people justify actions on the basis of *perceived* evidence, and physicians are not notably exceptional in that regard. When it first appeared in the medical literature, evidence-based medicine was described as de-emphasizing the "evidence" of *unsystematic clinical experience* in favor of clinical

*decision making [that] stresses the examination of evidence from clinical research.*

(Guyatt et al., 1992, p. 2420)

As such, evidence-based medicine was never about encouraging doctors to use evidence (there was never any doubt that they were already doing that), but a movement to define what constitutes *acceptable* evidence, namely, that which is "scientific" versus "clinical" or "personal."

The most widely cited definition of the approach states that evidence-based medicine

*is the conscientious, explicit, and judicious use of current best evidence in making decisions about the care of individual patients.*

(Sackett et al., 1996, p. 71)

Of critical importance, evidence, in this instance, refers to that which is *external* to physicians' individual practice. In particular, it refers to the *randomized controlled trial* or *clinical trial*. In place of the "striking variations" in behavior that had characterized clinical practice theretofore, the "ascendancy of the randomized trial" was heralded as the "fundamental shift" enabling doctors to employ uniformly valid and effective practices (Sackett and Rosenberg, 1995). Although evidence-based clinical practice has been widely disseminated and accepted, from its outset the movement elicited substantial skepticism and sometimes hostility—both of which continue to the present day. Much of that friction concerns physicians' obligations to deliver quality healthcare by adhering to uniform, accepted, and approved practices, and the countervailing right of physicians to exercise independent professional judgment.

The following quotes give a flavor of the diversity of opinion. The first quote is from an article published in 1983, titled *The End of Clinical Freedom*, which presaged the advent of evidenced-based medicine almost a decade before the term entered the medical literature:

*Clinical freedom is dead, and no one need regret its passing ... the right - some seemed to believe the divine right - of doctors to do whatever in their opinion was best for their patients ... is not good enough ... medical care must be limited to what is of proved value, and the medical profession will have to set opinion aside (p. 1237).*

The second quote, from an article published in 2011, titled *The Need for Clinical Freedom*, is diametrically opposite in content to the earlier article, stating that:

> *proscriptive guidelines, mechanistic doctors and financial control have come together to contribute to the demise of the responsibility that doctors used to have for individual patients. We need to change medical culture [and] return to clinical freedom (p. 852).*

Apart from the divergent views they express, these two quotes are notable for an additional fact. Both articles were authored by the same person, British cardiologist John Hampton. The quotes not only illustrate the dramatic change in Hampton's personal beliefs, but also the diversity of entrenched opinion that has permeated the medical community before and since evidence-based medicine gained dominance.

## 10.2.2   Internal Versus External Validity

Much of the continuing debate about evidence-based medicine has centered on the very epistemic point the movement aimed to settle, namely, the truth claims of different kinds of knowledge. Proponents of evidenced-based medicine have devoted much effort to the task of disseminating guidelines for evaluating *internal validity*, which is a property of a scientific study concerning the confidence with which causation may be inferred from observations made within the study. Evidence guidelines frequently focus on types of threats to internal validity, especially the likelihood of methodological bias (systematic error) as might arise, for example, if participants have not been assigned strictly at random to intervention and control conditions. With the focus on estimations of the internal validity of clinical trials, attempts have been made to delineate gradations of medical knowledge, which are then usually arranged hierarchically.

Often represented pictorially as a pyramid, as in Figure 10.1, evidence hierarchies typically show practitioner experience at the base (lowest ranking type of knowledge), randomized controlled trials toward the apex, and systematic reviews and meta-analyses of randomized controlled trials at the apex (highest ranking). Such characterizations are founded on the presumed scientific integrity of the clinical trials. That presumption, however, is undermined by extensive revelations, discussed in Part 2 (especially, Chapter 8), about manipulation of outcomes in clinical trials conducted by the pharmaceutical companies who conduct most of them. In that context, locating evidence from clinical trials at the apex of a "truth" hierarchy does little to instill confidence in the effectiveness and safety of contemporary biomedical healthcare. Apart from dubious scientific integrity, a further problem concerning clinical trials is that many practitioners consider the trials to be largely irrelevant to clinical medicine as ordinarily practiced. That is, whereas internal validity has been a preoccupation

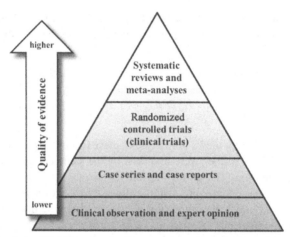

**FIGURE 10.1**  A simplified hierarchy of evidence used to guide evaluations of the *internal validity* of research.

of proponents of evidence-based medicine, skeptics have focused on the ***external validity*** of clinical trials, with the result that the opposing camps have tended to argue at cross-purposes.

To safeguard against threats to internal validity, clinical trials generally need to be highly specified with respect to patient characteristics and the clinical interventions examined. On the other hand, *external validity* refers to the confidence with which results of a precisely specified clinical trial can be generalized or extended to other circumstances, including other patients and other clinical settings. In practice, the priority given to the internal validity of clinical trials entails some sacrifice (usually unquantified) of external validity. Consequently, mindful of the diversity of circumstances encountered in clinical *practice* (as distinct from the comparatively uniform conditions in clinical *trials*), physicians have raised questions about how informative clinical trial results truly are for delivering medical care in the "real" world. Early skeptics expressed their concerns, since shown to be shared by many, as follows:

> *Derived almost exclusively from randomized trials and meta-analyses, the data [underpinning evidence-based medicine] do not include many types of treatments or patients seen in clinical practice; and the results show comparative efficacy of treatment for an "average" randomized patient, not for pertinent subgroups [of patients varying in] severity of symptoms, illness, comorbidity, and other clinical nuances.*

> (Feinstein and Horwitz, 1997, p. 527)

The problem was not unforeseen by the founders of evidence-based medicine. From the outset, advocates encouraged physicians to contextualize clinical-trial results by taking account of relevant clinical realities in order to arrive at nuanced interpretations of research findings. Leaders of the movement had this to say:

> *The practice of evidence based medicine means integrating individual clinical expertise with the best available external clinical evidence from systematic research ... External clinical evidence can inform, but can never replace, individual clinical expertise, and it is this expertise that decides whether the external evidence applies to the individual patient at all and, if so, how it should be integrated into a clinical decision. Similarly, any external guideline must be integrated with individual clinical expertise in deciding whether and how it matches the patient's clinical state, predicament, and preferences, and thus whether it should be applied.*

(Sackett et al., 1996, pp. 71-72)

Although it appears never to have been the intention of the founders of evidence-based medicine to usurp the independence of practitioners, the implementation of evidence-based practice may have followed a different course than the one the founders envisioned. Clinical guidelines have proliferated, and the explicit and implied *shoulds* and *should-nots* in the recommendations they contain often may leave practitioners feeling that they have little scope for exercising independent clinical judgment. In the minds of many practitioners, doubts concerning the relevance of clinical trials have not abated (e.g., Muth et al., 2014; Rees, 2013; Steel et al., 2014). In the opinion of critics, lack of clinical trial external validity is a problem that has never been satisfactorily addressed, and that opinion is well-founded. As discussed in the next section, a potential solution to the problem of external validity was long ago suggested, but has lain largely ignored for decades.

## 10.2.3 The Cochrane Program

The person most credited with elevating the status of the randomized controlled trial and for paving the way for the advent of evidence-based medicine is British clinical epidemiologist Archie Cochrane (1909-1988). During captivity in World War II, Cochrane served as sole medical officer for 20,000 inmates of a German camp for prisoners of war (Greenhalgh, 2004). During his time in the camp, four people died, three of whom were shot by camp guards. From his overall experience as a physician working under difficult conditions, and taking account of the low rate of mortality from "natural" causes, Cochrane became convinced that illness is mostly self-limiting and that medical intervention at that time, at least, was largely incidental to recovery. After the war, he made it his mission to rescue patients from ineffectual interventions. He called for an international register of randomized controlled trials, and for the development of explicit criteria for assessing the quality of published research. Neither goal was achieved in his lifetime, but today the register that bears his name, the Cochrane Controlled Trials Register, has several hundreds of thousands of entries, and there exists a global network, the Cochrane Collaboration (http://www.cochrane.org), dedicated to producing high-quality reviews of synthesized health-related evidence.

Not to be confused with the hierarchy of evidence illustrated in Figure 10.1, Cochrane's work gave rise to a separate hierarchy referred to as Cochrane's *ladder of evidence*, illustrated in Figure 10.2. Whereas the hierarchy of evidence (Figure 10.1) is used to guide evaluations of internal validity, Cochrane's ladder of evidence (Figure 10.2) can be used to guide evaluations of external validity. The ladder, which summarizes the essential criteria for assessing the adequacy of healthcare interventions in clinical practice, is comprised of three rungs, each of which asks a specific question: The first rung of the ladder asks, *Can* it (the drug or other intervention) work? This is the *efficacy* question. The second rung asks, *Does* it work in ordinary clinical practice? This is the *effectiveness* question. The third and final rung of the ladder asks, Is it *worth it?* This is the *cost-effectiveness* question. Conceptually, the hierarchy of Figure 10.1 and the ladder of Figure 10.2 overlap. Specifically, the first rung of Cochrane's ladder of evidence asks about efficacy, and the hierarchy of evidence illustrated in Figure 10.1 is a response to that question in the form of recommendations intended to assist evaluations of efficacy. Importantly, however, Cochrane's ladder of evidence asks two additional questions that are not part of the hierarchy of evidence depicted in Figure 10.1, which is solely concerned with efficacy. The additional questions in Cochrane's ladder of evidence relate to issues of great concern to practitioners and health policy makers, and they are issues that should also concern patients. Those questions are about the effectiveness of interventions when applied in usual clinical settings and the cost-effectiveness of those interventions.

Evidence of efficacy is only the first step in the process of assessing whether proposed healthcare interventions are suitable for wider clinical use. To date,

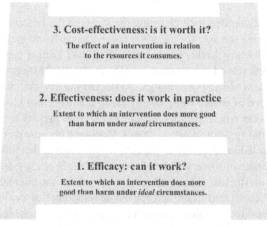

**3. Cost-effectiveness: is it worth it?**
The effect of an intervention in relation
to the resources it consumes.

**2. Effectiveness: does it work in practice**
Extent to which an intervention does more good
than harm under *usual* circumstances.

**1. Efficacy: can it work?**
Extent to which an intervention does more
good than harm under *ideal* circumstances.

**FIGURE 10.2** Cochrane's ladder of evidence to guide evaluations of the *external validity* of clinical research.

extraordinary effort and resources have been dedicated to the task of conducting internally valid randomized controlled trials to evaluate the efficacy of existing and new interventions (Järvinen et al., 2011). Understandably, great care typically goes into the selection of patients in order to maximize the likelihood of obtaining favorable clinical results. For example, trial participants are usually chosen on the basis of strictly defined diagnostic criteria, and those with other illnesses that could undermine treatment efficacy are excluded. Consequently, compared to published clinical-trial results, there are many factors at play in usual clinical settings that could dilute or interfere with treatment effects, including patient misdiagnosis, additional diagnoses (comorbidity), and effects from other drugs that patients may have been prescribed for other conditions. Additionally, in clinical trials, it is usual to employ dedicated personnel to maximize the quality of the intervention, and to use detailed protocols to maximize adherence by both patients and physicians to the prescribed treatment. Whereas 90% patient compliance is common in efficacy trials, patients in usual clinical settings typically take less than half of prescribed treatments (Yeaw et al., 2009).

Given the effort usually made to ensure that circumstances in randomized controlled trials are as near to ideal as possible, it would be wrong, as critics of evidence-based medicine rightly argue, to assume that efficacy results faithfully represent outcomes in usual clinical practice. In that respect, we might be comforted in the knowledge that Cochrane's ladder of evidence identifies efficacy studies as the *first* of three rungs. It would be reasonable, therefore, to expect that, in the rigorous and rule-bound world of evidence-based biomedical healthcare, that interventions accepted for common use would also have been evaluated for effectiveness and cost-effectiveness, which occupy the second and third rungs of Cochrane's evidence ladder, respectively. However, as often happens in the world of biomedical healthcare, reasonable expectations disappoint. In reference to Cochrane's ladder of evidence, a Finnish-based team of researchers came to the startling conclusion that

> *few therapies have made the second rung, and we know of none that have alighted the third.*

(Järvinen et al., 2011, pp. 1006-1007)

With specific reference to widely used statins (for lowering cholesterol), antihypertensives (for lowering blood pressure), and bisphosphonates (used in the treatment of osteoporosis to prevent loss of bone mass), the team concluded that

> *there are no valid data on the effectiveness, and particularly the cost effectiveness, in usual clinical care (p. 1007).*

Although largely ignored, considerations of effectiveness and cost-effectiveness are crucial to rational healthcare policies concerning the adoption and dissemination of healthcare practices, especially when demand is high. To illustrate, the Finish team quoted above examined the use of drugs for osteoporosis from the point of view of their cost-effectiveness in preventing hip fractures (Järvinen et al., 2011). In the United States, it is recommended that osteoporosis drugs should be prescribed for patients who, according to a WHO risk calculator, have a 10-year risk of hip fracture of 3% or over. On the optimistic assumption that drug treatment would be effective in lowering risk of fracture by 50%, Järvinen et al. (2011) estimated that 667 patients would need to be treated for 1 year to prevent one hip fracture. Based on the additional assumption that the average total cost of one hip fracture is about USD27,000, Järvinen et al. were able to estimate the cost-effectives of osteoporosis treatment for the prevention of hip fracture. They found that none of the available drugs was cost-effective for that purpose. Specifically, depending on which drug was used, ranging from the cheapest generic drug to the most expensive branded alternative, avoidance of a single hip fracture would cost between USD48,000 and USD5.21 million, between almost twice and almost 200-times, respectively, the cost of treating the one hip fracture likely to have been avoided.

In a further illustration of errors frequently made when efficacy data are used to estimate cost-effectiveness, a joint British and Dutch team compared cost-effectiveness using data from clinical trials versus usual clinical practice (van Staa et al., 2009). The comparison was for selective Cox-2 inhibitors (*coxibs*) used in the treatment of chronic pain. When introduced, coxibs were heavily marketed as a safer alternative to conventional nonsteroidal anti-inflammatory drugs (*NSAIDs*), including widely used compounds such as aspirin and ibuprofen, which can cause gastrointestinal bleeding and ulcers. Clinical-trial data suggested that switching from conventional NSAIDs to coxibs would (in "principle") cost between USD16,000 and USD20,000 on average for each instance of gastrointestinal bleeding that would be avoided. However, based on *actual* clinical usage involving more than one million patients in usual clinical care, cost-effectiveness analysis estimated a greater than fivefold cost of USD104,000 for each gastrointestinal bleeding event avoided. The researchers concluded:

> The published cost-effectiveness analyses of coxibs lacked external validity, did not represent patients in actual clinical practice, and should not have been used to inform prescribing policies. External validity should be an explicit requirement for cost-effectiveness analyses.

(van Staa et al., 2009)

As it happens, despite initial enthusiasm fuelled by heavy marketing, the pattern of coxib usage has changed not because of cost-ineffectiveness but because evidence, not revealed in earlier clinical trials, that their use is associated with a substantially increased risk of life-threatening cardiovascular events (Trelle et al., 2011).

Evidence-based claims in clinical medicine are widely assumed to possess high truth value. In fact, it appears that the expert panels responsible for preparing clinical guidelines routinely presume effectiveness and cost-effectiveness analyses based on efficacy results alone. Thus, the critics of evidence-based medicine have been vindicated. Almost all of the evidence (such as it is, considering the compromising influence of industry sponsorship) concerning widely used prescription drugs relates to efficacy when used with selected patients in well-resourced and closely monitored clinical trials. There is little evidence concerning how the same drugs actually perform when used in everyday clinical practice, and even less to indicate whether those interventions represent good value for money.

## 10.2.4   Where the Middle Ground Lies

There is no doubting the good intentions of the founders of evidence-based medicine to promote consistent and sound clinical practice. At times, the response of opponents has bordered on self-righteous indignation underpinning expressions of seemingly unbounded confidence in the veracity of clinical judgment. However, for humbler practitioners, it is obviously no easy matter to reconcile individual judgment, with its intrinsic subjectivity, with clinical practice ostensibly informed by objective science. What constitutes one practitioner's clinical judgment may appear imprudent to another equally experienced practitioner. This, indeed, appears to be the nub of ongoing debate and rancor. Whereas few would argue that practitioners should follow clinical guidelines unwaveringly, it could be said that failing to so do takes us back to where the matter of evidence-based guidelines started in the first place; the need, according to Hampton's (1983) initial thoughts, to *set opinion aside*.

The future appears uncertain for evidence-based medicine as currently construed. Its message has been disseminated to saturation level, but the movement appears to be faltering. Part of the problem is that the arguments of proponents are largely quantitative, statistical, and reputedly universal, whereas those of critics tend to be qualitative, anecdotal, and local. Although the critics have a weak epistemological foundation, their position is strong because their criticisms have gone largely unanswered. Evidence-based medicine has made great strides in addressing concerns about internal validity, whereas the questions that most concern clinical practitioners are related to external validity, and those questions remain largely unanswered. In short, deciding the relevance of findings from randomized controlled trials for the care of individual patients ultimately depends on myriad (largely unsystematic) clinical judgments pertaining to the particulars of each case. In that event, it is difficult to say what precisely has been achieved by the great campaign that has been waged under the banner of evidence-based medicine.

In the end, all sides may find themselves battling for lost causes, due to their collective failure to address the overarching problem of medical commercialism. To what cause are advocates of evidence-based medicine actually committed? The much-lauded evidence is tarnished by pharmaceutical companies who sponsor most trials and routinely bias, sometimes fraudulently, clinical-trial results in order to exaggerate the efficacy and safety of their products. Conversely, how is the practitioner meant to exercise sound judgment? Challenged to address the diverse and individual needs of patients, the practitioner is confronted with "scientific" evidence that cannot be wholly believed and personal judgment that cannot be wholly validated.

Hampton's (1983, 2011) personal journey of approximately 3 decades from vocal advocate of evidence-based medicine to damning critic seems to parody the story of evidence-based medicine itself. On one hand, *un*evidenced-based medicine can claim no authority; it is mere charlatanism. On the other hand, given the dissimilarity between the imperfect and convoluted world of clinical practice and the allegedly pristine world of clinical trials tainted by self-serving bias, slavish adherence to evidence-based guidelines would be doltish. The proponents of evidence-based medicine have reasonable arguments to buttress their cause, but so too do skeptics. Probably, most practitioners seek solace in what they perceive to be a middle ground between the two opposing camps. However, who can say where the true middle ground lies, or even if there is a middle ground containing the truth of what works and what does not in healthcare? Given that whatever middle ground of truth there may be must lie somewhere between potentially biased clinical-trial evidence of efficacy on one hand, and little or no evidence of effectiveness in usual clinical practice on the other hand, the inescapable conclusion is that some, and probably a lot, of what happens in biomedical healthcare does not work, or at least not very well.

## 10.2.5   Escalating Healthcare Expenditure

A perversity of modern biomedical healthcare is that level of expenditure is *not* a good measure of healthcare outcomes, and may possibly indicate the opposite. Neither differences in expenditure between regions nor increasing expenditure over time have much to do with personal or population health outcomes. Consequently, cost, and especially increasing cost, is a major source of pessimism in healthcare. Satisfying perceived healthcare needs is an increasingly formidable challenge even for the richest countries. At almost USD8500 per capita per annum, the United States has the highest total expenditure on health of any country, nearly 18% of GDP (WHO, 2014). As a key indicator of population health, life expectancy at birth in the United States is 79 years, 34th in the world and, as shown in Figure 10.3, equal to Columbia, Costa Rica, Cuba, Nauru, and Qatar. As can be seen from the figure, all of the countries with which the United States shares equal life expectancy have much lower healthcare expenditures. The next highest spending country in that group, Qatar, and the lowest spending country, Nauru, have annual health expenditures of less than one-fifth and about one thirty-fifth, respectively, of that of the United States. Japan, which has a life expectancy at birth of 84 years, the longest of any country, has a per capita total

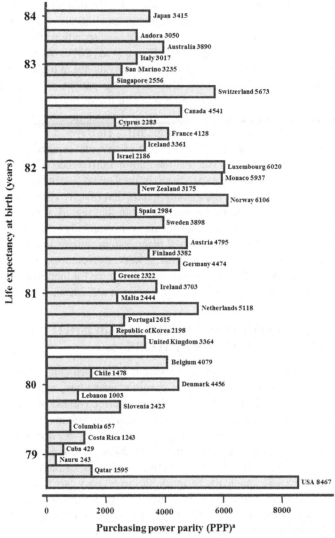

**FIGURE 10.3** National health expenditure grouped by average life expectancy at birth. [a]PPP = adjusted USD to standardize the cost of marketed goods (purchasing power) between countries. *(Derived from WHO (2014).)*

expenditure on health well below half of that of the United States. Among the group of countries with the next longest life expectancy after Japan, health expenditure in all except Switzerland is less than half of that of the United States. Indeed, the large majority of the 33 countries that enjoy longer life expectancy than the United States have achieved that result despite having healthcare expenditures less than half, and sometimes much less, than the United States.

The disjunction between the cost of biomedical healthcare and health outcomes is also evident in the pattern of healthcare expenditure *within* the United States, where there are marked differences in regional spending after adjustment for relevant health-related demographics (Fisher et al., 2009). Almost one-third of the difference in expenditure is explained by physicians in higher-cost, compared to lower-cost, regions more often recalling patients for repeat visits (Sirovich et al., 2008) and making more frequent use of diagnostic procedures and other resource-intensive interventions that produce doubtful benefit while exposing patients to risk of harm (Brody, 2010). However, increasing cost is only partly due to the ceaseless development of new and reputedly successful procedures and technologies. Much of the regional difference and much of the overall increase in healthcare expenditure in the United States is explained by physician behavior, namely, prescribing habits that involve excessive use not only of new but also of existing technologies. In that regard, it is evident that "clinical" judgment in biomedical healthcare is substantially influenced by self-serving incentives in for-profit oriented practice—incentives that reward physicians for delivering more rather than fewer interventions. The incessant stream of new technologies merely increases the scope for lucrative discretionary practices (Grady and Redberg, 2010; Fisher et al., 2009).

## 10.3 UNWELCOME DISCOVERIES: ESTABLISHED INTERVENTIONS THAT DON'T WORK

As discussed, reasons for doubting the extent of biomedicine's effectiveness include the compromising influence of commercial culture, and the dearth of trials to verify the clinical effectiveness and cost-effectiveness of interventions used in routine practice. In addition to those concerns, biomedical healthcare includes interventions of longstanding that have never been evaluated, and therefore are of unknown efficacy or effectiveness. In some instances, even when evaluated and found to be ineffective, interventions have continued to be practiced. The widespread use of antibiotics in primary care for the treatment of acute viral upper respiratory tract infection (Kenealy and Arroll, 2013; Heikkinen and Järvinen, 2003), discussed in Chapter 6, is an illustrative case in point. Similarly, there is wide acceptance (or at least tolerance) within biomedical healthcare of the use of over-the-counter medicines which are sold to the public without prescription for treatment of coughs and colds. Comprehensive reviews have consistently found that, of the many such medicines on sale, none is more effective than placebo in relieving symptoms of acute cough (Schroeder and Fahey, 2002; Paul, 2012). Despite consistently negative evaluations, especially of over-the-counter medicines for coughs and colds intended for use with children, biomedical researchers have been consistently reluctant to advise against use because so doing "may seem unacceptable to many parents and lay people" (Schroeder and Fahey, 2002, p. 174). Moreover, while acknowledging safety concerns, use of such medications is sometimes actually endorsed (even if grudgingly) with the advice that recommendations "should be restricted to less expensive preparations" (Schroeder and Fahey, 2002, p. 174).

In a study to assess the effectiveness of two specific commonly used over-the-counter medications for treating nocturnal cough and sleep difficulty associated with upper respiratory infections in children, assessments were also made of whether parents had improved sleep quality when their children received the medications (Paul et al., 2004). The treatments were found to be not superior to placebo in either respect. Noting "the potential for adverse effects, and the individual and cumulative costs," the authors questioned whether the "medications have a place in the treatment of these illnesses for children." At the same time, they also noted the "desire to ease symptoms is strong for both parents and clinician," and concluded by advising healthcare practitioners to "consider [the negative consequences] of the drugs before *recommending* them to families" (p. e89, italics added). Such recommendations, even if hesitant, are not consistent with biomedicine's claim to being evidence based. Given that findings have consistently shown over-the-counter cough and cold medicines to be no more effective than placebo or taking honey in solution (Oduwole et al., 2014; Paul et al., 2007), widespread medical-profession complicity in the use of over-the-counter cough and cold medicines poses a threat to child safety. Moreover, notwithstanding lack of efficacy and potential for harm when used correctly, evidence indicates that the likelihood of harm is exacerbated due to widespread misuse of over-the-counter cough and cold medicines (Lokker et al., 2009).

## 10.3.1 Surgery

If considered in relation to dimensions such as required level of expertise and involvement of technology in medical care, self-administered over-the-counter medicines are more-or-less at the opposite end of the spectrum from surgery. Surgery is widespread and increasing, due in part to the introduction of new technologies such as **endoscopy** and **laparotomy** which are less invasive than *open* surgery. However, unlike pharmaceuticals, formal evaluations of the safety and efficacy of surgical procedures generally are not mandated by regulatory authorities (McCulloch et al., 2009). To a substantial degree, surgical intervention and innovation proceed by stages of trial and error. The trial of new practices builds on experience gained from existing practice. However, most of this occurs in the absence of the systematic evaluations (such as they are) undertaken when randomized controlled trials are used to evaluate the efficacy and safety of new drugs. Although uncommon, randomized controlled trials of surgical procedures have been conducted. In those relatively few instances, the *control arm* of the trial has usually been placebo intervention, with the surgical procedure being the *experimental arm*. All too often when such trials have been conducted, surgery has been found to be no better than placebo and sometimes worse.

In a recent study, relevant databases were searched for reports of randomized controlled trials in which any surgical intervention was compared with placebo and only 53 such trials were located (Wartolowska et al., 2014). Most of the trials were concerned with non-life-threatening medical conditions, where the main aim of surgery was to improve function, symptoms, and quality of life; to reduce pain; or to remove the need for pharmacotherapy. In half of the trials, patients who received surgery had no better outcomes than patients who received placebo intervention, but even in the trials in which surgery was superior the magnitude of the effect was generally small.

One recurring argument in the surgery literature is that it is unethical to conduct placebo-controlled trials of surgery because of risk to safety from denial of treatment for patients receiving placebo. However, that argument is based largely on opinion and ignores evidence of medical harm, discussed in Chapter 6, indicating that surgery is a particularly hazardous form of medical intervention. In their review, Wartolowska et al. (2014) found little difference between surgery and placebo in relation to incidence of harm measured either as lack of improvement or the development of complications. As such, harm is not an inevitable result of being denied surgery and therefore it is not necessarily a valid reason for eschewing placebo-controlled trials. Wartolowska et al. concluded that more controlled trials of surgery are needed to prevent continued use of ineffective and potentially hazardous interventions.

### 10.3.1.1    Osteoarthritic Knee Surgery

In a seminal controlled clinical trial of a surgical procedure, Moseley et al. (2002) examined patients with *osteoarthritis* (degenerative joint disease) of the knee who were randomly assigned to receive one of three interventions: *arthroscopic debridement* (smoothing rough or worn bone surfaces that interfere with free movement of the joint), *arthroscopic lavage* (cleaning of loose "debris" such as fragments of cartilage or calcium phosphate crystals reputedly responsible for pain), or placebo (sham or pseudo) surgery. Patients in the sham-surgery group received skin incisions that simulated standard surgery without insertion of the arthroscope (a type of endoscope). Although this type of study design has attracted charges of unethicality for withholding treatment from needy patients, it should be noted that all patients in the trial consented to participate knowing that they might receive sham surgery which in all likelihood would be of no benefit to their knee arthritis.

Patients were assessed intermittently for knee function and pain for 24 months following intervention, and both patients and those conducting the assessments were blind to the intervention group to which patients had been assigned. A total of 165 patients completed the trial, and the results were unambiguous. At no point did either of the standard-surgery groups report better function or less pain than the sham-surgery group. At the time the study was conducted, 650,000 arthroscopic debridement and lavage procedures for osteoarthritis of the knee were being conducted annually in the United States at a cost of about USD3.5 billion. Initially, however, there was only reluctant acceptance of the conclusion that arthroscopic surgery is ineffective for the treatment of moderate-to-severe osteoarthritis of the knee, and for a time the procedure continued to be widely used despite direct evidence of inefficacy.

Similarly negative results were reported in a more recent study for patients with osteoarthritis of the knee who were assigned at random to one of two groups to receive either surgical lavage and arthroscopic debridement, plus physical and drug therapy as indicated, or to a control treatment consisting of physical and drug therapy without surgery (Kirkley et al., 2008). Again, all patients gave *informed consent* and were assessed intermittently for 2 years by personnel who were blind to the intervention group to which patients had been assigned. As in the earlier study by Moseley et al. (2002), patients who received arthroscopic surgery showed no additional benefit compared to those who received the control intervention. Faced with this additional evidence, there has at last been a substantial decline in the number of arthroscopic surgical procedures to treat knee osteoarthritis.

Concurrent with the decline in arthroscopic surgery for knee osteoarthritis, there was an increase of approximately 50% in arthroscopic partial meniscectomies (partial removal of cartilage in the knee) to about 700,000 annually in the United States at a cost of approximately USD7 billion (Kim et al., 2011). Again, growth in the popularity of that procedure occurred in the absence of controlled trials of clinical efficacy. Reminiscent of the rise and fall in popularity of arthroscopic surgery for knee osteoarthritis, recent clinical trials have been unfavorable to the cause of surgeons eager to perform ever more arthroscopic partial meniscectomies. In one multicenter American study, 351 symptomatic patients aged 45 years and older with a meniscal (cartilage) tear and imaging evidence of mild-to-moderate osteoarthritis were randomly assigned either to arthroscopic partial meniscectomy with postoperative physical therapy or to physical therapy alone (Katz et al, 2013). At 6 and 12 months, there were no significant differences in functional status and pain between the groups assigned to surgery or physical therapy without surgery.

Similarly, a recent multicenter Finish study (Sihvonen et al., 2013) involving 146 patients aged 35-65 years with symptoms of a degenerative meniscus tear without osteoarthritis were randomly assigned either to arthroscopic partial meniscectomy or sham surgery. At 12 months, both groups showed improvement in function and pain across a variety of tests, with neither group being significantly more improved than the other. Notably, however, there appears to be a large audience whose faith in the efficacy and safety of surgery is unresponsive to objective contrary evidence. In the study by Katz et al. (2013), despite evidence of inefficacy, 6 months into the trial, 30% of patients in the control arm opted to receive surgery and by 12 months the number had grown to 35%.

## 10.3.1.2 Spinal-Fusion Surgery

Controversy over ineffective surgery is not limited to the knee. Spinal-fusion surgery (to prevent movement between adjacent vertebrae) has increased rapidly, despite doubts about many of its applications (Deyo et al., 2004). Although there is wide acceptance of the efficacy of spinal-fusion surgery for *some* conditions in *some* patients, the rapid increase in the rate of such surgery, high rates of reoperation, and high rates of complications have led to concerns that the procedure is probably being overused. In particular, evidence of efficacy is lacking for the most common indications for which the procedure is being used. Specifically, given that back pain and disk degeneration are both nearly universal with aging, use of spinal fusion for those problems is not merely common but potential candidates for surgery are essentially limitless. However, following a thorough search of the literature, Mirza and Deyo (2007) succeeded in finding only four randomized trials suitable for systematic review of the efficacy of spinal fusion compared to nonsurgical interventions for low back pain (Mirza and Deyo, 2007). Unfortunately, there was a lack of uniformity between studies in the use of nonsurgical interventions, which included physical therapy supplemented with information and education, and *cognitive behavioral therapy*. Nevertheless, all four trials were consistent in suggesting little or no advantage of surgery over nonsurgical options.

Although the randomized controlled trial is not much favored by surgeons, failure to embrace it as an essential foundation for the practice of surgery is an abnegation of core professional responsibilities required of evidence-based medicine. Admittedly, randomized trials for surgery are challenging for practitioners, healthcare institutions, and patients alike. Trials are expensive and require elaborate coordination within and between healthcare facilities, and for patients it is difficult to allow the choice between having surgery or not to be determined by the toss of a coin. None of those challenges, however, is beyond solution (Hare et al., 2014; Kim et al., 2005; Wente et al., 2003). Failure to systematically evaluate widely-used surgical interventions has the certain outcome of unremitting patient harm.

## REFERENCES

Brody, H., 2010. Medicine's ethical responsibility for health care reform—the top five list. N. Engl. J. Med. 362, 283–285.

Davison, G.C., 1998. Being bolder with the Boulder model: the challenge of education and training in empirically supported treatments. J. Consult. Clin. Psychol. 66, 163–167.

Deyo, R.A., Nachemson, A., Mirza, S.K., 2004. Spinal-fusion surgery—the case for restraint. Spine J. 4, S138–S142.

Eccles, R., 2005. Understanding the symptoms of the common cold and influenza. Lancet Infect. Dis. 5, 718–725.

Feinstein, A.R., Horwitz, R.I., 1997. Problems in the "evidence" of "evidence-based medicine". Am. J. Med. 103, 529–535.

Fisher, E.S., Bynum, J.P., Skinner, J.S., 2009. Slowing the growth of health care costs—lessons from regional variation. N. Engl. J. Med. 360, 849–852.

Grady, D., Redberg, R.F., 2010. Less is more: how less health care can result in better health. Arch. Intern. Med. 170, 749–750.

Greenhalgh, T., 2004. Effectiveness and efficiency: random reflections on health services. Br. Med. J. 328, 529. http://dx.doi.org/10.1136/bmj.328.7438.529.

Guyatt, G., Cairns, J., Churchill, D., et al., 1992. Evidence-based medicine: a new approach to teaching the practice of medicine. JAMA 268, 2420–2425.

Hampton, J.R., 1983. The end of clinical freedom. Br. Med. J. 287, 1237–1238.

Hampton, J., 2011. Commentary: the need for clinical freedom. Int. J. Epidemiol. 40 (4), 849–852.

Hare, K.B., Lohmander, L.S., Roos, E.M., 2014. The challenge of recruiting patients into a placebo-controlled surgical trial. Trials 15, 1–5. http://dx.doi.org/10.1186/1745-6215-15-167.

Heikkinen, T., Järvinen, A., 2003. The common cold. Lancet 361, 51–59.

Humphrey, N., Skoyles, J., 2012. The evolutionary psychology of healing: a human success story. Curr. Biol. 22, R695–R698.

Järvinen, T.L., Sievänen, H., Kannus, P., et al., 2011. The true cost of pharmacological disease prevention. Br. Med. J. 342, 1006–1008.

Jubb, J., Bensing, J.M., 2013. The sweetest pill to swallow: how patient neurobiology can be harnessed to maximise placebo effects. Neurosci. Biobehav. Rev. 37, 2709–2720.

Kafka, F.A., 1919. A country doctor. Translated by Ian Johnston. Retrieved on 14 May 2014 from http://www.kafka-online.info/a-country-doctor.html.

Katz, J.N., Brophy, R.H., Chaisson, C.E., et al., 2013. Surgery versus physical therapy for a meniscal tear and osteoarthritis. N. Engl. J. Med. 368, 1675–1684.

Kenealy, T., Arroll, B., 2013. Antibiotics for the common cold and acute purulent rhinitis. Cochrane Database Syst. Rev. 6, 1–60. http://dx.doi.org/10.1002/14651858.CD000247.pub3.

Kim, S.Y., Frank, S., Holloway, R., et al., 2005. Science and ethics of sham surgery: a survey of Parkinson disease clinical researchers. Arch. Neurol. 62, 1357–1360.

Kim, S., Bosque, J., Meehan, J.P., et al., 2011. Increase in outpatient knee arthroscopy in the United States: a comparison of National Surveys of Ambulatory Surgery, 1996 and 2006. J. Bone Joint Surg. 93, 994–1000.

Kirkley, A., Birmingham, T.B., Litchfield, R.B., et al., 2008. A randomized trial of arthroscopic surgery for osteoarthritis of the knee. N. Engl. J. Med. 359, 1097–1107.

Kirmayer, L.J., 2004. The cultural diversity of healing: meaning, metaphor and mechanism. Br. Med. Bull. 69, 33–48.

Lokker, N., Sanders, L., Perrin, E.M., et al., 2009. Parental misinterpretations of over-the-counter pediatric cough and cold medication labels. Pediatrics 123, 1464–1471.

McCulloch, P., Altman, D.G., Campbell, W.B., et al., 2009. No surgical innovation without evaluation: the IDEAL recommendations. Lancet 374, 1105–1112.

Mirza, S.K., Deyo, R.A., 2007. Systematic review of randomized trials comparing lumbar fusion surgery to nonoperative care for treatment of chronic back pain. Spine 32, 816–823.

Moseley, J.B., O'Malley, K., Petersen, N.J., et al., 2002. A controlled trial of arthroscopic surgery for osteoarthritis of the knee. N. Engl. J. Med. 347, 81–88.

Muth, C., Kirchner, H., van den Akker, M., et al., 2014. Current guidelines poorly address multimorbidity: pilot of the interaction matrix method. J. Clin. Epidemiol. 67, 1242–1250.

Oduwole, O., Meremikwu, M.M., Oyo-Ita, A., Udoh, E.E., 2014. Honey for acute cough in children. Evid. Based Child Health 9, 401–444.

Paul, I.M., 2012. Therapeutic options for acute cough due to upper respiratory infections in children. Lung 190, 41–44.

Paul, I.M., Yoder, K.E., Crowell, K.R., et al., 2004. Effect of dextromethorphan, diphenhydramine, and placebo on nocturnal cough and sleep quality for coughing children and their parents. Pediatrics 114, e85–e90. http://dx.doi.org/10.1542/peds.114.1.e85.

Paul, I.M., Beiler, J., McMonagle, A., et al., 2007. Effect of honey, dextromethorphan, and no treatment on nocturnal cough and sleep quality for coughing children and their parents. Arch. Pediatr. Adolesc. Med. 161, 1140–1146.

Pearsall, J., Trumble, B. (Eds.), 1995. The Oxford English Reference Dictionary, second ed. Oxford University Press, Oxford, UK.

Rees, J., 2013. Why we should let "evidence-based medicine" rest in peace. Clin. Dermatol. 31, 806–810.

Sackett, D.L., Rosenberg, W.M., 1995. The need for evidence-based medicine. J. R. Soc. Med. 88, 620–624.

Sackett, D.L., Rosenberg, W.M., Gray, J.A., et al., 1996. Evidence based medicine: what it is and what it isn't. Br. Med. J. 312, 71–72.

Schroeder, K., Fahey, T., 2002. Should we advise parents to administer over the counter cough medicines for acute cough? Systematic review of randomised controlled trials. Arch. Dis. Child. 86, 170–175.

Sihvonen, R., Paavola, M., Malmivaara, A., et al., 2013. Arthroscopic partial meniscectomy versus sham surgery for a degenerative meniscal tear. N. Engl. J. Med. 369, 2515–2524.

Sirovich, B., Gallagher, P.M., Wennberg, D.E., Fisher, E.S., 2008. Discretionary decision making by primary care physicians and the cost of US health care. Health Aff. 27, 813–823.

Steel, N., Abdelhamid, A., Stokes, T., et al., 2014. A review of clinical practice guidelines found that they were often based on evidence of uncertain relevance to primary care patients. J. Clin. Epidemiol. 67, 1251–1257.

Szasz, T.S., 1974. The Second Sin. Doubleday, Garden City, NY.

Trelle, S., Reichenbach, S., Wandel, S., et al., 2011. Cardiovascular safety of non-steroidal anti-inflammatory drugs: network metaanalysis. Br. Med. J. 342, 1–11. http://dx.doi.org/10.1136/bmj.c7086.

van Staa, T.P., Leufkens, H.G., Zhang, B., Smeeth, L., 2009. A comparison of cost effectiveness using data from randomized trials or actual clinical practice: selective cox-2 inhibitors as an example. PLoS Med. 6, 1–10. http://dx.doi.org/10.1371/journal.pmed.1000194.

Wartolowska, K., Judge, A., Hopewell, S., et al., 2014. Use of placebo controls in the evaluation of surgery: systematic review. Br. Med. J. 348, 1–15. http://dx.doi.org/10.1136/bmj.g3253.

Wente, M.N., Seiler, C.M., Uhl, W., Büchler, M.W., 2003. Perspectives of evidence-based surgery. Dig. Surg. 20, 263–269.

WHO, 2014. World Health Statistics 2014. World Health Organization, Geneva. Retrieved 31 March 2014 from, http://apps.who.int/iris/bitstream/10665/112738/1/9789240692671_eng.pdf?ua=1.

Yates, J.S., Mustian, K.M., Morrow, G.R., et al., 2005. Prevalence of complementary and alternative medicine use in cancer patients during treatment. Support. Care Cancer 13, 806–811.

Yeaw, J., Benner, J.S., Walt, J.G., et al., 2009. Comparing adherence and persistence across 6 chronic medication classes. J. Manag. Care Pharm. 15, 728–740.

# Chapter 11

# Placebo and the Therapeutic Process

## Contents

An innovative feature of the studies of knee surgery by Moseley et al. (2002) and Sihvonen et al. (2013) discussed in Chapter 10 was the use of sham surgery as a placebo-control condition, and in both studies conventional surgery was no more effective than the sham. Nevertheless, the implications of the results of those studies are less straightforward than might appear. Although clinical improvement was probably due to processes common to both conventional and sham surgery, the nature of those common processes is unclear because evidence of improvement in both groups was based solely on before-and-after comparisons. While it is possible that the common process was placebo (a usual interpretation), the results are ambiguous in that regard. This is due to the fact that, as is true for almost all controlled trials of surgery, neither of the aforementioned studies of knee surgery included a *no-treatment control* group (sometimes referred to as an *observational* group).

The Health of Populations. http://dx.doi.org/10.1016/B978-0-12-802812-4.00011-4

## 11.1 THE PLACEBO: POWERFUL, POWERLESS, OR PASSÉ?

In the absence of a no-treatment group, it is impossible to know whether clinical improvement following conventional and sham surgery was due to placebo processes or to something else that would also have happened without the "placebo" intervention of sham surgery. That is, whereas the clinical improvement that occurred with both conventional and placebo surgery suggests that both were equally *effective*, it is also possible that both were equally *ineffective* and that the same clinical result would have been achieved if nothing at all had been done with the patients. Improvement could have occurred due to self-healing processes that occur with the passage of time and are part of the *natural progression* (or *natural history*) of the condition. Inclusion of a no-treatment condition in such studies is necessary to better differentiate effects directly attributable to placebo from effects that would have occurred in the absence of any intervention at all, whether conventional or placebo. From this, we might deduce that important additional information is provided when, irrespective of the type of intervention, efficacy studies include a no-treatment control as well as conventional and sham (placebo) interventions. So doing would help to isolate the *specific* effects of conventional interventions from effects attributable to *placebo* and other more general (*nonspecific*) effects.

### 11.1.1 Placebo Versus No Treatment

The distinction between placebo intervention and no treatment is important for clarifying some differences of opinion and confusion about the efficacy of placebos. In an influential article titled *The Powerful Placebo*, Beecher (1955) concluded that placebos have a "high degree of therapeutic effectiveness," producing "satisfactory" outcomes for 35% of patients. Beecher's widely-cited article appears to be at least partly responsible for encouraging sometimes hyperbolic claims about the benefit placebo interventions would have were they to be harnessed as part of routine clinical practice. However, the opposite view has also been argued by Hróbjartsson and Gøtzsche (2001, 2004, 2010) in a series of articles published in a number of influential outlets. Based on a review of over 200 clinical trials, Hróbjartsson and Gøtzsche (2010) concluded that placebo interventions generally produce "no major health benefits." However, the apparent gulf between Beecher's position and that of Hróbjartsson and Gøtzsche may not be as great as first appears after taking account of certain particulars of the clinical trials they examined.

To begin with, it should be noted that although Beecher based his conclusions on studies of a variety of medical conditions, he consistently reported larger effects on what he called "the subjective side" of those conditions, including pain, nausea, and mood (e.g., anxiety) than on "objective signs." As such, his conclusion of satisfactory outcomes for about one-third of patients may not be too far removed from that of the most recent of the three reviews by

Hróbjartsson and Gøtzsche. In their two earlier articles, they reported that placebo had a "small" effect on patient-reported outcomes, which they stated could not be distinguished from reporting bias due, for example, to patients wishing to "please the investigators" (Hróbjartsson and Gøtzsche, 2001, 2004). Later, however, they revised that earlier conclusion stating that although the effect of placebo on *observer-reported* outcomes was "small and uncertain," effects were generally "modest" for *patient-reported* outcomes such as pain and nausea (Hróbjartsson and Gøtzsche, 2010).

Possibly more important, however, is the fact that Beecher's review was based on estimates of the difference between before and after treatment in the placebo arm of randomized trials. Conversely, Hróbjartsson and Gøtzsche's (2010) expressly reviewed randomized placebo trials that also included a no-treatment control group allowing for direct comparison between placebo and no treatment. As explained above, before and after treatment comparisons do not provide an indication of the specific efficacy of placebo free of other (non-placebo) influences. For that reason, we should infer that Beecher (1955) may have overestimated the general efficacy of placebo, and that Hróbjartsson and Gøtzsche's (2010) conclusions may be more reliable. However, at this juncture, it is important to try to specify more precisely what is meant by the term *placebo*.

## 11.1.2    What is Placebo and What is It Not?

There is no universally-agreed definition of placebo. Indeed, a single all-encompassing definition may not be feasible due to the likelihood of different mechanisms being involved in placebos that differ in seemingly subtle but potentially important respects. The common idea that a placebo is an inactive agent that produces a therapeutic effect is meaningless because it is self-contradictory. Something that is inactive can have no effect. An important development in recent study is the realization that a placebo is not a *thing* but a *process* activated by a setting that *simulates* or *mimics* aspects of the therapeutic environment. That is, placebos have the appearance of conventional interventions but do not possess the specific content believed to be responsible for the therapeutic effect of the conventional intervention. A starch-containing capsule (*dummy* or *sugar pill*) identical in appearance to a pharmacologically-active prescribed drug is a case in point, as is sham surgery of the kind described in Chapter 10 in studies of the efficacy of arthroscopic surgery for knee osteoarthritis.

### 11.1.2.1    What Placebo is Not

Figure 11.1 illustrates the typical two-group placebo design in which intervention efficacy is estimated by subtracting the effect observed following placebo intervention from the effect observed following a conventional intervention. However, as mentioned above, the reputed placebo effect in such instances could be due to other nonspecific effects, the most likely of which are regression

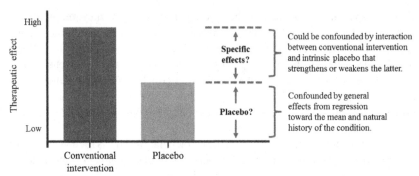

**FIGURE 11.1**   Schematic representation of the traditional placebo-controlled trial with reference to potential sources of confounding.

toward the mean and the natural history of the condition. *Regression toward the mean*, as mentioned in Chapter 6, relates to the natural variation in phenomena. Thus, irrespective of type of disorder, whether somatic or psychological, symptomatology typically waxes and wanes around an average or mean level within a natural range for a given condition, and it is likely that patients seek care when symptoms noticeably worsen. If so, part or all of any subsequent improvement in symptoms that might appear to be caused by an intervention (including placebo) could be accounted for by regression toward the mean.

The *natural history of the condition* refers to the course a disorder follows over time when left unattended. Although some conditions may worsen when untreated, much ill health is self-limiting and subject to spontaneous healing inclusive of the self- and natural-healing processes discussed in Chapter 10. Additionally, as discussed above, placebo-controlled trials frequently do not include a no-treatment control group. In such instances, as depicted in Figure 11.1, it is not possible to know with certainty whether any improvement in the disorder following placebo intervention can be attributed to a placebo effect *per se* or whether improvement was due to other influences, including regression toward the mean and healing reflected in the natural history of the condition.

Furthermore, as illustrated in Figure 11.1, almost all placebo-controlled trials assume that placebo response rates in the conventional intervention arm are equal to those in the placebo arm. It follows from this assumption that the "true" effect of an intervention can be determined by subtracting the placebo response in the placebo arm from the response in the arm containing the conventional intervention. Unfortunately, however, the underlying assumption that the placebo response is of equal size in both the conventional and placebo arms has never been adequately tested (Enck et al., 2011). Specifically, when combined with the efficacious content of conventional interventions it is not known whether placebo effects are unaffected (the usual assumption), weakened, or strengthened.

In the terminology of scientific methodology, such an effect (in this instance, a possible weakening or strengthening of placebo when combined with conventional intervention) is referred to as an *interaction effect*. Despite extensive study of the placebo phenomenon, it is essentially unknown whether placebo effects interact with the specific effects of interventions. Therefore, the usual practice of subtracting effects in the placebo arm from effects in the treatment arm may be an oversimplification. In that case, it must be admitted that it remains largely unknown what proportion of the benefit from conventional interventions is actually attributable to intrinsic and interactive placebo processes.

### 11.1.2.2   The Placebo as Simulated Therapy

"Genuine" placebo interventions involve deceit, wherein the patient and possibly (though rarely) the practitioner believe that a conventional intervention has been given when a placebo has been used. There is much debate about the ethicality of deceit in clinical practice, even if used in the patient's interests. In research, most placebo-controlled trials overcome the main ethical objection by informing the parties that either placebo or conventional intervention will be given on a random basis, with neither patient nor practitioner being aware which was given until after the trial. Patients frequently agree (i.e., give *informed consent*) to this arrangement in the interests of medical science, especially if they are promised the conventional intervention (assuming it is found to be effective and they were in the placebo arm) after the trial has been completed. When the patient alone is unaware of which treatment has been given the trial is said to be *single blind*, and when both patient and practitioner are unaware the trial is said to be *double blind*.

Importantly, the characterization of placebo as *simulated therapy* shifts the focus from "inactive" to *active* features that placebos share with conventional interventions. Although the critical psychological mechanisms responsible for placebo effects have proven difficult to characterize precisely, they are usually presumed to involve patient (and possibly practitioner) expectations, beliefs, desires, and hopes of therapeutic benefit. Thus, the comparing of placebo and conventional interventions is aimed at isolating the efficacy of the specific content of conventional interventions from more general effects of the therapeutic environment. However, from a scientific perspective, informing participants in randomized trials that they *may* receive either conventional intervention or placebo is less than ideal. Knowing that the conventional intervention may *not* be received could lessen its effectiveness (if received) due to a possible lowering of overall expectations of benefit. Similarly, knowing that one *may* receive placebo is not the same as falsely believing that conventional intervention has been delivered. Thus, the informed consent procedure may weaken the contrast between the conventional intervention and placebo arms of randomized controlled trials, thereby undermining stated purposes.

Comparison between the conventional intervention and placebo arms is potentially further undermined by the possibility of participants guessing to which arm they have been assigned. In drug trials, for example, being able (or believing, correctly or not, that one is able) to identify specific effects of a drug may suggest to participants that they have been assigned to the drug arm, and absence of discernible effects may suggest assignment to the placebo arm. This is likely to be a fairly frequent occurrence in clinical trials, where patients or practitioners or both, know or suspect that the drug under examination produces a variety of discernible symptoms or side effects. Thus, randomized controlled trials probably achieve considerably less control over expectations than is often conceded, an obvious disadvantage given that expectations are believed to be intrinsic to the placebo processes that the trials attempt to control.

In response to such problems, *balanced placebo designs*, that incorporate four rather than two conditions, are sometimes used in experimental studies. The two-by-two table in Figure 11.2 summarizes a balanced placebo design in which half of the participants are told that they will be given the conventional intervention, and then half of those are given the intervention while the other half are given placebo. Similarly, the remaining participants are told that they will be given placebo, half of whom do receive placebo while the other half are given the conventional intervention. By achieving stricter control over participants' expectations than is likely in the usual two-group placebo-controlled trial, the balanced design is more revealing of "true" drug and placebo effects. However, although the balanced design may be suitable for non-clinical

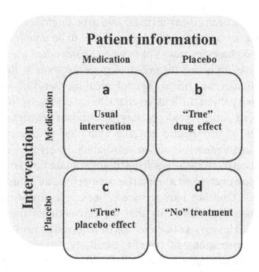

a: patients are *truthfully* told they are being administered *medication* which is given.
b: patients are *falsely* told they are being administered *placebo* but they are given medication.
c: patients are *falsely* told they are being administered *medication* but they are given placebo.
d: patients are *truthfully* told they are being administered *placebo* which is given.

**FIGURE 11.2** Balanced placebo design for experimentally investigating placebo effects.

experiments, use of deceit of that kind in clinical trials would be seen by many (including patients) as an abuse of patient trust. There are other novel designs for investigating placebos (Enck et al., 2011), but no single one is capable of addressing all relevant practical, methodological, and ethical challenges, which may explain why the placebo has remained elusive.

Among the many ingenious experimental arrangements that have been used to characterize placebos, the *open-hidden paradigm* has helped to elucidate aspects of placebo analgesia of clinical pain (Price et al., 2008). In the *open* arrangement of the paradigm, the patient, who may be recuperating postoperatively, receives a treatment in the standard manner, wherein analgesia is administered openly with the physician present and in full view of the patient. In the *hidden* arrangement, however, the patient is unaware of receiving analgesia which may be delivered intravenously by a preprogramed drug infusion pump without any personnel present. As shown in Figure 11.3, the difference between open and hidden arrangements represents the difference in the efficacy of the intervention when the patient does and does not know that the intervention has been administered. Several studies using the open-hidden paradigm have shown greater analgesia during the open arrangement than during the hidden arrangement, with the difference between the two sometimes being referred to as the placebo analgesic effect (Amanzio et al., 2001; Colloca et al., 2004).

One advantage of the open-hidden paradigm is that no participants are denied the active intervention, thereby enhancing its ethical suitability for clinical settings. A methodological strength of the design is the assurance it gives that the hidden intervention is free of placebo and other influences of the therapeutic environment such as compassionate caring. However, as explained by Colloca et al. (2004), it is misleading to refer to the difference in therapeutic effects between open and hidden interventions as "placebo," because no placebo has been given. Rather than a direct test of placebo effects, the design is revealing of the potential for active interventions to be enhanced by a

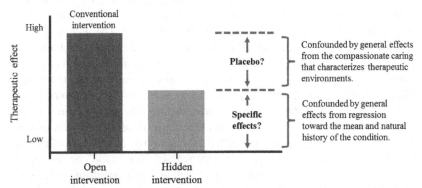

**FIGURE 11.3** Schematic representation of the "open-hidden" paradigm for examining placebo effects with reference to potential sources of confounding.

combination of knowledge that treatment has been received and active involve-ment of caregivers. It is likely that aspects of this combination of influences may be placebo, notably, expectations of improvement prompted by knowledge that active intervention has been received. However, as discussed below, it is likely that compassionate caregiver is capable of initiating healing processes that are independent of expectations.

That is, it may be an oversimplification to attribute to placebo all therapeutic outcomes not due to the specific content of interventions. As discussed above, confusion already exists in relation to failure to discriminate between placebo and seemingly quite distinct processes of regression toward the mean and nat-ural history of the condition. Additionally, it may be meaningful to differentiate between placebo effects involving expectations of improvement prompted by the therapeutic environment, and effects that are due to the empathic and com-passionate caring characteristic of therapeutic environments.

### 11.1.2.3   Practitioner Expectations

In placebo-controlled drug trials, identical preparations (pills or capsules) are used to keep patients and physicians (double) blind in relation to the conven-tional and placebo arms of the trial. For other kinds of clinical intervention, however, blinding, especially of physicians, is usually more challenging. When practitioners know which condition is conventional and which is placebo, it is possible they might unwittingly favor one or other intervention and thereby bias trial outcomes. As for the relatively few placebo-controlled trials of surgery, a single-blind arrangement has been typical. That is, patients do not know whether they have been assigned to receive conventional or placebo surgery (having been informed that they may receive either), but the surgeon is not blind as to which intervention was administered. As an added control, however, steps can be taken to ensure that independent observers, rather than the surgeons themselves, conduct post-intervention blind assessments of outcomes.

In an ingenious study of the efficacy of double-blinded surgery, Roberts et al. (1993) examined interventions once considered to be efficacious by their proponents but which later trials found to be ineffective. The ingenuity of the study derives from the fact that all of the chosen interventions, which included medical or surgical interventions for asthma, herpes simplex viral infection, and duodenal ulcer, are now considered to be devoid of any specific therapeutic effect. Consequently, prior use of those interventions by practitioners, and patients who believed them to be effective, can now be said to have involved double-blind administration of functional placebos. Based on a review of stud-ies conducted when the interventions were thought to be effective, Roberts et al. found that proponents had reported an average of 70% "good" or "excellent" outcomes for the five interventions combined, and concluded that heightened expectation of efficacy at the time the interventions were conducted was respon-sible for the high level of "success." However, it would be premature to attribute *all* of the improvements reported in those studies to placebo. Since none of the studies reviewed included a no-treatment control condition, some or all of the reported improvements may have been due to factors other than expectancy of improvement (e.g., natural progression of the condition).

### 11.1.3    The Biopsychosocial Nature of Placebos

The mechanisms responsible for placebo effects are not distinguished by whether the simulated intervention is biomedicine or psychotherapy, although the latter may involve more potential for confusion over what is placebo and what is therapy. For example, with a psychological intervention such as cognitive behavioral therapy clinical goals may include changes in patient beliefs and expectations, and those are not necessarily easily differentiated from changed beliefs and expectations due to placebo. As simulated therapy comprised of patient and practitioner enjoined in the mutual endeavor of producing beneficial patient outcomes, the placebo environment is intrinsically psychosocial. Whether interventions are somatic (e.g., pharmacotherapy or surgery) or psychological (e.g., behavioral or cognitive therapies), psychosocial aspects of the therapeutic environment common to both are responsible for placebo effects. The tangible reality, both biological and psychological, of placebo effects has been proven in numerous well-controlled experimental studies, the findings of which have often converged with clinical observations. Studies of pain and analgesia have been particularly illuminating of placebo processes, especially the role of *expectancy* and *learning* (Price et al., 2008).

Expectancy of benefit is reliably associated with biobehavioral symptom relief. In one study, patients received standard analgesic medication for relief of postoperative pain combined with intravenous infusion of saline accompanied by one of three different explanations (Pollo et al., 2001). One group was told nothing about the saline infusion (no-treatment or natural history group), a second was told that the infusion could be either a potent analgesic or a placebo (classic double-blind condition because both the patient and the person administering the infusion were blinded), and the third was told that the infusion was a potent analgesic (deceptive placebo condition). Placebo effects were measured by the amount of analgesic medication requested by patients over the three postoperative days of the trial. Compared to patients in the natural history condition, patients in the double-blind placebo condition requested and received about 20% less analgesic and patients in the deceptive placebo condition about one-third less analgesic. It seems, therefore, that expectation served as placebo analgesic.

It is reasonable to assume that the influence of placebo expectations derives from prior experience of expectations being confirmed. Certainly, placebo effects can be acquired (i.e., *learned*) as a result of prior experience, as has been demonstrated in studies of *associative learning* or *classical conditioning*. The processes of classical conditioning were first elucidated by the Russian physiologist Ivan Pavlov (1849-1936), who, in 1904, was awarded the Nobel Prize in Physiology or Medicine for his work on the physiology of digestion, especially his investigations of gastric functions in dogs. As an illustration of the role of classical conditioning in placebo, an individual's learning history may include repeated episodes of pain relief from an analgesic compound taken as a capsule. The analgesic relief that is experienced may lead to capsules of similar size and color, which contain no pharmacologically-active compound, coming by association to acquire analgesic properties.[1]

Classically-conditioned placebo analgesia has been extensively investigated and the effect has been shown to involve the endogenous opioid system (e.g., Benedetti, 2008). That is, the placebo acts as a stimulus to elicit release of an endogenous (i.e., produced naturally in the body) opiate-like substance (e.g., *endorphin*). Involvement of endorphin (a portmanteau of *endo*genous and mor-*phin*e) has been confirmed by studies in which *naloxone*, an opioid antagonist capable of blocking opioid-induced analgesia, also diminishes the effect of placebo analgesia. Because both expectancy-based and classically-conditioned placebos require prior experience with therapeutic settings, it is widely accepted that overlapping and potentially common mechanisms are involved. However, whereas expectancy involves conscious cognition, learning by classical conditioning can occur without conscious awareness. For any given placebo (simulated therapy) setting, either or both mechanisms may be involved (Stewart-Williams, 2004). This is not to say that expectancy and classical conditioning are necessarily the basis of all placebo effects. Notably, however, placebo cannot be explained away, as had been suggested by Hróbjartsson and Gøtzsche (2001, 2004), as mere response bias limited to subjective experience (e.g., pain perception).

### 11.1.4 Is the Placebo Effect in the Brain?

Interest has grown rapidly in the use of imaging and other technologies for observing blood flow and neurotransmitter release in the brain during adminis-tration of placebos (e.g., Enck et al., 2008; Jubb and Bensing, 2013; Meissner et al., 2011; Tausk et al., 2013). A recurring theme in the narrative describing those technologies is that neurobiological studies are unique in being able (at last) to confirm that placebo effects are "real." Moreover, according to that

---

1. The science of classical conditioning employs precise terminology to label various elements of the learning process. Although those details are not germane to the argument in this case, for the sake of completeness the elements in the illustrative example would be labeled as follows: the original analgesic compound is an *unconditioned stimulus* and its analgesic effect is an *unconditioned response*. Additionally, the capsule containing no pharmacologically-active compound is a *conditioned stimulus* and the analgesic effect it comes to acquire is a *conditioned response*. Classical con-ditioning (also known as *respondent learning*) is one of two main learning processes, the other being operant (or *instrumental learning*) mentioned in Chapter 2. Whereas classical conditioning is *learn-ing by association*, operant conditioning is *learning by consequences*.

reasoning, the effects *must* be real because we can see something happening in the brain. Such argumentation is further illustration of *biological exceptionalism*, as discussed in Chapter 10, involving the belief that biology is uniquely important and takes precedence over all other explanations, including those that posit behavioral and psychosocial processes and mechanisms. For example, according to Enck et al. (2008), placebo effects "*stem* from highly active processes in the brain" (p. 195, emphasis added). Similarly, according to Tausk et al. (2013) the "advent of detailed studies and modern imaging techniques have provided *the* basis to understand *the* underlying mechanisms of the placebo effect" (p. 86, emphasis added).

However, it is naïve to equate placebo with events in the brain. Placebos involve interactions between a host of subjective, behavioral, and biological processes that are far from wholly understood but certainly do not *stem* from the brain. Rather, the brain processes in question are activated by (and therefore stem from) particular conditions of the psychosocial environment characterized variously as "simulated therapy" and the "healing situation." *The* basis for understanding placebo is multifaceted and includes interacting environment-behavior-brain processes and mechanisms. To ascribe unique importance to one set of mechanisms (in this instance, neurobiological) that are but part of a wider set of interacting mechanisms only serves to impede scientific enquiry.

The hyperbole present in some of the current neurobiological narrative is evidenced in the claim that "*the* neurobiological pathways of placebo have been *uncovered*" (Tausk et al., 2013, p. 90, emphasis added). On that basis, it has been predicted that the new knowledge will enable physicians to "tap into the neurobiology of the patient's own brain [which] could truly revolutionize modern medicine" (Jubb and Bensing, 2013, p. 2717). In reality, a major disjunction exists between what is observed in imaging studies and claims about what those observations can achieve. It is as if the aforementioned commentators believe that the observed brain activity *causes* placebo outcomes when the one is merely a *correlate* of the other. The fact that there is brain activity is, in itself, trivial. All psychosocial processes, placebo or otherwise, involve brain activation. How viewing aspects of that activity through the aid of imaging technology will lead to novel clinical interventions that work is the challenge that has yet to be explained.

Intractable uncertainties persist about the size and extent of placebo effects. The cumulative evidence gives little reason to believe that there is scope for increasing placebo efficacy much beyond that which is already achievable when interventions are delivered by empathic and caring practitioners to willing and receptive patients. Accordingly, there is no reason to attach much confidence to optimistic forecasts, by neurobiologists or others, about future research unleashing hitherto unprecedented healthcare exploitation of the placebo. If further optimization of placebo efficacy is to be achieved, it is substantially more likely to come from studies of the behavioral and psychosocial dimensions of the therapeutic environment than visual representations of isolated events in the brain.

## 11.1.5    The Future of the Placebo

Few generalizations can be made about whether placebo effects are powerful or powerless. If interventions were judged powerful on the basis of generality of applicability, the placebo would qualify as powerful for that reason alone, and would be considered more so than most biomedical interventions. That is, whereas essentially all maladies are subject to some degree of placebo improvement, drugs and other biomedical interventions are, by design, applicable to a restricted range of conditions. Nevertheless, few studies, especially those conducted in clinical settings, can claim to have succeeded in isolating "pure" placebo effects. Typically, effects claimed to be placebo have been contaminated either by general effects unrelated to the therapeutic environment, including regression toward the mean and natural history of the condition, as in Figure 11.1, or by general effects of the therapeutic environment inclusive of the effects of compassionate healing, as in Figure 11.3. Furthermore, it appears that placebo efficacy varies with the type of intervention, the particular clinical outcomes, and individual differences between patients. Given that past attempts to isolate it in clinical settings have only been partially successful, it would be premature to dismiss the placebo as passé. On the other hand, there is little reason to believe, as some commentators apparently do, that the placebo offers powerful opportunities for greatly enhancing clinical outcomes in healthcare. Major challenges remain in trying to estimate, even approximately, the contribution of placebo to current biomedical practice. Trying to gauge to what extent, if at all, its contribution to healing can be enhanced is an even greater challenge, not least, because of ethical obstacles concerning the use of deception.

Continued scientific and clinical explorations of placebo effects are justified. At the same time, it is possible that widespread fascination with the placebo has distracted researchers from realizing a greater prize, namely, the exploitation of non-placebo general effects of the therapeutic environment. The healing that comes not from the specific content of interventions, but from the therapeutic environment created by genuine practitioner commitment to patient welfare, expressed through caring that is accepting, compassionate, and sincere, may exceed by a wide margin the potential therapeutic benefits of placebo. Moreover, practices that optimize general benefits of the therapeutic environment, unlike the use of placebo interventions, pose no ethical dilemmas. Thus, whereas there is continuing need for clinical efficacy trials to try to separate benefits attributable to the specific content of interventions from effects attributable to placebo, a greater need possibly exists for closer consideration of the health-enhancing potential of therapeutic processes other than placebo.

## 11.2    THE THERAPEUTIC PROCESS

A usual assumption of healthcare practice, whether somatic or psychological, is that any therapeutic benefit that accompanies the interventions practitioners employ is a *specific* outcome of those interventions. A key objective of evidence-based medicine has been to test that assumption, and thereby establish beyond reasonable doubt that the interventions employed by physicians work and that they work for the reasons claimed. Thus, when a drug is prescribed it should produce measureable benefit and that benefit should be due to specific pharmacological actions of the drug. The same should be true for surgery and the myriad other interventions that comprise modern biomedical practice. For biomedicine to claim its place of dominance in healthcare it must be shown not merely to have measurable effects but those effects must also be shown to be attributable *specifically* to the interventions. In that regard, evidence-based medicine has achieved comparatively little. As discussed in Chapter 10, the evidence-based movement has generally failed to confirm the *effectiveness* of interventions incorporated into routine clinical practice. Therefore, for any given patient, it is reasonable to suspect that therapeutic benefit from biomedical healthcare is due only partially, and possibly often not at all, to the specific interventions employed.

## 11.2.1 General Therapeutic Effects

It has been customary to refer to outcomes not attributable to the specific content of interventions as placebo effects or as *nonspecific* effects. However, the latter designation is even less satisfactory than the former. To begin with, the implication that so-called "nonspecific" effects do not have specific causes is wrong for epistemological reasons, because all specific outcomes *must* have specific causes. The mere fact that some causes of therapeutic outcomes may not be well-understood (yet) does not justify them being designated less specific than other causes that have been better characterized. Furthermore, the designation of nonspecific to mean the opposite of specific is grammatically wrong, because the antonym *of specific* is *general*, not *nonspecific*. Therefore, in the text that follows, the term *general therapeutic effects* (or simply *general effects*) is used in preference to *nonspecific* to refer to all outcomes not attributable to the specific content of interventions.

Figure 11.4 is a schematic representation of the therapeutic process, ranging from "disease" to "cure" and inclusive of specific and general effects. Rather than implying a hierarchy of effects, Figure 11.4 depicts an overarching process: a "path" from compromised to improved health that subsumes complementary subprocesses, each contributing to therapeutic change. Perhaps the first thing to note in the figure is that, irrespective of the intervention, whether somatic or psychological, conventional or "alternative," general therapeutic effects include but are not synonymous with placebo effects. The custom of labeling as placebo any outcome that is not attributable to the specific content of interventions ignores the diversity of processes that underpin therapeutic change. Moreover, although diverse lines of evidence support the conclusion that expectancy (or an expectancy-like process, such as classical conditioning)

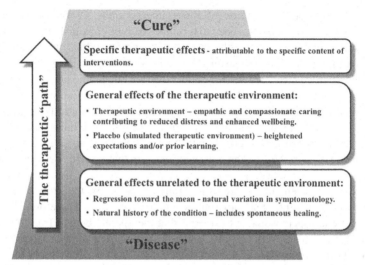

**FIGURE 11.4** The therapeutic "path": a schematic representation of healing processes.

is central to the placebo effect, expectation of improvement is not a defining feature of therapeutic benefit even if it may be facilitative.

For example, the spontaneous outpouring of feelings to a therapist, family member, friend, or even stranger by a person experiencing debilitating grief, while not dissolving the loss that is the cause of the grief, may nevertheless be therapeutic. The exchange may not have been motivated by any thought or desire of benefit, and benefit may not have been expected. Yet, after the encounter, the grieved person may feel less bereft and may show fewer signs of distress, leaving little doubt that the encounter was indeed therapeutic. Depending on the particular circumstances, effects could range from transient to transformational. Therapeutic benefit could include improvements in emotional and behavioral state, and these could be reflected in biological indicators, such as reduced levels of cardiovascular activity (e.g., heart rate and blood pressure) and lower levels of stress-related biomarkers (e.g., cortisol and $\alpha$-amylase), all signifying and potentially contributing to improved somatic and psychological health and wellbeing. In this instance, where there was no prior expectancy of benefit, actual benefit should not be thought of as placebo. Rather, as depicted in Figure 11.4, the benefit belongs to a category of general effects that derive from the non-placebo therapeutic environment.

Thus, Figure 11.4 depicts three main outcomes of the therapeutic process comprised of *specific effects* and two categories of *general effects*. At the base of the "path," though not conveying any necessary order of priority or chronology, there are general therapeutic effects that are independent of the intervention (including none). These include regression toward the mean and natural history of the condition, as described above. A second category of general effects refers to benefits that derive from the therapeutic environment. These include *placebo effects* stemming from expectancies of improvement created by interventions that mimic conventional interventions. Additionally, there are general effects of the *therapeutic environment* that are not placebo, because they do not depend on expectancy of improvement to be therapeutic. The existence of these effects possibly accounts for the universality of psychosocial healing practices discussed in Chapter 10. Although frequently occurring conjointly with expectancy of improvement, this class of general effects derives from the interpersonal encounter, and especially the environment of empathic and compassionate care that helps to decrease distress and contributes to enhanced feelings of wellbeing.

Healing, especially that which occurs in familiar healthcare settings, takes place in the social context of interpersonal interaction between patient and healthcare practitioner. It has long been believed that the social dimensions at the core of this *therapeutic environment* possess potent therapeutic potential (e.g., Frank and Frank, 1993). Some of that reputed benefit may be attributed to placebo, especially in so far as patients experience heightened expectancy of improvement. Other aspects of the experience, however, are likely to be related more to reductions in the intensity of emotions such as anxiety, worry, and despair than to cognitively-based expectancy. That is, the therapeutic environment bounded by the patient-practitioner relationship may be seen as incorporating both placebo and other less-well characterized processes that are largely emotion based. Accordingly, while acknowledging that placebo effects are created by the therapeutic environment, Figure 11.4 distinguishes between benefits attributable to expectancy of improvement and benefits facilitated by important practitioner characteristics such as empathy and compassionate caring as sources of healing and enhanced wellbeing.

Thus, to assess the potential benefit of the specific content of interventions, it is necessary first to take account of the two broad categories of general effects summarized in Figure 11.4, namely: effects that are *unrelated* (regression toward the mean and natural history of the condition) and effects that are *related* (placebo and compassionate caring) to the therapeutic environment. Specific therapeutic effects are the objects of discovery in clinical efficacy trials. However, the adequacy with which clinical trials take account of general effects, both unrelated and related to the therapeutic environment, is open to question. Considering the variety of sources of therapeutic benefit that may be unleashed during ordinary physician-patient encounters, genuinely specific effects may occur less often than is presumed. As discussed earlier in this and the preceding chapter, many interventions once thought to be effective have been shown to be without specific therapeutic benefit, and many more that continue to be widely practiced are of unknown benefit because relevant efficacy and/or effectiveness trials, as specified in Cochrane's ladder (Figure 10.2), have not been conducted.

## 11.3 WHY DO PRACTITIONERS USE INTERVENTIONS THAT HAVE NO SPECIFIC BENEFIT?

If, as argued here, many of the interventions of routine biomedicine possess little or no *specific* benefit for patients, we should ask, why do practitioners use them? Knowledge of the therapeutic process as outlined in Figure 11.4 offers a likely answer to that question. The key point is that even if much of what happens in biomedical healthcare is of little or no specific benefit, many and possibly most such interventions have some benefit. Patients consult practitioners of every description, including physicians, alternative practitioners, psychotherapists, and many others. Interventions are administered and in the large majority of instances there is improvement, if not full recovery. Neither patient nor practitioner is any the wiser when recovery has little or nothing to do with the specific content of the interventions that been employed. Essentially all interventions, barring those that are patently harmful, delivered by an understanding and compassionate practitioner to a responsive patient in need of care, will deliver a variety of general therapeutic effects that leave the patient improved compared with how they were before the consultation.

Regression toward the mean and the natural history of the condition each play a part, and neither of those sources of benefit owes anything to intervention. Additionally, placebo and other effects of the therapeutic environment play their part, which combined may be considerable. Only then is there opportunity for specific therapeutic effects to produce additional benefits. For much of the time, however, it is likely that patient recovery equates with the cumulative benefit of diverse general therapeutic processes, with little or no additional benefit attributable to the specific content of interventions. Physicians and patients persist in their use of interventions that have no specific benefit because it is benefit *per se* that counts most for physicians and patients alike, not whether the benefit is specific to the interventions that have been used. Substantial benefit to health undoubtedly comes from general effects of the therapeutic environment. Such benefit is probably sufficient to engage both physicians and patients indefinitely, despite only occasional benefit from the specific content of interventions.

On one hand, biomedical healthcare claims to have an extensive array of interventions that possess specific potency against common healthcare problems. On the other hand, evidence is lacking for the effectiveness of much of what happens in biomedical healthcare, whereas extensive evidence exists of harm. Of the benefit, much is probably due to general therapeutic effects and not the specific content of interventions. On the whole, it is reasonable to conclude that biomedicine's dominance in healthcare far exceeds the limited specific efficacy of its procedures, and a sizable proportion of patients possibly perceive the situation in those terms. One of the clearest signs of biomedicine's lack of specific efficacy is the popularity of therapies not usually considered to be part of conventional medicine.

## 11.3.1    Complementary and Alternative Medicine

Complementary and alternative medicine (CAM) includes a diverse array of therapies that do not yield easily to being organized into groups or categories. A substantially incomplete list includes alternative medicine (e.g., homeopathic medicine, naturopathic medicine), mind-body interventions (e.g., meditation, prayer, art therapy), biologically-based therapies (e.g., dietary supplements, herbal products), and manipulative and physical methods (e.g., chiropractic, massage, acupuncture) (Busato and Künzi, 2010; Yates et al., 2005). Without reference to particular interventions, there is wide agreement that in recent decades there has been a marked increase in the popularity of CAM therapies (MacArtney and Wahlberg, 2014). Although population estimates are complicated by the diversity of CAM, measured as at least one instance of use in the preceding 12 months, usage has been estimated to be 40% of adults and more than 10% of children in the United States (Barnes et al., 2008). Such popularity has been a source of puzzlement and occasional ire among conventional scientists and practitioners (MacArtney and Wahlberg, 2014; Moynihan, 2012). Some of that consternation, however, may stem from a double standard involving a higher level of skeptical enquiry being applied to CAM than to the claims of effectiveness and safety in conventional medicine.

Therapies used instead of conventional medicine are usually referred to as *alternative*, whereas therapies used alongside conventional medicine are said to be *complementary*. Considering the diversity of available therapies, it is not surprising that patients give diverse reasons for using CAM (Oldendick et al., 2000). In descending order of prevalence, reported benefits of CAM include: improved natural history of disease, disease prevention, promotion of general wellbeing, avoidance of side effects, personal control over health, symptom relief, boosting the immune system, emotional support, holistic care, improving quality of life, relief of side effects from conventional medicine, and good therapeutic relationship (Ernst and Hung, 2011). Some of these expectations reflect directly upon patients' perceptions of biomedical healthcare, including avoidance of side effects and relief of side effects from conventional medicine. All, however, reflect directly or indirectly patients' perceptions of shortcomings in biomedical healthcare, including perceived ineffectiveness, failure to take account of the "whole person," and risk of harm.

The fact that patients choose CAM partly to avoid perceived risk of harm from conventional biomedicine reflects badly on the latter, and is a situation that warrants due consideration. Although CAM is not free from adverse events, including serious and fatal outcomes (e.g., Ernst et al., 2011; Lim et al., 2011), CAM risk has generally been regarded as relatively low and below that of many common biomedical interventions (White, 2004). Regarding effectiveness, while care should be taken to avoid glib generalizations, numerous studies involving diverse therapies for a wide range of conditions have reported CAM to be no better than placebo control (e.g., Coulter and

Willis, 2004; Ernst, 2012; Lee et al., 2012; Milazzo et al., 2006; Smith et al., 2014; Yun et al., 2013). Again, considering the popularity of CAM, these facts reflect badly on conventional biomedicine. In that regard, however, as discussed above, CAM is not easily distinguishable from the effectiveness of much of modern biomedical healthcare. Accordingly, the same answer can be given in response to questions about the popularity of CAM as was given above in answer to questions concerning physician and patient persistence in the use of conventional medical care of uncertain effectiveness. As with conventional physicians and patients, it is likely that CAM practitioners and patients often persist in the use of CAM therapies because of benefit to health from the therapeutic environment (general therapeutic effects) rather than benefit from the specific content of interventions (which evidence generally suggests there is little or none in the case of CAM).

### 11.3.2 Physician-Patient Relationship

Good physician-patient communication has long been considered important in clinical practice and as having a potentially important influence on patient outcomes (Stewart, 1995). Common complex diseases, in particular, often require patients to assume considerable responsibility for implementing prescribed treatment plans. Consequently, the success of an intervention may depend on physician communication skills and ability to establish a strong bond of trust. Whereas such skills are necessary if patients are to be well-informed and suitably engaged as active participants in their medical care, studies of the quality of interpersonal communication in biomedicine suggest that requisite skills may often be lacking. In particular, interpersonal communication in biomedical healthcare seems often to involve physicians asking questions and patients answering them. An early study of recordings of physician-patient communications found that in about one-quarter of encounters patients were not given the opportunity to complete their opening statement of concerns (Beckman and Frankel, 1984). In more than two-thirds of the encounters, physicians interrupted patients and directed questions toward specific concerns of the physician. On the other hand, functional status, blood pressure, and blood glucose levels have all been found to be improved when physician-patient communication consists of patients being given more rather than less opportunity to contribute to the exchange and physicians reciprocating by providing more rather than less information in response to patient queries (Kaplan et al., 1989).

Length of biomedical healthcare consultations is a perennial concern for those interested in improving the quality of the physician-patient relationship. In a study involving six European countries, almost 200 primary practitioners, and nearly 4000 patients, average consultation time was less than 11 min (Deveugele et al., 2002). Notably, differences between countries in length of consultation coincided with prevailing marketplace incentives in each country. Incentives encouraging high patient loads in Germany and Spain, involving on average more than 200 consultations per week, were associated with shorter consultation times (about 7 min). An "open market" in Belgium and Switzerland, characterized by direct patient access to more than one physician, appears to encourage doctors in those countries to "invest time" in patient satisfaction, thereby contributing to longer consultations times (about 15 min). Organized primary healthcare in the Netherlands and the United Kingdom, characterized by restricted patient lists, gatekeeping, and government paid capitation (fee) per patient, was associated with intermediate consultation length (about 10 min).

Patient surveys suggest that CAM users are generally satisfied with their CAM experience and often may be more satisfied with the quality of practitioner-patient communication in CAM than in the conventional medical consultations they have experienced (Busato and Künzi, 2010; Stewart et al., 2000; Wong et al., 2010). Consequently, lack of perceived quality of physician-patient communication in biomedical healthcare could be an important reason CAM is popular, and this points to an area in need of research. Much of the scholarly narrative about CAM centers on questions of efficacy, with critics and proponents alike often calling for randomized controlled clinical trials of CAM (e.g., Fischer et al., 2014). As part of that endeavor, however, an important point has generally been missed, namely, the need for head-to-head comparisons between CAM and biomedical healthcare involving patients with diverse presenting problems. If, as predicted here, much of the benefit from routine biomedical healthcare is due to general effects of the therapeutic environment rather than the specific content of interventions, the efficacy of biomedicine and CAM often may not differ greatly or not at all. With regard to patient safety, the latter could even prove superior for some patients and problems.

### 11.3.3    Integrative Medicine

The fact that CAM is popular, despite on average appearing to be either no better than placebo or of unknown efficacy, should be taken as evidence of failure of conventional biomedical healthcare, and of evidence-based medicine in particular. If evidence-based medicine had succeeded in ensuring the dissemination of uniformly *effective* biomedical practice, as envisioned by the founders of the movement, there would be little cause for large numbers of patients to turn to CAM. Admittedly, in turning to CAM, most patients have not abandoned conventional medicine, but use CAM as a supplement to biomedical healthcare. Nevertheless, biomedicine seeks to provide comprehensive healthcare, which it evidently is not succeeding in doing in the opinion of the substantial and apparently growing number of patients who use CAM. That biomedicine is perceived by many lay people to be less than comprehensively successful in meeting personal and population healthcare needs is one thing, but it is quite another matter altogether that substantial numbers of physicians apparently share that perception as evidenced by the rise of *integrative medicine*.

Integrative medicine aims to assimilate elements of CAM into the methods of diagnosis and treatment of conventional biomedicine (Gorski, 2014). Typically, advocates draw heavily on the CAM narrative by emphasizing "wellness of the entire person… as primary goals…in the context of a supportive and effective physician-patient relationship" (Bell et al., 2002). In that regard, Coulter and Willis (2004) have raised an important point of criticism. They argue that for conventional medicine and CAM to be integrated the knowledge bases of the two approaches should be commensurable, that is, they should not be logically inconsistent. Yet, avoidance of logical inconsistency may be impossible.

To take homeopathy as an illustrative example, a core proposition of that popular CAM therapy relates to *infinitesimal dilutions* and *succussion* (vigorous shaking) of substances (Milgrom, 2009), wherein the therapeutic properties of substances are alleged to be greatly increased when diluted and succussed over and again until only a few molecules, or even none at all, of the original substance remain, by which time the solution is reputed to be imbued with the property known as *memory of water*. Such ideas are incompatible with the scientific underpinning of conventional biomedical healthcare. Obvious incompatibilities may possibly be resolved by the selective incorporation of CAM elements into conventional biomedicine. However, in so doing, practitioners of integrative medicine are likely to fail to achieve meaningful "integration," while managing to create a third indistinct and ill-defined form of healthcare that is neither conventional medicine nor CAM.

The integrative-medicine movement is to be applauded for recognizing major failings of conventional biomedical practice. However, an air of desperation permeates the movement's attempt to "humanize" biomedicine by incorporating, either wholesale or selectively, CAM theories, practices, and philosophies. Rather than looking to be rescued by a body of practices as unfounded as those of CAM, the cause of integrative medicine and the welfare of the patients it seeks to serve might be better supported by less reckless disregard for scientific principle. More could be achieved by adopting systematic and empirical means for addressing the major shortcomings of conventional biomedicine that the proponents of integrative medicine evidently perceive, including poor therapeutic environment, commercial culture, and dearth of evidence of effectiveness and cost-effectiveness. The advent of integrative medicine was a response to failures of conventional medicine, and is further evidence of the recurring incapacity of conventional biomedical healthcare to remedy fundamental problems of its own making. By courting pseudoscience, integrative medicine as currently construed runs the risk of compounding past failure. If the aim of integrative medicine is to humanize biomedical practice, a more successful course might have been to seek integration with psychological science, where respect for empirical evidence is as unwavering as in the biological and physical sciences that are the current foundations of medical science and practice.

# REFERENCES

Amanzio, M., Pollo, A., Maggi, G., Benedetti, F., 2001. Response variability to analgesics: a role for non-specific activation of endogenous opioids. Pain 90, 205–215.

Barnes, P.M., Bloom, B., Nahin, R.L., 2008. Complementary and alternative medicine use among adults and children: United States, 2007. National health statistics reports, 12. National Center for Health Statistics, Hyattsville, MD.

Beckman, H.B., Frankel, R.M., 1984. The effect of physician behavior on the collection of data. Ann. Intern. Med. 101, 692–696.

Beecher, H.K., 1955. The powerful placebo. J. Am. Med. Assoc. 159, 1602–1606.

Bell, I.R., Caspi, O., Schwartz, G.E., et al., 2002. Integrative medicine and systemic outcomes research: issues in the emergence of a new model for primary health care. Arch. Intern. Med. 162, 133–140.

Benedetti, F., 2008. Mechanisms of placebo and placebo-related effects across diseases and treatments. Annu. Rev. Pharmacol. Toxicol. 48, 33–60.

Busato, A., Künzi, B., 2010. Differences in the quality of interpersonal care in complementary and conventional medicine. BMC Complement. Altern. Med. 10, 1–15. http://dx.doi.org/10.1186/1472-6882-10-63.

Colloca, L., Lopiano, L., Lanotte, M., Benedetti, F., 2004. Overt versus covert treatment for pain, anxiety, and Parkinson's disease. Lancet Neurol. 3, 679–684.

Coulter, I.D., Willis, E.M., 2004. The rise and rise of complementary and alternative medicine: a sociological perspective. Med. J. Aust. 180, 587–590.

Deveugele, M., Derese, A., van den Brink-Muinen, A., et al., 2002. Consultation length in general practice: cross sectional study in six European countries. Br. Med. J. 325, 472.

Enck, P., Benedetti, F., Schedlowski, M., 2008. New insights into the placebo and nocebo responses. Neuron 59, 195–206.

Enck, P., Klosterhalfen, S., Weimer, K., et al., 2011. The placebo response in clinical trials: more questions than answers. Philos. Trans. R. Soc. Lond. B Biol. Sci. 366, 1889–1895.

Ernst, E., 2012. Homeopathy for eczema: a systematic review of controlled clinical trials. Br. J. Dermatol. 166, 1170–1172.

Ernst, E., Hung, S.K., 2011. Great expectations: what do patients using complementary and alternative medicine hope for? Patient 4, 89–101.

Ernst, E., Lee, M.S., Choi, T.Y., 2011. Acupuncture: does it alleviate pain and are there serious risks? A review of reviews. Pain 152, 755–764.

Fischer, F.H., Lewith, G., Witt, C.M., et al., 2014. High prevalence but limited evidence in complementary and alternative medicine: guidelines for future research. BMC Complement. Altern. Med. 14, 1–9. http://dx.doi.org/10.1186/1472-6882-14-46.

Frank, J.D., Frank, J.B., 1993. Persuasion and healing: a comparative study of psychotherapy, third ed. Johns Hopkins University Press, Baltimore, MD.

Gorski, D.H., 2014. Integrative oncology: really the best of both worlds? Nat. Rev. Cancer 14, 692–700. http://dx.doi.org/10.1038/nrc3822.

Hróbjartsson, A., Gøtzsche, P.C., 2001. Is the placebo powerless? An analysis of clinical trials comparing placebo with no treatment. N. Engl. J. Med. 344, 1594–1602.

Hróbjartsson, A., Gøtzsche, P.C., 2004. Is the placebo powerless? Update of a systematic review with 52 new randomized trials comparing placebo with no treatment. J. Intern. Med. 256, 91–100.

Hróbjartsson, A., Gøtzsche, P.C., 2010. Placebo interventions for all clinical conditions. Cochrane Database of Syst. Rev 1, 1–448. http://dx.doi.org/10.1002/14651858.CD003974.pub3.

Jubb, J., Bensing, J.M., 2013. The sweetest pill to swallow: how patient neurobiology can be harnessed to maximise placebo effects. Neurosci. Biobehav. Rev. 37, 2709–2720.

Kaplan, S.H., Greenfield, S., Ware Jr., J.E., 1989. Assessing the effects of physician-patient interactions on the outcomes of chronic disease. Med. Care 27, S110–S127.

Lee, M.S., Choi, J., Posadzki, P., Ernst, E., 2012. Aromatherapy for health care: an overview of systematic reviews. Maturitas 71, 257–260.

Lim, A., Cranswick, N., South, M., 2011. Adverse events associated with the use of complementary and alternative medicine in children. Arch. Dis. Child. 96, 297–300.

MacArtney, J.I., Wahlberg, A., 2014. The problem of complementary and alternative medicine use today eyes half closed? Qual. Health Res. 24, 114–123. http://dx.doi.org/10.1177/1049732313518977.

Meissner, K., Bingel, U., Colloca, L., et al., 2011. The placebo effect: advances from different methodological approaches. J. Neurosci. 31, 16117–16124.

Milazzo, S., Russell, N., Ernst, E., 2006. Efficacy of homeopathic therapy in cancer treatment. Eur. J. Cancer 42, 282–289.

Milgrom, L.R., 2009. The eternal closure of the biased mind? The clinical and scientific relevance of biophysics, infinitesimal dilutions, and the memory of water. J. Altern. Complement. Med. 15, 1255–1257.

Moseley, J.B., O'Malley, K., Petersen, N.J., et al., 2002. A controlled trial of arthroscopic surgery for osteoarthritis of the knee. N. Engl. J. Med. 347, 81–88.

Moynihan, R., 2012. Assaulting alternative medicine: worthwhile or witch hunt? Br. Med. J. 344, 1–2. http://dx.doi.org/10.1136/bmj.e1075.

Oldendick, R., Coker, A.L., Wieland, D., et al., 2000. Population-based survey of complementary and alternative medicine usage, patient satisfaction, and physician involvement. South. Med. J. 93, 375–381.

Pollo, A., Amanzio, M., Arslanian, A., et al., 2001. Response expectancies in placebo analgesia and their clinical relevance. Pain 93, 77–84.

Price, D.D., Finniss, D.G., Benedetti, F., 2008. A comprehensive review of the placebo effect: recent advances and current thought. Annu. Rev. Psychol. 59, 565–590.

Roberts, A.H., Kewman, D.G., Mercier, L., Hovell, M., 1993. The power of nonspecific effects in healing: implications for psychosocial and biological treatments. Clin. Psychol. Rev. 13, 375–391.

Sihvonen, R., Paavola, M., Malmivaara, A., et al., 2013. Arthroscopic partial meniscectomy versus sham surgery for a degenerative meniscal tear. N. Engl. J. Med. 369, 2515–2524.

Smith, P.J., Clavarino, A., Long, J., Steadman, K.J., 2014. Why do some cancer patients receiving chemotherapy choose to take complementary and alternative medicines and what are the risks? Asia Pac. J. Clin. Oncol. 10, 1–10.

Stewart, M.A., 1995. Effective physician-patient communication and health outcomes: a review. Can. Med. Assoc. J. 152, 1423–1433.

Stewart, D., Weeks, J., Bent, S., 2000. Utilization, patient satisfaction, and cost implications of acupuncture, massage, and naturopathic medicine offered as covered health benefits: a comparison of two delivery models. Altern. Ther. Health Med. 7, 66–70.

Stewart-Williams, S., 2004. The placebo puzzle: putting together the pieces. Health Psychol. 23, 198–206.

Tausk, F., Ader, R., Duffy, N., 2013. The placebo effect: why we should care. Clin. Dermatol. 31, 86–91.

White, A., 2004. A cumulative review of the range and incidence of significant adverse events associated with acupuncture. Acupunct. Med. 22, 122–133.

Wong, L.Y., Toh, M.P., Kong, K.H., 2010. Barriers to patient referral for complementary and alternative medicines and its implications on interventions. Complement. Ther. Med. 18, 135–142.

Yates, J.S., Mustian, K.M., Morrow, G.R., et al., 2005. Prevalence of complementary and alternative medicine use in cancer patients during treatment. Support. Care Cancer 13, 806–811.

Yun, Y.H., Lee, M.K., Park, S.M., et al., 2013. Effect of complementary and alternative medicine on the survival and health-related quality of life among terminally ill cancer patients: a prospective cohort study. Ann. Oncol. 24, 489–494.

Chapter 12

# Prevention and Control
# of Disease

*[N]early all the major improvements in national health have been due to preven-*
*tion, mostly resulting from population-wide changes.*

(Rose, 1992, p. 103)

## Contents

The realization that health and disease are determined more by psychosocial factors than by biology has profound implications for healthcare. Both sets of broad determinants, biological and psychosocial, are important, but psychosocial determinants explain the greater portion of the distribution of personal and population health. In essence, it is human habits and habitats rather than biological endowment that most determine who is healthy and who is sick. Moreover, whereas biology is an important mediator of disease and disease outcomes are commonly defined biologically, the consequences of disease are profoundly behavioral and social. Accordingly, instead of being overwhelmingly biomedical in orientation, as now, healthcare needs to reorient to incorporate a judicious balance of psychosocial and biological aims and strategies.

The Health of Populations. http://dx.doi.org/10.1016/B978-0-12-802812-4.00012-6

Healthcare for optimal population health must first be directed at promoting health and preventing disease by minimizing exposure to risk factors. When disease occurs, a balance is required consisting of risk factor minimization augmented by effective and cost-effective biomedical intervention to control disease progression.

The essence of these ideas, arguments, and conclusions are not new. They existed in the "medicine" of Ancient Greece (see Box 4.1), and have persisted ever since, though mostly in nascent form. Three decades ago, in a seminal article titled *Sick Individuals and Sick Populations*, epidemiologist Geoffrey Rose (mentioned in Chapter 9) foresaw more clearly than most what was, and is, needed to maximize the health of populations (Rose, 1985). The recurring refrain of habits and habitats in the present book echoes Rose's (1992) belief that for individuals and populations:

> The scale and pattern of disease reflect the way that people live and their social, economic, and environmental circumstances (p. 1).

Compared with what he saw as the "rescue operation" of individualized clinical medicine, population-wide prevention addresses the "essential determinants" of health and disease (Rose, 1992). Unfortunately, biological exceptionalism, faith in technology, and gargantuan commercial incentives have conspired to keep healthcare mired in a biomedical rescue operation.

Only recently do there appear to be real prospects for the evolution of genuinely new approaches to healthcare. Recognition of the need for radical change in healthcare is being propelled by the global crisis in noncommunicable diseases. Confronted with the current crisis, the predominantly biomedical approach that has characterized the last half-century of developments in healthcare has proven unfit for purpose and is unsustainable. The current global burden of disease is evidence of what has long been suspected: biomedical dominance produces healthcare of disappointing efficacy, dubious effectiveness, and ever-diminishing cost-effectiveness.

## 12.1 THE POWER OF PREVENTION: REORIENTING HEALTHCARE

> The doctor of the future will give no medicine but will interest his patients in … the cause and prevention of disease.
> (1902-1903, attributed to Thomas Edison, 1847-1931, American inventor and business man, http://www.snopes.com/quotes/futuredoctor.asp)

Rose (1981, 1992) delineated two contrasting but complementary approaches for responding to disease, especially common complex diseases: the *high-risk approach* and the *population approach*. These approaches are distinguished primarily by their respective aims. The high-risk approach, which characterizes clinical medicine, aims to *control* disease in "conspicuously vulnerable" (sick) individuals, whereas the population approach aims to reduce the incidence of disease through *prevention*. While proposing that healthcare should incorporate both approaches, Rose (1981) argued that prevention is the more potent contributor largely because biomedicine typically comes too late to make any real difference to the health of populations. Population-wide prevention is effective because it addresses the causes of health and disease in advance of the development of disease processes (Rose, 1992). Thus, even in the minority of instances when biomedicine succeeds in curing disease, the disease and its consequences have already occurred, which is an obvious lesser outcome compared to disease prevention.

Optimal healthcare depends on risk factor reduction from prebirth to end of life ("womb to tomb") complemented by strategic use of biomedical healthcare when indicated. The benefits of lifelong risk factor reduction include maximization of longevity and minimization of morbidity. Compared with current patterns of death and disease, successful population-wide preventive healthcare promises to compress morbidity leading to personal and population survival curves tending to become more rectangular in the manner described in Chapter 3. In consequence, there is less lifelong need for medicine and less end-of-life suffering. Despite continuing need for medical "rescue operations" after injury and the tendency toward increased need during later stages of life, life-course risk factor reduction leads to less need for biomedicine during all stages of life.

## 12.1.1    Incidence and Prevalence

The epidemiological concepts of incidence and prevalence help to clarify important differences between preventive healthcare and biomedical healthcare. *Incidence* refers to the *number of new cases of disease in a defined population within a specified time period*. Incidence is expressed as a rate, reflecting the fact that it refers to the number of events per unit of time. *Prevalence* is an indicator of the population burden of disease, and refers to the *proportion of the population that has the disease at a given moment in time (point prevalence)* or *at some time during a given period (period prevalence)*. Incidence rate and prevalence of disease are linked, wherein prevalence is approximately equal to *incidence times duration*. Thus, when the duration of disease is brief, because the natural time course is brief (those afflicted recover quickly or die) or interventions are available and produce rapid cure, prevalence tends to be lower. On the other hand, when disease is chronic (i.e., is of long duration), the pool of people afflicted necessarily expands, and prevalence is higher.

Preventive healthcare addresses incidence, whereas the disease control focus of clinical medicine mostly addresses prevalence. One benefit of prevention is that, by thwarting incidence, prevalence is also curbed. That is, the number of cases of disease at any time is kept low (comparatively speaking). Conversely, to the extent that biomedicine responds to disease after it has occurred, incidence of that disease is unaffected. Although biomedicine aims to curb prevalence, the endeavor is limited because cure, especially of common complex diseases, is comparatively rare. High and rising prevalence of common complex diseases, a reflection of high incidence and chronicity, is proof of the failure of biomedical healthcare. That the failure is of immense proportions is evidenced by the fact that common complex diseases are the major contributors to the current global burden of disease. The situation is certain to worsen unless action is taken to redirect healthcare priorities. Current healthcare practice, which gives priority to the clinical management of disease, is ill-fated because it does little to reduce incidence of disease and is only modestly effective in limiting prevalence. Conversely, success in prevention will lower incidence directly, which necessarily contributes to lower prevalence and lower overall burden of disease.

## 12.1.2  What Is Prevention and What Is It Not?

Individualized clinical medicine having little or no effect on disease incidence is synonymous with clinical medicine having little or no disease-preventing impact. Yet, it is routinely claimed that biomedical healthcare "prevents" disease. The reason for this apparent contradiction is that *prevention* in the health-science literature has multiple definitions, some of which are counterintuitive and others contradictory. Sadly, considering the importance of the concept, little effort has been made to clarify the confusion, part of which may be due to the fact that prevention is used to describe both actions and outcomes. For example, if a local water supply becomes contaminated with microbial pollutants, the responsible public health authority may issue advice to householders that water should be boiled before being consumed. Both the action, including health-authority advice and householder compliance with that advice, as well as the outcome (avoidance of infection), are likely to be described as prevention. However, when thinking about prevention, precedence should be given to outcomes. Thus, if the particular microbe in question is resistant to boiling, and disease is not prevented, the action cannot be said to have been preventive. Unfortunately, discussions about actions that are intended or claimed to be preventive often make no reference to, or proceed without much knowledge of, outcomes.

A major theme of the present book is that prevention of exposure to risk factors can contribute far more to population health than clinical medicine. Accordingly, the reader is encouraged to exercise the same caution in respect of that claim as should be exercised in respect of claims for disease prevention in any other context, such as personalized medicine and over-zealous population screening (discussed in Chapter 9) or individualized preventive primary care (discussed below). That is, claims in respect of prevention should be assessed on the basis of outcomes showing actual improvements in health and decreases in disease, not merely on the basis of actions *intended* to have preventive outcomes. Expectation of preventive benefit from reduced exposure to identified risk factors is well-founded for some risks but less so for others. For example, the health benefits of decreased tobacco use are well-founded, but the benefits of increased consumption of fruit and vegetables are less-well understood. In either case, and in all such cases, it is important to note that evidence of reduced rate of smoking or presumed improvement in diet is not sufficient to support claims of disease prevention. Only the measurement of outcomes showing improvements in health and decreases in disease can confirm whether changes in particular habits and habitats intended to be preventive are so in fact.

Notwithstanding possible confusion due to prevention being used to describe actions as well as outcomes, still greater confusion arises due to efforts to differentiate between different preventive actions. An early account of prevention distinguished between actions "that avert the occurrence of disease (*primary prevention*) and interventions that halt or slow the progression of a disease ... (*secondary prevention*)" (Starfield et al., 2008, p. 580). Later accounts largely retained the earlier meaning for primary prevention but distinguished between secondary prevention involving early-stage disease and *tertiary prevention* to halt the progression and limit the consequences (e.g., pain) of manifest disease (Starfield et al., 2008). More recent accounts have included additional concepts and further distinctions. Just as primary prevention is generally targeted at asymptomatic (healthy) populations, a newer concept, *primordial prevention*, refers to intervention for asymptomatic populations not yet exposed to a targeted risk factor (Mensah et al., 2005).

For example, many in the field of prevention studies are comfortable using the term prevention to refer to anti-tobacco campaigns that encourage the healthy population not yet smoking (e.g., adolescents) not to start while also using prevention to refer to a program that encourages healthy existing smokers to quit. The concept of primordial prevention, however, distinguishes between two such interventions. Encouraging non-smokers (not yet exposed to the health risks of smoking) not to start is primordial prevention, whereas encouraging asymptomatic smokers (those already exposed to the risk factor) to quit is primary prevention. Within that framework, the recent call for a "tobacco-free world" (Beaglehole et al., 2015) is a primordial prevention initiative. Yet another addition to the prevention vocabulary, but one with inconsistent uses, is *quaternary prevention*. On one hand, quaternary prevention has been described as the prevention of medical harm (Kuehlein et al., 2010). On the other hand, the same term has been used to describe interventions for persons whose condition is deemed to be "severe" (Mensah et al., 2005).

### 12.1.2.1   Primary and Secondary Prevention

Notwithstanding the many and varied attempts to define distinct types of prevention, delineations often are arbitrary and subject to confusing overlap. In that context, it is important to preserve the essential meaning of prevention as *disease non-occurrence*. In particular, it is important not to conflate the meaning of *prevention of disease incidence* with ideas about clinical *control of disease progression*. To avoid confusion, the present text distinguishes between primary and secondary prevention only, and eschews use of the remaining varieties (primordial, tertiary, and quaternary). Here, *primary prevention* refers to prevention or postponement of disease (and pathophysiological precursors of disease) due to reduced exposure to risk factors in substantially asymptomatic populations. *Secondary prevention* refers to prevention or postponement of manifest disease in individuals already showing signs of pathophysiology (e.g., elevated blood pressure or blood glucose levels). Even so, the distinction between "substantially asymptomatic" and "signs of pathophysiology" is somewhat arbitrary because presence or absence of most disease is part of a continuum that has no precise demarcation between health and disease.

A further basis for distinguishing primary and secondary prevention concerns the nature of the preventive action. Primary prevention usually involves behavioral and social interventions (e.g., changed patterns of food choices and regulatory controls on advertising junk food to children), whereas secondary prevention often involves biomedical intervention (e.g., medication for elevated blood pressure). In addition, primary prevention reduces incidence of disease and in turn prevalence of early- and late-stage disease. Secondary prevention, on the other hand, does not affect the population incidence of early-stage disease processes, although it limits the incidence and therefore the prevalence of manifest disease. Unfortunately, although the usage proposed here is broadly representative, it is not universal. For example, pharmacotherapy for raised blood pressure in individuals who are asymptomatic (i.e., are not experiencing evident symptoms) but are showing early signs of pathophysiology, which according to the definitions proposed here would be an instance of secondary prevention, is sometimes described simply as "prevention" but also sometimes as "primary prevention."

### 12.1.3  Prevention Versus Control

Concepts of disease prevention have come to encompass such a confusing array of interrelated ideas and aspirations that one critic has suggested that "the concept of prevention has lost all practical meaning" (Starfield et al., 2008, p. 508). The point is well taken, especially in relation to attempts to merge prevention with clinical practice, as implied by *preventive medicine*. Rose (1992), the great advocate of prevention, must shoulder some responsibility for this contradictory terminology. His seminal work on prevention included the phrase "preventive medicine" in the title. It is curious that he should have chosen that phrase because his espousal of prevention was in direct opposition to the very thing that medicine most emphasizes, namely, the treatment of extant disease in "high risk" individuals. Criticism of *preventive medicine* and similarly fused terms, such as *clinical prevention* and *lifestyle medicine*, could be countered on the basis that biomedical procedures are sometimes used with healthy populations for the purpose of preventing disease (e.g., population screening and immunization). Such interventions should indeed be considered preventive (when they work), but they should be considered preventive *healthcare* not medicine. Using "medicine" as a proxy for "healthcare" perpetuates the misleading conflation of medicine wherein the two terms are treated as if they were synonymous.

Moreover, the fusion of prevention and medicine suggests that a comfortable bond exists between the two, when the linkage is distinctly uneasy. Indeed, on the premise that anything that seeks to relieve suffering is preventive at some level, linking the two terms encourages the view that *all* clinical practice is intrinsically preventive. However, the need for clinical relief of manifest disease is precisely what preventive healthcare seeks to eliminate to the greatest extent possible. As such, a firm distinction can and should be made between *prevention of disease incidence* and clinical/biomedical *control of manifest disease prevalence*. Table 12.1 summarizes the main differences between the two approaches. Biomedical healthcare focusses on individuals at high risk due to manifest disease, whereas prevention focuses on asymptomatic populations assumed to be healthy and at "low risk." Biomedicine comes late in the time course of disease progression and aims to arrest disease by treating pathophysiology. Preventive healthcare, on the other hand, seeks to preempt the need for biomedical intervention by reducing exposure to the risk factors that cause disease.

## 12.1.4    The Prevention Paradox

*A population-wide preventive measure may offer a disappointingly trivial benefit to individuals, but yet its cumulative benefit for the population as a whole can be unexpectedly large*

(Rose, 1992, p. 102)

**TABLE 12.1** Summary of differences between conventional biomedical healthcare and population-wide health promotion and disease prevention

| | Healthcare strategy | |
| --- | --- | --- |
| **Dimension** | **Biomedicine** | **Prevention** |
| Focus | "High-risk" individual | "Low-risk" population |
| Intervention target | Pathophysiology | Health risk factors |
| Timing of intervention | After disease onset | Before evidence of disease |
| Objective of intervention | Alter pathophysiology | Reduced exposure to risk factors |
| Intended outcome | Arrest disease progression | Prevent or postpone disease onset |

While arguing that the "mass approach is inherently the only ultimate answer to the problem of a mass disease" (p. 1850), Rose (1981) acknowledged a limitation of preventive intervention such that "however much it may offer to the community as a whole, it offers little to each participating individual" (p. 1850). He illustrated his concern by referring to immunization against diphtheria, which may save the life of one child for each of many hundreds who derive no benefit because they would not have contracted the disease anyway. Similarly, he described a lifetime of wearing seatbelts as being of no benefit for the majority of drivers, although the minority who suffer a traffic accident may benefit greatly by being spared death or serious injury. As both examples illustrate, the population benefit may be great (cumulatively, fewer deaths among infants and fewer traffic-related deaths and injuries) even if the majority of individuals would in any event have been spared those outcomes. Rose was particularly concerned that despite greater overall health benefit from prevention than biomedicine, the ratio of benefit wherein an intervention "applied to many will actually benefit few" would be a barrier to the adoption of preventive healthcare.

Notwithstanding the challenges of this dilemma, which Rose (1981, 1985, 1992) labeled the *prevention paradox*, it is arguable that his concern was disproportionate. In the examples he gave of immunization and the wearing of car seatbelts, it may not be strictly correct to say that all who in any event would not have experienced illness, injury, or death experienced *no* benefit. To begin with, the effective immunization of one child who would otherwise have contracted diphtheria might prevent spread of the infection to many others (though typically not everyone). Additionally, if not the children themselves, then certainly many parents, will have been comforted by the presumption that immunization meant that their child was unlikely to contract diphtheria. Likewise, many car drivers probably derive similar psychological benefit from comfort in knowing that their chances of avoiding serious injury and death are substantially improved by the wearing of a seatbelt.

Moreover, the illustrations Rose used were dichotomous, either one is immunized or not and does or does not contract diphtheria, and one wears a seatbelt or not and does or does not have an accident. However, as discussed in the following section, most prevalent diseases are not dichotomous nor are their causes. A downward shift in the population prevalence of smoking means that those who continue will, on average, smoke less and many others will quit. All who have reduced are likely to benefit, as will many who have never smoked due to lower levels of exposure to secondhand smoke. To paraphrase Rose, the power of prevention often derives from the shifting of population norms of behavior. When that happens, it is usual for benefits of one kind or another to accrue to everyone.

## 12.1.5    Disease Is Continuous Not Dichotomous

With relatively few exceptions, comprised largely of congenital disorders determined by a dominant gene with high penetrance, "diseases, whether genetic or

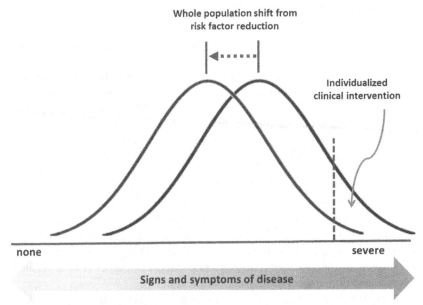

Whole population shift from
risk factor reduction

Individualized
clinical intervention

severe

Signs and symptoms of disease

**FIGURE 12.1**   Schematic representation of the relative impact on population health of individualized clinical intervention and whole-population risk factor reduction. *(Note. The vertical dashed line represents the intervention threshold, which is the consensually agreed cut-off above which clinical intervention is recommended. Whole-population shifts to the left of the intervention threshold produce comparatively large reductions in the incidence of disease and the accompanying need for clinical intervention.)*

acquired, come in all sizes" (Rose, 1992, p. 9). This fact provides a major point of contrast between prevention and clinical medicine. Whereas "the traditional principle of medical diagnosis [holds that] the world falls into just two classes, namely those who have [disease] and those who do not," the prevention approach exploits the fact that disease "has no natural definitions" (Rose, 1992, pp. 6-8). In particular, common complex diseases, including cardiovascular diseases, cancers, chronic respiratory diseases, and diabetes, are not single entities but disease processes comprising a continuum ranging from more-or-less healthy to more-or-less diseased, as illustrated in Figure 12.1.

Despite disease generally consisting of a continuum of severity from few or no signs and symptoms to confirmed pathology, clinical medicine demands diagnostic categorization. Although such categorization is entirely consistent with the logic of medicine, it imposes artificial distinctions between those who have disease and those who do not, indicated by the vertical line in Figure 12.1. Thus, routine clinical decision making is intrinsically binary in nature ("To treat or not to treat?"), wherein positive diagnosis signifying presence of disease is the basis for intervention. Absence of positive diagnosis implies absence of disease, necessitating acceptance of the default option, namely health, not requiring intervention. It is no wonder, then, that medicine falls short of being an optimal strategy for maximizing the health of populations. Besides espousing rudimentary ideas about health (e.g., health is the absence of disease), medicine is founded on a fundamentally flawed concept of the nature of disease.

An illustration of the continuity of disease is provided by the relationship between blood pressure level and the population distribution of cardiovascular diseases such as coronary heart disease and stroke. Physicians frequently behave as if a dichotomous relationship exists between blood pressure and disease, wherein "low" blood pressure denotes health (the default for absence of disease) and "high" blood pressure denotes disease or risk of disease sufficiently high to warrant intervention. The approach is internally consistent because of the binary (yes/no) nature of the decisions that physicians are typically required to make when assessing the suitability of a patient for a given intervention. The physician is guided in the clinical decision-making process by guidelines that incorporate thresholds (the vertical line in Figure 12.1) representing consensus agreements among experts about when the benefits of intervention are likely to outweigh harm, including the negative side effects and other risks associated with the use of therapeutic drugs and medical devices.

Notwithstanding internal consistency, much of the clinical decision-making logic of biomedicine is not consistent with reality. In the real world, blood pressure level does not exist in a binary or dichotomous relationship to disease. With the exception of pathologically very low and very high levels, the relationship between blood pressure and disease is primarily linear in that disease risk increases progressively and more-or-less evenly (forming an approximate straight line on a graph) as blood pressure increases (Prospective Studies Collaboration, 2002). Heart attack and stroke, for example, can and do occur at any level of blood pressure, although their likelihood increases as blood pressure increases. In that sense, everyone is simultaneously more-or-less healthy and more-or-less diseased. For that reason, prevention (Rose's "population approach") is potentially good for everyone, irrespective of how healthy or how sick they may appear to be. Population-wide prevention leads to whole-population shifts in risk factors, benefiting high-risk patients in receipt of treatment while simultaneously reducing the number in need of treatment.

The First International Conference on Health Promotion, convened by the World Health Organization (WHO) and held in Ottawa, Canada, in 1986, highlighted the need for healthcare services to refocus away from preoccupation with control of manifest disease toward greater emphasis on disease prevention (WHO, 1986). The Conference produced an agreement, known as the Ottawa Charter, which confirmed the strong consensus among the international community that priority should be given to disease prevention in order to deliver "health for all" by the end of the millennium. The Ottawa Charter failed in its main objective of delivering health for all by 2000, and even now, almost three decades after its signing, no substantial reorienting of healthcare has occurred. Healthcare remains primarily remedial in focus, with few signs of being refocussed away from traditional preoccupations with rescue operations that seek to limit disaster after disease has occurred toward preventive interventions that avoid disaster. If anything, healthcare is less preventive than ever, as evidenced by commercial intensification and increased public expenditure on marketable technologies in the form of drugs, devices, and genomic-inspired personalized medicine.

### 12.1.5.1 Prevention Lag Time

The fact that disease is generally distributed in the population as a continuous function, ranging from few or no signs and symptoms to confirmed pathology, has implications for the speed with which disease prevention has detectable benefits. It is often assumed that the lag time is necessarily long when the targeted diseases involve common complex diseases that develop slowly after decades of exposure to risk factors. Therefore, it is reasoned, any benefits from interventions that reduce exposure to known risk factors will not emerge until decades later. This presumed relation is sometimes proffered as an explanation for the lack of political commitment to disease prevention strategies that have the potential to deliver substantial population benefits, which, it is said, will not be enjoyed until sometime, probably decades, in the future. Thus, it is argued, there is a lack of incentive for action as any politician devoted to the task of disease prevention is likely to be out of office by the time measurable benefits occur. Such reasoning, however, is flawed. Population-wide risk factor reduction can bring major benefits to health almost immediately.

For example, studies of alcohol consumption and disease have consistently shown short lag time between changes in consumption and health benefits. This is not surprising when it occurs in relation to acute outcomes such as alcohol-related injury (accidents and violence), which indeed show *no* noticeable lag between reduced alcohol consumption and reduced incidence of injury (Rehm and Shield, 2014). However, even in relation to slowly progressing diseases, such as liver cirrhosis and heart disease, evidence shows that short lag time is the rule. The reason for this relates to the fact that disease is continuous not dichotomous. Individuals in whom disease processes are already close to the threshold of an index event (e.g., death) are likely to be immediately responsive to changes in exposure, wherein the event is either triggered if exposure continues or postponed if exposure ceases. In other words, it is essential to take account of prior history of population exposure when considering the potential effects of changes in exposure that occur at a given time and to remember that disease is continuous not dichotomous (Norström and Skog, 2001).

Staying with the example of alcohol, although 15-20 years of excessive drinking may be needed to bring an individual to imminent death from liver damage, a population in which drinking has been prevalent for at least that period of time will contain a constant *reservoir* of such individuals. Theoretical expectations are confirmed by empirical evidence in showing abrupt population reduction in mortality from liver cirrhosis and heart disease following per capita reduction in alcohol consumption (Holmes et al., 2012). Thus, a reduction in consumption at a particular time due, for example, to an increase in the price of alcohol, reliably postpones looming death for some drinkers, whereas a fall in price has the opposite effect. Similarly, immediate benefits have been reported for tobacco-related events following the introduction of smoking bans. For example, substantial reductions in hospital admissions for cardiovascular and/or respiratory diseases were found to occur within months following the introduction of public smoking bans in Italy (Cesaroni et al., 2008), Scotland (Pell et al., 2008), Switzerland (Humair et al., 2014), the United States (Juster, et al., 2007; Sargent et al., 2004), and other regions where the effects of bans have been implemented and the benefits assessed (Meyers et al., 2009). Importantly, health benefits have consistently been found to accrue due to decreases in both active and passive smoking.

Somewhat more surprising perhaps is the fact that changes in diet may also produce rapid benefits, as indicated in a number of "natural experiments." In Poland, for example, rates of coronary mortality fell steeply shortly after Poland became democratic in 1989. Prior to that, rates had been rising steadily, but a reversal in trend occurred that is attributed to the loss of communist subsidies for meat and animal fats and an associated influx of more healthy but cheap vegetable oils (Zatonski and Willett, 2005). A similar sequence of events occurred for apparently the same reasons in other central European countries (Capewell and O'Flaherty, 2011a). Elsewhere, Cuba experienced a severe economic crisis in the early 1990s as a consequence of the collapse of the Soviet Union. During this period, although nutritional sufficiency was unaffected, there was a marked decrease in per capita caloric intake and associated population-wide weight loss accompanied by increased physical activity due to a marked fall in private transport (Franco et al., 2007). These changes were associated with simultaneous and rapid falls in mortality due to coronary heart disease, stroke, diabetes, and all causes. The rapid population-wide impact on health suggested by these naturally-occurring events is supported by findings from randomized controlled trials. Reductions in fatal cardiovascular events (e.g., heart attack and stroke) have been repeatedly reported to occur within months of random assignment to intervention arms of clinical trials involving relatively modest changes in diet (e.g., increased consumption of fatty fish or dietary supplements of omega-3 fatty acids; Capewell and O'Flaherty, 2011a,b).

### 12.1.6   Life-course Health

*The child is father of the man.*
(From the poem My Heart Leaps Up, 1802, William Wordsworth, 1770-1850)

The concept of prevention sits comfortably within theoretical frameworks that encourage a life-course perspective on health. The preventive benefits of decreased exposure to disease risk factors are generally greater when intervention comes earlier in life than later. Never smoking, for example, is undeniably more favorable to health than smoking, and the earlier age at which smoking begins the greater the likelihood of harm. For those who have started smoking, measurable benefit occurs from quitting whenever that occurs, but the earlier the better. Rarely, however, will removing exposure to a disease risk factor reverse manifest disease. Therefore, stopping smoking after contracting tobacco-related disease is not likely to reverse the disease process, although it may slow disease progression and contribute to the efficacy of biomedical intervention.

Evidence indicates that opportunities to promote health and prevent disease exist at all stages from pre-conception to end of life (Halfon et al., 2014). For example, it has long been accepted that maternal smoking is harmful to the developing fetus, causing long-term harm to child development (Butler and Goldstein, 1973). In addition to maternal smoking, prenatal as well as postnatal exposure to secondhand tobacco smoke contributes to numerous adverse consequences for fetal and child development (Zhou et al., 2014). Alcohol consumed during pregnancy crosses the placenta, resulting in approximately equal concentrations of alcohol in the mother and fetus, and maternal exposure to alcohol, both prior to and during pregnancy, increases the risk of adverse birth outcomes (Nykjaer et al., 2014). Similarly, fetal nutrition, which depends on the transfer across the placenta of nutrients in the maternal circulation to fetal circulation, can have profound consequences for fetal, as well as long-term child, development. In particular, under-nutrition during pregnancy, which continues to be prevalent in some developing countries, can lead to prematurity, low birth weight, increased risk of mortality, and irreversible harm to later physical and cognitive development (Black et al., 2008).

Even in the absence of obvious malnutrition, it appears that aspects of maternal diet have the potential to influence long-term health in offspring. For example, in Motherwell, Scotland, from the early 1950s to the mid-1970s, pregnant women were advised to eat a high-meat low-carbohydrate diet, as this was thought to facilitate healthy pregnancy (Shiell et al., 2001). When followed-up three decades later, the offspring of women who were successful in implementing the diet were found to have clinically-significant raised blood pressure. However, neither high protein nor low carbohydrate during pregnancy is necessarily implicated since the women who followed the recommended diet might also have adopted other potentially important dietary changes, such as increasing their intake of salt or fatty acids, none of which were measured. Nevertheless, the findings illustrate the potential importance for later adult health of relatively subtle differences in uterine environment, independent of other pregnancy-related variables known to affect fetal and child development (e.g., maternal smoking and alcohol consumption).

The "long shadow" cast by early life experience over health in later life is as evident for maternal and child mental health as it is for physical health. Moreover, effects are reciprocal, revealing that both maternal physical and mental health have implications for the physical and mental health of offspring during childhood, adolescence, and adulthood. For example, studies show that maternal exposure to psychological *stress* during pregnancy can contribute to lower birth weight. Stress in this context is a broad concept that can include any event or experience that the individual perceives as threatening to their wellbeing and the emotions that accompany such appraisals. The Yugoslav Wars of the 1990s appear to be illustrative. Belgrade experienced 3 months of bombing in 1999, and although it is not possible to exclude a variety of negative influences on health (e.g., compromised nutrition due to disruption of food supplies), the population, including pregnant women, were exposed to high levels of psychosocial stress (Maric et al., 2010). Compared to periods before and after the bombing, infants born to mothers who experienced the bombing had a lower birth weight.

Even stress experienced prior to conception may have the potential to harm offspring. In a recent representative study of American mothers, prior exposure to stressful life events, including, for example, the death of a close family member, divorce, or separation, was associated with low birth weight independently of stressors experienced during the pregnancy (Witt et al., 2014). In fact, a dose-response relationship was observed such that cumulative life stress of mothers-to-be was associated with progressively lower infant birth weight. High maternal family stress during pregnancy has also been found to be potentially detrimental. For example, poorer early-childhood verbal and nonverbal cognitive development has been reported for the offspring of pregnant women who answered affirmatively to questions such as, "There are lots of bad feelings in our family," versus "In times of crisis we turn to each other for support" (Henrichs et al., 2011). Furthermore, a distinction can be made between stress associated with events and circumstances, on one hand, and *distress* in the form of mental health problems characterized by more pervasive and persistent negative emotions, on the other hand. In that regard, particular attention has been given to maternal depression, which has consistently been found to be associated with negative child outcomes, including emotional problems (e.g., anxiety and depression) and conduct disorders (e.g., antisocial behavior and delinquency) which generally are evident by middle childhood (Goodman et al., 2011).

Although consistent associations have been found between the behavior and life experience of mothers and the physical and mental health of their offspring, the strength of associations is often modest, indicating that there is considerable variation between individuals. Additionally, however, postnatal and early-childhood experiences may also contribute to outcomes in adulthood, with the effects sometimes appearing to be pronounced. For example, in New Zealand's South Island, most young adults with a mental disorder were found to have had diagnosable problems stemming from childhood (Kim-Cohen et al., 2003). Specifically, half of the adults with a mental disorder met criteria for a diagnosable disorder by age 15 years, and by the late teens the figure approached 75%. This suggests that many adult mental health problems may be thought of as extensions of juvenile difficulties (Maughan and Kim-Cohen, 2005) and that interventions to promote mental health should begin early and continue throughout the life-course.

A life-course approach must by definition range across the life span. In addition to early life, a second main focus of research attention has been at the opposite end of the age continuum, namely, advanced age. There is extensive evidence of improved *healthy aging*, both physical (Loef and Walach, 2012) and cognitive (Qiu, 2014) when established habits of living include fewer behavioral risk factors, especially not smoking, not drinking alcohol to excess, being physically active, and consuming a nutritious diet. Although midlife has attracted less attention than earlier and later life stages, evidence of behavioral and neural plasticity throughout the life-course suggests the potential for benefit from preventive interventions that are directed at the transition from younger to older age. In fact, midlife has been described as a pivotal stage in the life-course. Addressing early-life risk factors through preventive interventions delivered during midlife could be important for the promotion of healthy aging (Lachman et al., 2015).

## 12.1.7  Is Primary Care Preventive?

A large proportion of healthcare is dispensed by individual community-based *primary-care physicians*, also known as *general practitioners* or *family physicians*, who serve as the point of first contact for individuals in need. Generally speaking, the process of healthcare begins when patients self-diagnose illness. Feeling unwell, the patient contacts a physician, who conducts a formal diagnosis to determine what intervention, if any, to administer. Because of extensive deployment throughout the community, primary care is widely regarded as being a major resource for promoting population health and preventing disease (Starfield et al., 2005). Primary-care physicians are likely to have ties to the local communities in which they live and work, and extensive knowledge of the personal lives and experiences of the patients in their care. However, it cannot be assumed that community deployment and extensive personal ties necessarily are sufficient to ensure an effective preventive role. The core training for primary care is individualized clinical medicine, which should not be confused with population disease prevention even if the medicine that is dispensed occurs in a community context. Evidence indicates that there are major impediments to primary care, as usually practiced, serving as a resource for population-wide disease prevention.

Despite the obvious fact that the delivery of preventive services requires time, there has been relatively little appreciation of the actual time required to perform even routine preventive services when delivered individually in the primary care setting. The *Guide to Clinical Preventive Services* of the United States Preventive Services Task Force (1996) contains recommendations for about 90 prevention targets related to cardiovascular health, cancers, drug abuse, diabetes, dementia, injury, mental health, and sexual health. These preventive interventions have been designed to be used specifically in primary care for individuals of all age groups and risk categories, and include health screening, counseling, immunization, and prophylactic medication. However, when examined from the perspective of the time required to implement the actions, it was found that primary-care physicians would have to commit almost 90% of their full-time working hours to doing nothing other than serving in a preventive role if they are to satisfy guideline recommendations (Yarnall et al., 2003). Thus, by reason of limited opportunity due to insufficient time, there is little prospect of primary-care physicians fulfilling, other than in a minor way, the role widely envisioned for them as agents for improving population health through disease prevention.

In view of evidence of limited opportunity, there may be little point in asking whether the role of primary-care physicians in disease prevention is cost-effective. In fact, cost-effectiveness studies have been conducted, and the findings, at least with respect to disease prevention through increased physical activity (Gulliford et al., 2013) and improved diet (Gulliford et al., 2014), have not been encouraging. In view of evidence indicating modest impact, it is of interest to know that physicians' views of themselves suggest that they have insight into their limited role in preventive healthcare. Of several major healthcare professions, including dieticians, midwives, occupational therapists, physiotherapists, and psychologists, hospital- and community-based physicians reported least willingness among those professions to refocus their traditional clinical role in the direction of preventing disease (Johansson et al., 2010).

When primary-care physicians try to encourage patients to make lifestyle improvements related to key risk factors such as diet, physical activity, smoking, and consumption of alcohol, their attempts reflect the traditions of their profession. It is a characteristic of the biomedical model that healthcare is dispensed individually to each patient. In that context, it is tempting to think of exposure to risk factors as something that is the result of individual lifestyle choices. Thus, in principle, those choices should be amenable to change in response to authoritative physician recommendations. In reality, individual choices (habits) are greatly influenced by social and economic circumstances (habitat). Consequently, although it is admirable that physicians attempt to promote health and prevent disease as part of individual patient consultations, the effectiveness of those efforts generally compares poorly to population-wide interventions. With regard to all major considerations (i.e., opportunity, effectiveness, cost-effectiveness, and practitioner interest), primary care is an unpromising platform for delivering preventive healthcare.

### 12.1.7.1 Routine Health Checks

Among the array of biomedical services intended to prevent disease, the general "health check," especially that performed by primary-care physicians, is possibly the most common. The history of periodic health examinations for healthy individuals dates from the 1860s (Han, 1997). Although professional enthusiasm for the practice has waxed and waned, patients appear in general to believe in the value of periodic health checks. Evidence suggests that patients who perceive their health check to have been "thorough," as inferred from more rather than less extensive physical examinations and laboratory tests performed, whether needed or not, the more comforted patients feel about their health and their physician (Laine, 2002). Conversely, patients who are not examined or tested to the extent they expected are less trusting of their physician and are more likely to seek care elsewhere.

Consistent with patient intuition, general health checks are widely believed to be an important feature of *preventive medicine*, but unfortunately patient intuition and popular belief are not supported by fact. A systematic review conducted at the Nordic Cochrane Centre, an affiliate of the Cochrane Collaboration, found that general health checks do not reduce overall mortality or morbidity (Krogsbøll et al., 2011). The finding included health checks for cardiovascular disease and cancer, often thought to be among the conditions most likely to benefit from routine checking. Conversely, health checks were associated with more diagnoses and more medical treatment, especially for hypertension. However, as this did not improve mortality or morbidity, the additional treatment was considered to be harmful rather than beneficial (Krogsbøll et al., 2011).

Similar conclusions were drawn from a more recent review, which also found no mortality benefit from health checks conducted in primary-care settings (Si et al., 2014). The study considered surrogate endpoints, including blood pressure, total cholesterol, body mass index, and smoking status, and these showed statistically significant, albeit "clinically small," improvements associated with health checks. On the other hand, health checks were associated with the clinical endpoint of higher mortality due to cardiovascular disease, which is suggestive of harm due to overtreatment. As such, the routine health check, which forms part of the gatekeeping role of primary care, while contributing to greater use of clinical medicine, appears to be largely ineffective in preventing disease.

Although routine health checks evidently function as a reassuring ritual for patients, such reassurance is misplaced if the practice leads to overdiagnosis, overtreatment, and unnecessary expense. However, patients should not be the ones held responsible for having unrealistic expectations about practices that in fact deliver little benefit and may involve harm. Rather, physicians must take the lead by replacing unnecessary health checks with improved patient education about the limited benefit and potential risks of routine examination. So doing could also help to relieve the problem of overwork that primary-care physicians sometimes complain about. For example, in a recent article concerning "general practice in crisis," it was claimed that physicians in the United Kingdom sometimes see "60-70 patients a day at 10-minute intervals" (Wilkinson,

2014, p. 295). That situation, which was understandably described as "not sustainable or safe" (presumably for both patients and practitioners), could be partially relieved were physicians to perform fewer unnecessary tasks such as the routine health examinations of healthy patients.

## REFERENCES

Beaglehole, R., Bonita, R., Yach, D., et al., 2015. A tobacco-free world: a call to action to phase out the sale of tobacco products by 2040. Lancet 385, 1011–1018.

Black, R.E., Allen, L.H., Bhutta, Z.A., et al., 2008. Maternal and child undernutrition: global and regional exposures and health consequences. Lancet 371, 243–260.

Butler, N.R., Goldstein, H., 1973. Smoking in pregnancy and subsequent child development. Br. Med. J. 4, 573–575.

Capewell, S., O'Flaherty, M., 2011a. Rapid mortality falls after risk-factor changes in populations. Lancet 378, 752–753.

Capewell, S., O'Flaherty, M., 2011b. Can dietary changes rapidly decrease cardiovascular mortality rates? Eur. Heart J. 32, 1187–1189.

Cesaroni, G., Forastiere, F., Agabiti, N., et al., 2008. Effect of the Italian smoking ban on population rates of acute coronary events. Circulation 117, 1183–1188.

Franco, M., Ordunez, P., Caballero, B., et al., 2007. Impact of energy intake, physical activity, and population-wide weight loss on cardiovascular disease and diabetes mortality in Cuba, 1980–2005. Am. J. Epidemiol. 166, 1374–1380.

Goodman, S.H., Rouse, M.H., Connell, A.M., et al., 2011. Maternal depression and child psychopathology: a meta-analytic review. Clin. Child. Fam. Psychol. Rev. 14, 1–27.

Gulliford, M.C., Charlton, J., Bhattarai, N., et al., 2013. Impact and cost-effectiveness of a universal strategy to promote physical activity in primary care: population-based Cohort study and Markov model. Eur. J. Health Econ. 14, 341–351. http://dx.doi.org/10.1007/s10198-013-0477-0.

Gulliford, M.C., Bhattarai, N., Charlton, J., Rudisill, C., 2014. Cost-effectiveness of a universal strategy of brief dietary intervention for primary prevention in primary care: population-based cohort study and Markov model. Cost Eff. Resour. Alloc. 12, 1–9. http://www.resource-allocation.com/content/12/1/4.

Halfon, N., Larson, K., Lu, M., et al., 2014. Lifecourse health development: past, present and future. Matern. Child Health J. 18, 344–365.

Han, P.K., 1997. Historical changes in the objectives of the periodic health examination. Ann. Intern. Med. 127, 910–917.

Henrichs, J., Schenk, J.J., Kok, R., et al., 2011. Parental family stress during pregnancy and cognitive functioning in early childhood: the Generation R Study. Early Child. Res. Q. 26, 332–343.

Holmes, J., Meier, P.S., Booth, A., et al., 2012. The temporal relationship between per capita alcohol consumption and harm: a systematic review of time lag specifications in aggregate time series analyses. Drug Alcohol Depend. 123, 7–14.

Humair, J.P., Garin, N., Gerstel, E., et al., 2014. Acute respiratory and cardiovascular admissions after a public smoking ban in Geneva, Switzerland. PLoS One 9, 1–6. http://dx.doi.org/10.1371/journal.pone.0090417.

Johansson, H., Stenlund, H., Lundström, L., Weinehall, L., 2010. Reorientation to more health promotion in health services-a study of barriers and possibilities from the perspective of health professionals. J. Multidiscip. Healthc. 3, 213–224.

Juster, H.R., Loomis, B.R., Hinman, T.M., et al., 2007. Declines in hospital admissions for acute myocardial infarction in New York State after implementation of a comprehensive smoking ban. Am. J. Public Health 97, 2035.

Kim-Cohen, J., Caspi, A., Moffitt, T.E., et al., 2003. Prior juvenile diagnoses in adults with mental disorder: developmental follow-back of a prospective-longitudinal cohort. Arch. Gen. Psychiatry 60, 709–717.

Krogsbøll, L.T., Jørgensen, K.J., Gøtzsche, P.C., 2011. General health checks in adults for reducing morbidity and mortality from disease: Cochrane systematic review and meta-analysis. Br. Med. J. 345, e7191.

Kuehlein, T., Sghedoni, D., Visentin, G., et al., 2010. Quaternary prevention: a task of the general practitioner. Prim. Care 10, 350–354.

Lachman, M.E., Teshale, S., Agrigoroaei, S., 2015. Midlife as a pivotal period in the life course balancing growth and decline at the crossroads of youth and old age. Int. J. Behav. Dev. 39, 20–31.

Laine, C., 2002. The annual physical examination: needless ritual or necessary routine? Ann. Intern. Med. 136, 701–703.

Loef, M., Walach, H., 2012. The combined effects of healthy lifestyle behaviors on all cause mortality: a systematic review and meta-analysis. Prev. Med. 55, 163–170.

Maric, N.P., Dunjic, B., Stojiljkovic, D.J., Britvic, D., Jasovic-Gasic, M., 2010. Prenatal stress during the 1999 bombing associated with lower birth weight: a study of 3,815 births from Belgrade. Arch. Womens Ment. Health 13, 83–89.

Maughan, B., Kim-Cohen, J., 2005. Continuities between childhood and adult life. Br. J. Psychiatry 187, 301–303.

Mensah, G.A., Dietz, W.H., Harris, V.B., et al., 2005. Prevention and control of coronary heart disease and stroke – nomenclature for prevention approaches in public health: a statement for public health practice from the Centers for Disease Control and Prevention. Am. J. Prev. Med. 29, 152–157.

Meyers, D.G., Neuberger, J.S., He, J., 2009. Cardiovascular effect of bans on smoking in public places: a systematic review and meta-analysis. J. Am. Coll. Cardiol. 54, 1249–1255.

Norström, T., Skog, O.J., 2001. Alcohol and mortality: methodological and analytical issues in aggregate analyses. Addict 96, 5–17.

Nykjaer, C., Alwan, N.A., Greenwood, D.C., et al., 2014. Maternal alcohol intake prior to and during pregnancy and risk of adverse birth outcomes: evidence from a British cohort. J. Epidemiol. Community Health 68, 542–549. http://dx.doi.org/10.1136/jech-2013-202934.

Pell, J.P., Haw, S., Cobbe, S., et al., 2008. Smoke-free legislation and hospitalizations for acute coronary syndrome. N. Engl. J. Med. 359, 482–491.

Prospective Studies Collaboration, 2002. Age-specific relevance of usual blood pressure to vascular mortality: a meta-analysis of individual data for one million adults in 61 prospective studies. Lancet 360, 1903–1913.

Qiu, C., 2014. Lifestyle factors in the prevention of dementia: a life course perspective. In: Leist, A.-K., Kulmala, J., Nyqvist, F. (Eds.), Health and Cognition in Old Age: From Biomedical and Life Course Factors to Policy and Practice. Springer, Switzerland, pp. 161–175.

Rehm, J., Shield, K.D., 2014. Alcohol and mortality: global alcohol-attributable deaths from cancer, liver cirrhosis, and injury in 2010. Alcohol Res.: Curr. Rev. 35, 174–183.

Rose, G., 1981. Strategy of prevention: lessons from cardiovascular disease. Br. Med. J. 282, 1847–1851.

Rose, G., 1985. Sick individuals and sick populations. Int. J. Epidemiol. 14, 32–38.

Rose, G., 1992. The Strategy of Preventive Medicine. Oxford University Press, Oxford, UK.

Sargent, R.P., Shepard, R.M., Glantz, S.A., 2004. Reduced incidence of admissions for myocardial infarction associated with public smoking ban: before and after study. Br. Med. J. 328, 977–980.

Shiell, A.W., Campbell-Brown, M., Haselden, S., et al., 2001. High-meat, low-carbohydrate diet in pregnancy: relation to adult blood pressure in the offspring. Hypertension 38, 1282–1288.

Si, S., Moss, J.R., Sullivan, T.R., et al., 2014. Effectiveness of general practice-based health checks: a systematic review and meta-analysis. Br. J. Gen. Pract. 64, e47–e53.

Starfield, B., Shi, L., Macinko, J., 2005. Contribution of primary care to health systems and health. Milbank Q. 83, 457–502.

Starfield, B., Hyde, J., Gérvas, J., Heath, I., 2008. The concept of prevention: a good idea gone astray? J. Epidemiol. Community Health 62, 580–583.

United States Preventive Services Task Force, 1996. The Guide to Clinical Preventive Services. Agency for Healthcare Research and Quality, Rockville, MD.

WHO, 1986. The Ottawa Charter for Health Promotion. World Health Organization, Geneva. Retrieved 31 March 2014 from, http://www.who.int/healthpromotion/conferences/previous/ottawa/en.

Wilkinson, E., 2014. UK general practice in crisis: time for a rethink? Lancet 384, 295–296.

Witt, W.P., Cheng, E.R., Wisk, L.E., et al., 2014. Maternal stressful life events prior to conception and the impact on infant birth weight in the United States. Am. J. Public Health 104, S81–S89.

Yarnall, K.S., Pollak, K.I., Østbye, T., et al., 2003. Primary care: is there enough time for prevention? Am. J. Public Health 93, 635–641.

Zatonski, W.A., Willett, W., 2005. Changes in dietary fat and declining coronary heart disease in Poland: population based study. Br. Med. J. 331, 187–188.

Zhou, S., Rosenthal, D.G., Sherman, S., et al., 2014. Physical, behavioral, and cognitive effects of prenatal tobacco and postnatal secondhand smoke exposure. Curr. Probl. Pediatr. Adolesc. Health Care 44, 219–241.

Chapter 13

# Associated Prevention Concepts and Models

*There is no quality in human nature which causes more fatal errors in our conduct than that which leads us to prefer whatever is present to the distant and remote.*

(Hume, 1739, p. 358)

## Contents

With reference to the essentials of prevention summarized in Table 12.1, preventive interventions are often incorporated into other theoretical frameworks, the burgeoning number of which suggests growing dissatisfaction

The Health of Populations. http://dx.doi.org/10.1016/B978-0-12-802812-4.00013-8

with the manifold failures of contemporary biomedical healthcare. Reference has been made throughout this book to habits and habitats as a succinct abstraction of the myriad individual and environmental factors that determine and define human health. That depiction echoes various concepts and models in the biological and health sciences that seek to encapsulate complexity in dynamic systems. Many such proposals incorporate *systems biology*, which examines biological processes in context in order to understand interactions within and between processes that are embedded, often hierarchically, in larger biological systems (Ahn et al., 2006; Ideker et al., 2001; Kitano, 2002).

The tradition has been to explain complex biological phenomena through detailed *analysis* of fundamental constituent parts, and that tradition of *reductionism* has been enormously successful. Systems biology, on the other hand, seeks to *synthesize* knowledge of intact functioning entities in the service of *holistic* explanation believed to be more than the mere sum of parts. For example, investigations of individual genes or proteins, one at a time, over the past 30 years has produced great advances in knowledge. Systems biology builds on that tradition by examining the relationships between many such multiple elements within larger functioning biological systems, including whole organisms (Ideker et al., 2001). Of necessity, systems biology is interdisciplinary in approach, drawing on principles, knowledge, and techniques from biology, physics, chemistry, computer science, and engineering (Medina, 2013).

As the term suggests, systems biology is primarily concerned with biological processes. However, as emphasized throughout this book, biological systems do not exist in isolation and biological knowledge alone does not provide an adequate understanding of health. In common with systems biology, *ecological models* emphasize *systems thinking* which has roots in ancient philosophical traditions that focus on cyclical rather than linear cause-and-effect relations (Skyttner, 1996). In the health sciences, systems thinking is often identified with Austrian-born biologist Ludwig von Bertalanffy (1951, 1972). His formulation of *general systems theory* is the application of systems thinking that is transdisciplinary, extending beyond biology to include behavior and the environment. A key objective of general systems theory is to illuminate reciprocal influences and interactions between *systems*, defined as sets of "elements standing in interrelation among themselves and with the environment" (von Bertalanffy, 1972, p. 417).

## 13.1    ECOLOGICAL MODELS OF HEALTH AND THE EMERGENCE OF HEALTH ACTIVISM

Ecological perspectives on health have long emphasized the environment as a source of influence and as a target for intervention for health promotion and disease prevention (e.g., Bronfenbrenner, 1977; McLeroy et al., 1988; Sallis et al., 2008). *The environment*, however, is subject to varying interpretation, and that has led to an assortment of overlapping ecological models of health. Of these, perhaps the longest established is *public health*, which has itself undergone several reputedly discernible "waves" of development (Davies et al., 2014; Hanlon et al., 2011). Throughout human history, practices, both secular and religious, have been promulgated with the aim of promoting health and wellbeing among groups and entire populations. However, the humanitarian and health-related social reforms that emerged in the wake of nineteenth century European industrialization are often taken as the beginnings of modern public-health practice (Hanlon et al., 2011). As discussed in Chapter 1, transformational improvements in population health were achieved by addressing key aspects of the natural and built environment, including the provision of clean water, sewage disposal, and slum eradication. Those initiatives have been described as representing the first wave of development in public health where the focus was on the *structural* environment, followed by subsequent waves, each with differing biomedical, social, and cultural emphases (Davies et al., 2014).

Increasingly, public health, most of which falls within the ambit of governments, both local and national, has been extended to include international relations in the form of *global health* (or *global public health*). In an early depiction, global health was taken to refer to

> *Health problems, issues, and concerns that transcend national boundaries, may be influenced by circumstances or experiences in other countries, and are best addressed by cooperative actions and solutions.*

> (Institute of Medicine, 1997, p. 2)

It is reasonable to assume good alignment between the priorities of global health and the need to address the diseases that contribute most to the global burden of disease (see Table 2.1 and Figures 2.1 and 2.2). However, until recently, this was not necessarily the case. A comparison of World Health Organization (WHO) biennial budgetary allocations and the burden of disease found that as recently as 2006-2007, the WHO allocated almost 87% of its total budget to infectious diseases, 12% to noncommunicable diseases, and less than 1% to injuries (Stuckler and McKee, 2008). That pattern of funding is substantially misaligned with the prevailing global distribution of diseases discussed in previous chapters, wherein noncommunicable diseases, in particular, have come to be the leading cause of death and disability. Only relatively recently, with the adoption of a new framework for the prevention and control of noncommunicable diseases, discussed in Chapter 14, have WHO priorities been brought into closer alignment with the actual burden of disease in most countries and regions (WHO, 2012a).

Over recent decades, increased attention has been given to the implications for health of the diverse effects of human activity on the natural environment. This has led to a variety of ecological models of health, including *health ecology*, *ecotoxicology*, and *public health ecology*, all of which emphasize interdependence between aspects of the natural environment, human activity, and health (Coutts et al., 2014). More recently still, the progression in ecological models of health from public health to global and the concomitant growth in concern over environmental integrity has led to the emergence of *planetary health*. Here, interest in the role of the environment, intrinsic to ecological models, has led to attention being focussed on threats to human health from global environmental degradation, especially climate change, signified by global warming, ice melt, rising sea levels, and ocean acidification. The logic is compelling. Efforts to foster human health may count for little if threats to the integrity of the planet's biosphere are ignored. The recently proposed planetary health *manifesto* envisions:

> *A planet that nourishes and sustains the diversity of life with which we coexist and on which we depend.*

(Horton et al., 2014, p. 847)

The potential consequences of planetary environmental degradation are many, and include under-nutrition due to decreased food security, increased infectious diseases, diminished sources of freshwater, poverty, population displacement, conflict, and "ultimately collapse of our civilization" (Haines et al., 2014).

Despite differing specific emphases, human-environment (habit-habitat) interdependence is a prominent concern of most ecological models of health, which are increasingly assuming the character of a movement in the form of *health activism*. In that regard there are parallels with modern environmentalist movements. However, whereas *environmental activism* is a familiar term in everyday discourse, *health activism* is less so (Zoller, 2005). This is not to say that health activism is new. Many of the nineteenth-century health-related social reforms, now often identified as early instances of public health, should also be viewed as instances of health activism. Crucially, health activism, in common with activism in general, involves challenging an existing order of entrenched interests (Berridge, 2007; Laverack, 2013; Zoller, 2005). In that regard, there is a distinction to be made between advocacy and activism (Brown et al., 2004). Health advocacy usually involves the promulgation of information for purposes of education within the existing order, whereas health activism is characterized by the use of direct action to challenge and disrupt entrenched interests.

Typically, health activism is undertaken by or on behalf of groups which, by reason of social position, culture, ethnicity, or other socially-constructed category, are perceived to be denied a level of health advantage that is afforded to others. That is, the direct action of health activists is often intended not merely to promote health and prevent disease, but to achieve those objectives in the service of social equity and justice. Corporate interests related to the marketing of pharmaceuticals, chemicals, tobacco, foods, and beverages are among the entrenched interests targeted by health activists. However, just as corporations have proved adept at health advocacy, either by infiltrating existing groups or creating their own such groups (discussed in Chapter 7), it is to be expected that transnational corporations will defend against challenges to corporate interests by sponsoring their own health "activist" causes (Labonté, 2013).

## 13.2  HEALTH INEQUALITY AND THE SOCIAL DETERMINANTS OF HEALTH

In addition to highlighting environmental issues, ecological models of health tend to be concerned with problems of *health inequality*, especially in relation to systematic difference in the distribution of health status and the distribution of health risk factors within and between countries. *Inequality* in health, however, has different shades of meaning (Whitehead, 1991). Some health inequality may entail difference in a purely mathematical sense. For example, the biological predispositions responsible for the gender-linked diseases of testicular and prostate cancers in men and cervical and ovarian cancers in women cause gender-based "inequalities" in the distribution of those diseases. However, rather than mere difference, health inequality is more often used to refer to differences in health status that are contextualized as historical, social, or cultural in origin. In that more usual but narrower sense, health inequality has moral connotations to do with *fairness* and *justice*, and when used in that way the term is essentially synonymous with *health inequity* (Braveman, 2006; Braveman and Gruskin, 2003). Indeed, the two terms are often used more-or-less interchangeably, and as a traditional concern of public health, health inequality/inequity has carried forward to remain a key consideration of global health (Beaglehole and Bonita, 2008) and planetary health (Horton et al., 2014).

Unless otherwise stated, health inequality in the present text refers to socially- or economically-determined differences in health status or differences in exposure to health risk factors. Such differences are usually considered to be avoidable (in principle) and to be of a nature that fairness and justice demand *should* be avoided. Health inequality may be illustrated by citing the following instances: Referring to between-country inequalities, the rate of infant death among live births is more than 60-times higher in Mozambique than in Iceland, and the lifetime risk of maternal death during or shortly after pregnancy is more than 2000-times higher in Afghanistan than in Sweden (WHO, 2014a). Within countries, the rate of infant mortality in Bolivia is more than halved for mothers who have secondary education compared to those with no education; life expectancy at birth in Australia is 17 years longer for the general population than for indigenous Australians; and life expectancy at birth for men in the Lenzie neighborhood of Glasgow is 28 years longer than for men in Calton, a few kilometers away (WHO, 2014a).

These examples are illustrative of innumerable health inequalities that comprise the *social gradient* in health: the fact that health throughout the world is heavily influenced by income and social position (Marmot, 2005). The social gradient runs from top to bottom of the socioeconomic spectrum; is present in low-, middle-, and high-income countries; and is evident for all population age categories from infancy to old age. The renowned Whitehall studies conducted with British civil servants were important in revealing the ubiquity and nuanced character of the social gradient in health (Marmot and Brunner, 2005). The first of two major studies began in the late 1960s and showed a steep inverse gradient between employment level and rate of death. For grades of employment ranging from senior administrative staff to intermediate professional and executive levels to clerical and support staff, the likelihood of premature death increased progressively for men in lower employment grades compared to those in higher grades. Overall, for men aged 40-64 years, the lowest employment grade had an approximate threefold increased rate of mortality compared to that of those in the highest grade (Marmot and Shipley, 1996).

The initial Whitehall findings of a social gradient in health were extended with Whitehall II, which began in the mid-1980s and included both men and women and a wider range of health outcomes, including heart disease, cancer, chronic lung disease, gastrointestinal disease, depression, suicide, back pain, and self-reported health (Ferrie, 2004). Whereas an association between poverty and poorer health has long been recognized, the Whitehall findings were seminal for the reason that they demonstrated a social gradient in disease even among white-collar civil servants in stable employment in a high-income country. The findings, confirmed by studies in other countries (e.g., Adler et al., 2008; Braveman and Gottlieb, 2014; Mackenbach et al., 2008) illustrate the important role of *social determinants of health*. Evidence of a pronounced social gradient in health that is independent of the material deprivations of poverty is illustrative of the need for policies to address broad social, economic, and political inequalities. Thus, as well as tackling material poverty where it exists, there is great need locally and globally to address social exclusion and diverse environmental and cultural harmful influences on health in everyday life, education, and work across the life-course (Marmot, 2013a).

## 13.2.1   Causes Proximal and Distal

More than a half-century ago, in an influential account of causation in biology, Mayr (1961) distinguished proximal and distal causes. For Mayr, *proximal* causes are immediate influences such as the physiology of an organism, whereas *distal* causes (which he called "ultimate") are evolutionary influences such as natural selection. Moreover, whereas proximal causes are likely to operate at the level of the individual, distal causes are more likely to operate at the group level (or, as in the case of evolution, an entire species). The broad concept of nearness versus distance in causation applies not only to biology but can be applied to psychosocial processes and the environment, and to the combination of all three (biological, psychosocial, and environmental). With reference to the Whitehall studies, in addition to demonstrating a social gradient in disease outcomes, a similar gradient was also evident for a variety of precursors of disease.

These included biological determinants such as elevated plasma cholesterol and blood pressure, and behavioral determinants such as tobacco use and physical inactivity (Ferrie, 2004). The diversity and chronological pattern of the many potential biological and social determinants of health suggests the need to consider *causes of the causes* (Rose, 1992).

The distinction between proximal and distal causes is sometimes conveyed in the form of a parable, known variously as *the river story* and *the public health parable*. Although the story can be embellished to suit any of a number of specific purposes, it usually begins with the reader or audience being asked to imagine standing on the bank of a river. Suddenly, the cries are heard of a drowning person, who may be man, women, or child. Without hesitation, you (the reader or audience member) plunge into the river, struggle against the current, reach the drowning person, who is brought to shore, bedraggled but alive. Just then, the cries are heard of a second person, who too is rescued. Then, the cries of a third person are heard and that person is rescued, then there is a fourth, and so on. Nearing exhaustion, you realize that you have no hope of rescuing a never-ending succession of drowning individuals. With that realization, you make haste upstream to find out what is causing people to fall into the river. Thus, the problem is solved by addressing the root cause. In similar fashion, each of the leading "causes" of mortality and morbidity has multiple causes that can be rank ordered on a continuum representing proximity to the appearance of disease comprised of proximal "downstream" causes and distal "upstream" causes.

The proximal cause of death from myocardial infarction (heart attack) could be accumulation of atherosclerotic plaque occluding one of the coronary arteries. Less proximal causes could include a history of individual behavior such as cigarette smoking, poor diet, physical inactivity, or any combination of those. Somewhat more distal causes could comprise a wide variety of factors within the social, cultural, and political milieu. These could include unregulated tobacco advertising, policies regarding workplace smoking, and pricing controls on tobacco products. Since all causes of disease, proximal and distal, offer opportunities for intervention, it is essential that attention be given to the relative effectiveness and cost-effectiveness of different options in order to arrive at the best balance of intervention types (e.g., biomedical intervention, health-related behavior change, and regulatory policy).

As mentioned in Chapter 3, one of the most difficult challenges in healthcare is ensuring that attention to immediate urgent need does not lead to neglect of needs that appear less urgent but are potentially more important (McGinnis and Foege, 2004). Most current healthcare policy focuses on the provision of proximal biomedical care for people with diagnosed disorder. However, common complex diseases have been shown to be substantially avoidable, due in large part to the fact that distal behavioral and social determinants are amenable to change in advance of more proximal pathophysiological causes of disease. Thus, with reference to cardiovascular disease, population health would benefit considerably by reordering priorities to shift the focus from proximal biomedical rescue operations intended to save people with manifest disease toward greater emphasis on preventing disease by addressing distal behavioral and social causes relevant to, for example, cigarette smoking, poor diet, and physical inactivity.

Given that the same disease may have multiple distal and proximal causes, each particular cause requires different strategies and level of focus on the individual, group, or population. Progressing from the distal to the proximal, one causal pathway for heart attack might be: absence of public walkways and cycle paths—sedentary habits—atherosclerotic deposits in coronary arteries—angina—occluded artery—heart attack. Moreover, whereas biomedical intervention to control manifest disease (e.g., heart attack) addresses that specific disease, remedying individual distal causes tends to have multiple beneficial health outcomes. Therefore, reduced exposure to cigarette smoking, poor diet, and physical inactivity each individually contributes to reduced incidence not merely of heart attack and other cardiovascular diseases (e.g., stroke) but also other leading diseases, including cancers, chronic respiratory diseases, and diabetes.

Importantly, whereas specific proximal causes (e.g., occluded coronary arteries) tend to contribute to specific disease outcomes (e.g., heart attack), a single distal cause (e.g., smoking) may contribute to multiple disease outcomes (e.g., cardiovascular disease, cancer, and diabetes), and multiple distal causes (e.g., physical inactivity, diet, and smoking) may contribute to a common proximal cause (e.g., atherosclerosis). The temporal character of causality is for distal (upstream) causes to cause proximal (downstream) causes and not vice versa. Given that health policy is generally aimed at minimizing population mortality and morbidity across the life-course, the current preoccupation on proximal biological events is illogical and needs to be redirected toward greatly increased focus on the diverse and modifiable distal behavioral and social determinants.

## 13.3 HEALTH IN ALL POLICIES

*Medicine is a social science and politics is nothing else but medicine on a large scale.*

(Attributed to Rudolf Virchow, 1821-1902, Prussian-born physician, anthropologist, and polymath; Ashton, 2006)

A shared objective of ecological models of health and health activism is to affect the decisions of health policy makers, and recent international developments in health promotion have been encouraging in that regard. The Eighth Global Conference on Health Promotion, held in Helsinki in 2013 had the theme "health in all policies," which reaffirmed and extended an approach to global health promotion that can be traced to the origins of the WHO, the constitution of which states that "Governments have a responsibility for the health of their peoples which can be fulfilled only by the provision of adequate health and social measures" (WHO, 1946, p. 1). The most recent affirmation of this long tradition is the *Health in All Policies Framework for Country Action*, which recognizes that governments must contend with wide ranging responsibilities that compete for priority and may sometimes conflict with population health objectives (WHO, 2014b).

The Health in All Policies framework espouses a theme of this book that the main determinants of personal and population health have environmental origins that lay mostly outside the direct influence of the healthcare sector in the personal, social, cultural, and economic lives of people. Consequently, population health is influenced by policies and decisions across all spheres of government, and not merely the ministries that have an explicit health remit. By calling on governments to make health a stated priority in *all* decisions, the health in all policies framework seeks to "health proof" all government policies. A major objective of the framework is to avoid the unintended negative consequences that can occur when policy decisions outside the health portfolio are made in the absence of health being expressly identified as a priority. The World Health Organization has defined health promotion as

> *The process of enabling people to increase control over their health and its determinants, and thereby improve their health.*

(WHO, 2005, p. 1)

The Health in All Policies framework offers important opportunities for addressing the social determinants of personal and population health and especially the economic and social conditions responsible for health inequalities.

The 53 countries that comprise the WHO's European region provide a "rich natural laboratory" showing the crucial role of public policy for promoting health and removing health inequalities between and within countries (Mackenbach et al., 2013a). After World War 2, there was substantial convergence of improved life expectancy in Western Europe in contrast to the countries of Central and Eastern Europe. Following initial improvements, the former communist countries and the former Soviet Union experienced stagnation, and in some instances falls, in life expectancy from 1960 until the collapse of the communist bloc and dissolution of the Soviet Union in 1991. Exit from the communist bloc was followed by almost immediate improvements in life expectancy in Poland, the former East Germany, and what had been Czechoslovakia. These events support the view that optimal personal and population health are profoundly influenced by broad social, economic, political, and cultural factors that influence ways of living (Marmot, 2013b). In addition, evidence indicates that a host of specific health policies, including tobacco and alcohol control, road traffic safety, and food policy, have contributed to the comparatively favorable health outcomes for Western Europe and former communist-bloc countries since the collapse of communism.

Notably, even within Western Europe, differences in specific policies have contributed to health inequalities (Mackenbach et al., 2013a). For example, in contrast to Denmark, where smoking controls have been delayed in the face of success by the tobacco industry in promoting the view that smoking is an expression of individual freedom, Sweden has implemented wide-ranging smoking-control policies. Consequently, the death rate from lung cancer in Denmark is twice that of Sweden, and Danish women have the highest rate of lung-cancer death in Western Europe. Similarly, following implementation of wide-ranging policies to limit alcohol consumption, France has experienced a drop in the rate of death from cirrhosis of the liver, which is now less than a third of what it was in 1970 when it was one of the highest in Western Europe (Mackenbach et al., 2013a). In the same period, death from liver cirrhosis has increased nearly fourfold in the United Kingdom, where alcohol consumption has increased in response to a policy of deregulation which made alcohol easier to access.

At the same time, it should be noted that a portion of the superior health outcomes of Western Europe compared to Central and Eastern Europe is likely to have come from biomedical healthcare, including, for example, improved perinatal and maternal health, and detection and treatment of hypertension and cancer (Mackenbach et al., 2013a). Considering those gains, it is important to ask what has been the relative contribution of biomedical healthcare and public policies aimed at population-wide health promotion and disease prevention. That question was examined in a study of overall rates of mortality and morbidity in the Netherlands from 1970 to 2010 (Mackenbach et al., 2013b). The study found that health benefits from policies to promote population-wide risk factor reduction were three-times greater than the benefit attributable to biomedical healthcare. That estimate, it may be remembered from Chapter 4, is similar in magnitude to estimates of the relative contribution of risk factor reduction and biomedical intervention to improvements in cardiovascular disease that have occurred in some countries.

Taking account of the extent of past benefits from relatively ad hoc applications of public policies that address social determinants of health, Health in All Policies represents an important development for global health promotion, disease prevention, and reduction of health inequalities. This is not to say that the framework represents a major advance in knowledge. Nor is the idea new, because, as mentioned above, it has long infused WHO thinking. Rather, adoption of the framework promises greater consolidation of future effort and the likelihood of existing knowledge being brought to bear more effectively on a global basis (Van den Broucke, 2013). Among existing examples, Finland is notable for having developed a national level Health in All Policies approach that has been in operation for four decades (Melkas, 2013). After making public health a political priority in Finland, it was recognized that success in addressing key determinants of health required the involvement of sectors beyond the health portfolio. Policies on nutrition, smoking, and accident prevention were among the main priorities. Success was indicated by improvements in a range of indices, including increases in average life expectancy of almost 11 years for men and 9 years for women.

The hope is that individual instances of Health in All Policies will not prove to be overly context-specific, and that generalizable preventive health-care will become dominant worldwide (Mulgan, 2010). Importantly, health in all policies should have the dual purpose of promoting health gains in the population as a whole while also reducing health inequalities (Benach et al., 2013). To achieve both objectives, policies need to address the entire social gradient of health, while considering carefully the proportionate impact at different points along the gradient. In the aforementioned example of Finland, the average level of health for the population improved, but contrary to aims and expectations, health inequalities between groups did not diminish (Melkas, 2013). Failure to address health inequality in that instance, especially between income groups, appears to have been due at least in part to the adoption of neo-liberal economic policy in response to a recession in the 1990s, a specific decision that was made without reference to Health in All Policies. In accordance with an approach known as *proportionate universalism* (Marmot and Bell, 2012), to be considered truly effective, the impact of Health in All Policies must be proportionately greater for those who are most disadvantaged if the gap between those at the top and bottom of the gradient is to be narrowed.

The Health in All Policies initiative, currently being pursued in the international healthcare arena, offers important possibilities for developing countries to learn from mistakes made in high-income counties and to arrive at a more effective and sustainable balance between preventive healthcare and conventional biomedical healthcare. Having to contend with many of the same challenges that currently exist in high-income countries, such as an aging demographic and increased burden of noncommunicable diseases (Beard and Bloom, 2014; Bloom et al, 2014; Mathers et al., 2015), the economies and infrastructure of developing countries are likely to be particularly challenged. However, those countries have one important potential advantage in being able to avoid the over-investment in technological "solutions" that is already heavily entrenched in high-income countries. In particular, by employing Health in All Policies, developing countries have the potential to develop healthcare based on legislative and policy initiatives largely outside the usual remit of health ministries, and thereby exceed high-income countries in achieving equitable, effective, and cost-effective population-wide life-course health promotion and prevention of disease.

## 13.4 WELFARE PROVISION AND THE "NANNY" STATE: NUDGE, FUDGE, SMUDGE, GRUDGE, OR BUDGE

There is a strong consensus that one of the factors responsible for variation in health between countries is level of state welfare provision (Bambra, 2007; Marmot et al., 2012; Muntaner et al., 2011). Welfare arrangements appear capable of addressing important social determinants of health, evidenced by comparatively more generous and universal welfare state regimes being associated with better health.[1] A review of data on social welfare spending collected by the Organization for Economic Cooperation and Development (OECD)— with age-standardized all-cause mortality in 15 countries of the European Union over 25 years—found that each additional USD100 per capita increase in social welfare spending was associated with a 1.2% drop in mortality (Stuckler et al., 2010). However, despite apparently contributing to health, welfare provision remains controversial, and claimed benefits are sometimes contested. In particular, more generous welfare regimes are sometimes criticized for being overly protective (the *nanny* state) and unduly interfering with personal choice.

Thus, advocates argue for state welfare provision to promote health and wellbeing throughout the life-course, whereas opponents argue that welfare provision promotes a culture of dependency that stifles individual initiative to the eventual detriment of health and wellbeing. In recent times, the latter view appears to have gained momentum. Governments worldwide are exploring options for reducing levels of welfare provision in response to global economic volatility and the perceived need to create "efficiencies" which others might be more inclined to regard as austerity measures. In that context, the political philosophy of *libertarian paternalism* (Thaler and Sunstein, 2008) has become influential among policymakers, most prominently in the United Kingdom and the United States, but also in Denmark, France, and Sweden (e.g., Blumenthal-Barby and Burroughs, 2012; Bonell et al., 2011; Marteau et al., 2011; Oliver, 2013; Vallgårda, 2012). The approach is predicated on the presumption that most people value their health, yet many persist in ways that undermine it. The term *libertarian* seems intended to modify *paternalism* with a view to signifying that the approach aims to encourage patterns of behavior that benefit the individual (paternalism) but do so without imposing the punitive regulation or impediments to liberty (libertarianism) that some associate with welfare regimes.

That people sometimes act in self-defeating ways is thought by advocates of libertarian paternalism to reflect two different thinking styles, a reflective deliberative style that uses knowledge about facts and values and a second automatic nonrational style involving decisions and choices influenced by impulse and environmental cues (Strack & Deutsch, 2004). The prospects of exploiting the latter style to unobtrusively influence health-related behavior gained attention following popularization of *nudge* by American behavioral economists Thaler and Sunstein (2008). Nudge entails a variety of interventions (Blumenthal-Barby and Burroughs, 2012), one of which, the default strategy, is exemplified by encouraging people to save for their retirement by manipulating

---

1. Although better welfare provision is consistently associated with overall improvements in population health, the effectiveness of welfare arrangements for reducing health inequalities is less clear (e.g., Bambra, 2011; Mackenbach et al., 2008; Muntaner et al., 2011; Popham et al., 2013).

choice through a particular arrangement of "opt-in" versus "opt-out" alternatives. Specifically, greater saving is encouraged by replacement of an opt-in system, which requires people to make a deliberate choice to set aside a portion of their salary, with an opt-out system in which savings are set aside by default. People are free to choose either alternative, but experience shows that many automatically accept the default option, which in this instance results in more people saving for retirement.

Although perceived by some to be new, nudge draws on decades-old theory, findings, and practices from behavioral psychology, economics, and social psychology (Marteau et al., 2011; Oliver et al., 2011; Vallgårda, 2012). Perceived newness is attributable in part to public-relations efforts by government aimed at winning approval for this reputedly cost-effective and more people-sensitive approach. However, premature attempts to win public support could backfire, as evidenced by commentaries that have labeled political endorsement of the approach as "smudge" (Bonell et al., 2011) and "fudge" (Wise, 2011), implying that true purposes have been hidden. One such purpose may be that nudge is believed by governments to offer opportunities for reducing welfare, because nudging is believed to be capable of achieving some of the social and health benefits that accompany welfare provisions. Thus, there is concern that a hidden purpose of government could be the introduction of nudge coupled with simultaneous reduction in welfare.

Whereas general acceptance of nudge may depend on its being accepted as libertarian and paternalistic as claimed, critics have argued that it may be neither (Hausman and Welch, 2010; Vallgårda, 2012). Nudges may be paternalistic if they benefit the person whose behavior is influenced. The nudge technique of *priming*, for example, might involve the arrangement of nutritious foods on supermarket shelves in an attractive and convenient manner to encourage their selection in favor of less-nutritious foods that are inconveniently placed. However, it is doubtful whether manipulations of that kind can be considered libertarian if they work by coercing particular choices at the expense of others. Moreover, the exact same nudge, used by food manufacturers to promote selection of foods and drinks that have little nutritional value but yield good profit, would be considered neither paternalistic nor libertarian, but exploitative. While commercial exploitation of that comparatively subtle kind appears to be widely tolerated, people may be less tolerant and more grudging if the same methods are used by the governments they elect.

In theory, people are at liberty to ignore a nudge if they so wish. However, the theory appears self-contradictory in that regard because, if it is true that many routine but important choices are unconscious, it cannot also be argued that those (nudged) choices are easily ignored. Indeed, the reliance of nudges on reputedly automatic unconscious processes seems problematic in the context of public policy, since it is questionable whether the business of democratic governments should include (the conscious) use of covert manipulation to influence the unconscious choices of citizens. Enthusiasm for nudging is to be welcomed to the extent that it encourages policymakers to understand the role of the physical and social environment in shaping health-related behavior. However, evidence of the effectiveness of nudge policies is limited, including the absence of evidence as well as evidence of little or no effect either for garnering public support for health promotion or for actually improving personal and population health (Marteau et al., 2011).

Rather than nudging *citizens*, it has been argued that the attention of government should instead be focussed on the activities of *corporations* that in many instances are directly responsible for creating circumstances that encourage people to make automatic nonrational choices that are harmful to health (Oliver, 2013). More specifically, it has been proposed that, in addition to overt regulation, government should reserve its use of nudge to *budge* the private sector, including the food, drink, and tobacco industries that employ nudge and other strategies to maximize profits to the detriment of health. Additionally, it could be argued that while governments are elected to implement the wishes of the electorate, governments should also initiate and lead. The legislative and regulatory authority of government is crucial in that respect. Governmental leadership in the use of legislation to regulate behavior combined with explicit enforcement, including penalties for noncompliance, is sometimes initially unpopular, but in due course the targeted behavior becomes an accepted social norm. Workplace smoking bans and the wearing of seatbelts are two such instances. Whereas nudge may in some circumstances complement overt legislative forms of regulation, it is unlikely, despite current optimism in some quarters, to offer an effective replacement.

### 13.4.1 The Logic and Ethics of Prevention Versus Biomedical Intervention

A widely-acknowledged peculiarity of contemporary somatic and mental healthcare is the imbalance of expenditure on biomedicine compared with preventive interventions. In the United States, for example, medical care attracts approximately 95% of total spending on healthcare, with just 5% going to population-wide preventive interventions (McGinnis et al., 2002). Similarly, in the United Kingdom, only 4% of National Health Service funding is spent on prevention (Marmot and Bell, 2012). Yet, as exemplified in Table 4.2 and Figures 4.6–4.8, most health benefit comes from preventive activities not biomedical care, and most benefit occurs in the general population and not among patients where most healthcare expenditure is concentrated.[2] Whereas logic and economics demand increased use of population-wide preventive interventions, healthcare worldwide continues to be dominated by individually-focussed biomedicine.

---

2. Notwithstanding this substantial imbalance in direct healthcare expenditure, the crucial role of social determinants means that personal and population health are also influenced by public expenditure in diverse other areas, including education, social welfare, economic and employment policy, justice, and human rights (Marmot et al., 2012). Indeed, the Health in All Policies framework is predicated on that fact (WHO, 2014b).

There are many reasons for this perverse distribution of emphases and resources, as described below.

## 13.5 WHY DO SUCCESSES IN PREVENTION NOT ATTRACT THE ACCOLADES THEY DESERVE?

It is in the nature of prevention that even extraordinary contributions to humanity go largely unnoticed. Were such achievements made in other ways, for example, a new medical device or procedure, it is likely that the persons or group responsible would be feted. Conversely, successes in prevention rarely attract accolades, and part of the reason appears to be the fact that prevention relates to threats to health that were averted. That is, prevention concerns events that did *not* happen. In countries where successful action has been taken to reduce road fatalities and injuries to normatively low levels, large numbers of people are alive who would have died, and larger numbers still have remained healthy who would have been disabled. Yet, those remarkable successes go largely uncelebrated. Rose (1985) opined, "Grateful patients are few ... where success is marked by a non-event" (p. 38). In general, it appears that humans possess a relative incapacity to appreciate the significance of things that have not happened.[3]

As one minor illustration of the point, there are many (although arguably not enough) city precincts where it is possible to "cross" a multilane highway by going under it. In each instance, anonymous planners and engineers in a municipal office had the foresight to separate people and cars by building a pedestrian tunnel under the road. Very likely, there has been no celebration to commemorate this type of contribution to preventive healthcare. Why? Probably, the impossibility of knowing precisely who has been spared victimhood is the barrier to our appreciation of the importance of the death and injury that did not happen. The importance of what happened (e.g., the construction of an underpass) is directly proportional to that which was prevented (human deaths and injuries). Yet, direct material evidence of what was prevented cannot be shown, and therefore is difficult to comprehend and even harder to celebrate. Who are the living healthy individuals who would have died or been injured? There is no answer to that question, no "human face" to represent those who did not die and those who were not injured.

Where, for example, is the child, who, in a moment of inattention, would have skipped onto the road into the path of a passing lorry? We see no image

---

3. Ironic acknowledgement of this conundrum can be found in several countries where, at a particular location, a plaque may be found declaring "On this site in [a particular year] nothing happened." One such example is in the West of Ireland. On the road from Galway to Clifden, at the town of Recess (which consists of a short row of small shops on one side of the road), there is a monument with the inscription "On this site in 1897 nothing happened." Working the irony, the storekeeper across the road from the monument claims to receive frequent enquiries from curious passers-by wanting to know "What was it that *didn't* happen in 1897".

of the smiling would-be victim who did not become one. There are no accompanying tearful expressions of heartrending relief of parents, forever grateful for the life of the child who has been permitted the opportunity to grow to adulthood. The underpass spared the child, but neither the child nor the parents nor anyone can know that specific would-be fact. The closest we can come to such awareness is by inferring it from statistics. However, announcements of statistics showing, for example, "a fall in road deaths," may catch the attention of a few people, while going more-or-less unnoticed by most and almost certainly forgotten by everyone soon thereafter. Given our cognitive and emotional architecture, we do not have the capacity to feel anything but abstract and ephemeral relief over tragedies that would have happened but did not, while being ever primed to be intensely moved emotionally by tragedies that do happen.

The psychological dimensions alluded to in the preceding paragraphs are related to a phenomenon known as the *identifiable victim effect*, which refers to the tendency for people to care more about identifiable than statistical victims (Small and Loewenstein, 2003; Västfjäll et al., 2014). For example, in a field trial, participants contributed more to a charity when their contributions benefitted a family that had already been selected from a list than when told that the family *would* be selected, even when the list from which the family was/would be selected was the same in both instances (Small and Loewenstein, 2003). Preventive initiatives such as the pedestrian underpass outlined above are commonplace, and similar examples could be cited from communities everywhere. Despite the success, large or small, such initiatives may have in preventing death and injury, they share a disappointing fate. It is the fate of not being valued in correct proportion to other interventions that produce less benefit but have identifiable beneficiaries. Evidence suggests that action aimed at saving identifiable victims is valued intuitively, if illogically, over actions intended to benefit victims who are no less real but are anonymous, even when the latter are in far greater numbers. Future success in optimizing personal and population health may depend on overcoming this idiosyncrasy of human cognition.

## 13.6 WHY HAS HEALTHCARE NOT ADOPTED PREVENTION AS THE PREDOMINANT APPROACH?

> ...the prospect of personal benefits to health provides only a weak motivation to accept a change, since it is neither immediate nor substantial, and an individual's health next year is likely to be much the same, regardless of whether that person accepts or rejects the proffered advice.

> (Rose, 1992, p. 105)

Who would not prefer to avoid disease than to suffer it, even if medical interventions exist to limit suffering? On the face of it, preventive healthcare seems the more appealing alternative. Nevertheless, many appear willing to opt for healthcare that prioritizes rescue over prevention, even if rescue is only partially successful and sometimes harmful. According to Rose (1992), part of the explanation is that illness is personal and not collective: "it happens to individuals." Collective action against underlying causes is not perceived as addressing personal concerns, and "among doctors, public, and governments alike the natural focus for [healthcare] is action for individuals" (Rose, 1992, p. 29). Moreover, at the level of the individual, it is a well-established psychological principle that although behavior is powerfully influenced by consequences (Skinner, 1969), immediate rather than delayed outcomes tend to have greater influence even when the latter are more highly valued (Rachlin, 2009).[4] As such, individual health-promoting behavior tends to be undermined because the benefit of improved health lies in the future, and threats of possible future suffering are by definition uncertain in both intensity and duration.

The fact remains, however, that treating individual cases of disease has little impact on the population burden of disease and often is of limited benefit even for the individuals treated. Conversely, whole-population risk factor reduction reduces the risk of disease for everyone, evidenced by increased health overall and fewer cases of manifest disease. Nevertheless, Rose (1985, 1992) suggested that the reputed economic benefits of prevention, which some cite as a main benefit, may be more apparent than real. As everyone dies eventually, disease prevented is at best disease postponed, with the associated costs also being postponed rather than avoided. However, the reader may recall from Chapter 3 that there is much conjecture about patterns of morbidity during the latter stages of life, with competing hypotheses about expansion versus compression of morbidity.

Disease postponed is likely to produce economic benefits if the lives that are prolonged remain mostly healthy, as predicted by the compression of morbidity hypothesis (Fries et al., 2011). Conversely, if postponement of death is accompanied by prolonged chronic illness, as predicted by the expansion of morbidity hypothesis (Olshansky et al., 1991), the economic burden of increased longevity may escalate due to increased demand for healthcare. As mentioned in Chapter 3, there is concern about population-wide expansion of morbidity from the "failure of success" of biomedical healthcare which may be keeping many people alive in poor health (Gruenberg, 1977). Whereas individualized biomedicine has the potential to contribute to expansion of morbidity, compression of morbidity, the preferred scenario for both humane and economic reasons, is a more likely outcome of success in population-wide preventive healthcare.

## 13.7   WHAT ABOUT THE ETHICAL IMPERATIVE OF PROVIDING HELP FOR SUFFERING INDIVIDUALS?

All healthcare systems are confronted with the challenge of having to fund diverse healthcare needs from finite budgets. Given the impossibility of

---

4. Notably, the quote at the beginning of this chapter suggests that Scottish philosopher, empiricist, and essayist David Hume (1711–1776) had similar insight almost three centuries earlier.

providing unlimited funding for every healthcare need, priorities must be set and choices made. Compared to most countries, the United Kingdom is well-advanced in addressing healthcare resource allocation problems. Since 1999, systematic, transparent, and evidence-based processes for resolving resource allocation dilemmas for England and Wales have been led by the National Institute for Health and Clinical Excellence (NICE) (Shah et al., 2013). Cost-effectiveness of interventions is determined using quality-adjusted life years (QALYs) (NICE, 2013), which, as may be remembered from Chapter 3, is a metric incorporating both length and quality of life that enables quantitative comparisons to be made between diverse health outcomes from diverse interventions for diverse conditions.

In 2009, NICE announced that its various committees would henceforth give extra weight to health gains from life-extending end-of-life treatments (Raftery, 2009). That decision to adjust policy appears to have been largely in response to a political predicament involving public concern about reputed unnecessary suffering among seriously-ill individuals. In addition, there appears to have been industry lobbying against the withholding of free NHS provision of expensive (i.e., cost-ineffective) drugs that could help to extend life modestly for some terminally-ill cancer patients. One of the main attractions of the QALY metric is that its universality encourages fairness, which the new NICE policy breached. The making of an exception in the form of preferential weighting for life-extending treatment for terminally-ill patients not only set a precedent for other groups, but in the context of fixed budgets necessarily leads to some patient groups receiving less preferential treatment (Round, 2012). Recently, British health economist Richard Cookson (2013) examined the problem with reference to eleven potential justifications and concluded that none yielded a coherent ethical basis for the new policy.

The events that prompted the revised NICE policy, and reactions to the revision, highlight the controversial nature of decision making in healthcare. Given those events, it seems reasonable to infer that substantial challenges are likely to confront decision makers who seek to strengthen the priority of preventive interventions that promote *future* health over biomedical interventions for *present* suffering, even when preventive interventions are known to produce greater overall gains for personal and population health. For that reason, it is instructive to consider Cookson's (2013) framework of eleven potential justifications in the context of the current need for healthcare to be reoriented away from ever-increasing commitment to biomedical interventions for existing diseases toward greater emphasis on population-wide preventive healthcare. The subheadings in the concluding sections of this chapter refer to Cookson's eleven justifications, and each is followed by a statement of the key argument and a brief discussion of limitations. The discussion extends Cookson's analysis of end-of-life treatments to consider limitations from the standpoint of each argument being proffered as justification for continuing the current healthcare *status quo* characterized by biomedical dominance at the expense of preventive healthcare.

## 13.7.1 Rule of Rescue

*Argument*: Society has a duty to rescue all identifiable citizens in immediate peril, regardless of cost.

*Limitations*: The moral imperative that is the *rule of rescue* is most obvious in unusual emergency situations where innocent individuals are at great peril. Cookson (2013) cites the Chilean mine disaster of 2010, which prompted the Chilean government and the international community to take immediate action to mount the best possible rescue effort without consideration for the financial cost. Conversely, the rescue operations of routine biomedical practice are not one-off self-contained events but an ever-present and perpetual daily occurrence, which force upon us the need to consider resource limitations. As such, the moral imperative of the rule of rescue prompted by circumscribed and imminent tragedies is necessarily not of the same order when responding to routine tragedies of clinical practice, especially when alternatives to rescue are available in the form of preventive healthcare. Moreover, the moral imperative claimed by biomedical intervention is diminished by the fact that the personal and population benefits for quality and length of life from disease prevention exceed those from biomedical rescue.

Admittedly, the specific individuals who benefit from preventive interventions, and the extent of their individual benefit, are not likely to be known even to the beneficiaries. As such, the identifiable victim effect discussed above comes into play, involving the human tendency to be moved more by the plight of known victims than that of unknown victims. This dilemma raises awkward questions which demand answers, especially in light of the manifold limitations of biomedicine: Should the welfare of identifiable victims that biomedical healthcare seeks to rescue necessarily take precedence over the greater number who can be spared the need for such rescue but who happen to be unidentifiable? Logic and humanitarian concern indicate the need to counter the familiar human bias that favors identifiable over unidentifiable victims. Action is needed to achieve a better balance than currently exists between priorities for addressing the important (i.e., interventions that deliver health to many) and the perceived urgent biomedical rescue operations of limited efficacy that tend to postpone mortality at the cost of extended morbidity.

## 13.7.2 Fair Chances

*Argument*: All patients should have a fair chance of receiving the best possible treatment.

*Limitations*: The "fair chances" argument is usually invoked in relation to decisions about which patient should receive a resource of fixed quantity, such as an organ transplant. The argument is not usually used for deciding resource allocations of varying amounts for diverse healthcare purposes, and not at all for deciding the allocation of resources between preventive and remedial purposes.

Application to the latter circumstance would presumably lead to an immediate reallocation of resources in favor of population-wide prevention, because as discussed above preventive interventions currently attract something in the region of 5% or less of total expenditure on healthcare with the remainder going to individualized biomedical healthcare (McGinnis et al., 2002; Marmot and Bell, 2012). In particular, fair chances would militate against the use of some current "best possible" treatments for chronic debilitating or terminal illnesses that confer comparatively little individual health benefit but at considerable expense.

Extending the fair chances argument to health promotion and disease prevention would mean extending fair chances to all, including patients and non-patients. In that case, a fair chance of receiving *the best possible treatment* would have to be re-phrased as a fair chance of receiving *the best possible health*. Again, any such extension would indicate the need for an immediate reallocation of resources in favor of prevention, because evidence shows that more emphasis on prevention (e.g., in smoking cessation, limiting harmful consumption of alcohol, improving dietary practices, and encouraging more physical activity) would produce substantial improvements in personal and population health and lower incidence of common complex diseases.

### 13.7.3    Ex Post Willingness to Pay

*Argument*: After individuals are diagnosed with serious and especially terminal illness they will be willing to pay a high amount for a chance to receive even a small improvement in quality of life and especially a small extension of life. *Limitations*: Individual willingness to pay *after* diagnosis is of little relevance, because most healthcare is publicly funded or subsidized. Even in privatized healthcare systems, major public subsidies exist in the form of healthcare infrastructure, including clinical services and healthcare research and development. Therefore, willingness of individuals to pay for some high-cost healthcare does not equate to full-cost when healthcare is embedded within an existing publicly-subsidized infrastructure. Moreover, there is effectively no limit to the ingenuity of pharmaceutical and medical device companies in using public subsidies to invent new products that *may* benefit health and longevity for some patient groups. Therefore, *willingness to pay* actually requires the willingness of *everyone* to pay essentially unlimited higher taxes and higher insurance premiums to provide some individuals access to products many of which will have little therapeutic value.

### 13.7.4    Caring Externality

*Argument*: Family members gain important non-health benefits from improvements to the health of seriously-ill loved ones, including life extension for those who are terminally ill.

*Limitations*: If the argument applies at all, it should apply to health gains of all kinds and not merely to serious illness, and it should apply equally to both preventive and biomedical interventions. As such, the argument offers no justification for claiming special privilege for biomedical interventions. Additionally, the argument is contestable for reasons of relevance because it is predicated on benefit to others (namely, loved ones of sick individuals), whereas biomedical healthcare is usually intended to benefit sick individuals. To the extent that healthcare benefits others, it should be for the good of society in *general* rather than specific *others*. Furthermore, the argument is difficult to sustain, because it is potentially discriminatory. It implies that those who are more loved have more right to healthcare, and by logical extension those who have outlived loved ones may have no right.

## 13.7.5   Financial Protection

*Argument*: Individuals put high value on protection against the prospect of needing expensive interventions at some time in the future because of the threat of catastrophic financial consequences compared to bearable financial loss in the event of need for less-expensive interventions.

*Limitations*: This is largely irrelevant because financial protection is not a health benefit. People seeking to protect themselves against high out-of-pocket costs for future cost-ineffective healthcare have other options, and calling upon the healthcare budget in lieu of those options is unjustified. For example, people may take private health insurance commensurate with their personal choice to prioritize future high-cost healthcare, they may pursue wealth-creating paths (e.g., career choices and investments) to bolster their financial security, and in the event of absolute poverty they may avail themselves of social welfare provisions in countries where those are available. Crucially, poverty should not be a barrier to accessing reasonable healthcare. All things considered, desire to avoid potential high-cost from future biomedical healthcare argues for immediate priority to be given to population-wide preventive healthcare because of the known greater efficacy and cost-effectiveness of prevention over clinical rescue.

## 13.7.6   Symbolic Value

*Argument*: Expensive biomedical healthcare symbolizes the special value society attaches to human life.

*Limitations*: Healthcare is not unique in this regard. A wide range of public and charitable institutions symbolize the special value society places on human life and quality of life. For example, emergency rescue services give priority to life-and-death situations, police and criminal justice systems give priority to protecting against endangerment of life, and safety regulations in the workplace and the community give priority to protecting human life. Therefore, the benefits of specific actions should be weighted in light of community benefit and opportunity cost. Compared to preventive healthcare, biomedical interventions consume disproportionate resources for the benefit of comparatively few in the population. Moreover, public expenditure on expensive biomedical technologies is influenced by lobbying from influential vested interests, including corporations whose main priority is profit not human welfare.

### 13.7.7   Diminishing Marginal Value of Future Life Years

*Argument*: Clinical rescue undertaken later in life has priority over early-life prevention, because the more future life years one has in prospect the less valuable is an additional life year.

*Limitations*: This is based on the idea that the value of an additional unit of a particular good may depend on how much of the good one already possesses. For example, the young who have an abundance of health may assign low priority to disease prevention, whereas the elderly may assign high priority to high-cost biomedical technologies in the hope of having some of their waning health restored. Both groups, however, are responding to their respective immediate circumstances, which are independent of the relative value of preventive and biomedical healthcare for society as a whole. Thus, lack of concern among the young for more health should lead them to assign low priority for healthcare in general, whether preventive or biomedical. The elderly, however, may perceive that opportunities for disease prevention have passed, leaving them little choice other than potential biomedical remedies. Public priorities should reflect expected benefits from the perspective of society rather than subgroups whose points of reference are likely to change over time. Individuals and society benefit most when there is less need for high-cost biomedical healthcare to address (albeit imperfectly and often not at all) age-related expansion of morbidity (Olshansky et al., 1991). Given that the whole of society is best served by people living long healthy lives characterized by compression of morbidity (Fries et al., 2011), prevention should have priority over clinical rescue.

### 13.7.8   Concentration of Benefits

*Argument*: Large health benefits for a small number of patients are more valuable than small health benefits for a large number of other patients, even if the aggregate of the latter is greater than the former.

*Limitations*: The values expressed in this argument may be held by some people, but they are not necessarily shared by everyone. However, irrespective of personal values, the argument is unsustainable because low-cost preventive interventions yield larger (not smaller) benefits for more people than high-cost biomedical interventions yield for fewer people (Rose, 1992). Therefore, healthcare realities dictate that prevention should take precedence over biomedical intervention.

## 13.7.9   Dread

*Argument*: The high prevalence of noncommunicable diseases, including those that are dreaded (e.g., cancer), means that most people will encounter them personally or among loved ones. Therefore, remedial interventions for these conditions should attract the highest priority.

*Limitations*: There is no obvious proportionality between the health impact of a particular disease and the amount of public anxiety or dread it elicits. For example, there would be little justification in giving high priority to a rare condition that has come to elicit anxiety, fear, and dread among the public due to intense attention it may happen to have received in the public media, while simultaneously ignoring an equally rare and dreadful condition that has not come to the attention of the general public. In any event, if dread of disease were proportional, it should lead to increased priority being given to preventive healthcare. Compared to clinical rescue, preventive intervention optimizes avoidance and postponement of noncommunicable diseases (e.g., WHO, 2012a,b) including those that are most dreaded.

## 13.7.10   Time to Set One's Affairs in Order

*Argument*: The maintenance of life involving expensive biomedical intervention offers the important benefit of allowing people time to set their affairs in order.

*Limitations*: Setting one's affairs in order (e.g., updating one's will, organizing financial matters and possessions, and fare-welling family and friends), as with the caring externality argument above, benefits others and the relevance of that for prioritizing healthcare is open to question. Moreover, the argument violates the principle that healthcare should be deployed on the basis of balance between healthcare needs and benefits, and not on the basis of the individual wealth, status, or role. That is, the argument gives priority to people who have affairs to set in order over those who have no such affairs. Moreover, the argument does not recognize those with affairs who have the foresight to set them in order as preparation for an unanticipated personal health crisis. Thus, the argument is intrinsically dubious, while also being of little relevance to decisions about relative resource allocation to preventive and biomedical healthcare.

## 13.7.11   Severity of Illness

*Argument*: Compared to health promotion and prevention among the healthy, sick individuals are worse off and in greater need and therefore should be given priority.

*Limitations*: A common ethical argument for the setting of healthcare priorities is that healthcare resources should be allocated on the basis of need and that more severely-ill patients have greater need. In the absence of other considerations, that argument would give automatic priority to life-threatening conditions irrespective of prognosis. In practice, healthcare priorities reflect anticipated health benefit as well as need. This is justified, as there would be little justification in prioritizing resource-intensive interventions that confer marginal extension to life of low quality over interventions that promote long life of high quality. Indeed, empirical studies show that people put higher value on interventions that improve quality of life over interventions that merely extend life (Pinto-Prades et al., 2014; Shah et al., 2014). Thus, proportionality of benefit, and not merely need, is and should continue to be a priority of healthcare.

## 13.8 THE ETHICAL FOUNDATIONS OF PREVENTION OVER CURE

In conclusion, Cookson's (2013) examination of ethical justifications for giving special priority to life-extending end-of-life treatments over other biomedical interventions can be generalized to a wider examination of the ethicality of current priorities that consistently favor biomedical over preventive interventions. Just as Cookson found that none of eleven potential justifications provides an ethical basis for prioritizing end-of-life treatments, so the present analysis finds no ethical justification for continuing current biomedical dominance of healthcare. On the contrary, several major ethical considerations provide formidable grounds for the immediate and radical reordering of priorities away from biomedical intervention in favor of population-wide life-course preventive healthcare.

## REFERENCES

Adler, N., Singh-Manoux, A., Schwartz, J., et al., 2008. Social status and health: a comparison of British civil servants in Whitehall-II with European-and African-Americans in CARDIA. Soc. Sci. Med. 66, 1034–1045.

Ahn, A.C., Tewari, M., Poon, C.S., Phillips, R.S., 2006. The limits of reductionism in medicine: could systems biology offer an alternative? PLoS Med. 3, 1–5. http://dx.doi.org/10.1371/journal.pmed.0030208.

Ashton, J.R., 2006. Virchow misquoted, part-quoted, and the real McCoy. J. Epidemiol. Community Health 60, 671.

Bambra, C., 2007. Going beyond the three worlds of welfare capitalism: regime theory and public health research. J. Epidemiol. Community Health 61, 1098–1102.

Bambra, C., 2011. Health inequalities and welfare state regimes: theoretical insights on a public health "puzzle" J. Epidemiol. Community Health 65, 740–745.

Beaglehole, R., Bonita, R., 2008. Global public health: a scorecard. Lancet 372, 1988–1996.

Beard, J.R., Bloom, D.E., 2014. Towards a comprehensive public health response to population ageing. Lancet 383, 65–68.

Benach, J., Malmusi, D., Yasui, Y., Martínez, J.M., 2013. A new typology of policies to tackle health inequalities and scenarios of impact based on Rose's population approach. J. Epidemiol. Community Health 67, 286–291.

Berridge, V., 2007. Public health activism. Br. Med. J. 335, 1310–1312.

Bloom, D.E., Chatterji, S., Kowal, P., et al., 2014. Macroeconomic implications of population age-ing and selected policy responses. Lancet 384, 649–657. http://dx.doi.org/10.1016/S0140-6736 (14)61464-1.

Blumenthal-Barby, J.S., Burroughs, H., 2012. Seeking better health care outcomes: the ethics of using the" nudge". Am. J. Bioeth. 12, 1–10.

Bonell, C., McKee, M., Fletcher, A., Haines, A., Wilkinson, P., 2011. Nudge smudge: UK Govern-ment misrepresents "nudge" Lancet 377, 2158–2159.

Braveman, P., 2006. Health disparities and health equity: concepts and measurement. Annu. Rev. Public Health 27, 167–194.

Braveman, P., Gottlieb, L., 2014. The social determinants of health: it's time to consider the causes of the causes. Public Health Rep. 129 (suppl 2), 19–31.

Braveman, P., Gruskin, S., 2003. Defining equity in health. J. Epidemiol. Community Health 57, 254–258.

Bronfenbrenner, U., 1977. Toward an experimental ecology of human development. Am. Psychol. 32, 513–531.

Brown, P., Zavestoski, S., McCormick, S., et al., 2004. Embodied health movements: new approaches to social movements in health. Sociol. Health Illn. 26, 50–80.

Cookson, R., 2013. Can the NICE "end-of-life premium" be given a coherent ethical justification? J. Health Polit. Policy Law 38, 1129–1148.

Coutts, C., Forkink, A., Weiner, J., 2014. The portrayal of natural environment in the evolution of the ecological public health paradigm. Int. J. Environ. Res. Public Health 11, 1005–1019.

Davies, S.C., Winpenny, E., Ball, S., et al., 2014. For debate: a new wave in public health improve-ment. Lancet 383, 1889–1895. http://dx.doi.org/10.1016/S0140-6736(13)62341-7.

Ferrie, J.E. (Ed.), 2004. Work Stress and Health: The Whitehall II Study. Public and Commercial Services Union, London.

Fries, J.F., Bruce, B., Chakravarty, E., 2011. Compression of morbidity 1980–2011: a focused review of paradigms and progress. J. Aging Res. 1–10. http://dx.doi.org/10.4061/2011/261702.

Gruenberg, E.M., 1977. The failures of success. Milbank Mem. Fund Q. Health Soc. 55, 3–24.

Haines, A., Whitmee, S., Horton, R., 2014. Planetary health: a call for papers. Lancet 384, 479–480.

Hanlon, P., Carlisle, S., Hannah, M., et al., 2011. Making the case for a "fifth wave" in public health. Public Health 125, 30–36.

Hausman, D.M., Welch, B., 2010. Debate: to nudge or not to nudge. J. Polit. Philos. 18, 123–136.

Horton, R., Beaglehole, R., Bonita, R., et al., 2014. From public to planetary health: a manifesto. Lancet 383, 847.

Hume, D., 1739. A treatise of human nature. Reprinted from the Original Edition in three volumes and edited, with an analytical index, by L.A. Selby-Bigge. Oxford: Clarendon Press, 1896. Indi-anapolis, IN: Online Library of Liberty. Retrieved on 31 May 2014 from http://people.rit.edu/wlrgsh/Treatise.pdf.

Ideker, T., Galitski, T., Hood, L., 2001. A new approach to decoding life: systems biology. Annu. Rev. Genomics Hum. Genet. 2, 343–372.

Institute of Medicine, 1997. America's vital interest in global health: protecting our people, enhanc-ing our economy, and advancing our international interests. Institute of Medicine Board on International Health, National Academies Press, Washington, DC.

Kitano, H., 2002. Systems biology: a brief overview. Science 295, 1662–1664.

Labonté, R., 2013. Health activism in a globalising era: lessons past for efforts future. Lancet 381, 2158–2159.

Laverack, G., 2013. Individualism or activism for health in hard times? J. Public Health 21, 385–386.

Mackenbach, J., Stirbu, I., Roskam, A., et al., 2008. Socioeconomic inequalities in health in 22 European countries. N. Engl. J. Med. 358, 2468–2481.

Mackenbach, J.P., Karanikolos, M., McKee, M., 2013a. The unequal health of Europeans: successes and failures of policies. Lancet 381, 1125–1134.

Mackenbach, J.P., Lingsma, H.F., van Ravesteyn, N.T., Kamphuis, C.B., 2013b. The population and high-risk approaches to prevention: quantitative estimates of their contribution to population health in the Netherlands, 1970–2010. Eur. J. Pub. Health 23, 909–915. http://dx.doi.org/10.1093/eurpub/cks106.

Marmot, M., 2005. Social determinants of health inequalities. Lancet 365, 1099–1104.

Marmot, M., 2013a. Health inequalities in the EU: Final report of a consortium. European Union: European Commission Directorate-General for Health and Consumers, http://dx.doi.org/10.2772/34426.

Marmot, M., 2013b. Europe: good, bad, and beautiful. Lancet 381, 1190–1191.

Marmot, M., Bell, R., 2012. Fair society, healthy lives. Public Health 126, S4–S10.

Marmot, M., Brunner, E., 2005. Cohort profile: the Whitehall II study. Int. J. Epidemiol. 34, 251–256.

Marmot, M.G., Shipley, M.J., 1996. Do socioeconomic differences in mortality persist after retirement? 25 year follow up of civil servants from the first Whitehall study. Br. Med. J. 313, 1177–1180.

Marmot, M., Allen, J., Bell, R., et al., 2012. WHO European review of social determinants of health and the health divide. Lancet 380, 1011–1029.

Marteau, T.M., Ogilvie, D., Roland, M., Suhrcke, M., Kelly, M.P., 2011. Judging nudging: can nudging improve population health? Br. Med. J. 342, 263–265.

Mathers, C.D., Stevens, G.A., Boerma, T., et al., 2015. Causes of international increases in older age life expectancy. Lancet 385, 540–548. http://dx.doi.org/10.1016/S0140-6736(14)60569-9.

Mayr, E., 1961. Cause and effect in biology. Science 134, 1501–1506.

McGinnis, J.M., Foege, W.H., 2004. The immediate vs the important. J. Am. Med. Assoc. 291, 1263–1264.

McGinnis, J.M., Williams-Russo, P., Knickman, J.R., 2002. The case for more active policy attention to health promotion. Health Aff. 21, 78–93.

McLeroy, K.R., Bibeau, D., Steckler, A., Glanz, K., 1988. An ecological perspective on health promotion programs. Health Educ. Behav. 15, 351–377.

Medina, M.Á., 2013. Systems biology for molecular life sciences and its impact in biomedicine. Cell. Mol. Life Sci. 70, 1035–1053.

Melkas, T., 2013. Health in all policies as a priority in Finnish health policy: a case study on national health policy development. Scand. J. Public Health 41 (11 suppl), 3–28.

Mulgan, G., 2010. Health is not just the absence of illness: health in all policies and 'all in health policies'. In: Kickbusch, I., Buckett, K. (Eds.), Implementing Health in All Policies. Government of South Australia, Adelaide, pp. 39–48.

Muntaner, C., Borrell, C., Ng, E., Chung, H., Espelt, A., Rodriguez-Sanz, M., Benach, J., O'Campo, P., 2011. Politics, welfare regimes, and population health: controversies and evidence. Sociol. Health Illn. 33, 946–964.

NICE, 2013. Guide to the Methods of Technology Appraisal. National Institute for Health and Clinical Excellence, London. Retrieved 29 December 2014 from, http://publications.nice.org.uk/pmg9.

Oliver, A., 2013. From nudging to budging: using behavioural economics to inform public sector policy. J. Soc. Policy 42, 685–700.

Oliver, A., Ryner, G., Lang, T., 2011. Is nudge an effective public health strategy to tackle obesity? Yes. Br. Med. J. 342, 1–2. http://dx.doi.org/10.1136/bmj.d2168.

Olshansky, S.J., Rudberg, M.A., Carnes, B.A., et al., 1991. Trading off longer life for worsening health: the expansion of morbidity hypothesis. Aging Health 3, 194–216.

Pinto-Prades, J.L., Sánchez-Martínez, F.I., Corbacho, B., Baker, R., 2014. Valuing QALYS at the end of life. Soc. Sci. Med. 113, 5–14.

Popham, F., Dibben, C., Bambra, C., 2013. Are health inequalities really not the smallest in the Nordic welfare states? a comparison of mortality inequality in 37 countries. J. Epidemiol. Community Health 67, 412–418.

Rachlin, H., 2009. The Science of Self-Control. Harvard University Press, Cambridge, MA.

Raftery, J., 2009. NICE and the challenge of cancer drugs. Br. Med. J. 338, 271–272.

Rose, G., 1985. Sick individuals and sick populations. Int. J. Epidemiol. 14, 32–38.

Rose, G., 1992. The Strategy of Preventive Medicine. Oxford University Press, Oxford, UK.

Round, J., 2012. Is a QALY still a QALY at the end of life? J. Health Econ. 31, 521–527.

Sallis, J.F., Owen, N., Fisher, E.B., 2008. Ecological models of health behavior. In: Glanz, K., Rimer, B.K., Viswanath, K. (Eds.), Health Behavior and Health Education: Theory, Research, and Practice, fourth ed. Wiley, San Francisco, CA, pp. 465–486.

Shah, K.K., Cookson, R., Culyer, A.J., Littlejohns, P., 2013. NICE's social value judgements about equity in health and health care. Health Econ. Policy Law 8, 145–165.

Shah, K.K., Tsuchiya, A., Wailoo, A.J., 2014. Valuing health at the end of life: an empirical study of public preferences. Eur. J. Health Econ. 15, 389–399.

Skinner, B.F., 1969. Contingencies of Reinforcement: A Theoretical Analysis. Appleton-Century-Crofts, New York.

Skyttner, L., 1996. General systems theory: origin and hallmarks. Kybernetes 25, 16–22.

Small, D.A., Loewenstein, G., 2003. Helping a victim or helping the victim: altruism and identifiability. J. Risk Uncertain. 26, 5–16.

Strack, F., Deutsch, R., 2004. Reflective and impulsive determinants of social behavior. Pers. Soc. Psychol. Rev. 8, 220–247.

Stuckler, D., McKee, M., 2008. Five metaphors about global-health policy. Lancet 372, 95–97.

Stuckler, D., Basu, S., McKee, M., 2010. Budget crises, health, and social welfare programmes. Br. Med. J. 341, 77–79.

Thaler, R.H., Sunstein, C.R., 2008. Nudge: Improving Decisions About Health, Wealth, and Happiness. Yale University Press, New Haven.

Vallgårda, S., 2012. Nudge: a new and better way to improve health? Health Policy 104, 200–203.

Van den Broucke, S., 2013. Implementing health in all policies post Helsinki 2013: why, what, who and how. Health Promot. Int. 28, 281–284.

Västfjäll, D., Slovic, P., Mayorga, M., Peters, E., 2014. Compassion fade: affect and charity are greatest for a single child in need. PLoS One 9, 1–10. http://dx.doi.org/10.1371/journal.pone.0100115.

von Bertalanffy, L., 1951. Theoretical models in biology and psychology. J. Pers. 20, 24–38.

von Bertalanffy, L., 1972. The history and status of general systems theory. Acad. Manag. J. 15, 407–426.

Whitehead, M., 1991. The concepts and principles of equity and health. Health Promot. Int. 6, 217–228.

WHO, 1946. Constitution of the World Health Organization. Supplement, October 2006, 45th ed. World Health Organization, Geneva. Retrieved 7 May 2014 from, http://www.who.int/governance/eb/who_constitution_en.pdf.

WHO, 2005. The Bangkok Charter for Health Promotion in a Globalized World. World Health Organization, Geneva. Retrieved 29 December 2014 from, http://www.who.int/healthpromotion/conferences/6gchp/bangkok_charter/en/.

WHO, 2012a. Prevention and Control of Noncommunicable Diseases. Sixty-fifth World Health Assembly. World Health Organization, Geneva. Retrieved 25 May 2013 from, http://apps.who.int/gb/ebwha/pdf_files/WHA65/A65_6-en.pdf.

WHO, 2012b. A Comprehensive Global Monitoring Framework, Including Indicators, and a Set of Voluntary Global Targets for the Prevention and Control of Noncommunicabale Diseases. Revised WHO discussion paper (Version dated 25 July 2012), World Health Organization, Geneva. Retrieved 24 November 2013 from, http://www.who.int/nmh/events/2012/discussion_paper3.pdf.

WHO, 2014a. Social Determinants of Health. World Health Organization, Geneva. Retrieved 28 December 2014 from, http://www.who.int/social_determinants/thecommission/finalreport/key_concepts/en/.

WHO, 2014b. Health in All Policies Framework for Country Action. World Health Organization, Geneva. Retrieved 26 April 2014 from, http://www.who.int/cardiovascular_diseases/140120HPRHiAPFramework.pdf?ua=1.

Wise, J., 2011. Nudge or fudge? doctors debate best approach to improve public health. Br. Med. J. 342, 1–2. http://dx.doi.org/10.1136/bmj.d580.

Zoller, H.M., 2005. Health activism: communication theory and action for social change. Commun. Theory 15, 341–364.

# Chapter 14

# Optimal Healthcare: Risk Factor Reduction and Adjunctive Biomedical Intervention

*The function of protecting and developing health must rank even above that of restoring it when it is impaired*

Attributed to Hippocrates.[1]

## Contents

---

1. Where precisely this statement appears in the works of Hippocrates seems to be a mystery. The quote is typically cited in text advocating preventive healthcare. However, none of many references to the statement that this author has viewed cites an original source other than "Hippocrates," suggesting that it may be misattributed.

The Health of Populations. http://dx.doi.org/10.1016/B978-0-12-802812-4.00014-X

History shows that the benefits of prevention on disease incidence can be transformative for personal and population health. Examples include the dramatic increases in average life expectancy that characterized the epidemiologic transition of the nineteenth century (Chapters 1-3) and the reductions in incidence of cardiovascular disease in many countries during the last decades of the twentieth century (Chapter 4). As is true of disease patterns in all eras, habits and habitats are primarily responsible for the current global crisis in noncommunicable diseases. These account for two-thirds of deaths worldwide and more than four-fifths of deaths in high-income countries (WHO, 2014a). The worldwide proportion of total deaths attributable to noncommunicable diseases is expected to increase due to the combined effects of population aging and trends in developing countries toward the higher rates characteristic of high-income countries.

Excluding mental health problems (discussed in Chapter 15), noncommunicable diseases are responsible for an annual loss of more than 500 million disability adjusted life years, approximately 60% of which are due to premature mortality and 40% due to morbidity (WHO, 2008). The projected growth in both mortality and morbidity over the next two decades will create a global economic burden of immense proportions. The projected cumulative worldwide loss of economic output for 2011-2030 has been estimated to be USD47 trillion, about one-third of which will be due to mental health problems and nearly half of which will be borne by developing countries (Bloom et al., 2012). This has raised concerns about economic instability, governmental fragility, and national security (Atun, 2014). Such threats demand alternatives to the *status quo* of biomedical dominance of healthcare.

Belated recognition by the international community of the need to address the growing crisis in noncommunicable diseases is evidenced by the fact that the United Nations has only ever twice held meetings of heads of state on an issue related to health (Hunter and Reddy, 2013). The first meeting, in 2001, concerned HIV/AIDS, and the second occurred a decade later in 2011 to discuss the global threat from noncommunicable diseases. It has been suggested that the mobilization of international effort to combat HIV offers lessons for action now needed to address the rise of common complex diseases (Atun, 2014). Strong advocacy from within secular society and affected communities, and from scientists, contributed to a comparatively well-organized, well-resourced, and substantially successful response to the spread of HIV. Although the comparative clinical diversity of common complex diseases poses challenges for their control, the fact that they share common behavioral and social determinants should advantage coordinated international preventive effort.

## 14.1   PROGRAM FOR PREVENTION AND CONTROL OF COMMON COMPLEX DISEASES INCORPORATING RISK FACTOR REDUCTION AND ADJUNCTIVE BIOMEDICAL CARE

*Realistically, many diseases will long continue to call for both [population preven- tion and clinical medicine]...Nevertheless, the priority of concern should always be the discovery and control of the causes of incidence*

(Rose, 1985, p. 38)

There is no single set of effective prevention strategies. Rather, the history of successful prevention shows that the strategies that work are many and varied and that success sometimes depends on synergy created by several strategies operating simultaneously. As discussed in Chapter 1, improved methods of agricultural production, storage, and distribution, as well as public engineer- ing works involving the provision of clean water and the disposal of sewage, were core elements of habitat changes responsible for the epidemiologic transition in nineteenth century England and Wales. As has been mentioned, Rose (1985, 1992) argued, in respect of habits, that the most effective strat- egies are often those that produce whole population shifts in social norms of behavior. Unfortunately, some recent whole-population shifts, notably, increased physical inactivity, poor diet, obesity, and diabetes are contributing to the increased global burden of common complex diseases, especially in developing countries.

Following formal adoption by the United Nations (2012) of the Political Declaration of the aforementioned 2011 meeting about prevention and control of noncommunicable diseases, the WHO undertook to develop

*a comprehensive global monitoring framework, including a set of indicators and voluntary global targets for the prevention and control of noncommunicable diseases.*

(WHO, 2012a)

The program specifies targets for 2025, with reviews scheduled for 2015 and 2020. The overall target is for a 25% reduction in premature mortality among adults aged 30-70 years from four targeted diseases: cardiovascular diseases (especially coro- nary heart disease and stroke), cancers, chronic respiratory diseases, and diabetes. The proposed overall target of "25 by 25" (i.e., 25% reduction in premature mor- tality by 2025) has a 15-year timeframe from 2010 to 2025 and is based on analysis of historical trends within a selected group of 81 member states. Trends in death rate in the best performing countries for the 30-year period from 1980 to 2010 were used to calculate the "achievable" specific targets as well as the overall target of 25% reduction in leading noncommunicable diseases by 2025.

The 25% target is to be measured as relative change using 2010 as the reference baseline for each country or region. Assuming that all countries and regions benefit as intended, the fact that one change-target of 25% has been set for all means that differences in burden of disease that existed between countries and regions at the start of the period are likely to be present at the end of the period. In other words, for the time being at least, priority is being given to *improved* health for all rather than *equality* of health. It is assumed that the more poorly performing countries, mostly in developing regions of the world, can emulate gains made by better performing countries, and that the better performers are also capable of further gains in health. The historic experience of the best performing countries over the period 1980-2010 shows that major gains in the prevention of premature death can be achieved through population-wide reductions in exposure to known risk factors. Additional, more modest, gains are also achievable by complementing risk factor reduction with affordable biomedical interventions for patients with manifest disease.

Table 14.1 summarizes the main actions of the WHO program for achieving the overall target for all countries of 25% reduction (relative to 2010 levels) in

**TABLE 14.1** Adaptation and extension of the WHO global program for prevention (risk factor reduction) and control (biomedical intervention) of common complex diseases

**Program for prevention and control of common complex diseases**

| Overall outcomes (to be achieved by 2025)[a] | Outcome target |
|---|---|
| Premature mortality among adults aged 30-70 years from cardiovascular diseases, cancers, chronic respiratory diseases, and diabetes | 25% reduction |

**1. Prevention of disease through risk factor reduction to reduce disease incidence**

| Risk factor | Outcome target | Change targets |
|---|---|---|
| **1.1. Primary (behavioral) risk factor reduction** | | |
| Tobacco use | 30% reduction | Prevalence of tobacco use among adolescents and adults aged 18+ years |
| Harmful consumption of alcohol | 10% reduction | 1. Reduction in total alcohol per capita consumption (aged 15+ years) <br> 2. Reduced prevalence of heavy episodic drinking among adolescents and adults |
| Dietary factors: Salt/sodium intake | 30% reduction | 1. Mean population intake of salt (sodium chloride) with the long-term aim of achieving the recommended level of less than 5 g per day |

*Continued*

**TABLE 14.1** Adaptation and extension of the WHO global program for prevention (risk factor reduction) and control (biomedical intervention) of common complex diseases—cont'd

### 1. Prevention of disease through risk factor reduction to reduce disease incidence—Con'd

| Risk factor | Outcome target | Change targets |
|---|---|---|
| Saturated fatty acid intake | 15% reduction | 2. Mean proportion of total energy intake from saturated fatty acids, with the aim of achieving the recommended level of less than 10% of total energy intake. Includes replacement of saturated fatty acids by polyunsaturated fatty acids |
| Trans fatty acids | Complete or near-complete removal | 3. Replacement with polyunsaturated fatty acids to eliminate partially hydrogenated vegetable oils from the food supply |
| Fruit and vegetable intake | Replacement of processed foods | 4. Replacement of high-energy foods, especially processed foods high in fats and sugars, by the recommended level of five total servings (400 g) of fruit and vegetables per day |
| Marketing to children | Remove | 5. Remove exposure to marketing of foods and drinks high in saturated fats, trans fatty acids, free sugars, or salt |
| Physical inactivity | 10% reduction | 1. Reduced prevalence of physical inactivity (less than 60 min of moderate to vigorous intensity activity per day) for adolescents<br>2. Reduced prevalence of physical inactivity (less than 150 min of moderate intensity activity per week) for adults aged 18+ years |

### 1.2. Secondary (biological) risk factor reduction

| | | |
|---|---|---|
| Raised blood pressure | 25% reduction | Reduction in population prevalence of systolic blood pressure $\geq$140 mm Hg and/or diastolic blood pressure $\geq$90 mm Hg |
| Overweight and obesity | No increase | No increase in prevalence of persons overweight (body mass index $\geq$25) and obese (body mass index $\geq$30) |
| Raised blood glucose | No increase | No increase in prevalence of fasting blood glucose concentration $\geq$7.0 mmol/l (126 mg/dl) |

### 2. Biomedical intervention to control disease progression

| Disease category | Outcome target | Change targets |
|---|---|---|
| All major noncommunicable diseases | An 80% availability | Population proportion covered by "affordable basic technologies and essential medicines," including generics, for treating |

*Continued*

**TABLE 14.1** Adaptation and extension of the WHO global program for prevention (risk factor reduction) and control (biomedical intervention) of common complex diseases—cont'd

2. Biomedical intervention to control disease progression—Con'd

| Disease category | Outcome target | Change targets |
|---|---|---|
| | | major noncommunicable diseases in both public and private facilities |
| Cardiovascular diseases | At least 50% of eligible people | Proportion of eligible people receiving drug therapy and counseling, including glycemic control, to prevent heart attacks and stroke. Eligibility includes individuals with existing cardiovascular disease and anyone aged 40 years and older with an estimated 10-year cardiovascular risk of $\geq 30\%$ |
| Terminal disease | Palliative care | Access to palliative care including the use of strong opioid analgesics to relieve end-of-life suffering |
| Vaccination against infectious cancers: Human papillomavirus (cervical cancer) | Screening and vaccination | All women aged 30-49 years screened at least once, and vaccination against HPV16 and HPV18.[b] |
| Hepatitis B virus (liver cancer) | Vaccination | Universal infant immunization involving three doses of hepatitis B vaccine |

[a]Outcomes at 2025 are to be measured as relative change since 2010, which is the reference baseline for each country or region.
[b]Whereas mass screening for human papillomavirus is uncontroversial, the proposal to mass vaccinate against the virus has been criticized (Carlos et al., 2014).
Adapted from WHO (2011, 2012b, 2013b) and other sources cited in the text

common complex diseases by 2025. That table and the text below are divided into two main sections: 1. *Prevention of Disease through Risk Factor Reduction to Reduce Disease Incidence*, and 2. *Biomedical Intervention to Control Disease Progression*. The first section is concerned with prevention of disease by reducing exposure to primary and secondary risk factors. Primary risk factors are behaviors, including tobacco use, harmful consumption of alcohol, dietary factors, and physical inactivity. Secondary risk factors are biological, namely, raised blood pressure, overweight and obesity, and raised blood glucose. Reduced exposure to primary and secondary risk factors is achievable by intervening to influence habits and habitats. For each risk factor, the table summarizes the *outcome target* (the proposed targeted reduction) and relevant *change targets* for a specific health-risk behavior or particular age cohort. The second section of Table 14.1 concerns adjunctive biomedical interventions and summarizes outcomes and change targets for several disease categories.

In principle, much of the program for health promotion and disease prevention summarized in Table 14.1 can be implemented by individuals without recourse to biomedicine or other institutional support, including *all* elements of the program that relate to primary and secondary risk factors. Moreover, the importance of risk factor reduction does not cease when biomedical intervention begins, because clinical outcomes for most common complex diseases are improved when patients are also less exposed to risk factors such as smoking, poor diet, and physical inactivity. Accordingly, given favorable circumstances, individuals possess the means for optimizing their own health through largely autonomous and independent action. In practice, the health of populations is invariably enhanced when individual action is facilitated by supportive social institutions, including the family, local community, school, workplace, and central government. Thus, appropriate public policy is crucial for creating physical and social habitats that minimize exposure to risk factors and foster health-promoting habits.

## 14.2   PREVENTION OF DISEASE THROUGH RISK FACTOR REDUCTION TO REDUCE DISEASE INCIDENCE

### 14.2.1   Primary (Behavioral) Risk Factor Reduction

> *[The] superficial [response is] simply to treat the cases and the conspicuously vulnerable individuals [but] they represent the manifestation of the problem, not its roots*
>
> (Rose, 1992, p. 100)

As discussed in previous chapters (especially Chapter 2), common complex diseases comprise five main clusters: cardiovascular diseases, cancers, chronic pulmonary diseases, digestive diseases, and diabetes, which account for almost 90% of deaths from noncommunicable diseases (see Figure 2.2). Notably, these diverse diseases are associated with just four major behavioral pathways: tobacco use, harmful alcohol consumption, poor diet, and lack of physical activity. The prevailing pattern of disease, and associated risk factors, is strikingly consistent worldwide. Individual susceptibility to disease is primarily caused by what people *do*, and specifically what they eat and drink, whether they smoke, consume alcohol (and other drugs, especially pharmaceuticals) to excess, and the amount of physical activity they engage in while at work and at leisure.

Moreover, in reference to *causes of the causes* discussed in the previous chapter, individual behavior patterns are relatively distal causes whereas biological factors are proximal causes of disease. Behavior patterns are in turn influenced by the social milieu, including family, friends, school, workplace, and neighborhood. These, too, are shaped by still broader social determinants involving the economy, access to education, social harmony, and political stability both locally and nationally. The temporality of causes supports the chronology of different interventions which are delivered earlier or later in the process of disease progression.

As discussed in the preceding chapter (and illustrated in Figure 4.5 with respect to cardiovascular disease), preventive interventions focus on relatively distal population behavioral and social causes and aim to avoid or postpone disease, whereas biomedical interventions focus on proximal causes of manifest disease and generally aim to control rate of disease progression. It is for these reasons that population health is best served by healthcare comprising risk factor reduction as the principal and crucial element, with biomedical intervention serving in an adjunctive role.

Effort to change for the better the health-related behavior of individuals is often met with a measure of success. However, change in individual behavior may be resisted by more distal influences in the social environment that are indifferent or antagonistic to individual efforts at health promotion. It is not easy, for example, to exercise healthier food choices if other family members persist with less healthy dietary habits. In turn, families face challenges to healthy dietary habits in a nutritional environment saturated with promotional appeals for unhealthy foods carefully engineered to be attractive and palatable for consumers and profitable for industry. In that context, population-wide intervention that targets substantially distal causes of disease is necessary for encouraging optimal personal and social dynamics leading to sustainable reductions in exposure to common risk factors.

As discussed in Chapter 12, whole-population shifts in behavior such as illustrated schematically in Figure 12.1 tend to be associated with changed social norms. When that happens, primary prevention of disease tends to be long-lasting, as Rose (1985) explained:

> *Once a social norm of behavior has become accepted…the maintenance of that situation no longer requires effort from individuals*

(Rose, 1985, p. 37)

Successful preventive interventions are characterized by reduced population-wide individual exposure to risk factors complemented by changed social norms that maintain the new health-promoting disease-avoiding habits. Thus, unlike biomedical intervention, the need to revisit the same problems in succeeding generations tends to be obviated by preventive interventions. The exception is when earlier positive social change is undone by policy reversals or marked social upheaval, as evidenced by events discussed below that took place in Europe during the latter half of the twentieth century.

### 14.2.1.1 Tobacco Use

The harmful effects of smoking are extensive and well-known (Alberg et al., 2014). One in ten deaths worldwide is attributable to the 10 million cigarettes sold every minute (Mucha et al., 2006). There is a strong population-wide dose-response relation, which has no harm-free lower limit. Earlier uptake, longer duration, and heavier rate of smoking are all associated with increased mortality and morbidity (Edwards, 2004). Whereas the causal relation between smoking and disease is strongest for lung cancer and chronic respiratory diseases, smoking is also a leading cause of cardiovascular and other noncommunicable diseases, including cancers at virtually all sites. The prevalence of smoking is not evenly distributed across the globe, being more prevalent in developing than in high-income countries. Uneven geographic distribution has more to do with greater acceptance of smoking as a norm of behavior than it does with people in some regions having less knowledge or less understanding of the dangers of smoking than people in regions where prevalence of smoking has declined.

Progressive shifts in norms pertaining to the acceptability of smoking behavior in high-income countries have been responsible for enormous saving of life and avoidance of suffering over several decades. Drawing upon the vast material and political resources at its disposal, the tobacco industry (*Big Tobacco*) has actively resisted those changes. Although success in resisting change in high-income countries has been limited, industry has been successful in opposing the development of antismoking norms in many developing countries. A WHO expert committee reported that, for many years, the tobacco industry has operated with the

*purpose of subverting the efforts of the World Health Organization to address tobacco issues [and that the] attempted subversion has been elaborate, well financed, sophisticated and usually invisible*

(Zeltner et al., 2000, p. 18)

Consequently, whereas tobacco use has fallen in many high-income countries, it has increased in developing countries (Yach, 2014).

Historically, smoking was more prevalent among men, but in recent years prevalence among women has increased and in some regions approximates if not surpasses that for men (Mucha et al., 2006). Maternal smoking presents particular risks, including premature delivery, low-birth-weight infants, sudden infant death syndrome, and long-term threats to child development (Butler and Goldstein, 1973; Fleming and Blair, 2007; Horta et al., 1997; Zhang and Wang, 2013). In addition, some common risks associated with smoking may be disproportionately higher in women than in men. For example, the risk of myocardial infarction due to smoking has been found to be 50% greater in women than in men (Prescott et al., 1998). Because the smoking epidemic does not appear to have peaked for women, extensive smoking-related avoidable female premature mortality is likely to worsen in the absence of action to reverse the trend (Janssen et al., 2015).

For both men and women, stopping smoking has immediate and long-term benefits, including reduced risk of death, which continues to fall for at least 10-15 years after smoking cessation (Edwards, 2004). The excess risk of myocardial infarction attributable to smoking reduces by as much as 50% within the first year after quitting, and within 15 years the absolute risk of myocardial infarction is almost the same as in people who have never smoked. For lung cancer, the risk falls to about 30-50% over 10 years compared to continuing smokers, and, for those who quit before age 30 years, 90% of the lifelong risk of lung cancer is removed. Overall, for those who quit smoking before age 35 years, long-term survival approximates that for nonsmokers.

Most smokers attempt to quit but most self-initiated attempts to quit end in failure. Studies show that, in a given year, anywhere between one-quarter and one-half of all smokers make an attempt to quit (Vangeli et al., 2011), but only 3-5% succeed at remaining abstinent for as long as 6-12 months (Hughes et al., 2004). Such is the low rate of success of self-quitting attempts that quit-smoking intervention is considered "successful" if as few as 5-10% of participants become long-term abstinent. Yet, stopping smoking is likely to be the most important health-promoting action in the life of any smoker. While the health benefit of successful quitting is proportionally greater the earlier it happens, quitting at any age and in any state of health has measurable benefits for the quitter. Notably, the remarkable reductions in tobacco use in high-income countries have been due to population-wide public policies, rather than self-initiated quit attempts or individual- and group-oriented smoking-cessation interventions. As such, reduced population exposure to tobacco smoke is a leading example of the crucial role of government in promoting health and preventing disease.

A striking example of the superior benefit of risk factor reduction over biomedicine has been found in smokers who suffer a myocardial infarction. Not surprisingly, patients who stop smoking after their infarction have a better rate of survival than those who continue to smoke. Additionally, however, it has been estimated that the improvement in survival for quitters is greater than the life-saving benefit that both groups, continuing smokers and quitters, derive from clinical intervention appropriate to their condition (Wilson et al., 2000). Unfortunately, it has also been found that the threat of serious health consequences does not in general greatly influence rate of quitting. For example, measuring lung function and informing "healthy" smokers about the demonstrable adverse effects that their smoking is having on them improves their likelihood of quitting only modestly (Parkes et al., 2008). This is further evidence that long-term abstinence rates stemming from individual- or physician-supported initiatives remain poor compared to population-wide initiatives.

Rose's (1992) axiom about success in prevention often depending on shifts in social norms has possibly been most consistently confirmed in relation to population patterns of smoking. For example, the introduction of smoke-free workplaces not only reduces smoking in the workplace but also facilitates the adoption of broader social norms that encourage overall population-wide reductions in smoking (Fichtenberg and Glantz, 2002a). Conversely, less comprehensive measures, such as limiting smoking to designated areas, have little desired effect. In other words, smoking in the workplace, even when limited to designated areas, contributes to a culture of acceptance toward smoking in general, whereas workplaces that are entirely smoke-free encourage the opposite, namely, the adoption of generalized antitobacco social norms. Before these facts were established in the public scientific literature, they were known to tobacco-industry scientists from internal research which was not made public (Heironimus, 1992). Industry understood that the introduction of smoke-free workplaces would lead to lower population rates of smoking and loss of profits from tobacco sales. Delay in implementing that knowledge cost millions of lives.

The substantial positive impact on population health following the introduction of smoke-free workplaces demonstrates the potential for cumulative and generalized benefits to health from initiatives that appear to be specific in focus. Smoke-free workplaces discourage smoking on a number of different levels, including decreased uptake of smoking by youth (Farkas et al., 2000), decreased rate of smoking by existing smokers, and increased rates of quitting among long-term smokers (Fichtenberg and Glantz, 2002a). The effect on youth uptake is all the more notable given that laws intended to restrict cigarette purchasing by youth have been found to have comparatively little effect on the prevalence of adolescent smoking (Fichtenberg and Glantz, 2002b; Wakefield and Giovino, 2003). Furthermore, notwithstanding the life-saving importance of smoking restrictions for current and would-be smokers, estimated overall to be approximately 10 years of additional life (Jha, 2009), substantial additional benefit is obtained due to avoidance of *passive smoking* from inhalation by nonsmokers of harmful *secondhand* tobacco smoke when in the presence of smokers.

Inhalation of secondhand smoke by nonsmokers is estimated to be responsible for 1-in-10 smoking-related deaths worldwide (WHO, 2012b) or about 1.0% of total worldwide mortality (Öberg et al., 2011). Indeed, the occupational health and safety of *nonsmokers* has always been the main impetus behind the introduction of smoking bans. It is a bonus that the growth of antitobacco culture encouraged by mandated smoke-free workplaces has led to more generalized cultural intolerance of smoking. Evidence of this kind of *social contagion* effect, wherein norms pertaining to one setting generalize to others, is indicative of the power of prevention. Specifically, smoke-free workplace initiatives have been found to encourage adoption of voluntary restrictions on smoking in private areas, such as the home and motor vehicles, which are less amenable to mandated restrictions than are public spaces. Because children are at particular risk of suffering tobacco-related harm in households containing adult smokers (Been et al., 2014), a substantial burden of future disease due to years of exposure to environmental smoke is averted by changes in social norms that encourage the adoption of voluntary limits on smoking in private dwellings and transport.

A recent study compared levels of voluntary home-smoking bans in three Mexican populations resident in three cities: San Diego, California, located immediately adjacent to the border with Mexico; Tijuana, in the north of Mexico and adjacent to the border with California; and Guadalajara, in the central region of Mexico and comparatively distant from the border with California (Hovell et al., 2014). Compared to Mexico, California has a long-established and strong antitobacco culture. After controlling for demographics and other potential confounders, comparisons between the three cities showed that voluntary implementation by Mexican households of a ban on smoking in the home was proportional to geographic distance from California. That is, voluntary banning of smoking in Mexican households was most frequent in San Diego in California (91%), followed by Tijuana (66%), adjacent to California, and least likely in Guadalajara (38%), furthest from California. As such, the results suggest that "social contagion" of antitobacco social norms in California have encouraged adoption of similar norms by Mexican families in direct proportion to the level of exposure that those Mexican populations have had to Californian social norms.

Awareness about the threat of smoking in the home and in private transport has been further highlighted due to growing concern over harm due to ***third-hand smoke***. This refers to residual nicotine and other chemicals from tobacco smoke being deposited on surfaces whenever smoking occurs within enclosed spaces (Ferrante et al., 2013). Smoke residue that accumulates on furniture, walls, carpets, clothing, and skin, resists normal cleaning, and cannot be eliminated by airing-out. This residue is believed to react with common indoor pollutants to create toxic mixtures of chemicals that include several known human carcinogens (Sleiman et al., 2010). Consequently, smoking in enclosed areas is hazardous not merely because of inhalation of second-hand smoke while smoking is in progress, but also because of inhalation and ingestion of third-hand smoke, even at times when smoking is not in progress and may not have been for some time.

The WHO target of a 30% relative reduction in prevalence of smoking shown in Table 14.1 is based on reductions in prevalence of smoking achieved by some high- and middle-income countries that have implemented strong tobacco programs. The target is believed to be achievable by implementing the WHO Framework Convention on Tobacco Control (WHO, 2005), which includes the following actions:

- use of price and tax measures as a means for reducing tobacco consumption, particularly by the young;
- prohibition of the sale of tobacco products to under-age persons (typically, under 18 years);
- provision of protection from exposure to tobacco smoke in indoor workplaces, public transport, and indoor public places;
- inclusion of large, clear, visible, and legible health warning labels, including pictograms, on the packaging of tobacco products;
- development and dissemination of appropriate, comprehensive and integrated guidelines aimed at promoting the cessation of tobacco use;
- promotion of broad access to comprehensive educational and public awareness programs on the health risks of tobacco use, including passive exposure to tobacco smoke, and the benefits of cessation of tobacco use and a tobacco-free environment; and
- undertaking a comprehensive ban on all tobacco advertising, promotion, and sponsorship.

## e-Cigarettes—Industry Strikes Back!

The advent of e-cigarettes, which the WHO did not anticipate (Yach, 2014), threatens to undermine aspects of tobacco control policy. Containing no tobacco, e-cigarettes deliver nicotine as a vapor containing solvent and flavoring agents. They have been marketed for recreational use, as an aid to quitting smoking, and as a reputedly safer alternative to smoking tobacco. However, although the agents inhaled from e-cigarettes probably possess less immediate toxicity than cigarette smoke, there are concerns about their long-term safety. A particular concern is that *vaping* (the term for "smoking" an e-cigarette) a flavored product could be attractive to children and adolescents, and lead to long-term nicotine addiction followed by experimentation and subsequent recruitment to tobacco smoking (Harrell et al., 2014; Yamin et al., 2010).

Additionally, there are doubts about the likelihood of seasoned smokers of conventional cigarettes switching to e-cigarettes. If that did happen, there would probably be an overall reduction in the burden of disease from nicotine addiction. However, evidence is lacking that e-cigarettes are effective as an aid to quitting conventional smoking. Rather, current evidence indicates high levels of dual use of e-cigarettes with conventional cigarettes (Grana et al., 2014). In that regard, rather than serving as an aid to quitting, dual use may delay or deter quitting conventional cigarettes and thereby contribute to the long-term harmful effects of smoking.

There are further concerns that e-cigarettes may be used to circumvent smoke-free laws. If so, toxins in the exhaled vapor of e-cigarettes could pose a direct threat to the safety of nonsmokers (Yamin et al., 2010). Also, as mentioned above, total smoke-free policies have a wider positive health impact on society at large by encouraging antismoking norms of behavior. The use of e-cigarettes in smoke-free areas would not only expose nonsmokers to potential toxins but could also undo the wider benefit of smoking bans by encouraging a return to smoking as socially acceptable (Grana et al., 2014). As part of a wider policy framework to regulate e-cigarettes in the interests of public health, Grana et al. (2014) have recommended that e-cigarettes be:

- prohibited anywhere that use of conventional cigarettes is prohibited;
- prohibited from sale to anyone who cannot legally buy cigarettes or in any venues where sale of conventional cigarettes is prohibited;
- subject to the same level of marketing restrictions that apply to conventional cigarettes (including no television or radio advertising); and
- prohibited from being cobranded with conventional cigarettes or marketed in a way that promotes dual use.

### 14.2.1.2  Alcohol

Alcohol consumption is a risk factor for multiple noncommunicable diseases, injuries, and even infectious diseases such as tuberculosis (Ezzati and Riboli, 2012). Nearly 2 billion people worldwide consume alcohol regularly (Roswall and Weiderpass, 2015), and alcohol is responsible for more than 3 million deaths worldwide per year or approximately 6% of all deaths (WHO, 2014b). There is a substantial gender difference in alcohol-related harm, with the proportion of male deaths attributable to alcohol being almost twice that for females (approximately 8% and 4%, respectively). There is also geographic variation in the proportion of alcohol-attributable deaths, with a higher proportion generally occurring in high-income countries, where more alcohol is consumed per capita and fewer people abstain than in developing countries. Irrespective of per capita consumption, deprived and marginalized social groups in most countries are at greater risk of harm from alcohol consumption. In the absence of intervention, increased consumption of alcohol, and therefore increased associated burden of disease, are anticipated trends in developing countries.

Apart from per capita consumption, drinking pattern is an important additional consideration, especially the prevalence of heavy episodic drinking (Norström and Skog, 2001). Notably, a substantial proportion of the global burden of disease caused by the consumption of alcohol is due to harm caused to people other than the active drinker, including fetal harm from maternal alcohol consumption, bystander injury due to drink driving, and alcohol-related violence, both domestic and public (Rehm et al., 2009). The vulnerability of females to alcohol-related harm is a particular concern because alcohol consumption among women has increased steadily in recent decades, possibly influenced by greater economic independence and changing gender roles (Grucza et al., 2008). The higher burden of disease among men is largely attributable to men being less often abstainers, drinking more frequently, and drinking larger quantities (WHO, 2014b). As with smoking, alcohol-related harm is the end result of choices by individuals, choices that are influenced by habitat, especially the social environment. In the case of alcohol, consumption can be seen to reflect fluctuations in norms of behavior associated with broad social changes, including those that accompany political instability and turmoil.

In recent history, Europe has experienced dramatic fluctuations in alcohol-related harm arising from the impact of geopolitical developments that have influenced the availability and social acceptability of alcohol (Zatoński et al., 2010). Change in incidence of liver cirrhosis, which is a relatively specific alcohol-attributable outcome, has served as a reliable indicator of overall alcohol-related harm (Mann et al., 2003). For example, reduced availability due to the German seizure of alcohol during the occupation of France in World War II led to a marked reduction in the incidence of cirrhosis mortality. In Paris, alcohol consumption fell by 80% and cirrhosis mortality reduced by more than 50% after 1 year, and more than 80% in 5 years (Zatoński et al., 2010). Similar effects, involving increases as well as decreases in consumption and disease, have been observed more recently in the countries of North-eastern Europe in conjunction with upheavals surrounding the dissolution of the Soviet Union.

Reduced consumption of alcohol and associated alcohol-related harm coincided with restrictions on alcohol sales during the Gorbachev era of the mid-1980s (Leon et al., 1997). However, with relaxation of restrictions later in the decade, there was a return to earlier levels of consumption and associated harm. In the following decade, coincidental with the dramatic political and socioeconomic transformations in the region at that time, there were successive periods of increases and decreases in alcohol consumption and harm (Zatoński et al., 2010). In the early 1990s, following eventual dissolution of the Soviet Union, there were sharp increases, and these were followed by decreases in the late 1990s. Toward the end of the millennium, increases again occurred, even sharper than previously, resulting in unprecedentedly high levels for both consumption and harm, including liver cirrhosis and death due to alcohol poisoning, homicide, and suicide. Each successive fluctuation, all occurring within a relatively short period, represents a whole-population shift up or down the scale of per capita consumption. In those patterns, we see the potential power of prevention to deliver radical improvements in health when population shifts are in the direction of reduced exposure to risk factors, and the potential for great harm when positive trends are reversed.

A recent analysis of alcohol-related mortality in Russia again shows alcohol to be a major cause of premature death due to many individuals consuming extraordinarily large quantities of vodka exacerbated by episodic heavy drinking patterns (Rehm, 2014; Zaridze et al., 2014). One approach to dealing with the problem would be to target interventions at those high-risk individuals, since obvious benefit would be derived from reducing the size of the "tail" that is at high risk of harm. However, even if such attempts were successful, substantially greater benefit would accrue from success in encouraging a whole-population shift toward reduced per capita alcohol consumption. Many people would benefit among the majority who belong to that portion of the distribution that is not at discernibly high risk. Some, despite not being at obvious risk, are nevertheless likely to be drinking at a level that renders them susceptible to injury and disease, including cardiovascular diseases, liver cirrhosis, and cancers. With a whole-population shift toward lower levels of consumption, injury and disease for those individuals would be postponed or avoided. Moreover, many among those at little or no direct risk would be spared becoming victims of harm due to reduced alcohol-related accident and violence among higher-risk consumers.

The WHO 10% target relates to both relative reduction in per capita consumption and hazardous drinking (Table 14.1). The target is based on the performance of several countries in achieving substantial reductions over recent years, while also taking account of equally substantial increases in consumption by some other countries. The target is recommended to be pursued by implementing the *Global Strategy to Reduce the Harmful Use of Alcohol* (WHO, 2010a), which includes actions to address the following:

- drink-driving policies and countermeasures, including enforcement of limits on blood alcohol concentration, graduated licensing for novice drivers with zero-tolerance for drink-driving, check points and random breath-testing, and running mass media campaigns targeted at specific situations (e.g., holiday seasons) or audiences (e.g., young people);

- availability of alcohol, including establishing an appropriate minimum age for purchase or consumption of alcoholic beverages and other policies that create barriers to drinking by adolescents, and setting policies regarding drinking in public places;
- marketing of alcoholic beverages, including regulation of the content and the volume of marketing, especially to young people, direct and indirect marketing, and sponsorship activities that promote alcoholic beverages;
- pricing policies to reflect the fact that increasing the price of alcohol is one of the most effective interventions against harmful consumption, and action to counter the use of direct and indirect price promotions, discount sales, sales below cost, and other strategies that encourage unlimited drinking or volume sales;
- reducing the negative consequences of drinking and alcohol intoxication by regulating situations (e.g., entertainment events) to minimize violence and disruptive behavior where drinking occurs.

## What About the Reputed Cardioprotective Benefit of Moderate Alcohol Consumption?

As discussed above, liver cirrhosis is a serious outcome largely specific to harmful alcohol consumption, much as lung cancer is largely specific to cigarette smoking. Despite the scale of alcohol-attributable liver disease, greater harm derives from alcohol-attributable cardiovascular disease. Yet, moderate alcohol consumption (e.g., 2-3 standard drinks per day for men and 1-2 standard drinks per day for women) has long been considered beneficial to cardiovascular health as well as diabetes (Rehm et al., 2009; Rimm et al., 1996; Ronksley et al., 2011). Although the evidence concerning benefit is mostly correlational, it is supported by plausible biological mechanisms of action that have led to the belief that the relationship is causal (e.g., Brien et al., 2011). The idea that alcohol in low doses has a causal cardioprotective effect has been widely disseminated through the public media, and has often been presented as a truism in the scientific literature (e.g., Roswall and Weiderpass, 2015; WHO, 2014b), even forming the basis of recommendations to clinicians to "spread the message" that moderate alcohol consumption is beneficial to health (Artero et al., 2015).

Overall, then, there is general acceptance that a *J*-shaped curve describes the relationship between alcohol consumption and cardiovascular health. That is, moderate consumption is associated with better health than zero consumption, whereas amounts increasingly above those considered to be moderate are associated with progressively increasing levels of harm. However, whether the association is causal has always been the source of controversy (e.g., Ronksley et al., 2011). Among diverse evidence that the relationship is not causal, inadequate control of potential confounders in epidemiological research, especially the classification of "abstainers," has been a particular concern (Chikritzhs et al., 2015). For example, as people age and become increasingly unwell there is a tendency for alcohol to be avoided. At the population level, that process could lead to the appearance that abstaining is harmful to health. Moreover, since a proportion of those who quit alcohol are likely to have been consumers whose health has been compromised by previous higher levels of consumption, classifying those individuals as abstainers would contribute further to the likelihood of spurious inferences about little or no consumption causing poorer health and moderate regular consumption being health protective. The current state of overall evidence is such that Chikritzhs et al. (2015) concluded:

*The foundations of the hypothesis for protective effects of low-dose alcohol have now been so undermined [that] future estimates of the alcohol related burden of disease and national drinking guidelines should no longer assume any protective effects from low dose consumption...Health professionals should not recommend moderate alcohol consumption as a means of reducing cardiovascular risk for patients.*

(Chikritzhs et al., 2015, p. 2)

### 14.2.1.3  Nutritional Habits and Habitats

For much of human history, dietary concerns have been dominated by efforts to avoid undernutrition during periods of food scarcity. In addition to long-standing problems of undernutrition in developing countries, substantial risk due to nutritional imbalance as a cause of noncommunicable diseases has emerged in relatively recent times in countries where there is no food scarcity (WHO, 2009). Indeed, many more people now live in countries where mortality and morbidity due to overweight and obesity exceed that due to underweight. The high global prevalence of different forms of malnutrition, inclusive of undernutrition and nutritional imbalance, has been described as the "new normal" (IFPRI, 2014).

The rapid rise in overweight and obesity cannot be attributed to changes in the human genome, as the span of time involved has been too brief for the population gene pool to have changed appreciably. Rather, altered nutritional habits and habitats, in combination with high prevalence of physical inactivity (discussed below), are implicated as leading causes of the increased prevalence of overweight and obesity in recent decades and the associated global "epidemic" of obesity-related diseases, including cardiovascular disease, hypertension, and diabetes. The WHO program for a 25% reduction in premature mortality by 2025 includes several specific targets under the rubric of dietary factors, including reduced intake of salt/sodium, saturated fatty acids, and trans fatty acids, increased intake of fruits and vegetables, and regulatory action to restrict the marketing of food to children.

## Salt (Sodium)

There is extensive and consistent evidence that high dietary salt contributes significantly to a wide range of diseases, including raised blood pressure level and consequential cardiovascular disease, especially ischemic heart disease and stroke, gastric cancer, calcium-containing kidney stones, and osteoporosis (Aburto et al., 2013; Campbell et al., 2011; Lim et al., 2013). There is also evidence of a strong association between high dietary salt and obesity, supported by plausible pathophysiological mechanisms indicating a causal relation (He and MacGregor, 2008). Excess consumption of salt is mostly due to it being added by householders during cooking and by food manufacturers during production of processed foods of every description. Strong and direct evidence of the benefits of reduced salt intake comes from studies measuring 24 h urinary sodium levels and blood pressure. Even modest reductions in dietary salt cause blood pressure to fall significantly in both hypertensive and normotensive individuals, irrespective of gender and ethnicity (He et al., 2013).

Reduced salt intake is considered to be one of the most cost-effective measures ("best buys" in WHO terminology) for reversing the rising global burden of cardiovascular and other diseases. It has been estimated that well over a million deaths could be avoided each year through salt-reduction strategies costing less than USD0.50 per person per year in low-income countries and less than USD1.00 per person per year for higher-income countries (Asaria et al., 2007). The WHO target of 30% relative reduction in mean population intake is intended to represent a minimum (Table 14.1). Achievability of that target is supported by experience from a limited number of countries that have implemented salt/sodium reduction strategies (e.g., Finland, United Kingdom) showing that mean population intake can be reduced by up to 30% over a period of 7-10 years. The current global level of dietary salt is estimated to be 9-12 g per day (WHO, 2012b), which greatly exceeds the estimated physiological need of about 1 g per day (Brown et al., 2009).

Considering the ubiquity of salt in the nutritional environment, especially in processed foods, the role of government is vital for achieving reduction targets (Cappuccio et al., 2011; Webster et al., 2011). The WHO recommends that daily intake of salt should be less than 5 g (approximately 2 g of sodium), about half the current global per capita level. The main recommended strategies for achieving the more modest target of a 30% relative reduction from current levels by 2025 are:

- the introduction of mass media campaigns and labeling of foods to encourage consumers to make informed choices, and
- the setting of progressive targets requiring manufacturers to reduce the salt content of processed foods through product reformulation.

## Fat Intake

Fats are a large group of *lipid* compounds that are generally insoluble in water, and are referred to as *fats* if solid at room temperature and *oils* if liquid at room temperature. Fatty acids comprise the major form of dietary fat, which, depending on chemical composition, are classified as either *saturated* or *unsaturated*. Within these broad categories, individual fatty acids can have different influences on health and disease, and scientific attention has increasingly come to be concentrated on examining the effects of individual compounds. Overall, high dietary fat has been linked to increased risk of coronary heart disease, certain types of cancer, and obesity (WHO, 2012b).

Prevention of disease, especially cardiovascular diseases, by changing dietary fat has been a focus of attention since the 1950s, and has led to the view that the fatty acid *composition* of the diet is more important than total fat intake (Perk et al., 2012). Extensive knowledge has accumulated about the effects of the major classes of fatty acids, *saturated*, *monounsaturated*, and *polyunsaturated*, as well as specific fatty acids within classes (e.g., *trans fatty acids*). Replacement of saturated with polyunsaturated fatty acids has been found to reduce risk of coronary heart disease (Astrup et al., 2011). It is considered important that reduction of saturated fatty acids is by replacement with polyunsaturated fatty acids, because replacement with carbohydrates has no benefit and replacement with monounsaturated fatty acids has uncertain effects (Micha and Mozaffarian, 2010). Polyunsaturated fatty acids broadly comprise two subgroups: omega-6 fatty acids, mainly from plant foods, and omega-3 fatty acids, mainly from fish.

Trans fatty acids are a subclass of unsaturated fatty acids and are formed during the partial hydrogenation of vegetable oils, which is the basis of margarine production. Partially hydrogenated vegetable oils are used extensively in the food industry because of their long shelf-life, stability during deep-frying, and ability to enhance the palatability of baked goods and sweets. Major sources of trans fats are deep-fried fast foods, bakery products, and packaged snack foods. Although small amounts are present in dairy products and in meat from ruminant animals such as cows and sheep (due to the action of bacteria in the stomachs of those animals), almost all dietary trans fats are consumed in commercially processed foods. The consumption of trans fats is believed to contribute substantially to increased risk of cardiovascular disease while conferring no apparent nutritional benefit (Mozaffarian et al., 2006). Consequently, the consensus recommendation is the complete or near-complete avoidance of industrially produced trans fats. Ruminant trans fatty acids cannot be removed entirely from the nutritional environment, but their consumption is already low and has not been found to contribute to increased risk of disease (Mozaffarian et al., 2009).

As summarized in Table 14.1, the WHO target for dietary fat is for a 15% relative reduction in mean proportion of total energy intake from saturated fatty acids, with the aim of achieving the recommended level of total energy intake comprising less than 10% from saturated fatty acids. Those targets are based on success in a number of countries in reducing saturated fat consumption largely by replacing whole-milk with low-fat dairy products, reformulated margarine containing little or no trans fat, and reducing the use of highly-saturated types of cooking oil. The WHO policy is to encourage continued replacement where it has begun and the initiation of similar replacement where it has yet to occur.

## Cholesterol Controversies

Cholesterol is produced by the body as well as being consumed in foods, and is involved in several important functions, including maintenance of healthy cell membranes and the production of some hormones and vitamin D. Raised total cholesterol is associated with increased disease risk, especially cardiovascular disease (e.g., Ezzati et al., 2002; Lim et al., 2013). Therefore, lowering cholesterol level is widely considered desirable. A common recommendation, supported by the WHO (2012b, 2013), is a level of total serum cholesterol less than 5.0 mmol/l (90 mg/dl). That target, however, does not require observance of specific recommendations concerning intake of dietary cholesterol because cholesterol consumed in the diet (e.g., animal fat, cheese, and eggs) has only a modest effect on the level of serum cholesterol in the body. By comparison, fatty acid consumption has a greater effect (Perk et al., 2012). Accordingly, a lower serum cholesterol level can be achieved by reducing dietary fat, especially saturated fatty acids combined with increased consumption of fruits and vegetables, without necessitating additional specific attempts to reduce intake of cholesterol in the diet.

Cholesterol is transported through the body by lipoproteins (literally, biochemical assemblies comprising lipids bound to protein) in blood. Lipoproteins include low-density lipoproteins (LDL; "bad" cholesterol) and high-density lipoproteins (HDL; "good" cholesterol). Increased risk of cardiovascular events (e.g., heart failure and stroke) is associated with raised concentrations of LDL cholesterol, which can be reduced through diet and exercise. Prescription medication, primarily statins, can also be used to lower LDL cholesterol, but as discussed in earlier chapters, there is controversy about the safety of widespread use of statins (Ridker, 2014). Raised HDL cholesterol is associated with decreased risk of cardiovascular disease, giving rise to the possibility of a causal relationship. However, the hypothesis that HDL cholesterol is protective against atherosclerosis has become increasingly difficult to defend, and there is increasing doubt about the likely safety and clinical utility of drugs that are currently under clinical development for HDL cholesterol (Rader and Hovingh, 2014). Raised plasma triglycerides, involved in bidirectional transference of fat and glucose, has also been considered a cardiovascular risk factor similar to raised

LDL cholesterol. However, despite the absence of evidence from randomized-controlled trials supporting the benefit of triglyceride-lowering for cardiovascular health, new triglyceride-lowering drugs of uncertain, and possibly unpromising, clinical utility are currently being developed (Nordestgaard and Varbo, 2014).

## Fruit and Vegetables

Fruit and vegetables are important sources of nutrients and fiber in the human diet. The *nutrition transition*, whereby traditional plant-based diets, rich in complex carbohydrates, are being replaced in all but the poorest countries by animal fat and *free sugar*, has focussed attention on the role of fruit and vegetables in human health (Lock et al., 2005). Interest in the potential preventive effects of fruit and vegetables strengthened in the 1990s, with a particular focus on possible prevention of cancer. However, the cumulative evidence since then has more consistently indicated prevention of cardiovascular disease, including heart disease (e.g., Crowe et al., 2011; Dauchet et al., 2006; He et al., 2007) and stroke (e.g., He et al., 2006), whereas a substantial body of research suggests only weak associations for cancer (e.g., Boffetta et al., 2010; Key, 2011).

Isolating the effects of fruit and vegetables from other components of diet is complicated by the fact that fruits and vegetables comprise an extraordinarily diverse array of foods, and obtaining reliable measurements of amounts consumed is challenging. Moreover, given that people who are conscious about consuming fruit and vegetables might be more mindful about their health in general, studies routinely try to control for the effects of confounding from other health-related variables, including age, gender, social class, education, body mass index, smoking status, alcohol consumption, and physical activity. Despite such challenges to precision, there is consistent and robust evidence of an inverse association between the consumption of fruit and vegetables and the incidence of disease (i.e., higher dietary fruit and vegetables is associated with lower disease incidence).

In a recent comprehensive review of literature, it was concluded that there is "convincing" evidence that consumption of fruit and vegetables reduces the risk of hypertension, coronary heart disease, and stroke, and evidence of a "probable" risk-reducing influence of increased fruit and vegetable consumption for cancer (Boeing et al., 2012). Evidence of a protective effect was found to be less certain but "possible" for a wide variety of other conditions, including dementia, overweight, rheumatoid arthritis, osteoporosis, the lung diseases of asthma and chronic obstructive pulmonary disease, and the eye diseases of macular degeneration and cataract. Consistent with those findings, a recent study of a large nationally representative sample in England reported a strong inverse relationship between fruit and vegetable consumption and all-cause mortality (Oyebode et al., 2014). Additional analyses showed that the observed protective effect was apparent for both cardiovascular disease and cancer, with the consumption of vegetables and salad being of greater benefit than similar quantities of fruit. Importantly, no upper limit on benefit was observed. That is, within reasonable limits, the more fruit and vegetables the better.

A key objective of the WHO program, as summarized in Table 14.1, is for fruit and vegetable consumption to replace processed foods, with a recommended minimum consumption of five servings, each of 80 g (total of 400 g), per day. In fact, this has been a recommendation of the WHO since 1990 (WHO, 1990), which was followed by the introduction of "5-a-day" fruit and vegetable campaigns in several European countries and the United States. The recent introduction in Australia of the "Go-for-2 + 5" campaign in 2005, which advises higher levels of daily consumption, comprising two portions of fruit of 150 g per portion and five portions of vegetables of 75 g per portion (total of 675 g of fruit and vegetables per day), is possibly more consistent with current scientific evidence than the WHO's current recommendation.

However, despite widespread knowledge of recommendations to consume more fruit and vegetables, adherence to recommendations has often been found to be poor (Oyebode et al., 2014). For example, the United States Centers for Disease Control and Prevention has estimated that most Americans (75%) consume less fruit and vegetables than recommended by the 5-a-day program (CDC, 2009). Consequently, recommending higher levels, which, if achievable, would be health promoting, could be premature and counterproductive. In the meantime, the conservative levels recommended by WHO with programs such as the 5-a-day might represent a more realistic and achievable intermediate target. If, in due course, those conservative targets are met and associated health benefits are confirmed, attention could be given to the possibility of achieving additional gains by promoting further increases in consumption of fruit and vegetables.

## Marketing of Foods to Children

In May 2010, the WHO endorsed a set of recommendations encouraging new policies and the strengthening of existing policies concerning the marketing of food to children (WHO, 2010b). The purpose of the recommendations is to ensure that children are protected against the impact of marketing of foods high in saturated fats, trans-fatty acids, free sugars, or salt (WHO, 2010c). Evidence shows that advertising directed at children, especially television advertising, influences children's food preferences, purchase requests, and consumption patterns, with implications for long-term health. For example, surveys in Australia, several European countries, and the United States have found the prevalence of overweight children to be associated with advertising on children's television (Lobstein and Dibb, 2005). Specifically, a higher number of advertisements per hour on children's television promoting consumption of energy-dense low-nutritious foods is associated with a higher prevalence of overweight children, and a higher number of advertisements promoting healthier food choices is also, though less strongly, associated with a lower prevalence of overweight children.

While television remains an important medium, it is being complemented by an increasingly multifaceted mix of marketing communications that focus on branding and building relationships with consumers, including *product placement*, celebrity endorsement, use of brand mascots or characters popular with children, gifts of toys, posting of messages on web sites, use of email and text messaging, philanthropic activities tied to branding opportunities, and communication by word-of-mouth and through *viral marketing*. Globalization of trade, including massive markets in India, South America, and China, has greatly extended the reach of food marketing by transnational corporations and contributed to the use of marketing strategies honed in developed countries being increasingly used in developing countries (Cairns et al., 2009). As well as being effective in influencing the habits of the children who are targeted, the marketing of food to children can spearhead wider cultural change, especially in developing countries. For example, it has been found that worldwide distribution of fast-food chains, involving the marketing of hamburgers, fried chicken, pizzas, and other fast-food products having great appeal for children, serves as a bridgehead to influence the dietary habits of generations of older people, including parents and grandparents, who were not exposed as children to similar promotional pressures to consume highly-processed fast foods.

It is too early to say what effect if any WHO (2012c) initiatives summarized in Table 14.1 will have in limiting harm from the marketing of energy-dense low-nutritious foods to children. Experience from similar initiatives in other fields, notably, tobacco and alcohol, suggests that the food industry will take counteraction (Brownell, 2012). Growing concern about children's advertising led to the introduction of restrictions in several countries (e.g., Australia, Norway, Canada, Sweden, and the United Kingdom) even before the WHO initiatives were promulgated. However, results from evaluations have been disappointing, showing little impact of restrictions on children's overall level of exposure to food advertisements that encourage unhealthy food choices (Adams et al., 2012; Kelly et al., 2007). In that context, the WHO initiatives should be regarded as being at an early stage of development. While conclusions regarding effectiveness of the initiatives await continued monitoring and evaluation, the need for further development of the regulatory environment should be anticipated.

In the meantime, WHO (2012c) has begun the process of encouraging governments worldwide to assume leadership in developing policies that:

- remove the impact on children of marketing of foods high in saturated fats, trans-fatty acids, free sugars, or salt;
- ensure no such marketing occurs in settings where children gather, including nurseries, schools, school and preschool centers, playgrounds, family and child clinics, and during sporting or cultural activities;
- ensure cooperation between states to reduce the influence of cross-border marketing (in-flowing and out-flowing);

- adopt enforcement mechanisms, including systems for monitoring compliance, reporting complaints, and imposing sanctions;
- adopt systems to evaluate the impact and effectiveness of the policy aims.

### 14.2.1.4 Physical Activity

*...eating alone will not keep a man well; he must also take exercise.*

(Hippocrates, 460-370 BC)

Beliefs about the benefits of physical activity have vacillated over time. It is obvious from the forgoing quote, and similar views expressed by the physician, Galen, a successor to Hippocrates, that lack of physical activity was considered in ancient times to be detrimental to health (Berryman, 2010). Similar views about the health benefits of physical activity were also prevalent in ancient China (Lee and Skerrett, 2001). However, during the early part of the twentieth century the view emerged that physical activity, especially when vigorous, could be harmful (Lee et al., 2012). It was not until the 1950s and afterwards that extensive population studies confirmed what the ancients had believed (Morris and Crawford, 1958). The predominant contemporary view, supported by voluminous scientific evidence accumulated over the past several decades, is that a substantial proportion of the current global burden of disease can be attributed to insufficient physical activity (e.g., Heath et al., 2012). Today, physical inactivity is considered to be a leading cause of death and one of the most important public health problems of the twenty-first century (Bauman, 2004; Blair, 2009; Kohl et al., 2012).

Despite incontrovertible evidence of the benefits of physical activity for health, life-long physical *inactivity* is common in most regions of the world (Bauman et al., 2012). It was not until relatively recently, however, that the scale of the problem was quantifiable. Until a little more than a decade ago, suitable measurement protocols were absent, making it impossible to conduct systematic comparisons of patterns of physical activity between countries and regions (Hallal et al., 2012). International effort to develop standardized measurement began in the late 1990s and has progressed to the point where data are now available for approximately two-thirds of countries worldwide. Those data reveal the extent of physical inactivity worldwide, showing for the first time that more than 30% of the global adult population is inactive, defined as not meeting any of the following three criteria:

- 30 min of moderate-intensity physical activity on at least 5 days every week (150 min per week),
- 25 min of vigorous-intensity physical activity on at least 3 days every week (75 min per week), or
- an equivalent combination of activity achieving 600 metabolic (*MET*) equivalent minutes per week.[2]

---

2. One MET refers to the energy spent per minute when seated quietly. Only nonsedentary activity is taken into account when measuring physical activity level.

Moderate-intensity activity (e.g., casual walking) is the approximate equivalent of 4 MET per minute and vigorous-intensity activity (e.g., running) is equivalent to 8 MET per minute. Thus, walking 30 min a day (120 MET) for 5 days yields 600 MET. For assessment purposes, physical activity is inclusive of activity associated with leisure-time, occupation, and transportation (e.g., walking or cycling). The inclusion of housework for purposes of assessing physical activity level is controversial, as that aspect of physical activity may not be a good predictor of key health outcomes (Sabia et al., 2012). Studies show that inactivity increases with age, is higher among women than men, and is higher in higher-income and more urbanized countries, although higher-income groups within countries tend to be more physically active than lower-income groups (e.g., Bauman et al., 2012).

Box 14.1 summarizes health benefits of physical activity for which there is strong evidence. It has been estimated that worldwide physical inactivity is responsible for up to 10% of the common complex diseases of coronary heart disease, type 2 diabetes, and breast and colon cancers (Lee et al., 2012; Warburton et al., 2010). By becoming moderately active, the majority of people who are currently inactive would gain an estimated 2-4 years of life, and benefit from improved health while alive. From among the diverse proposed strategies for stemming the currently increasing global trend in common complex diseases, increased physical activity is possibly the most cost-effective.

---

**BOX 14.1 Health benefits of physical activity for which there is strong evidence**

**Reduced incidence of:**
- all-cause mortality
- raised blood pressure
- metabolic syndrome
- breast cancer
- osteoporosis
- falls
- coronary heart disease
- stroke
- type 2 diabetes
- colon cancer
- depression

**Improved function in the following:**
- cardiorespiratory and muscular fitness
- bone health
- cognitive function
- body mass and composition
- functional health

---

*Adapted from Lee et al. (2012) and Warburton et al. (2010).*

There is an overall dose-response relationship between physical activity and health, wherein increasing levels of physical activity generally confer at least modest additional increased benefit for both physical health (Lee and Skerrett, 2001; Sattelmair et al., 2011; Warburton et al., 2010) and mental health (Hamer et al., 2009). Hesitations about engaging in physical activity due to concerns about potential dangers, particularly in relation to vigorous activity, are generally not well-founded. Just as life-threatening events and deaths sometimes occur when people sleep or sit quietly, these events also sometimes occur in proximity to physical activity. However, even vigorous-level physical activity is almost always beneficial (Lee and Paffenbarger, 2000; Paffenbarger et al., 2001). Despite evidence of transient increased risk of cardiovascular-related events during or immediately after physical activity, with vigorous activity posing higher risk, the overall risk to health of vigorous activity is markedly lower than for physical inactivity for both patients and asymptomatic persons across the lifecourse (Warburton et al., 2011).

Compared with improvements in other major behavioral risk factors, such as reduced smoking and improved diet, physical activity not only contributes to extended life but may also be important for gaining life-years free from morbidity. It was noted in Chapter 3 that reduced smoking and improved diet are considered to have been responsible for marked decreases in cardiovascular disease in some developed countries in the latter part of the twentieth century (e.g., Ünal et al., 2005a,b). Remarkably, those improvements occurred against a background of increased levels of physical inactivity, which is regarded as having contributed to the negative trends for diabetes and obesity that occurred during the same period. As such, it is evident that even larger reductions in cardiovascular disease would have occurred had physical activity shown the same positive trends as smoking and diet. Chapter 3 also discussed encouraging findings from the Global Burden of Disease Study, which reported recent increases in life expectancy globally due to postponement of mortality (Salomon et al., 2012). However, the same findings were simultaneously disappointing because there has been an expansion of morbidity worldwide (Olshansky et al., 1991), wherein longer life has been accompanied by increased years living with disability.

Level of physical activity may be the key to achieving population compression of morbidity (Fries et al., 2011) comprising postponement of mortality *and* postponement of morbidity. That is, whereas health-related habits such as not smoking and nutritious diet may be essential for longer life, physical activity throughout the lifecourse may be essential for longer life free of disability. At the international level, care is needed to ensure that actions that postpone mortality do not contribute to increased human and economic burdens due to expansion of morbidity. This is typified by the expansion of morbidity due to the "failure of success" attributed to medical interventions that extend life without removing disability (Gruenberg, 1977). Expansion of morbidity, in particular, is contributing to current alarm over the potential unsustainability of rising healthcare costs. Increased physical activity, on the other hand, has the potential not merely to contribute to postponement of mortality but to simultaneously contribute to postponement of morbidity, thereby contributing to increased healthcare cost-effectiveness.

## Physical Activity Among Children

While adult inactivity presents major health concerns, inactivity among children is an even greater concern. The determination of inactivity for children differs from that for adults. For children, inactivity is defined by doing fewer than 60 min of daily physical activity of moderate to vigorous intensity (approximately 360 MET). By that measure, 80% of children worldwide in their mid-teens are inactive, a finding that has major implications for future global health. Physical activity is associated with numerous health benefits in school-aged children and youth (Janssen and LeBlanc, 2010; Strong et al., 2005). As for adults, there is a dose-response relationship for children, wherein more physical activity is associated with greater health benefit. Although modest amounts of physical activity are beneficial for children who are at higher health risk due to obesity or high blood pressure, habitual moderate and preferably vigorous physical activity are needed to achieve optimal health benefits.

Unlike adults, for whom moderate and vigorous activities are considered exchangeable (e.g., 60 min activity of moderate-intensity is regarded as equivalent in health benefit to 30 min activity of vigorous-intensity), activities of vigorous intensity may be particularly beneficial for children. Unfortunately, however, evidence suggests that habitual physical activity, and especially vigorous-intensity activity, among young people is declining in proportion to the increasing incidence of childhood overweight and obesity (Belton et al., 2014). Additionally, girls are consistently found to be more inactive than boys (Hallal et al., 2012). The likely consequence of declining levels of physical fitness among women of childbearing age has implications for the health of offspring and therefore of future generations. Incorporating moderate-to-vigorous physical activity as part of regular school activities is one way of encouraging adequate physical activity among young people. Unfortunately, increased focus on educational goals, especially academic achievement, appears to have become a barrier to the promotion of physical activity in schools (Hills et al., 2015).

Evidence does not support the suspicion that the incorporation of physical activity as part of the school day may simply lead to same-day "compensation" in the form of reduced physical activity after school (Long et al., 2013). The provision of neighborhood parks also encourages physical activity among children and youths, with more vigorous activity being observed in parks that offer a greater number of amenities (Coughenour et al., 2014). Furthermore, level of physical activity is influenced by level of *independent mobility*, which for children refers to the freedom "to travel around their own neighborhood or city without adult supervision" (Stone et al., 2014, p. 2). Greater independent mobility implies more time spent on foot and less time being chaperoned in the family car. Evidence indicates that independent mobility in children has been declining worldwide since the 1970s, apparently largely in response to parental concerns about threats to child safety. Ironically, for many, the threat to health from being physically inactive almost certainly exceeds the threat to security from being more independent.

### Physical Activity Targets

Notice that the risk factor change targets summarized in Table 14.1 specify decreases in population physical inactivity as distinct from increases in physical activity. The distinction is important because an increase in overall physical activity does not necessitate change in the population level of physical inactivity. This would be achieved, for example, if the active population increased their level of physical activity without the inactive population changing theirs. Thus, despite evidence showing likely benefits to health from increased activity by everyone, including those who are already active, the principal physical-activity targets identified by the WHO are expressly focussed on the population that is at greater risk, namely, the population defined as inactive.

The focus on inactivity reflects growing concern about the high population prevalence of sedentary time which has increased substantially over recent decades (Kohl et al., 2012; Owen et al., 2014). Extended periods of continuous sitting have become ubiquitous throughout the lifecourse due to varied economic, social, environmental, and technological influences, especially increased use of motorized transport and screen-based communication and entertainment devices. The specific outcome target summarized in Table 14.1 is for a decrease of 10% in physical inactivity by 2025, where inactivity is defined:

- for children and adolescents, as less than 60 min daily of moderate-to-vigorous intensity activity; and
- for adults, as less than 150 min of moderate-intensity activity (or equivalent) per week.

These targets are deemed achievable by the WHO on the basis of a history of success of the order of 1% change per year in a number of high- and middle-income countries where national physical-activity programs have been implemented (WHO, 2012b). Specific actions that have proven effective include:

- the promotion of physical activity through mass media;
- environmental design to increase accessibility to walking, cycling, sports, and other recreational activities;
- involvement of multiple settings, including schools, workplaces, and neighborhoods; and
- involvement of multiple sectors, including transport, education, and environmental planning.

## 14.2.2    Secondary (Biological) Risk Factor Reduction

*...the enormous difficulty for medical personnel [is] to see health as a population issue and not merely as a problem for individuals.*

(Rose, 1985, p. 38)

An important feature of primary (behavioral) and secondary (biological) risk factors is that the latter are largely caused by the former, but not the reverse. For example, tobacco use is a major cause of raised blood pressure, as is high consumption of salt, but neither tobacco use nor high salt intake is caused by raised blood pressure. Generally, then, the most effective way to reduce population secondary risk factors is to reduce exposure to primary risk factors. Hence, reduced exposure to the primary risk factors of tobacco use, harmful consumption of alcohol, poor diet, and physical inactivity is effective in preventing the secondary risk factors of raised blood pressure, overweight and obesity, and raised blood glucose. Prevention of both primary and secondary risk factors is in turn effective in preventing common complex diseases, including cardiovascular diseases, cancers, chronic respiratory diseases, and diabetes.

When present, secondary risk factors can be reduced by reducing either exposure to primary risk factors or by biomedical intervention, or both. However, using biomedical intervention to reduce secondary risk factors is less safe, overall, less effective, and less cost-effective than reducing exposure to primary risk factors before secondary risk factors have developed. Furthermore, as discussed in Chapter 9, whereas the secondary risk factors of raised blood pressure and blood glucose are responsive to biomedical intervention, there are no viable biomedical interventions for safely reducing population-level primary risk factors. Here, again, we see the limited capacity of biomedicine to improve the health of populations. Moreover, whereas everyone is a potential beneficiary of population-wide preventive behavioral and social interventions to reduce primary and secondary risk factors, biomedical interventions, when effective, benefit only those relatively few individuals who are at high risk.

### 14.2.2.1    Raised Blood Pressure

During each heartbeat, blood pressure alternates between a maximum (systolic) and a minimum (diastolic) pressure. Concerns about raised blood pressure often focus on hypertension, defined as an average systolic blood pressure of 140 mm Hg or greater, or diastolic blood pressure of 90 mm Hg or greater, or on use of antihypertensive medication. By that definition, more than a quarter of the global adult population has hypertension, and this is predicted to increase to about 30% by 2025 due in part to negative trends in physical activity levels (Kearney et al., 2005). However, risk of cardiovascular disease is not restricted to those who are hypertensive, but increases or decreases progressively (i.e., as a *monotonic function*) with variation in mean level. Hence, downward shifts in blood pressure level, to a low of at least 115 mm Hg systolic and 75 mm Hg diastolic, are associated with monotonic falls in disease prevalence and vice versa (Prospective Studies Collaboration, 2002). Specifically, for systolic pressure, which tends to be a better predictor of cardiovascular risk than diastolic pressure, a decrease in the population average of 5 mm Hg would result in an approximate 20% reduction in stroke mortality and 15% reduction in mortality from heart disease.

Worldwide, mean systolic pressure is about 128 for men and 124 mm Hg for women, and tends to be higher in developing countries than in developed countries (Danaei et al., 2011a). Notwithstanding cardiovascular disease risk being highest for people with blood pressure levels in the hypertensive range, there is a higher absolute *number* (despite lower *rate*) of cardiovascular events in the normotensive than in the hypertensive population. This is due to the relative difference in size of the two populations, with the normotensive population being the larger (Cappuccio et al., 2011). Therefore, even relatively modest shifts in population exposure to primary risk factors that influence blood pressure can have major effects on population incidence of disease. For example, it has been estimated that systolic blood pressure is decreased by about 1 mm Hg for every 1-g reduction in daily salt intake (He et al., 2013). Consequently, in respect of this one risk factor, success in reducing the current global salt intake of 9-12 g per day to the WHO target of less than 5 g daily would be expected to produce an average reduction in population systolic blood pressure by 4-7 mm Hg, leading to approximately 20% or more reductions in global cardiovascular mortality and morbidity.

The WHO target for raised blood pressure is a reduction of 25% by 2025 in the population prevalence of blood pressure levels in the hypertensive range. The interventions of choice (*first-line interventions*) for achieving that target include all of the strategies listed above and in Table 14.1 under the rubric of *primary (behavioral) risk factors*. In addition to those strategies, adjunctive use of blood pressure-lowering drugs is recommended for people with persistent hypertension (WHO, 2012b). It may be remembered that evidence discussed in Chapter 4 and summarized in Figure 4.8 shows that considerably more overall benefit derives from risk factor reduction than biomedical intervention. Thus, risk factor reduction is not merely the best approach for achieving optimally low population levels of blood pressure, but it is also the best first-line intervention for hypertension. At the same time, the adjunctive use of biomedical intervention for hypertensive patients is supported by evidence indicating an approximate 50% reduction in mortality rate after 5 years for patients receiving pharmacotherapy compared to placebo (Sundström et al., 2014). However, the extensive body of research involving drugs for hypertensive patients has rarely examined the optimal balance of pharmacotherapy and risk factor reduction (Wilson et al., 2014). Consequently, in the treatment of hypertension, drugs are routinely used *instead of* rather than as an *adjunct to* risk factor reduction.

## 14.2.2.2 Overweight and Obesity

Worldwide prevalence of overweight and obesity has been increasing for the past 30 years (Ng et al., 2014). Notwithstanding the immensity of the problem of undernutrition, with more than one-tenth of the world's population being chronically undernourished (FAO, 2014), more than 2 billion people, almost 30% of the global population, is either overweight, *body mass index (BMI)* of 25 kg/m$^2$ or greater, or obese, BMI of 30 kg/m$^2$ or greater. A systematic review of multiple data sources showed that obesity rates are increasing in both high-income and developing countries, with rates generally higher among women than men (Ng et al., 2014). Whereas the increases in obesity that began in the 1980s in high-income countries have slowed during the past decade, increases are likely to continue in the developing world where almost two-thirds of obese people live. Excess bodyweight is a risk factor for mortality and morbidity from cardiovascular diseases, cancers, diabetes, and musculoskeletal disorders. Total mortality attributable to overweight and obesity is estimated to be nearly 3 million deaths a year (Finucane et al., 2011). Most of that mortality is attributable to cardiovascular diseases (Lim et al., 2013), but increased BMI is also a contributing factor for many different cancer types (Bhaskaran et al., 2014).

Attempts to explain the large increases in obesity over the past 3 decades have focussed on several possible pathways, including increased calorie consumption, changes in dietary composition, and decreased physical activity (Ng et al., 2014). The thrifty gene hypothesis, discussed in Chapter 2, represents one attempt to provide a genetic explanation for the rise in overweight and obesity not involving evolutionary change in the genome (Neel, 1962). More recently, genome-wide association studies and other approaches have identified several obesity-associated gene variants that could account for increased numbers of people being susceptible to overweight when exposed to the contemporary nutritional environment (Loos, 2009). However, just as evidence of thrifty genes has been lacking, known gene variants have low predictive value and explain only a small proportion of the population incidence of overweight and obesity. Possibly more promising are studies of the influence of early-life environment on later phenotypic expression. In particular, it has been hypothesized that poor maternal nutrition could induce in offspring a phenotype that is matched to a postnatal environment in which nutrients are scarce (Burdge and Lillycrop, 2014).

More specifically, a maternal diet consisting largely of energy-dense low-nutritious foods could result in a deficiency in the supply of nutrients crossing the placenta during pregnancy. In that event, activation of a fetal phenotype suited to nutrient scarcity would be mismatched with the nutrient-rich environment into which the infant is born. As such, nutrition-related genetic and epigenetic processes induced during early critical stages of development could influence later susceptibility to an *obesogenic* habitat consisting of an abundance of energy-dense foods accessible with a minimum of physical effort. This is consistent with the more general proposal that common complex diseases in adulthood are linked to the nutritional quality of the uterine environment during fetal development (Barker and Thornburg, 2013). Whereas babies born in high-income countries may risk being inadequately nourished due to poorly-balanced maternal nutrition, babies born in low-income countries may be malnourished because their mothers are chronically undernourished (Barker, 2012). Thus, maternal behavior and nutritional environment before and after conception may contribute to susceptibility to obesity in offspring during childhood and adulthood. Diet and nutrition, perhaps more than any other health-related factors, support the importance of adopting a lifecourse approach for optimizing the health of populations.

The obesogenic habitat of palatable processed foods is a creation of commercial incentives operating in the market economy. Accordingly, attempts to reverse-engineer the environment to undo the past 3 decades of rising overweight and obesity must of necessity address market activity in ways that threaten profits for the global food industry (*Big Food*). Indeed, it has been claimed that "unhealthy commodities, their producers, and the markets that power them," are the main risk factors responsible for the noncommunicable diseases that are the leading global causes of mortality and morbidity (Stuckler et al., 2012). It has been shown that regulatory measures can be successful in reducing levels of salt, sugar, fat, and trans fat in processed foods marketed in high-income countries. However, the same transnational companies required by regulatory controls to produce healthier reformulations in some products have generally failed to apply the same nutritional improvements in developing countries. Consequently, *ultra-processed* foods, characterized by hyper-palatable, energy-dense, fatty, sugary, and salty ready-to-eat products, which continue to dominate food supplies in high-income countries, are now coming to dominate food supplies globally (Monteiro et al., 2013).

Mozaffarian et al. (2014) have argued that the comparative low prices at which processed foods can be sold while still generating high commercial profit do not reflect the true costs of those foods for society. Due to their high prevalence, diet-related diseases are responsible for a substantial proportion of healthcare expenditure, the continuing rise of which threatens to impose intolerable burdens on national economies. Conversely, healthy individuals, sustained by nutritionally balanced diets contribute to a lowering of expenditure on healthcare, have more productive lives, and in turn contribute more to tax revenue. Taking these factors into account, there are strong economic as well as humanitarian grounds for the introduction of national subsidies for healthy foods and increased taxes for unhealthy foods to encourage health-promoting dietary choices.

## Overweight and Obesity Targets

In view of the scale of preventable mortality and morbidity attributable to overweight and obesity, the WHO target of "zero increase" summarized in Table 14.1 could appear *unambitious*. In reality, given the relentless rise in prevalence of overweight and obesity and the relative absence of comprehensive interventions with demonstrated effectiveness (Popkin et al., 2012), success in merely halting further increases at this time would be genuine success. In response to evidence that overweight and obesity tend to follow a lifecourse pattern, the WHO recently resolved to encourage member states to take particular action to ensure zero increase among infants and young children (WHO, 2012d). Here again, that outcome target, although possibly seeming to be unambitious, may be all that can be realistically expected at this time. Considering the current relative lack of experience with policies and interventions to improve dietary balance and increase physical activity among children (Brennan et al., 2014), stemming further increases in childhood overweight and obesity would be success.

Effective intervention against overweight and obesity will require sustained action at multiple levels of the range and kind discussed above in relation to dietary risk factors and physical inactivity. At the same time, care is required to avoid exacerbating existing antiweight and antiobesity social bias that has been found for all ages and strata within society. For example, compared to children who are not overweight, those who are overweight and obese are more likely to experience social marginalization and bullying (Fitzgerald et al., 2013). In addition, studies of adults from the general public as well as physicians and other healthcare professionals have revealed a high prevalence of attitudes that associate overweight and obese individuals with negative traits and stereotypes such as being weak-willed, sloppy, and lazy (Sabia et al., 2012). It is evident, therefore, that the manifold problems of overweight and obesity require population-wide strategies that are supportive and non-stigmatizing. Action is needed that engages individuals, families, and communities in diverse settings including schools and workplaces. Public policies must encompass multiple sectors including education, transportation, and urban planning, as well as extensive regulation of private sector agriculture, food processing, and marketing (Gortmaker et al., 2011; Penhollow and Rhoads, 2014).

### 14.2.2.3 Raised Blood Glucose

Mean blood glucose level tested in healthy individuals while fasting is about 5.5 mmol/l (100 mg/dl). Raised blood glucose is referred to as *impaired glucose tolerance* if levels are higher than normal but less than the level required for a diagnosis of diabetes. Impaired glucose tolerance is a prediabetic state that may precede type 2 diabetes mellitus by many years and is associated with *insulin resistance* and increased risk of mortality and cardiovascular disease. Blood glucose levels and diabetes have been rising globally, and the number of adults with diabetes has more than doubled over the past 3 decades (Danaei et al., 2011b). Although the increased number is partly due to population growth and aging, other factors have also contributed, including the global increase in overweight and obesity. Adults who are obese and have signs of being metabolically unhealthy (e.g., impaired glycemic control or raised cholesterol) have an eightfold greater risk of developing type 2 diabetes than healthy normal weight adults, and even for metabolically healthy obese adults there is a four-fold increase in diabetes risk (Bell et al., 2014).

As with overweight and obesity, the WHO target for raised blood glucose is for no increase in prevalence of concentrations indicative of diabetes, defined as persistent fasting blood glucose levels $\geq 7.0$ mmol/l (126 mg/dl). Also, as with overweight and obesity, preventive strategies primarily involve control of the behavioral risk factors of physical activity and diet, especially the consumption of sugar and processed carbohydrates. Accordingly, actions to halt the increased incidence of raised blood glucose are essentially the same as those for over-weight and obesity, namely, multisectoral interventions including regulation of private sector marketing of unhealthy commodities.

## 14.3   BIOMEDICAL INTERVENTION TO CONTROL DISEASE PROGRESSION

*In chronic diseases the clinician's first contact with the patient comes late in the natural history of the disease.*

(Rose, 1981, p. 1850)

Substantial progress in reducing the current global burden of common complex diseases will only come from prevention of disease incidence through reduced exposure to known risk factors. When disease processes become established in individuals, with or without prior risk factor reduction, adjunctive biomedical intervention should be used in the interests of relieving discomfort and controlling disease progression. Accordingly, Section 2 of Table 14.1 summarizes key healthcare objectives for individuals who are at high risk due to manifest noncommunicable diseases. However, to reiterate an earlier point, biomedicine does not replace the benefits of risk factor reduction, which should not be abandoned when biomedical healthcare is used. That is, not smoking, avoiding harmful use of alcohol, a healthy diet, and physical activity all continue to be beneficial when accompanied by biomedical intervention for manifest disease and injury, and can often contribute more to stemming disease progression than biomedicine.

### 14.3.1   All Major Noncommunicable Diseases

Of all deaths from noncommunicable diseases, approximately three-quarters occur in developing countries, where basic healthcare is often not accessible. Therefore, a major objective of the WHO global program for prevention and control of disease is to encourage wider access to "affordable basic technologies and essential medicines" in developing countries. The WHO publishes a *model list of essential medicines* containing nearly 400 items (WHO, 2013a) and a separate (though substantially overlapping) list of nearly 300 medicines for children up to 12 years (WHO, 2013b). The listed medicines are selected as being the most efficacious, safe, and cost-effective. They are considered to be the minimum stock needed for problems that contribute most of the burden of disease, including all of the leading noncommunicable diseases. The WHO 2025 goal is for 80% availability of the recommended affordable basic technologies and essential medicines.

A further objective is to encourage technology transfer, including knowledge and skills, to facilitate production by developing countries of affordable, safe, and effective medicines and vaccines, diagnostics and medical technologies, information and electronic communication technologies (eHealth), and mobile and wireless devices (mHealth) (WHO, 2013c). Without diminishing the importance of the WHO recommendations on increased access to healthcare technologies in developing countries, it is important that sight is not lost of the vital role of population-wide reduction of the primary (behavioral) and secondary (biological) risk factors, the reduction of which greatly reduces the need for biomedical intervention. Given that access to biomedical technology is widespread in high-income countries (and, as discussed in earlier chapters, is characterized by oversupply in some countries), risk factor reduction is the leading and essential means for successfully reducing incidence of noncommunicable diseases in high-income and developing countries alike.

## 14.3.2   Cardiovascular Disease

The WHO is reputed to have stated in 1969 that cardiovascular disease

*…will result in coming years in the greatest epidemic mankind has faced unless we are able to reverse the trend.*

(De Backer, 2009, p. 343)

As discussed in Chapter 4, there was a reversal of trend in some high-income countries during the last decades of the twentieth century. That trend, however, appears to have stalled, possibly due to increased overweight and obesity and decreased physical activity. Confirming WHO warnings, cardiovascular disease is indeed at epidemic levels in high-income countries, where it is responsible for almost 40% of deaths from all causes (Table 2.1) and about half of all deaths from noncommunicable diseases. Although the proportion of total deaths attributable to cardiovascular diseases is lower in developing compared to high-income countries, the total burden of cardiovascular disease is substantially greater in developing countries. Specifically, developing countries account for almost 13 million of more than 17 million cardiovascular deaths annually (Table 2.1). To gain perspective, it is sometimes helpful to reference statistics against more familiar quantities. In that vein, it might be noted that the largest airliners from Boeing and Airbus are capable of carrying 500 passengers per flight. It would require 96 such aircraft fully laden with passengers to crash every day with no survivors, or one such crash every 15 min, to match the current global tally of deaths from cardiovascular disease.

The fact that some high-income countries experienced marked reductions in cardiovascular disease over several decades demonstrates that those diseases are substantially preventable. In that vein, the American Heart Association (Lloyd-Jones, et al., 2010) uses the concepts of *ideal health* and *cardiovascular health* as the basis for recommending increased focus on population risk factors associated with smoking, body mass index, diet, and physical activity for "promoting health rather than solely treating disease" (p. 588). The WHO target for cardiovascular disease, as summarized in Table 14.1, is to ensure by 2025 that at least 50% of "eligible" people worldwide receive appropriate drug therapy and counseling to reduce the risk of heart attacks and strokes. Eligibility includes individuals with existing cardiovascular disease and anyone at high risk of a cardiovascular event (WHO, 2012b). While capable of contributing modestly to reducing disease incidence, the recommended interventions are intended primarily to postpone mortality and limit morbidity for individuals with manifest disease.

### 14.3.3   Terminal Disease

The WHO (2013c) recommends the provision of palliative care, including the use of strong opioid analgesics to relieve end-of-life suffering and improve quality of life. More specifically, the WHO's (2015) definition of palliative care stresses that it:

- neither hastens nor postpones death,
- is life-affirming and accepting of the view that dying is a normal process,
- helps patients live as actively as possible until death, and
- provides support to assist families to cope during their loved one's illness and their own bereavement.

Advanced-cancer care has been a focus of research into the effectiveness of palliative care, and evidence supports its early provision for individuals suffering terminal disease (Bandieri et al., 2012; Zimmermann et al., 2014). Although palliative care appears to be most effective when delivered by a multidisciplinary care team, it need not necessarily be resource-intensive and is amenable to being delivered in tertiary-care facilities, community centers, and the home.

Success in relation to WHO recommendations regarding access to opioid analgesics would end needless suffering for millions of patients with terminal disease (WHO, 2012b, 2013d). In particular, access should be available for patients in moderate to severe pain in palliative care and patients for whom curative treatment is no longer an option. In all instances, the aim is to facilitate the highest possible quality of life. However, despite the high efficacy of opioid analgesics, good access to pain management is more the exception worldwide than the rule. It may be recalled from Chapter 6 that studies have shown a high incidence in the United States of death from overdose of opioids prescribed for pain relief (Bohnert et al., 2011; Paulozzi et al., 2012). That situation, however, is not representative of opioid access globally. Seya et al. (2011) reviewed consumption levels of relevant strong opioid analgesics in 188 countries and found that more than 80% of people worldwide live in countries with levels of access that are inadequate to meet estimated total population need for pain control.

A prominent barrier, among several that obstruct patient access to opioid analgesics in many countries, is *opiophobia*, which is fear of risk of overdose-related death (Atkinson et al., 2014). Additional barriers relate to concerns about dependence and abuse, and fears about prescribed opioids being diverted for illicit use. Such concerns have led many countries to introduce laws and public policies that restrict access without giving due regard to the important benefits of opioids for patients in pain. Such fears, moreover, are not justified given current evidence. Specifically, studies of long-term use of opioids for chronic pain show that risk of overdose death is low when rigorous clinical guidelines are observed (Bohnert et al., 2011). Incidence of drug dependence and abuse is also low (Noble et al., 2008), and largely irrelevant for terminal patients (Seya et al., 2011).

As for fears about diversion for illicit use, the facts are less than straightforward. Diversion of prescribed opioid analgesics for nonmedical uses has been reported to increase with increased clinical use of those drugs (Hall et al., 2008). Conversely, demand from patients and families suffering unbearably but without legal access to opioid analgesics is documented to create a market for illicit supply (Krakauer et al., 2010). Moreover, patients driven to obtain medicinal drugs illicitly face increased risks due to harm from inappropriate self-administration and absence of quality control over the drugs supplied.

## 14.3.4   Infectious Disease

Notwithstanding limited efficacy of some attempts at population immunization, such as was discussed in Chapter 1 in relation to the use of BCG vaccine for tuberculosis, a host of acute infectious diseases appear to be effectively controlled through mass vaccination. Several child vaccinations, in particular, are considered to be highly effective public health interventions (Batra, 2015; Wang et al., 2014). These include *MMR* against measles, mumps, and rubella, the three-in-one *DTaP* against diphtheria, tetanus, and pertussis (whooping cough), and the five-in-one *DTaP/IPV/Hib* which combines DTaP with vaccines against polio and *Haemophilus influenzae* type b (*Hib*). Although not free from potentially serious side-effects, adverse events tend to be rare, and the harm caused is greatly outweighed by benefits (Batra, 2015; Maglione et al., 2014). Infectious disease is also responsible for a substantial burden of *noncommunicable* disease, including human papillomavirus (*HPV*) which has a causal role in cervical cancer, and hepatitis B virus (*HBV*) which is a leading cause of liver cancer (Jemal et al., 2011).

### 14.3.4.1   Human Papillomavirus

More than 80% of cervical cancer occurs in low- and middle-income countries, where it is the most common cancer in women (de Martel et al., 2012). HPV refers to a group of more than 150 related viruses, many of which are spread sexually. HPV infection is responsible for most cervical cancers, with just two HPV types, HPV16 and HPV18, being responsible for about 70% of cases (Schiffman et al., 2007). As mentioned in Chapter 9, widespread cancer screening over the past 3 decades using the Pap smear is believed to have been responsible for substantially reducing the incidence of cervical cancer mortality in high-income countries. Consequently, the WHO proposes that all women worldwide aged 30-49 years be screened at least once (WHO, 2012b).

In addition, HPV vaccines have been approved for use in many countries, and these are recommended by the WHO as potentially effective in preventing infections with the two types of HPV that cause most precancerous cervical lesions and cervical cancer (WHO, 2012b, 2013b). However, Carlos et al. (2014) have warned that many issues surrounding mass HPV vaccination remain unsolved. In particular, they identified affordability and other barriers to adequate population coverage in the most vulnerable and affected countries, and uncertainties about long-term effectiveness, including possible replacement of the vaccine-targeted strains with other high-risk types of the virus.

Carlos et al. (2014) also cite conflict of interest as an additional concern, given that "available trials were sponsored by the manufacturers." They question the basis for current proposals to implement mass vaccination when well-known and comparatively effective preventive measures already exist and may be downgraded with the introduction of vaccination programs. Specifically, they refer to the success of screening in reducing incidence of HPV infection, and since infection rate is related to risky sexual behavior they recommend sexual education as an additional preventive measure in place of mass vaccination.

### 14.3.4.2 Hepatitis B Virus

Liver cancer has a comparatively high rate of fatality and is the third leading cause of cancer deaths worldwide (Jemal et al., 2010). More than 80% of liver cancers occur in less developed countries, with China alone accounting for more than 50% of the total. Variation between countries is largely attributable to variation in the distribution of chronic hepatitis B and C viruses, with HBV generally dominating. Recent decreases in rates of liver cancer in some Asian countries, including China and Korea, are believed to be partly attributable to reduced transmission of HBV due to improved hygienic and sanitary conditions, and reduced contamination of food due to improved food storage. In addition, infant hepatitis immunization programs implemented over the past 2 decades in those and other countries appear to have been responsible for substantial reductions in child and adolescent rates of infection (Chang et al., 2009). In view of such findings, the WHO recommends universal infant immunization involving three doses of hepatitis B vaccine, with the first dose delivered within 24 h of birth (WHO, 2012b).

## 14.3.5   Stakeholders in Personal and Population Health: Self-regulation, Public-Private Partnership, and Industry

Concern about conflict of interest has been a recurring theme of this book, and it is an issue deserving of further discussion in the context of the *WHO 25 by 25 Program* for the prevention and control of noncommunicable diseases. Progress in reducing the global burden of noncommunicable disease undoubtedly requires action from multiple sectors of society. Many practitioners, researchers, and policy makers believe that managing the complex challenges at hand requires partnership between public and private sectors. It may be recalled from Part 2 that collaboration in the form of public-private partnership has been the basis of much pharmaceutical entanglement with biomedical practice and research. However, as with pharmaceuticals, there is little evidence in the field of public health of dialog with industry being of much benefit to anyone's interests except those of industry. For example, the Food and Health Dialogue, established by the Australian Government in 2009 to create a healthier food environment (e.g., less salt, less fat, and less sugar in foods), has had no measurable success in lessening Australia's "unprecedented burden of disease attributable to poor diet" (Elliott et al., 2014, p. 95).

The science of the effects of corporate behavior on health is at an early stage of development. Much still remains to be learned about the extent of harm from unhealthy commodities marketed by transnational corporations and how best to prevent such harm. The record of partnership between the public sector and the *Big* consumer industries of *Big Pharma, Big Tobacco, Big Alcohol, Big Food,* and *Big Soda* is largely one of meager public benefit or worse (Angell, 2004; Monteiro et al., 2013; Moodie et al., 2013; Stuckler et al., 2012). Generally, self-regulation is industry's preferred form of "regulation," but from a public safety perspective that strategy is high-risk in principle and largely ineffective in practice (Brownell, 2012; Brownell and Warner, 2009; Ronit and Jensen, 2014; Sharma et al., 2010; Stuckler and Nestle, 2012).

National and international food and related markets are now largely controlled by a relatively small number of corporate *oligopolies* (Moodie et al., 2013). Recognition that the corporate purveyors of unhealthy commodities are themselves leading risk factors in the current global crisis in noncommunicable diseases has led to calls for a stronger response from governments, public health organizations, and civil society in support of greater regulation of the corporations whose products contribute to endemic disease. Based on an analysis of *profits and pandemics* created by the tobacco, alcohol, and ultra-processed food and drink industries, Moodie et al. (2013) concluded that corporations marketing unhealthy commodities should be permitted

> *no role in the formation of national or international policy for noncommunicable diseases*

(Moodie et al., 2013, p. 7)

## 14.3.6 Are the WHO 25 by 25 Targets Achievable?

Extensive evidence supports the conclusion that exposure to fewer behavioral risk factors (i.e., adopting a healthy lifestyle) is related to improved health. On the basis of a meta-analysis of relevant studies, Loef and Walach (2012) concluded that each of the following contributes individually and cumulatively to the promotion and maintenance of health: healthy nutrition (e.g., regular eating of fruit and vegetables and avoidance of ultra-processed foods), physical activity (the equivalent of a minimum of 150 min of moderate- to vigorous-intensity physical activity per week), optimal weight balance (estimated to be in the approximate range of 19-25 kg/m$^2$), no consumption of tobacco, and no or moderate consumption of alcohol. They found that individuals with four or more of these five healthy lifestyle factors had a mortality risk one-third of that for people who were unhealthy on all factors. By this reckoning, the WHO target of 25% reduction by 2025 in overall mortality from common complex diseases appears feasible. Conversely, Loef and Walach found that of the more than 500,000 participants included in their meta-analysis, all of whom were healthy at baseline and were followed-up on average for more than 13 years, fewer than one-quarter adhered to all factors.

In a more direct analysis of the WHO program for prevention and control of disease, Kontis et al. (2014) found that the prevalence of noncommunicable diseases could be reduced by nearly 20% (i.e., not far short of the target of 25%) by meeting just six of the population change targets summarized in Table 14.1. Those targets are the three primary (behavioral) risk factors of tobacco use (target of 30% reduction), harmful alcohol consumption (10% reduction), and salt intake (30% reduction) and the three secondary (biological) risk factors of raised blood pressure (25% reduction), obesity (no increase), and raised blood glucose (no increase). Given that Kontis et al. included only one of the five WHO dietary targets in their estimations, it follows that if the remaining dietary targets are also met (15% reduction in saturated fatty acid intake, complete or near-complete removal of trans fatty acids, replacement of processed foods by fruit and vegetables, no marketing of unhealthy foods to children), a result approximating or exceeding the overall target of a 25% decrease in noncommunicable diseases should be achievable.

While it is still too early say whether targets will be met, it is reassuring that relevant analyses indicate that major improvements in the health of populations can be achieved on the basis of relatively modest changes in the ways of living for most individuals. As discussed above, the WHO targets were not chosen arbitrarily, but were selected on the basis of improvements already made by the best performing countries. The hope is that the remaining countries of the world will be able to follow the lead of the best performers, and thereby arrest and reverse the increasing incidence of common complex diseases. A 25% reduction in deaths from noncommunicable diseases by 2025 compared to 2010 levels would mean approximately 10 million fewer deaths per year.

That would be a seminal achievement in the history of humankind, which would in turn pave the way for new prevention targets and further improvements in the health of populations worldwide.

## REFERENCES

Aburto, N.J., Ziolkovska, A., Hooper, L., et al., 2013. Effect of lower sodium intake on health: systematic review and meta-analyses. Br. Med. J. 346, 1–20. http://dx.doi.org/10.1136/bmj.f1326.

Adams, J., Tyrrell, R., Adamson, A.J., White, M., 2012. Effect of restrictions on television food advertising to children on exposure to advertisements for "less healthy" foods: repeat cross-sectional study. PLoS One. 7, http://dx.doi.org/10.1371/journal.pone.0031578.

Alberg, A.J., Shopland, D.R., Cummings, K.M., 2014. The 2014 Surgeon General's report: commemorating the 50th Anniversary of the 1964 report of the Advisory Committee to the US surgeon general and updating the evidence on the health consequences of cigarette smoking. Am. J. Epidemiol. 179, 403–412.

Angell, M., 2004. The Truth About the Drug Companies: How They Deceive Us and What to Do About It. Random House, New York.

Artero, A., Artero, A., Tarín, J.J., Cano, A., 2015. The impact of moderate wine consumption on health. Maturitas 80, 3–13.

Asaria, P., Chisholm, D., Mathers, C., et al., 2007. Chronic disease prevention: health effects and financial costs of strategies to reduce salt intake and control tobacco use. Lancet 370, 2044–2053.

Astrup, A., Dyerberg, J., Elwood, P., et al., 2011. The role of reducing intakes of saturated fat in the prevention of cardiovascular disease: where does the evidence stand in 2010? Am. J. Clin. Nutr. 93, 684–688.

Atkinson, T.J., Schatman, M.E., Fudin, J., 2014. The damage done by the war on opioids: the pendulum has swung too far. J. Pain Res. 7, 265–268.

Atun, R., 2014. Decisive action to end apathy and achieve $25 \times 25$ NCD targets. Lancet 384, 384–385.

Bandieri, E., Sichetti, D., Romero, M., et al., 2012. Impact of early access to a palliative/supportive care intervention on pain management in patients with cancer. Ann. Oncol. 23, 2016–2020. http://dx.doi.org/10.1093/annonc/mds103.

Barker, D.J.P., 2012. Developmental origins of chronic disease. Public Health 126, 185–189.

Barker, D.J.P., Thornburg, K.L., 2013. Placental programming of chronic diseases, cancer and life-span: a review. Placenta 34, 841–845.

Batra, N., 2015. Neurological disorders associated with measles-mumps-rubella vaccination: a review. Science 3, 81–86.

Bauman, A.E., 2004. Updating the evidence that physical activity is good for health: an epidemiological review 2000–2003. J. Sci. Med. Sport 7, 6–19.

Bauman, A.E., Reis, R.S., Sallis, J.F., et al., 2012. Correlates of physical activity: why are some people physically active and others not? Lancet 380, 258–271.

Been, J.V., Nurmatov, U.B., Cox, B., et al., 2014. Effect of smoke-free legislation on perinatal and child health: a systematic review and meta-analysis. Lancet 383, 1549–1560. http://dx.doi.org/10.1016/S0140-6736(14)60082-9.

Bell, J.A., Kivimaki, M., Hamer, M., 2014. Metabolically healthy obesity and risk of incident type 2 diabetes: a meta-analysis of prospective cohort studies. Obes. Rev. 15, 504–515. http://dx.doi.org/10.1111/obr.12157.

Belton, S., Wesley, O., Meegan, S., et al., 2014. Youth-physical activity towards health: evidence and background to the development of the Y-PATH physical activity intervention for adolescents. BMC Public Health 14, 1–12. http://www.biomedcentral.com/1471-2458/14/122.

Berryman, J.W., 2010. Exercise is medicine: a historical perspective. Curr. Sports Med. Rep. 9, 195–201.

Bhaskaran, K., Douglas, I., Forbes, H., dos-Santos-Silva, I., Leon, D.A., Smeeth, L., 2014. Body-mass index and risk of 22 specific cancers: a population-based cohort study of 5.24 million UK adults. Lancet 384, 755–765.

Blair, S.N., 2009. Physical inactivity: the biggest public health problem of the 21st century. Br. J. Sports Med. 43, 1–2.

Bloom, D.E., Cafiero, E., Jané-Llopis, E., et al., 2012. The Global Economic Burden of Noncommunicable Diseases (No. 8712). World Economic Forum, Geneva. Retrieved 30 December 2014 from, http://www.hsph.harvard.edu/pgda/working.htm.

Boeing, H., Bechthold, A., Bub, A., et al., 2012. Critical review: vegetables and fruit in the prevention of chronic diseases. Eur. J. Nutr. 51, 637–663.

Boffetta, P., Couto, E., Wichmann, J., et al., 2010. Fruit and vegetable intake and overall cancer risk in the European Prospective Investigation into Cancer and Nutrition (EPIC). J. Natl. Cancer Inst. 102, 529–537.

Bohnert, A.S., Valenstein, M., Bair, M.J., et al., 2011. Association between opioid prescribing patterns and opioid overdose-related deaths. J. Am. Med. Assoc. 305, 1315–1321.

Brennan, L.K., Brownson, R.C., Orleans, C.T., 2014. Childhood obesity policy research and practice: evidence for policy and environmental strategies. Am. J. Prev. Med. 46, e1–e16. http://dx.doi.org/10.1016/j.amepre.2013.08.022.

Brien, S.E., Ronksley, P.E., Turner, B.J., et al., 2011. Effect of alcohol consumption on biological markers associated with risk of coronary heart disease: systematic review and meta-analysis of interventional studies. Br. Med. J. 342, 1–15. http://dx.doi.org/10.1136/bmj.d636.

Brown, I.J., Tzoulaki, I., Candeias, V., Elliott, P., 2009. Salt intakes around the world: implications for public health. Int. J. Epidemiol. 38, 791–813.

Brownell, K.D., 2012. Thinking forward: the quicksand of appeasing the food industry. PLoS Med. 9, 1–2. http://dx.doi.org/10.1371/journal.pmed.1001254.

Brownell, K.D., Warner, K.E., 2009. The perils of ignoring history: big Tobacco played dirty and millions died. How similar is Big Food? Milbank Q. 87, 259–294.

Burdge, G.C., Lillycrop, K.A., 2014. Environment-physiology, diet quality and energy balance: the influence of early life nutrition on future energy balance. Physiol. Behav. 134, 119–122.

Butler, N.R., Goldstein, H., 1973. Smoking in pregnancy and subsequent child development. Br. Med. J. 4, 573–575.

Cairns, G., Angus, K., Hastings, G., 2009. The Extent, Nature and Effects of Food Promotion to Children: A Review of the Evidence to December 2008. World Health Organization, Geneva. Retrieved on 17 April 2014 from, http://www.who.int/dietphysicalactivity/Evidence_Update_2009.pdf.

Campbell, N., Correa-Rotter, R., Neal, B., Cappuccio, F.P., 2011. New evidence relating to the health impact of reducing salt intake. Nutr. Metab. Cardiovasc. Dis. 21, 617–619.

Cappuccio, F.P., Capewell, S., Lincoln, P., McPherson, K., 2011. Policy options to reduce population salt intake. Br. Med. J. 343, 1–8. http://dx.doi.org/10.1136/bmj.d4995.

Carlos, S., de Irala, J., Hanley, M., Martínez-González, M.Á., 2014. The use of expensive technologies instead of simple, sound and effective lifestyle interventions: a perpetual delusion. J. Epidemiol. Community Health 68, 897–904. http://dx.doi.org/10.1136/jech-2014-203884.

CDC, 2009. Prevalence and Trends Data: Fruits and Vegetables—2009. Centers for Disease Control and Prevention, Atlanta, GA. Retrieved 16 April 2014 from, http://apps.nccd.cdc.gov/brfss/list. asp?cat=FV&yr=2009&qkey=4415&state=All.

Chang, M.H., You, S.L., Chen, C.J., et al., 2009. Decreased incidence of hepatocellular carcinoma in hepatitis B vaccines: a 20-year follow-up study. J. Natl. Cancer Inst. 101, 1348–1355.

Chikritzhs, T., Stockwell, T., Naimi, T., et al., 2015. Has the leaning tower of presumed health benefits from 'moderate' alcohol use finally collapsed? Addiction 110, 726–727. http://dx.doi.org/10.1111/add.12828.

Coughenour, C., Coker, L., Bungum, T.J., 2014. Environmental and social determinants of youth physical activity intensity levels at neighborhood parks in Las Vegas, NV. J. Community Health 39, 1092–1096. http://dx.doi.org/10.1007/s10900-014-9856-4.

Crowe, F.L., Roddam, A.W., Key, T.J., et al., 2011. Fruit and vegetable intake and mortality from ischaemic heart disease: results from the European Prospective Investigation into Cancer and Nutrition (EPIC)-Heart study. Eur. Heart J. 32, 1235–1243.

Danaei, G., Finucane, M.M., Lin, J.K., et al., 2011a. National, regional, and global trends in systolic blood pressure since 1980: systematic analysis of health examination surveys and epidemiological studies with 786 country-years and 5.4 million participants. Lancet 377, 568–577.

Danaei, G., Finucane, M.M., Lu, Y., et al., 2011b. National, regional, and global trends in fasting plasma glucose and diabetes prevalence since 1980: systematic analysis of health examination surveys and epidemiological studies with 370 country-years and 2.7 million participants. Lancet 378, 31–40.

Dauchet, L., Amouyel, P., Hercberg, S., Dallongeville, J., 2006. Fruit and vegetable consumption and risk of coronary heart disease: a meta-analysis of cohort studies. J. Nutr. 136, 2588–2593.

De Backer, G.G., 2009. The global burden of coronary heart disease. Medicographia 31, 343–348.

de Martel, C., Ferlay, J., Franceschi, S., et al., 2012. Global burden of cancers attributable to infections in 2008: a review and synthetic analysis. Lancet Oncol. 13, 607–615.

Edwards, R., 2004. The problem of tobacco smoking. Br. Med. J. 328, 217–219.

Elliott, T., Trevena, H., Sacks, G., et al., 2014. A systematic interim assessment of the Australian Government's Food and Health Dialogue. Med. J. Aust. 200, 92–95.

Ezzati, M., Riboli, E., 2012. Can noncommunicable diseases be prevented? Lessons from studies of populations and individuals. Science 337, 1482–1487.

Ezzati, M., Lopez, A.D., Rodgers, A., et al., 2002. Selected major risk factors and global and regional burden of disease. Lancet 360, 1347–1360.

FAO, 2014. The State of Food Insecurity in the World 2014: Strengthening the Enabling Environment for Food Security and Nutrition. Food and Agriculture Organization of the United Nations, Rome, Italy. Retrieved 10 February 2015 from, http://www.fao.org/3/a-i4030e.pdf.

Farkas, A.J., Gilpin, E.A., White, M.M., Pierce, J.P., 2000. Association between household and workplace smoking restrictions and adolescent smoking. J. Am. Med. Assoc. 284, 717–722.

Ferrante, G., Simoni, M., Cibella, F., et al., 2013. Third-hand smoke exposure and health hazards in children. Monaldi Arch. Chest Dis. 79, 38–43.

Fichtenberg, C.M., Glantz, S.A., 2002a. Effect of smoke-free workplaces on smoking behaviour: systematic review. Br. Med. J. 325, 188–191.

Fichtenberg, C.M., Glantz, S.A., 2002b. Youth access interventions do not affect youth smoking. Pediatrics 109, 1088–1092.

Finucane, M.M., Stevens, G.A., Cowan, M.J., et al., 2011. National, regional, and global trends in body-mass index since 1980: systematic analysis of health examination surveys and epidemiological studies with 960 country-years and 9.1 million participants. Lancet 377, 557–567.

Fitzgerald, A., Heary, C., Roddy, S., 2013. Causal information on children's attitudes and behavioural intentions toward a peer with obesity. Obes. Facts 6, 247–257.

Fleming, P., Blair, P.S., 2007. Sudden infant death syndrome and parental smoking. Early Hum. Dev. 83, 721–725.

Fries, J.F., Bruce, B., Chakravarty, E., 2011. Compression of morbidity 1980–2011: a focused review of paradigms and progress. J. Aging Res. 1–10. http://dx.doi.org/10.4061/2011/261702.

Gortmaker, S.L., Swinburn, B.A., Levy, D., et al., 2011. Changing the future of obesity: science, policy, and action. Lancet 378, 838–847.

Grana, R., Benowitz, N., Glantz, S.A., 2014. E-cigarettes a scientific review. Circulation 129, 1972–1986.

Grucza, R.A., Bucholz, K.K., Rice, J.P., Bierut, L.J., 2008. Secular trends in the lifetime prevalence of alcohol dependence in the United States: a re-evaluation. Alcohol. Clin. Exp. Res. 32, 763–770.

Gruenberg, E.M., 1977. The failures of success. Milbank Mem. Fund Q. Health Soc. 55, 3–24.

Hall, A.J., Logan, J.E., Toblin, R.L., et al., 2008. Patterns of abuse among unintentional pharmaceutical overdose fatalities. J. Am. Med. Assoc. 300, 2613–2620.

Hallal, P.C., Andersen, L.B., Bull, F.C., et al., 2012. Global physical activity levels: surveillance progress, pitfalls, and prospects. Lancet 380, 247–257.

Hamer, M., Stamatakis, E., Steptoe, A., 2009. Dose–response relationship between physical activity and mental health: the Scottish Health Survey. Br. J. Sports Med. 43, 1111–1114.

Harrell, P.T., Simmons, V.N., Correa, J.B., et al., 2014. Electronic nicotine delivery systems ("e-cigarettes"): review of safety and smoking cessation efficacy. Otolaryngol. Head Neck Surg. 151, 381–393.

He, F.J., MacGregor, G.A., 2008. A comprehensive review on salt and health and current experience of worldwide salt reduction programmes. J. Hum. Hypertens. 23, 363–384.

He, F.J., Nowson, C.A., MacGregor, G.A., 2006. Fruit and vegetable consumption and stroke: meta-analysis of cohort studies. Lancet 367, 320–326.

He, F.J., Nowson, C.A., Lucas, M., MacGregor, G.A., 2007. Increased consumption of fruit and vegetables is related to a reduced risk of coronary heart disease: meta-analysis of cohort studies. J. Hum. Hypertens. 21, 717–728.

He, F.J., Li, J., MacGregor, G.A., 2013. Effect of longer term modest salt reduction on blood pressure: cochrane systematic review and meta-analysis of randomised trials. Br. Med. J. 346, 1–15. http://dx.doi.org/10.1136/bmj.f1325.

Heath, G.W., Parra, D.C., Sarmiento, O.L., et al., 2012. Evidence-based intervention in physical activity: lessons from around the world. Lancet 380, 272–281.

Heironimus, J., 1992. Impact of workplace restrictions on consumption and incidence. Internal correspondence to Louis Suwarna, Phillip Morris Company, 22 January 1992. Retrieved 26 March 2014 from, http://tobaccodocuments.org/landman/2023914280-4284.pdf.

Hills, A.P., Dengel, D.R., Lubans, D.R., 2015. Supporting public health priorities: recommendations for physical education and physical activity promotion in schools. Prog. Cardiovasc. Dis. 57, 368–374.

Hippocrates (c. 400 BCE). On Airs, Water, and Places, Part 1. Translated by Francis Adams. Retrieved on 9 January 2012 from, http://classics.mit.edu/Hippocrates/airwatpl.1.1.html.

Horta, B.L., Victora, C.G., Menezes, A.M., et al., 1997. Low birthweight, preterm births and intrauterine growth retardation in relation to maternal smoking. Paediatr. Perinat. Epidemiol. 11, 140–151.

Hovell, M.F., Adams, M.A., Hofstetter, C.R., et al., 2014. Complete home smoking bans and anti-tobacco contingencies: a natural experiment. Nicotine Tob. Res. 16, 186–196.

Hughes, J.R., Keely, J., Naud, S., 2004. Shape of the relapse curve and long-term abstinence among untreated smokers. Addiction 99, 29–38.

Hunter, D.J., Reddy, K.S., 2013. Noncommunicable diseases. N. Engl. J. Med. 369, 1336–1343.

IFPRI, 2014. Global Nutrition Report 2014: Actions and Accountability to Accelerate the World's Progress on Nutrition. International Food Policy Research Institute, Washington, DC.

Janssen, I., LeBlanc, A.G., 2010. Review systematic review of the health benefits of physical activity and fitness in school-aged children and youth. Int. J. Behav. Nutr. Phys. Act. 7, 1–16.

Janssen, F., Rousson, V., Paccaud, F., 2015. The role of smoking in changes in the survival curve: an empirical study in 10 European countries. Ann. Epidemiol. 25, 243–249. http://dx.doi.org/10.1016/j.annepidem.2015.01.007.

Jemal, A., Center, M.M., DeSantis, C., Ward, E.M., 2010. Global patterns of cancer incidence and mortality rates and trends. Cancer Epidemiol. Biomark. Prev. 19 (8), 1893–1907.

Jemal, A., Bray, F., Center, M.M., et al., 2011. Global cancer statistics. CA Cancer J. Clin. 61, 69–90.

Jha, P., 2009. Avoidable global cancer deaths and total deaths from smoking. Nat. Rev. Cancer 9, 655–664.

Kearney, P.M., Whelton, M., Reynolds, K., et al., 2005. Global burden of hypertension: analysis of worldwide data. Lancet 365, 217–223.

Kelly, B., Smith, B., King, L., et al., 2007. Television food advertising to children: the extent and nature of exposure. Public Health Nutr. 10 (11), 1234–1240.

Key, T.J., 2011. Fruit and vegetables and cancer risk. Br. J. Cancer 104, 6–11.

Kohl, H.W., Craig, C.L., Lambert, E.V., et al., 2012. The pandemic of physical inactivity: global action for public health. Lancet 380, 294–305.

Kontis, V., Mathers, C.D., Rehm, J., et al., 2014. Contribution of six risk factors to achieving the $25 \times 25$ non-communicable disease mortality reduction target: a modelling study. Lancet 384, 427–437.

Krakauer, E.L., Wenk, R., Buitrago, R., et al., 2010. Opioid inaccessibility and its human consequences: reports from the field. J. Pain Palliat. Care Pharmacother. 24, 239–243.

Lee, I.M., Paffenbarger, R.S., 2000. Associations of light, moderate, and vigorous intensity physical activity with longevity: the Harvard Alumni Health Study. Am. J. Epidemiol. 151, 293–299.

Lee, I.M., Skerrett, P.J., 2001. Physical activity and all-cause mortality: what is the dose–response relation? Med. Sci. Sports Exerc. 33 (Supp), S459–S471.

Lee, I.M., Shiroma, E.J., Lobelo, F., et al., 2012. Effect of physical inactivity on major noncommunicable diseases worldwide: an analysis of burden of disease and life expectancy. Lancet 380, 219–229.

Leon, D.A., Chenet, L., Shkolnikov, V.M., et al., 1997. Huge variation in Russian mortality rates 1984–94: artefact, alcohol, or what? Lancet 350, 383–388.

Lim, S.S., Vos, T., Flaxman, A.D., et al., 2013. A comparative risk assessment of burden of disease and injury attributable to 67 risk factors and risk factor clusters in 21 regions, 1990–2010: a systematic analysis for the Global Burden of Disease Study 2010. Lancet 380, 2224–2260.

Lloyd-Jones, D.M., Hong, Y., Labarthe, D., et al., 2010. Defining and setting national goals for cardiovascular health promotion and disease reduction the American Heart Association's Strategic Impact Goal through 2020 and beyond. Circulation 121, 586–613.

Lobstein, T., Dibb, S., 2005. Evidence of a possible link between obesogenic food advertising and child overweight. Obes. Rev. 6, 203–208.

Lock, K., Pomerleau, J., Causer, L., et al., 2005. The global burden of disease attributable to low consumption of fruit and vegetables: implications for the global strategy on diet. Bull. World Health Organ. 83, 100–108.

Loef, M., Walach, H., 2012. The combined effects of healthy lifestyle behaviors on all cause mortality: a systematic review and meta-analysis. Prev. Med. 55, 163–170.

Long, M.W., Sobol, A.M., Cradock, A.L., et al., 2013. School-day and overall physical activity among youth. Am. J. Prev. Med. 45, 150–157.

Loos, R.J., 2009. Recent progress in the genetics of common obesity. Br. J. Clin. Pharmacol. 68, 811–829.

Maglione, M.A., Das, L., Raaen, L., et al., 2014. Safety of vaccines used for routine immunization of US children: a systematic review. Pediatrics 134, 325–337.

Mann, R.E., Smart, R.G., Govoni, R., 2003. The epidemiology of alcoholic liver disease. Alcohol Res. Health 27, 209–219.

Micha, R., Mozaffarian, D., 2010. Saturated fat and cardiometabolic risk factors, coronary heart disease, stroke, and diabetes: a fresh look at the evidence. Lipids 45, 893–905.

Monteiro, C.A., Moubarac, J.C., Cannon, G., et al., 2013. Ultra-processed products are becoming dominant in the global food system. Obes. Rev. 14 (Suppl. 2), 21–28.

Moodie, R., Stuckler, D., Monteiro, C., et al., 2013. Profits and pandemics: prevention of harmful effects of tobacco, alcohol, and ultra-processed food and drink industries. Lancet 381, 670–679.

Morris, J.N., Crawford, M.D., 1958. Coronary heart disease and physical activity of work. Br. Med. J. 2, 1485–1496.

Mozaffarian, D., Katan, M.B., Ascherio, A., et al., 2006. Trans fatty acids and cardiovascular disease. N. Engl. J. Med. 354, 1601–1613.

Mozaffarian, D., Aro, A., Willett, W.C., 2009. Health effects of trans-fatty acids: experimental and observational evidence. Eur. J. Clin. Nutr. 63, S5–S21.

Mozaffarian, D., Rogoff, K.S., Ludwig, D.S., 2014. The real cost of food: can taxes and subsidies improve public health? J. Am. Med. Assoc. 312, 889–890.

Mucha, L., Stephenson, J., Morandi, N., Dirani, R., 2006. Meta-analysis of disease risk associated with smoking, by gender and intensity of smoking. Gend. Med. 3, 279–291.

Neel, J.V., 1962. Diabetes mellitus: a "thrifty" genotype rendered detrimental by "progress"? Am. J. Hum. Genet. 14, 353–362.

Ng, M., Fleming, T., Robinson, M., et al., 2014. Global, regional, and national prevalence of overweight and obesity in children and adults during 1980–2013: a systematic analysis for the Global Burden of Disease Study 2013. Lancet 384, 766–781.

Noble, M., Tregear, S.J., Treadwell, J.R., Schoelles, K., 2008. Long-term opioid therapy for chronic noncancer pain: a systematic review and meta-analysis of efficacy and safety. J. Pain Symptom Manag. 35, 214–228.

Nordestgaard, B.G., Varbo, A., 2014. Triglycerides and cardiovascular disease. Lancet 384, 626–635.

Norström, T., Skog, O.J., 2001. Alcohol and mortality: methodological and analytical issues in aggregate analyses. Addiction 96, 5–17.

Öberg, M., Jaakkola, M.S., Woodward, A., et al., 2011. Worldwide burden of disease from exposure to second-hand smoke: a retrospective analysis of data from 192 countries. Lancet 377, 139–146.

Olshansky, S.J., Rudberg, M.A., Carnes, B.A., et al., 1991. Trading off longer life for worsening health: the expansion of morbidity hypothesis. Aging Health 3, 194–216.

Owen, N., Salmon, J., Koohsari, M.J., et al., 2014. Sedentary behaviour and health: mapping environmental and social contexts to underpin chronic disease prevention. Br. J. Sports Med. 48, 174–177.

Oyebode, O., Gordon-Dseagu, V., Walker, A., Mindell, J.S., 2014. Fruit and vegetable consumption and all-cause, cancer and CVD mortality: analysis of Health Survey for England data.

J. Epidemiol. Community Health 68 (9), 856–862. http://dx.doi.org/10.1136/jech-2013-203500.

Paffenbarger, R.S., Blair, S.N., Lee, I.M., 2001. A history of physical activity, cardiovascular health and longevity: the scientific contributions of Jeremy N Morris, DSc, DPH, FRCP. Int. J. Epidemiol. 30, 1184–1192.

Parkes, G., Greenhalgh, T., Griffin, M., Dent, R., 2008. Effect on smoking quit rate of telling patients their lung age: the Step2quit randomised controlled trial. Br. Med. J. 336 (7644), 598–600.

Paulozzi, L.J., Kilbourne, E.M., Shah, N.G., et al., 2012. A history of being prescribed controlled substances and risk of drug overdose death. Pain Med. 13, 87–95.

Penhollow, T.M., Rhoads, K.E., 2014. Preventing obesity and promoting fitness an ecological perspective. Am. J. Lifestyle Med. 8, 21–24.

Perk, J., De Backer, G., Gohlke, H., et al., 2012. European guidelines on cardiovascular disease prevention in clinical practice (version 2012). Eur. Heart J. 33, 1635–1701.

Popkin, B.M., Adair, L.S., Ng, S.W., 2012. Global nutrition transition and the pandemic of obesity in developing countries. Nutr. Rev. 70, 3–21.

Prescott, E., Hippe, M., Schnohr, P., et al., 1998. Smoking and risk of myocardial infarction in women and men: longitudinal population study. Br. Med. J. 316, 1043–1047.

Prospective Studies Collaboration, 2002. Age-specific relevance of usual blood pressure to vascular mortality: a meta-analysis of individual data for one million adults in 61 prospective studies. Lancet 360, 1903–1913.

Rader, D.J., Hovingh, G.K., 2014. HDL and cardiovascular disease. Lancet 384, 618–625.

Rehm, J., 2014. Russia: lessons for alcohol epidemiology and alcohol policy. Lancet 383, 1440–1442.

Rehm, J., Mathers, C., Popova, S., et al., 2009. Global burden of disease and injury and economic cost attributable to alcohol use and alcohol-use disorders. Lancet 373, 2223–2233.

Ridker, P.M., 2014. LDL cholesterol: controversies and future therapeutic directions. Lancet 384, 607–617.

Rimm, E.B., Klatsky, A., Grobbee, D., Stampfer, M.J., 1996. Review of moderate alcohol consumption and reduced risk of coronary heart disease: is the effect due to beer, wine, or spirits? Br. Med. J. 312, 731–736.

Ronit, K., Jensen, J.D., 2014. Obesity and industry self-regulation of food and beverage marketing: a literature review. Eur. J. Clin. Nutr. 68, 753–759.

Ronksley, P.E., Brien, S.E., Turner, B.J., et al., 2011. Association of alcohol consumption with selected cardiovascular disease outcomes: a systematic review and meta-analysis. Br. Med. J. 342, 1–13. http://dx.doi.org/10.1136/bmj.d671.

Rose, G., 1981. Strategy of prevention: lessons from cardiovascular disease. Br. Med. J. 282, 1847–1851.

Rose, G., 1985. Sick individuals and sick populations. Int. J. Epidemiol. 14, 32–38.

Rose, G., 1992. The Strategy of Preventive Medicine. Oxford University Press, Oxford, UK.

Roswall, N., Weiderpass, E., 2015. Alcohol as a risk factor for cancer: existing evidence in a global perspective. J. Prev. Med. Public Health 48, 1–9.

Sabia, S., Dugravot, A., Kivimaki, M., et al., 2012. Effect of intensity and type of physical activity on mortality: results from the Whitehall II cohort study. Am. J. Public Health 102, 698–704.

Salomon, J.A., et al., 2012. Healthy life expectancy for 187 countries, 1990–2010: a systematic analysis for the Global Burden Disease Study 2010. Lancet 380, 2144–2162.

Sattelmair, J., Pertman, J., Ding, E.L., et al., 2011. Dose response between physical activity and risk of coronary heart disease a meta-analysis. Circulation 124, 789–795.

Schiffman, M., Castle, P.E., Jeronimo, J., et al., 2007. Human papillomavirus and cervical cancer. Lancet 370, 890–907.

Seya, M.J., Gelders, S.F., Achara, O.U., et al., 2011. A first comparison between the consumption of and the need for opioid analgesics at country, regional, and global levels. J. Pain Palliat. Care Pharmacother. 25, 6–18.

Sharma, L.L., Teret, S.P., Brownell, K.D., 2010. The food industry and self-regulation: standards to promote success and to avoid public health failures. Am. J. Public Health 100, 240–246.

Sleiman, M., Gundel, L.A., Pankow, J.F., et al., 2010. Formation of carcinogens indoors by surface-mediated reactions of nicotine with nitrous acid, leading to potential thirdhand smoke hazards. Proc. Natl. Acad. Sci. U. S. A. 107, 6576–6581.

Stone, M.R., Faulkner, G.E., Mitra, R., Buliung, R.N., 2014. The freedom to explore: examining the influence of independent mobility on weekday, weekend and after-school physical activity behaviour in children living in urban and inner-suburban neighbourhoods of varying socioeconomic status. Int. J. Behav. Nutr. Phys. Act. 11, 5. http://www.ijbnpa.org/content/11/1/5.

Strong, W.B., Malina, R.M., Blimkie, C.J., et al., 2005. Evidence based physical activity for school-age youth. J. Pediatr. 146, 732–737.

Stuckler, D., Nestle, M., 2012. Big food, food systems, and global health. PLoS Med. 9, 1–4. http://dx.doi.org/10.1371/journal.pmed.1001242.

Stuckler, D., McKee, M., Ebrahim, S., Basu, S., 2012. Manufacturing epidemics: the role of global producers in increased consumption of unhealthy commodities including processed foods, alcohol, and tobacco. PLoS Med. 9, 1–8. http://dx.doi.org/10.1371/journal.pmed.1001235.

Sundström, J., Arima, H., Woodward, M., et al., 2014. Blood pressure-lowering treatment based on cardiovascular risk: a meta-analysis of individual patient data. Lancet 384, 591–598.

Ünal, B., Critchley, J.A., Capewell, S., 2005a. Modelling the decline in coronary heart disease deaths in England and Wales, 1981–2000: comparing contributions from primary prevention and secondary prevention. BMJ 331, 1–6. http://dx.doi.org/10.1136/bmj.38561.633345.8F.

Ünal, B., Critchley, J.A., Fidan, D., Capewell, S., 2005b. Life-years gained from modern cardiological treatments and population risk factor changes in England and Wales, 1981–2000. Am. J. Public Health 95, 103–108.

United Nations, 2012. Political declaration of the high-level meeting of the general assembly on the prevention and control of non-communicable diseases (A/RES/66/2). Retrieved 25 May 2013 from, http://daccess-dds-ny.un.org/doc/UNDOC/GEN/N11/458/94/PDF/N1145894.pdf.

Vangeli, E., Stapleton, J., Smit, E.S., et al., 2011. Predictors of attempts to stop smoking and their success in adult general population samples: a systematic review. Addiction 106, 2110–2121.

Wakefield, M., Giovino, G., 2003. Teen penalties for tobacco possession, use, and purchase: evidence and issues. Tob. Control 12 (Suppl. 1), i6–i13.

Wang, E., Clymer, J., Davis-Hayes, C., Buttenheim, A., 2014. Nonmedical exemptions from school immunization requirements: a systematic review. Am. J. Public Health 104, e62–e84.

Warburton, D., Charlesworth, S., Ivey, A., et al., 2010. A systematic review of the evidence for Canada's Physical Activity Guidelines for Adults. Int. J. Behav. Nutr. Phys. Act. 7, 39. http://www.ijbnpa.org/content/7/1/39.

Warburton, D.E., Gledhill, N., Jamnik, V.K., 2011. Evidence-based risk assessment and recommendations for physical activity clearance: consensus document 2011. Appl. Physiol. Nutr. Metab. 36 (Suppl. 1), S266–S298.

Webster, J.L., Dunford, E.K., Hawkes, C., Neal, B.C., 2011. Salt reduction initiatives around the world. J. Hypertens. 29, 1043–1050.

WHO, 1990. Diet, nutrition, and the prevention of chronic diseases: Report of a WHO Study Group (Vol. 797). Geneva: World Health Organization. Retrieved on 16 April 2014 from, http://

whqlibdoc.who.int/trs/WHO_TRS_797_(part1).pdf?ua=1 and http://whqlibdoc.who.int/trs/ WHO_TRS_797_(part2).pdf.

WHO, 2005. WHO Framework Convention on Tobacco Control. World Health Organization, Geneva. Retrieved 11 April 2014 from, http://whqlibdoc.who.int/publications/2003/ 9241591013.pdf?ua=1.

WHO, 2008. Global Burden of Disease 2004 Update. World Health Organisation, Geneva. Retrieved on 24 April 2014 from, http://www.who.int/healthinfo/global_burden_disease/ 2004_report_update/en/.

WHO, 2009. Global Health Risks: Mortality and Burden of Disease Attributable to Selected Major Risks. World Health Organization, Geneva. Retrieved 6 September 2012 from, http://apps.who. int/iris/bitstream/10665/44203/1/9789241563871_eng.pdf?ua=1.

WHO, 2010a. WHO Global Strategy to Reduce the Harmful Use of Alcohol. World Health Organization, Geneva. Retrieved 12 April 2014 from, http://www.who.int/substance_abuse/ msbalcstragegy.pdf.

WHO, 2010b. Marketing of food and non-alcoholic beverages to children. Sixty-third World Health Assembly (WHA63.14). World Health Organization, Geneva. Retrieved 16 April 2014 from, http://apps.who.int/gb/ebwha/pdf_files/WHA63/A63_R14-en.pdf.

WHO, 2010c. Set of Recommendations on the Marketing of Foods and Non-alcoholic Beverages to Children. World Health Organization, Geneva. Retrieved 16 April 2014 from, http://whqlibdoc. who.int/publications/2010/9789241500210_eng.pdf?ua=1.

WHO, 2011. Global Status Report on Alcohol and Health. World Health Organization, Geneva. Retrieved 13 April 2014 from, http://apps.who.int/iris/bitstream/10665/44499/1/9789241564151_eng.pdf.

WHO, 2012a. Prevention and control of noncommunicable diseases. Sixty-fifth World Health. World Health Organization, Geneva. Retrieved 25 May 2013 from, http://apps.who.int/gb/ ebwha/pdf_files/WHA65/A65_6-en.pdf.

WHO, 2012b. A Framework for Implementing the Set of Recommendations on the Marketing of Foods and Non-alcoholic Beverages to Children. World Health Organization, Geneva. Retrieved 16 April 2014 from, http://www.who.int/dietphysicalactivity/MarketingFramework2012.pdf.

WHO, 2012c. Maternal, infant and young child nutrition. Sixty-third World Health Assembly (WHA65.6). World Health Organization, Geneva. Retrieved 20 April 2014 from, http://apps. who.int/gb/ebwha/pdf_files/WHA65/A65_R6-en.pdf.

WHO, 2012b. A comprehensive global monitoring framework, including indicators, and a set of voluntary global targets for the prevention and control of noncommunicabale diseases. Revised WHO discussion paper (Version dated 25 July 2012). Geneva: World Health Organization. Retrieved 24 November 2013 from, http://www.who.int/nmh/events/2012/discussion_paper3. pdf.

WHO, 2013a. WHO Model List of Essential Medicines, eighteenth ed. World Health Organization, Geneva. Retrieved 13 February 2015 from, http://apps.who.int/iris/bitstream/10665/93142/1/ EML_18_eng.pdf?ua=1.

WHO, 2013b. WHO Model List of Essential Medicines for Children, fourth ed. World Health Organization, Geneva. Retrieved 24 April 2014 from, http://apps.who.int/iris/bitstream/10665/ 93143/1/EMLc_4_eng.pdf?ua=1.

WHO, 2013c. Global Action Plan for the Prevention and Control of Noncommunicable Diseases 2013–2020. World Health Organization, Geneva. Retrieved 7 April 2014 from, http://apps. who.int/iris/bitstream/10665/94384/1/9789241506236_eng.pdf?ua=1.

WHO, 2013d. Mental Health Action Plan 2013–2020. World Health Organization, Geneva. Retrieved 24 April 2014 from, http://apps.who.int/iris/bitstream/10665/89966/1/ 9789241506021_eng.pdf?ua=1.

WHO, 2014a. Global Health Estimates 2014 Summary Tables: Deaths by Cause, Age and Sex, by World Bank Region, 2000–2012. World Health Organization, Geneva. Retrieved 29 August 2014 from, http://www.who.int/healthinfo/global_burden_disease/en/.

WHO, 2014b. Global Status Report on Alcohol and Health 2014. World Health Organization, Geneva. Retrieved 9 February 2015 from, http://apps.who.int/iris/bitstream/10665/112736/1/9789240692763_eng.pdf.

Wilson, K., Gibson, N., Willan, A., Cook, D., 2000. Effect of smoking cessation on mortality after myocardial infarction: meta-analysis of cohort studies. Arch. Intern. Med. 160, 939–944.

Wilson, D.E., Van Vlack, T., Schievink, B.P., et al., 2014. Lifestyle factors in hypertension drug research: systematic analysis of articles in a leading Cochrane report. Int. J. Hypertens. 1–10. http://dx.doi.org/10.1155/2014/835716.

Yach, D., 2014. The origins, development, effects, and future of the WHO Framework Convention on Tobacco Control: a personal perspective. Lancet 383, 1771–1779. http://dx.doi.org/10.1016/S0140-6736(13)62155-8.

Yamin, C.K., Bitton, A., Bates, D.W., 2010. E-cigarettes: a rapidly growing Internet phenomenon. Ann. Intern. Med. 153, 607–609.

Zaridze, D., Lewington, S., Boroda, A., et al., 2014. Alcohol and mortality in Russia: prospective observational study of 151000 adults. Lancet 383, 1465–1473.

Zatoński, W.A., Sulkowska, U., Mańczuk, M., et al., 2010. Liver cirrhosis mortality in Europe, with special attention to Central and Eastern Europe. Eur. Addict. Res. 16, 193–201.

Zeltner, T., Kessler, D.A., Martiny, A., Randera, F., 2000. Tobacco Company Strategies to Undermine Tobacco Control Activities at the World Health Organization. World Health Organization, Geneva. Retrieved 23 April 2002 from, http://www.who.int/tobacco/en/who_inquiry.pdf.

Zhang, K., Wang, X., 2013. Maternal smoking and increased risk of sudden infant death syndrome: a meta-analysis. Legal Med. 15, 115–121.

Zimmermann, C., Swami, N., Krzyzanowska, M., et al., 2014. Early palliative care for patients with advanced cancer: a cluster-randomised controlled trial. Lancet 383, 1721–1730.

Chapter 15

# Mental Health

## Contents

The health of populations is only partially reflected in physical health. Although long neglected, recognition is increasing that mental health warrants priority in health care equal to that of physical health. The adage *no health without mental health*, which recently has come to be adopted by many local, national, and international bodies responsible for health policy, expresses the ambition to include mental health in the mainstream of health care (e.g., Mental Health Strategic Partnership, 2014; UK Department of Health, 2011; WHO, 2013). There is growing acceptance of the need to establish parity of esteem between services for people with mental and physical health problems and to integrate services for both.

It will be recalled from Chapter 1 that from its inception the WHO included *mental and social well-being* as integral elements of the definition of health

The Health of Populations. http://dx.doi.org/10.1016/B978-0-12-802812-4.00015-1
**429**

(WHO, 1948). Accordingly, all of the main issues and themes concerning problems of physical health discussed in preceding chapters are equally applicable to mental health, including the need for a population-wide life course approach, the benefits of risk factor reduction, the overall greater benefit of prevention over cure, the pronounced risk of harm from biomedical intervention, the role of social determinants, and the need for multisectoral involvement to optimize the mental health of populations. In preceding chapters, the term *noncommunicable disease* was used to refer exclusively to physical conditions, especially cardiovascular diseases, cancers, chronic respiratory diseases, and diabetes. *Mental disorders,*[1] however, represent an additional category of noncommunicable "disease."

Population surveys of mental health have been conducted ever since World War II, but it was not until the development in the 1980s of standardized psychiatric diagnostic interviewing that extensive international comparison became possible (Kessler et al, 2004). By the 1990s, it was apparent that mental disorders are not merely prevalent worldwide, but that their prevalence is among the highest of all noncommunicable conditions. The reported projected lifetime risk of mental disorders varies considerably between countries from about 1-in-5 to more than 1-in-2 of survey respondents (Kessler et al, 2007). Overall, rates tend to be lower in North and South East Asia and higher in English-speaking countries. With respect to depression, anxiety, and substance use disorders, it is usual for about 20% of survey respondents to satisfy relevant criteria for having experienced mental disorder during the preceding 12 months (Steel et al., 2014). Global prevalence of anxiety and mood disorders tends to be higher among women, whereas men tend to show higher prevalence of substance use disorders.

Mental disorders tend to have relatively early onset, usually in childhood or adolescence, are marked by chronicity, and incidence is associated with a number of socioeconomic indices (Whiteford et al., 2013). In particular, higher risk is associated with low income, low education, being unemployed, or not married. However, mental disorders go largely untreated, with between one- to two-thirds of people with serious mental disorder receiving no treatment (Bijl et al., 2003). Early onset, chronicity, and lack of treatment all contribute to mental disorders being a major cause of disability. Measured in terms of disability adjusted life years (DALYs), discussed in Chapter 3, mental disorders account for nearly 30% of disability worldwide from noncommunicable diseases and 13% from any cause (WHO, 2008). Among specific mental disorders, depression is cumulatively the most disabling, accounting for about one-third of total

---

1. The term *mental disorders* is used here not because it is the most suitable, but because of its wide currency. Among other shortcomings *disorder* in this context too readily and erroneously implies physical pathology. In fact, mental health problems are rarely attributable to identifiable physical pathology.

disability attributable to mental health problems, followed by alcohol use disorders and *schizophrenia*. Accounting for 4.3% of disability from all causes, depression is ahead of the leading physical noncommunicable diseases of ischemic heart disease (4.1%) and stroke (3.1%) (WHO, 2008).

Disability due to mental health is illustrative of the importance of behavioral outcomes in health, especially those discussed in Chapter 2 concerning the social roles of individuals and personal quality of life. The high value ascribed by lay people to behavioral function is evident from studies that have quantified the burden of disease using summary measures of health (e.g., Salomon et al., 2012). The sample of health conditions listed in Table 3.2 shows that mental health problems such as major depressive disorder, severe anxiety disorders, and schizophrenia are assigned morbidity weights comparable to or greater than serious physical diseases such as advanced-stage cancer, stroke, and end-stage renal disease.

## 15.1    MENTAL DISORDER: THE TRIPLE YOKE OF BIG PHARMA, PSYCHIATRY, AND THE BIOMEDICAL MODEL

As discussed in Part 1, the biomedical model of health is an all-pervasive influence in health care that seeks with limited success to restore to health the late-stage physical pathologies that define common diseases. More dogma and ideology than model, the extremist nature and intrinsic hazards of the biomedical model are most obvious in the context of mental health. In a seminal critique, American psychiatrist, Engel (1977), reflected on how the then (and still) dominant biomedical model encourages the view that both physical and mental health problems represent "disease...fully accounted for by deviations from the norm of measurable biological (somatic) variables" (p. 130). Over recent decades, the model of mental health as a biological phenomenon has become ever more firmly entrenched within psychiatry, evidenced by commonplace references to mental health problems as "brain disorders," "disorders of brain circuits," "neurobiological disorders," and "chemical imbalances in the brain" (Deacon, 2013). Acceptance of this dogma has spread beyond the medical community to become received wisdom in the general community.

The situation regarding intervention for mental health problems is paradoxical due to the simultaneous and widespread occurrence of both under- and overtreatment (Deacon, 2013). Despite large numbers of people worldwide who have mental disorders but receive little or no attention for their problems (*unmet need*), many others, especially in developed countries, are overtreated with prescribed psychiatric drugs (*met unneed*). The latter circumstance is attributed largely to the influence of the pharmaceutical industry, which has been successful in expanding the market for drugs by encouraging the creation of more and wider diagnostic categories of mental disorders. Poor servicing of mental health needs is also substantially attributable to limitations of the biomedical model of mental disorders, which focuses on biological correlates more aptly regarded as risk markers than risk factors (i.e., variables that are correlated with, but are not causally connected to, mental health). Preoccupation with biological risk markers has contributed to neglect of the behavioral and social causes and consequences of mental health problems (Moncrieff, 2007).

### 15.1.1   Stigma and the Biomedical Model of Mental Disorder

Due to decades of "awareness raising" by governments, patient advocacy groups, and the public media, there is now widespread acceptance of the proposition that mental disorders are diseases "like any other" and are responsive to biomedical intervention (e.g., Pescosolido et al., 2010; Schomerus et al., 2012). The extent of acceptance of those views is indicated by formal educational programs such as exist in the United States where children are explicitly taught as part of school-based mental health education that common mental health problems such as depression are "illnesses of the brain" (Watson et al., 2004). A commendable objective of such educational efforts is to reduce the stigma of mental disorders and to create greater social acceptance of persons with mental health problems. Such efforts, however, are not only largely ineffectual, but appear to have been counterproductive. Evidence indicates that the biomedical model of mental disorders does not merely fail to reduce stigma, but is stigmatizing.

A laboratory study of college students examined attitudes and behavior toward an individual (a confederate in the experiment), who was believed by different groups of participants to be either another typical student (control condition) or a student with a history of mental health problems (Mehta and Farina, 1997). Mental health history was in turn characterized for two separate groups of participants as either biological illness or as psychosocial problems. Not only did the biological view not improve attitudes, it provoked harsher behavior than both the psychosocial and control conditions. Behavior in this instance was in the form of penalties which participants, who had been assigned the role of "teacher," inflicted on the other "student" (always the confederate), who had been assigned the role of "learner" in a "learning task."

Another study examined results from a biennial survey of a population sample in the United States concerning public beliefs about major depression and schizophrenia (Pescosolido et al., 2010). Over the decade from 1996 to 2006, public endorsement of the disease model of mental disorders increased, but social stigma did not improve. Consistent with this and other studies, a recent extensive review of survey results from 10 geographically dispersed high-income countries observed an international trend in public acceptance of the biological model of mental disorders. However, as in other studies, the biological model was associated with either no change or change for the worse regarding social acceptance of people with mental health problems (Schomerus et al., 2012).

Notably, the tendency for social stigma to be exacerbated rather than ameliorated by biological explanations of mental disorders is not limited to the lay public, but has also been found to be prevalent in the professional mental health community. In a series of studies, psychiatrists, psychologists, and social workers serving as mental health professionals read vignettes describing patients with a variety of mental disorders (schizophrenia, major depression, obsessive-compulsive disorder, and social phobia) in which the patients' case histories were systematically manipulated to vary the degree of emphasis placed on biological versus psychosocial explanations (Lebowitz and Ahn, 2014). Consistent with similar speculation about lay attitudes, the researchers reasoned that clinicians' empathy for patients could *increase* when symptoms are construed biologically because biological causation implies reduced personal responsibility or blame. Conversely, they reasoned that clinicians' empathy for patients could *decrease* when symptoms are construed biologically, because traits and behavior that are biologically determined might foster perceptions of categorical dissimilarity from "normal," leading to dehumanization and fears of unpredictability. Using validated measurements of empathy, the studies indicated that biological explanations had a consistently negative effect on clinicians' feelings of empathy for patients who otherwise presented with the same symptomatology but which differed only by way of biological or psychosocial attribution.

## 15.1.2  Big Pharma, Psychiatry, and the Medical Model

The discovery in the early twentieth century that *general paresis*, a late-stage consequence of syphilis (involving dementia and other serious neurological problems), is caused by bacterial infection of the brain which later was found could be treated successfully with penicillin, fostered the view that biological causes and cures might be discovered for other mental disorders. Adoption of electroconvulsive therapy, lobotomy, and insulin coma therapy in the 1930s and 1940s was followed in the 1950s by the chance discovery of compounds that reduced overt symptoms of *psychosis*, depression, mania, anxiety, and hyperactivity. Shortly thereafter, accompanied by rapid escalation of entanglement between psychiatry and the pharmaceutical industry, mental disorders came increasing to be characterized as "brain diseases" caused by chemical imbalances purportedly treatable using prescription drugs.

Based on an extensive analysis of the validity and utility of the biomedical model of mental disorder, American psychologist, Deacon (2013) concluded that despite the atmosphere of "enthusiastic anticipation" that has surrounded biological psychiatry for decades, nothing approaching the predicted transformative advances has occurred. Upon examination, Deacon found that many of the central tenets of modern psychiatry, derived from the biomedical model of mental disorder, are without scientific support. The following summarizes Deacon's conclusions regarding specific biomedical claims and associated evidence:

| Biomedical claim | Evidence |
| --- | --- |
| • Mental disorders are brain diseases caused by neurotransmitter dysregulation, genetic anomalies, and defects in brain structure and function | • No biological cause of, or even reliable biomarker for, any major mental disorder has been identified |
| • Psychotropic medications work by correcting the neurotransmitter imbalances that cause mental disorders | • There is no credible evidence for the "chemical-imbalance" theory of mental disorders, and evidence is lacking that medicines work by "correcting" imbalances |
| • Advances in neuroscience have ushered in an era of safer and more effective pharmacological treatments | • Modern psychiatric drugs are generally no more safe or effective than those discovered by accident a half-century ago |
| • Biological psychiatry has made great progress in reducing the societal burden of mental disorder | • Mental disorders have become more chronic and severe, and the number of individuals disabled by their symptoms has risen steadily in recent decades |
| • Educating the public that mental disorders are biologically based medical diseases reduces stigma | • Despite widespread public acceptance that mental disorders are "diseases" with biological causes that are responsive to biological intervention, stigma has not improved and may have worsened |
| • Increased investment in neuroscience research will lead to diagnostic biological tests and curative pharmacological treatments | • The pharmaceutical industry has dramatically scaled back efforts to develop new psychiatric drugs due to the lack of promising molecular targets for mental disorders and the frequent failure of new compounds to demonstrate superiority over placebo |

Deacon (2013) describes how drugs that had formerly been known as *major tranquilizers* because of their generalized sedating effects came in time to be construed as *antipsychotic*, reputed to be capable of remediating specific disease processes. Similarly, *minor tranquilizers* were recast as agents having specific *antianxiety* actions. As part of this process of rebranding (essentially an exercise in marketing), negative actions of drugs such as generalized blunting of emotion (self-evidently implied in the original labeling of the same or similar drugs as "tranquilizers") came to be minimized as "side effects" and regarded as unfortunate yet acceptable. Despite popular acceptance of claims that common psychiatric drugs have precisely targeted effects, the fact is that most possess broad and poorly understood actions. Nevertheless, pharmacological intervention for mental disorders served the interests of psychiatry as a profession, helping to enhance its status by bolstering its longstanding aspiration for scientific legitimacy.

As a branch of medicine, psychiatry embraced the biomedical model of physical disease and extended it to characterize mental disorders as brain diseases reputedly caused by dysfunction in cerebral structures, faulty regulation of neurotransmitters, or other alleged specific neurological pathologies. The use of pharmaceuticals to tranquillize patients with mental disorders was a boon to the custodial endeavors of psychiatric practice, rendering patients docile and manageable (Allison and Moncrieff, 2014), but it produced a dismal record of therapeutic outcomes. Blunting of emotions is the principal effect of many psychiatric drugs, which when taken long-term exacerbate rather than cure serious mental disorders (Harrow et al., 2014). There is great need for reassessment of current approaches to the use of pharmaceuticals in the management of mental health problems. Rather than being the mainstay of intervention for mental disorders, pharmacotherapy should be used, if at all, sparingly and adjunctively to psychosocial interventions.

### 15.1.3  Schizophrenia

Rendering patients more manageable with the aid of psychiatric drugs is justifiable in exceptional circumstances such as when persons who by way of agitation and violent action are of potential harm to themselves or to others. Chemical restraint (also referred to as *chemical straitjacketing*), however, should not be mistaken for therapy. If antipsychotic drugs are effective therapeutically, as distinct from merely inducing docility, in what sense are they effective? More specifically, do they cure the underlying "disease," as is often implied and sometimes explicitly claimed? Evidence from multiple sources converges to indicate that these much-lauded "magic bullets" of biomedical innovation have little or no curative potential. Of particular interest are patients diagnosed with *schizophrenia* characterized by symptoms of *psychosis*, including *thought disorder*, *delusions*, and *hallucinations*.

Notwithstanding common beliefs about the effectiveness of antipsychotic medication, studies indicate that long-term outcomes for schizophrenia have improved little, if at all, from the preantipsychotic era (Deacon, 2013). In particular, the situation was *not* improved following introduction of the so-called *second-generation* antipsychotic medications, a classification applied to a heterogeneous variety of drugs and invented as a marketing strategy (Tyrer and Kendall, 2009). In reality, the newer drugs are consistently distinguishable from *first-generation* offerings of the 1950s, 1960s, and 1970s in no major respect other than higher cost (Leucht et al., 2009). Today, as in the past, approximately 1 in 7 individuals with schizophrenia recover, with experts in the field concluding that there is "no evidence to suggest that we are 'getting better' at getting our patients better" (Jääskeläinen et al., 2013, p. 1305). Additionally, despite becoming "better at detecting and treating the core symptoms of schizophrenia," overall health outcomes, including mortality, for patients with schizophrenia have worsened (Saha et al., 2007, p. 1129).

Figure 15.1 summarizes the findings for 64 patients with schizophrenia reassessed on six occasions over a period of 20 years following initial (baseline) assessment during the acute phase of hospitalization when the patients were

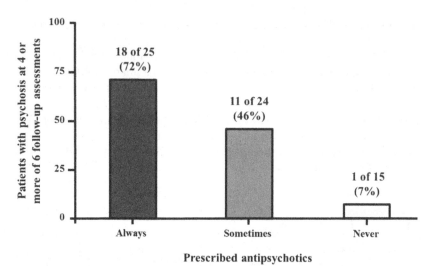

**FIGURE 15.1**   Longitudinal assessment of psychosis in schizophrenia patients ($N = 64$) prescribed antipsychotic medications after the acute phase always (15 patients), sometimes (24 patients), or never (15 patients) over a period of 20 years. *(Adapted from Harrow et al. (2014)).*

at an average age of 23 years (Harrow et al., 2014). The figure shows a dose-response relationship between amount of antipsychotic medication and the number of patients whose assessment indicated that they were experiencing psychotic activity on four or more of the six follow-up occasions. Notably, the association was in the *reverse* direction to that expected. Whereas only 1 of 15 patients (7%) who never took antipsychotic medication after the acute phase was persistently psychotic, the majority of patients (18 of 25; 72%) who always took medication were persistently psychotic and those who sometimes took medication (11 of 24; 46%) occupied a middle position. On first appearances, the results could be explained on the basis that patients whose disorder was more severe to begin with might have persisted for longer with their medication because of greater need than patients with less severe disorder. However, the results could not be explained on that basis. Rather, the authors concluded that long-term use of antipsychotics is not merely ineffective in eliminating or reducing psychosis in patients with schizophrenia, but may actually impede recovery for some.

    A recent Swedish study examined the relation between long-term use of antipsychotic drugs and violent crime (Fazel et al., 2014). In addition to schizophrenia, the study included other psychiatric diagnoses for which antipsychotics are prescribed, notably *bipolar disorder* and depression, and drugs other than antipsychotics, notably *mood stabilizers* (e.g., lithium for bipolar disorder) widely prescribed for patients with symptoms of psychosis. Rate of violent crime during a 4-year period among more than 80,000 patients who were prescribed medication was compared to the rate of crime in the same people when they were not prescribed medication. The overall rate of violent crime was found to have reduced by nearly two-thirds for patients with medication. Accordingly, it appears that medicating patients is a practice that can have important social benefits in the form of reduced rate of violent crime. At the same time, if the apparent crime-reducing benefit of medication is due to patient docility, questions of balance arise in relation to potential conflict between the rights of individual patients and the use (possible abuse) of psychiatry and pharmacology for purposes of social control rather than health benefit. At the very least, it should be acknowledged in such instances that the main purpose of chemical restraint is the common good rather than the usual purpose of health care which is to benefit those who are given the intervention.

It has long been known that psychiatric patients have a high rate of mortality compared to the general population (Waddington et al., 1998) and recent studies show that the gap remains high and may be widening (e.g., Reininghaus et al., 2014). That high risk of mortality persists despite increased use of antipsychotic medications is proof positive of intervention failure, especially given the suggestion that worsening outcomes may be due in part to the popularity of reputedly superior second-generation antipsychotics (Saha et al., 2007). Further evidence of therapeutic failure can be found in the common clinical practice of simultaneously prescribing multiple medications (polypharmacy) for patients in the hope of finding a combination that works (Ballon and Stroup, 2013; Gallego et al., 2012; Suokas et al., 2013; Weinmann et al., 2009). The practice extends even to the point of *off-label* combinations of drugs being recommended in the scientific literature, despite ethical and legal implications of such practice for clinicians and increased risk of side effects for patients (Baldwin and Kosky, 2007; Haw and Stubbs, 2005). Polypharmacy has not been discouraged by lack of supportive evidence (Ballon and Stroup, 2013; Weinmann et al., 2009) or by extensive evidence of drug combinations producing increased negative side effects and poorer outcomes (Gallego et al., 2012; Suokas et al., 2013).

### 15.1.4   Bipolar Disorder

Despite antipsychotic medications having a poor record when used to treat disturbances for which they are claimed to be specifically applicable, the same drugs have become widely marketed and widely used to treat conditions for which they were not intended. A striking example is the use of antipsychotics to treat bipolar disorder. The disorder, now relatively common, was formerly known as *manic depression*, a rare condition in which sufferers experience periods of deep depression punctuated by prolonged periods of pronounced hyperarousal and overactivity (*mania*). In the past, it was typical for the diagnosis to be made only after hyperarousal became so extreme and out of control as to lead to admission to a psychiatric unit, often compulsorily. Transition from the rare condition of manic depression to the common diagnosis of bipolar disorder has been premised on new "understanding" that strong emotions are manifestations of chemical imbalances signifying "abnormal" brain function.

The transition coincided with a period of increasing entanglement between psychiatry and the pharmaceutical industry. Creation of the modern concept of bipolar disorder (also known as *bipolar affective disorder*), recently extended to include *pediatric bipolar disorder*, has enabled drugs previously reserved for use with serious mental disorders to migrate into "the more profitable realm of everyday emotional problems" (Moncrieff, 2014). Thus, despite evidence supporting the use of psychosocial interventions, including psychoeducation, behavior therapy, and cognitive-behavioral therapy (Geddes and Miklowitz, 2013; Miklowitz and Scott, 2009), chemical modification of brain function has become the predominant strategy for managing psychological distress. Encouraged in part by clever direct-to-consumer marketing, physicians as well as patients increasingly make diagnoses such as bipolar disorder to "explain" problems arising from all manner of life stresses. Once disorder has been duly diagnosed, physicians and patients alike have come to believe that psychiatric medications offer realistic hope of relief.

This is precisely the direction taken by recent revisions to the *Diagnostic and Statistical Manual (DSM)*, an internationally respected (historically, at least) and widely used set of guidelines for diagnosing mental disorders, published by the American Psychiatric Association (2013). Recent revisions, the first in 20 years, have been criticized for contributing to diagnostic proliferation of mental disorders by expanding existing diagnoses and creating new ones, with the effect of medicalizing common problems of living (Pilgrim, 2014). New revisions have the effect of equating common misfortunes and ordeals such as grief and shyness with mental disorder, and of pathologizing common distresses and difficulties of childhood and adolescents. Challenging but essentially normal tribulations of life, when labeled *disorder*, become ripe for intervention using an ever expanding array of drugs of dubious applicability and effect.

Repeating what by now will be a familiar lamentation of this book, extensive revelations have been published concerning the many financial and other forms of pharmaceutical industry entanglement that have long existed among American Psychiatric Association task force members responsible for compiling the diagnostic and statistical manual, including earlier revisions and the current *DSM-5* (Cosgrove and Krimsky, 2012; Pilecki et al., 2011). An examination of financial ties covering a 15-year period to 2004 revealed that connections tended to be particularly strong in diagnostic areas where drugs are the first line of treatment for mental disorders (Cosgrove et al., 2006). Specifically, it was found that 100% of the members of the panels on *mood disorders* and *schizophrenia and other psychotic disorders* had financial ties, such as research funding and consultancies, to drug companies.

Incremental revisions to the guidelines have strongly aligned psychiatric diagnosis and practice with industry commercial interests. For example, shortly before expiry of the patent on its blockbuster antidepressant drug *Prozac*,

Eli Lilly rebranded and relicensed the drug (as *Sarafem*) for the treatment of *premenstrual dysphoric disorder* (*PMDD*). Inclusion of PMDD in *DSM* reified the disorder's existence as a "distinct clinical entity" encouraging treatment using antidepressants. The majority of the panel that decided PMDD's inclusion in *DSM* had financial ties to Eli Lilly (Cosgrove and Wheeler, 2013). Popularization of the chemical-imbalance theory of psychological disorder has been part of the general process of winning wider acceptance of the use of psychotropic drugs to "solve" problems of living. Over recent decades, the chemical-imbalance theory has been instrumental in disease mongering in the field of mental health and is substantially responsible for the gargantuan expansion of the market for prescribed psychotropic drugs.

Viewed charitably, the chemical-imbalance theory may be seen as benign fiction used to coax prospective patients into accepting drug taking as therapy, which without such coaxing may not appear self-evidently to be the best approach for solving personal problems. Viewed more frankly, the chemical-imbalance theory could be seen as fraud perpetrated for commercial gain against individuals rendered vulnerable by psychological distress. Worse, the fraud, if it is that, has been aided and abetted by the profession of psychiatry. It is arguable that industry entanglement over several decades has been the major factor in shaping the very nature of modern psychiatry, encouraging that profession into comprehensive adoption of the biomedical disease model of mental disorder.

Of various possible approaches for dealing with psychological distress and problems of living, including behavioral and psychosocial strategies known to be effective (Mohr et al., 2013), the medical model with its emphasis on pharmaceutical intervention for allegedly distinct categories of disorder seems to be the worst possible option, at least, from the perspective of patient well-being. A prominent feature that distinguishes the biomedical disease model of mental disorder from more suitable conceptualizations and forms of intervention is that it happens to be the option that is most amenable to commercial exploitation. With indispensable support from the profession of psychiatry, Big Pharma has been successful in creating for its own gain a form of psychiatric practice that is maximally attuned to private commercial profit and least congruent with actual mental health needs (Cartoon 15.1).

## 15.1.5 Depression

The relatively recent rise in diagnostic popularity of bipolar disorder parallels an earlier period of *disease mongering* of depression. Both diagnoses are characterized by intense marketing of ineffective and dangerous pharmaceuticals to ever larger numbers of people. As with bipolar disorder, expanded flexibility in the definition of depression has allowed for wide scope in its diagnosis and wide choice of pharmaceuticals. It has been found that physicians are especially willing to prescribe antidepressants if requested by patients, even if patients' symptoms

CARTOON 15.1    Caricature of the harmful (to patients) symbiotic entanglement of professional psychiatry and the pharmaceutical industry. *(Source: Accessed 7 October 2014 from http://www. naturalnews.com/021553_psychiatry_modern.html/ and reprinted with permission.)*

are not indicative of depression (Kravitz et al., 2005). However, marketing claims of efficacy for antidepressants, particularly selective serotonin reuptake inhibitors (*SSRIs*; e.g., *Prozac*), when used as first-line treatment for moderate and severe depression are not supported by relevant evidence. Meta-analyses show that the drugs have no clinically meaningful advantage over placebo in the treatment of adult depression and an even worse balance of outcomes when used with children (Moncrieff and Kirsch, 2005). Moreover, findings suggest that use of SSRIs may contribute to increased suicidality (Fergusson et al., 2005). Additionally, a recent critical review of nonsteroidal anti-inflammatory drugs (*NSAIDs*) for depression, when used either alone or adjunctively, found that they too possess "negligible" efficacy (Eyre et al., 2015).

Widespread publication bias has been revealed as a likely contributing factor to the common belief among clinicians and the public that antidepressants are effective. Turner et al. (2008) obtained reviews from the Food and Drug Administration (*FDA*) for 12 antidepressant drugs that had been assessed and approved for use in the United States. Drug companies are required by law to register *all* trials pertaining to drugs submitted for approval by the FDA. Accordingly, the researchers conducted a systematic literature search to identify matching publications in the scientific literature to compare with the FDA-registered trials. Of 74 registered studies, nearly one-third were not published in the usual publicly accessible scientific outlets. Notably, whether or not studies were published was found to be related to study outcomes. All except one of the studies viewed by the FDA as having positive results were published, whereas most of the studies viewed by the FDA as having negative or questionable results were not published. Of the negative studies that were published, almost all were misleadingly presented as having positive outcomes. Consequently, whereas only half of all trials that were conducted produced positive outcomes, the published literature available to scientists, clinicians, and the public gives the impression that almost all of the trials show antidepressants to be effective.

Similarly, after reviewing the relevant literature, Antonuccio and Healy (2012) identified a marked disjunction between the use of the term *antidepressant* and the actual effects of drugs that carry that label. They argued that to justify the label *antidepressant* a drug should be superior to placebo, should offer a risk/benefit balance that exceeds that of alternative treatments, should not increase suicidality, should not increase anxiety and agitation, should not interfere with sexual functioning, and should not increase the chronicity of depression. The authors concluded that the relevant evidence shows that so-called antidepressants "fall short" on all dimensions. They reasoned that although the label makes sense from a marketing perspective, it is misleading clinically and scientifically. A further point is that the term *antidepressant* risks breaching patients' informed consent. Few patients beginning antidepressant therapy are likely to be genuinely informed about the probable limited benefit and substantial risks involved with the drugs they are offered, and most are also likely to be uninformed about safer and more effective psychosocial interventions (e.g., Honyashiki et al., 2014).

## 15.1.6 Anxiety Disorders

Although depressive disorders are responsible for the largest cumulative burden of disability attributable to mental problems (WHO, 2008), anxiety disorders are the most prevalent of all mental health problems (Bystritsky et al., 2013). Benzodiazepines (e.g., *Valium*), discovered accidentally in the 1950s and initially claimed to be safe, effective, and nonaddictive, became widely used throughout the latter part of the twentieth century as first-line intervention for all anxiety disorders. In reality, benzodiazepines are associated with high incidence of negative side effects, including *physical dependence* and cognitive impairment. This has led to them being replaced as first-line treatments by antidepressants, especially SSRIs (Baldwin et al., 2005). Such is the parlous state of biomedical intervention for mental health problems that drugs (in this instance, SSRIs like *Prozac*) not initially intended for anxiety disorders are now preferred over other drugs (benzodiazepines) previously alleged to be specifically beneficial for anxiety. Moreover, the change in preferred drug was *not* predicated on the replacement having superior efficacy, but perceived less harmful side effects. Nevertheless, in contrast to benzodiazepines, which are fast-acting, antidepressants can take up to 4-6 weeks to begin relieving symptoms of anxiety, a disadvantage for patients but justification of sorts for manufacturers to recommend longer term use leading to consequent higher profits.

The fast-acting tranquilizing effect of benzodiazepines made them extremely popular with patients, but they are of little value therapeutically when used other than short-term. Despite guideline recommendations limiting use to 2-4 weeks, physicians worldwide commonly prescribed them for months and years (Ashton, 2005). Whatever efficacy benzodiazepines may possess in the short-term is lost in the long-term, with increased likelihood of serious negative emotional effects including, depression and (paradoxically) exacerbated feelings of anxiety. Cognitive harm includes negative effects on learning, memory, and attention. Although generally no longer considered to be first-line therapies for anxiety disorders, benzodiazepines are still commonly prescribed and misused. There is substantial leakage in supply of the drugs from general practitioner prescriptions onto the illicit market, where they are commonly used in combination with other illicit drugs (Ashton, 2005; McHugh et al., 2014).

While having a better side-effect profile than benzodiazepines, SSRIs are not free of side effects, including high incidence of patients reporting sleepiness, fatigue, nausea, sexual dysfunction, and weight gain (Anderson et al., 2012; Bet et al., 2013). Side-effect profiles for other antidepressant formulations used to treat anxiety disorders tend to be even worse. In addition, discontinuation of antidepressant drugs is associated with withdrawal symptoms, including rebound anxiety, dizziness, and nausea. Although such symptoms are indicative of withdrawal caused by physical dependence due to repeated drug use, a curious custom has become established in the relevant scientific literature wherein withdrawal effects associated with use of antidepressants are referred to as a *discontinuation syndrome* (Black et al., 2000; Nielsen et al., 2012). Granted, *addiction* is a term heavily ladened with emotive connotations and too equivocal in meaning to be warranted in this instance. However, there seems little reason scientifically or clinically not to refer to the negative effects that commonly accompany discontinuation of antidepressants by what they really are, namely, a *withdrawal syndrome* characteristic of drugs that produce physical dependence with prolonged use. Consequently, it is hard to avoid the conclusion that *discontinuation syndrome* is anything but a euphemism intended to evade the negative connotations of *physical dependence*, which would be implied if *withdrawal syndrome* were used.

The fact that antidepressants used in the treatment of anxiety can take several weeks to begin having desired effects is noteworthy, because psychosocial interventions (e.g., cognitive behavioral therapy) can show meaningful clinical results within the same or shorter time (e.g., Baker et al., 2009; Barlow, 2004). Indeed, behavioral and cognitive therapies have been found to be the most effective and cost-effective interventions for generalized anxiety disorder and panic disorder (Heuzenroeder et al., 2004). Pharmacological interventions alter biological processes via mechanisms that are poorly understood and have uncertain long-term consequences. By comparison, psychosocial interventions directly target the actual life problems anxious people experience (e.g., feelings of anxiety and associated avoidance behavior). In contrast, life problems are sometimes exacerbated by the drugs intended to bring relief, while also inflicting negative side effects, especially when used long term.

## 15.1.7    Attention-Deficit Hyperactivity Disorder

Attention-deficit hyperactivity disorder (ADHD) emerged in the twentieth century as the first formal mental disorder in children to be diagnosed and treated (Biederman and Faraone, 2005). Drug intervention studies were conducted in the 1930s and regulatory approval for use of stimulant drugs for children began in the 1960s. No single factor or consistent set of factors explains ADHD, and possible causes have long been controversial. In a recent review, Thapar et al. (2013) listed the possible causes as genes, pre- and postnatal risks (e.g., prematurity; low birth weight; maternal smoking, alcohol, and substance misuse; and maternal stress), dietary factors (e.g., nutritional deficiencies including zinc, magnesium, and polyunsaturated fatty acids and nutritional surpluses including sugar and artificial food colorings), psychosocial factors (e.g., family adversity and low income, severe early deprivation, family conflict, and parent-child hostility), and environmental toxins (e.g., organophosphate pesticides, polychlorinated biphenyls, and lead). Although ADHD shows heritability, genetic risk markers generally have small effect size and low predictive validity, adding little to diagnosis and prediction beyond that based on observation and family history (Thapar et al., 2013).

In light of the foregoing, it is not true, as is sometimes implied by some authoritative sources, that the causes of ADHD have been identified and are understood. For example, in answer to the question *What is attention deficit hyperactivity disorder?*, the ADHD homepage of the National Institute of Mental Health in the United States asserts that ADHD is a "brain disorder [wherein] the brain matures in a normal pattern but is delayed, on average, by about 3 years" (http://www.nimh.nih.gov/health/publications/attention-deficit-hyperactivity-disorder/index.shtml?rf=71264). The same site purports to identify the brain regions where alleged aberrant neurological delays are located. In reality, the *causes* that the site confidently describes are merely *associations*, the causal status of which is not known. At present, all that may confidently be said is that ADHD is probably caused by an interaction between inherited, psychosocial, and environmental factors. If that is stating the obvious, then that is indeed the current general state of the science of the causes of ADHD.

The worldwide pooled prevalence of mental disorders in children and adolescents has been found to be approximately 13% (Polanczyk et al., 2015). The most recent global estimate is that ADHD affects 3.4% of children and adolescents; being less prevalent than anxiety disorders, disruptive behavior disorders (e.g., temper tantrums, physical aggression, and stealing), which have high comorbidity with ADHD; and more prevalent than childhood depressive disorders (Polanczyk et al., 2015). Characterized by a pattern of persistent inattention, hyperactivity and behavioral impulsivity across settings, controversy has long surrounded the diagnosis and treatment of ADHD. Not necessarily easily distinguishable from high normal activity, controversy has been encouraged by questions about variability in estimates of prevalence reported in different studies; apparent higher prevalence in developed countries, especially the United States; and reports of increases in rates of the disorder over time. This variability has been attributed to differences in diagnostic method and perhaps readiness to diagnose, rather than to actual differences, either geographically or over time, in rates of "real" cases (Polanczyk et al., 2014).

Not unlike the situation pertaining to adult mental disorders, regarding unmet need and met unneed, there is evidence of underdiagnosis and undertreatment of child and adolescent mental disorders (Belfer, 2008; Morris et al., 2011), especially in developing countries, as well as evidence of overdiagnosis and overtreatment in some high-income countries (James et al., 2014; Visser et al., 2014). The motivation to treat ADHD in childhood stems in part from the fact that childhood ADHD is predictive of other mental health and life problems, including problematic academic and occupational performance, adolescent and adult substance use disorders, injury, and antisocial behavior (Biederman and Faraone, 2005). Interventions for ADHD fall into two broad categories, pharmacological and psychosocial. Pharmacological intervention, in turn, includes two broad classes of drugs, stimulants and "nonstimulants." Symptom relief from stimulant drugs (methylphenidate and amphetamine compounds) has led to their widespread use as first-line interventions (Watson et al., 2015). However, for the sizeable proportion (30%) of children who do not respond adequately to stimulant drugs, or who cannot tolerate the side effects of those drugs, the more recently developed nonstimulant drugs may be prescribed (Johnston and Park, 2015).

There is a tendency toward spontaneous reduction in symptom severity by adolescents and a large proportion of those diagnosed with ADHD who had been prescribed medication as children terminate their medication by the end of high school (Sibley et al., 2014). A probable contributing factor in this is the experience of side effects, the most common of which include headache, appetite loss, abdominal pain, and insomnia. Thus, despite broad agreement that pharmacological interventions for ADHD are generally effective in relieving symptoms, there are questions about the implications of long-term drug use (McCarthy et al., 2009; Molina et al., 2009; Watson et al., 2015). Moreover, the strength of the evidence base in support of pharmacological intervention as efficacious and safe is brought into question by extensive funding of the relevant clinical trials by the pharmaceutical industry and extensive associations between study authors and industry (e.g., Bushe and Savill, 2014; Faraone and Buitelaar, 2010; Newcorn et al., 2008).

Due to skilful marketing by the pharmaceutical industry, which routinely exaggerates the benefits and minimizes the risks of medication, rate of ADHD diagnosis and overtreatment has become particularly acute in the United States. The reported national diagnosis rate of 14% for ADHD in children and adolescents "exceeds all reasonable estimates of the true prevalence of the disorder" (Watson et al., 2013, p. 8). High rates of diagnosis and prescribing have persisted despite repeated communications about ADHD drug-related risks issued by the FDA, with particular reference to potentially serious cardiovascular events, sudden death, and suicidal ideation. Despite similar warnings from the American Heart Association and the American Academy of Pediatrics, a survey of a nationally representative sample found that there had been no discernible effect of the warnings on physician prescribing practices for ADHD (Kornfield et al., 2013).

Apart from potential harm, there are strong doubts about the extent to which pharmacological intervention can adequately address the range of behavioral and social problems associated with ADHD. There is a growing consensus that optimal therapeutic outcomes require psychosocial intervention, whether alone or in combination with medication (Fabiano et al., 2015; Johnston and Park, 2015; Sibley et al., 2014; Watson et al., 2015). The most effective psychosocial interventions for ADHD include teacher's use of classroom behavioral management methods, parent training to improve child management in out-of-school settings, and child and youth training in adaptive functional skills such as organizational and social skills. Overall, the evidence from meta-analyses indicates that the most appropriate first-line intervention for ADHD consists of psychosocial intervention with adjunctive use of low-dose medication (Johnston and Park, 2015; NICE, 2009; Sibley et al., 2014).

## 15.1.8  Pharmacotherapy as Adjunctive Intervention for Mental Disorders

The contemporary biomedical view that mental health problems are caused by disorders of the brain has been nourished by advances in biology, especially improved understanding of brain function founded on developments in neuroscience and new technologies such as brain imaging (e.g., functional magnetic resonance imaging; *fMRI*). Although there have been many genuine advances in these fields, scientific progress to date falls far short of justifying oft-repeated claims such as "depression is a chemical imbalance" or "schizophrenia is a brain disease" (Miller, 2010). The assumption (unfortunately pervasive) that events in the brain can be equated with psychological experience confuses two different logical categories. Brain events are *correlates* of behavioral and psychological events, but the former do not *explain* the latter. Mental health problems, which by definition involve behavioral and psychological events, require behavioral and psychological explanations and interventions. This is not to argue for wholesale abandonment of pharmacotherapy. There is, however, need for mental health research and practice to be reoriented toward greater emphasis on psychosocial interventions accompanied by occasional adjunctive use of pharmacotherapy.

There may be irony in the suggestion that pharmacotherapy should be considered a possible adjunct to psychosocial interventions, because it has long been customary in psychiatry to refer to these different therapies in precisely the opposite way (Geddes and Miklowitz, 2013 and Reinares et al., 2014 are illustrative recent examples). Nevertheless, a rational assessment of the evidence indicates that pharmacotherapy, if used at all, should be limited to the role of adjunct to behavioral and psychological interventions that target actual life challenges experienced by people who have mental health problems. Beliefs that psychological therapies are too labor intensive or slow acting to be considered mainstream interventions are ill-founded. Such interventions often are capable of producing clinically meaningful results within weeks of commencement. Moreover, as discussed below, cost-effectiveness is supported by the fact that psychosocial interventions are often able to be delivered effectively by trained nonspecialists under supervision. Even if that were not the case, there would be little justification for widespread use of drugs that are largely ineffectual or harmful. Fortunately, despite popular acceptance of exaggerated claims about mental health problems being explained by biology and continuing excessive use of psychiatric drugs, there has been considerable development of psychosocial interventions as first-line interventions for the promotion, prevention, and treatment of mental disorders.

## 15.2   PSYCHOSOCIAL MENTAL HEALTH PROMOTION, PREVENTION, AND INTERVENTION

As discussed in preceding chapters, the literature on health promotion and prevention customarily refers to interventions to reduce exposure to risk factors. In the field of mental health promotion there is a tradition of referring to protective factors as well as risk factors (CAMH, 2014). Mental health protective and risk factors both include an extensive range of characteristics of the individual, family circumstances, school or work context, life events, and diverse aspects of the community and culture. Examples of protective factors are positive self-regard, good social skills, family harmony, supportive social relationships, and freedom from discrimination. Conversely, mental health risk factors include labile temperament, harsh or inconsistent parenting, bullying, deviant peer group, and social exclusion.

There are several broad approaches for promoting mental health and for preventing mental health problems (Cuijpers, 2014). *Universal* prevention is population-wide primary prevention aimed at everyone within a defined population regardless of risk status. Examples include mass media campaigns and universal school-based programs. A second approach is *selective* prevention, which is primary prevention targeted at groups believed to be at increased risk of mental health problems (e.g., disadvantaged youth). A third approach is *indicated* prevention, which is secondary prevention aimed at people at high risk who have symptoms of disorder, but who do not meet full diagnostic criteria (e.g., family-based programs to address disruptive and aggressive child behavior).

Several decades of research have shown that promotion of mental health and prevention of mental health problems is most effective when focused on children, youth, and young adults (O'Connell et al., 2009). Emphasizing early mental health promotion and prevention is consistent with the pattern of early life course onset for many mental disorders. For example, one nationally representative face-to-face household survey of American adults found that more than half met the criteria for formal diagnosis of mental disorder sometime in their life (Kessler et al., 2005). For half of these, onset of the disorder was by age 14 years, and for three-quarters onset was by age 24 years (Kessler et al., 2005).

In a recent broad-ranging analysis, Sandler et al. (2014) provided an overview of 46 meta-analyses of more than 2000 separate trials of mental health promotion and prevention of mental health problems involving hundreds of thousands of children, adolescents, and young adults to age 26 years. The specific mental health problems that the meta-analyses examined were depression; anxiety; aggression, antisocial behavior, and violence; and alcohol and other substance use. Small-to-medium beneficial effects were reported overall, with anxiety being most responsive. The remaining problems benefitted less than anxiety, but to an approximate equivalent amount to one another. Medium beneficial effects were achieved in promoting healthy development using school-based, after-school, mentoring, and early childhood strategies, and interventions that targeted significant family disruptions, such as bereavement, divorce, and parental mental health problems.

Effects of psychosocial intervention generally appear to be well-maintained over time, with measurable benefits on many outcomes lasting one or more years (Sandler et al., 2014). Moreover, programs that targeted a specific problem outcome often reported benefits for other related outcomes, indicating that psychosocial approaches tend to have multiple or generalized preventive outcomes. Furthermore, the meta-analyses reviewed by Sandler et al. were consistent in showing that programs that emphasize skills building (e.g., provision of graduated skills-oriented homework) are more effective than those that are primarily didactic or educational in approach. Similarly, prevention of harmful alcohol and other substance use is more effective when skills are acquired in interpersonal communication, especially refusal skills, than when information and education about drugs is provided without skills training. For parenting programs, the learning of specific parenting skills, including the provision of feedback for parents practicing parenting skills with their own children, is more effective than educational programs that do not include skills acquisition and practice.

## 15.2.1   Going to Scale

Perceived need for psychosocial preventive interventions is likely to be inversely proportional to the effectiveness of interventions for manifest problems. That is, there would be less need for prevention if manifest problems could be treated effectively and cost-effectively. Therefore, just as exaggerated beliefs in the effectiveness of psychiatric drugs are likely to discourage use of psychosocial approaches as first-line interventions (Miller, 2010), the same exaggerated beliefs are probably also a barrier to greater policy commitment to interventions that emphasize promotion of mental health and prevention of mental health problems. However, given the limited effectiveness of drugs for major mental health disorders, the need for effective preventive interventions is not merely great, but is growing. The demonstrated efficacy of psychosocial preventive interventions provides strong justification for their wider application. Because most preventive interventions have been implemented and evaluated in high-income countries, a key challenge for mental health prevention is "going to scale" with existing evidence-based interventions to enable implementation worldwide (Catalano et al., 2012).

It has been argued that global health initiatives acquire momentum and priority when (1) political leaders publicly and privately express sustained concern for the issue, (2) organizations and political systems enact policies to address the problem, and (3) resources are allocated commensurate with the scale of the problem (Shiffman and Smith, 2007). Unfortunately, these conditions are currently largely unmet, especially in developing countries (Tomlinson and Lund, 2012). In particular, although mental health promotion and prevention *could* be implemented to good effect in school and community settings in developing countries (Barry et al., 2013), increased capacity and greater commitment to action are needed to achieve extensive and sustainable implementation (Hanlon et al., 2014). A general recommendation for developing countries, currently being supported by the WHO, is for the integration of mental health initiatives within existing mainstream health-care services (Collins et al., 2013; Kaaya et al., 2013; Ngo et al., 2013; Patel et al., 2013; Rahman et al., 2013).

### 15.2.2   WHO Framework for Mental Disorders

In 2013, the WHO launched the *Mental Health Action Plan 2013-2020*, designed to guide national initiatives in all countries with the aim of addressing worldwide unmet need for mental health promotion, prevention, and intervention. The action plan is intended for all countries for the express reason that there are current major shortfalls in mental health action worldwide. Although the shortfall is greatest in developing countries, where between 76% and 85% of people with severe mental disorders receive no treatment, a large shortfall of between 35% and 50% is also characteristic of high-income countries (WHO, 2013). Moreover, despite the fact that less than 2% of the health budget in most countries is allocated to mental health, about two-thirds of that already meager allocation goes to standalone mental hospitals. It is known, however, that such facilities are associated with poor health outcomes and frequent human rights violations (WHO, 2013). Therefore, even without increasing budgets, redirection of funding toward community-based services, including the integration of mental health into general health care, would create access to better and more cost-effective interventions for many more people than at present.

WHO initiatives for mental health have evolved against a background of increased awareness of the interdependence between physical health and mental health. Notably, some of the major risk factors for physical diseases are recognized as also being risk factors for mental disorders (e.g., physical inactivity predisposes to depression). Additionally, it is acknowledged that the simultaneous experience of physical and mental disorders is not merely a matter of comorbidity, but that considerable reciprocal causation also occurs between physical diseases and mental disorders. For example, depression predisposes to cardiovascular disease and cardiovascular disease increases the likelihood of depression. Consequently, strategies that decrease the incidence of physical health problems will inevitably also contribute to reduced incidence of mental disorders, and vice versa. In addition, both physical health and mental health are influenced by a wide range of shared psychosocial variables. For example, the social determinants that contribute to increased risk of harmful consumption of alcohol simultaneously contribute to a host of health problems, both physical (e.g., cardiovascular diseases) and psychosocial (e.g., depression and violence).

The WHO (2013) framework recognizes that physical and mental health are influenced by common social and economic determinants. Both benefit from higher per capita income; employment; education; material standard of living; and family cohesion. Both are also harmed by social exclusion; violations of human rights; and adverse life events including sexual violence, child abuse, and neglect. As such, physical and mental health are cross-cutting issues linking diverse personal and social processes. Inclusion of mental health in frameworks for action will help to achieve global priorities for development in such areas as poverty reduction, economic development, and protection for the most vulnerable in society (Eaton et al., 2014). Similarly, interventions that improve social and economic determinants benefit physical and mental health. Thus, the *Health in All Policies* framework (WHO, 2014) discussed in the preceding chapter, if applied generally would do much to enhance individual achievement, personal security, and social harmony, all of which underpin population mental health as well as physical health.

WHO recommendations urge early intervention with evidence-based psychosocial and other mostly nonpharmacological interventions to avoid medicalisation of mental health problems. Although these initiatives represent a landmark for global mental health, major ongoing effort will be necessary to achieve stated goals, especially in relation to population-wide implementation of preventive interventions and access to individualized psychosocial interventions when needed. Success in those endeavors depends on discouraging the use of medications as first-line interventions for mental health problems, and ensuring that when pharmacological intervention is indicated that it is time-limited and adjunctive to psychosocial intervention.

The WHO Mental Health Action Plan 2013-2020 is predicated on a *vision* of the world in which mental health is valued, those with mental health problems are supported in exercising the full range of human rights, and culturally appropriate care is available and accessible for those in need (WHO, 2013). The overall *goal* of the framework is to promote mental health and well-being; prevent mental disorders; provide care to foster recovery from mental health disorders;

promote human rights; and reduce mortality, morbidity, and disability associated with mental disorders. The action plan has the following *objectives*:

- to strengthen effective leadership and governance for mental health;
- to provide comprehensive, integrated, and responsive mental health and social care services in community-based settings;
- to implement strategies for promotion and prevention in mental health; and
- to strengthen information systems, evidence, and research for mental health.

The action plan builds upon but does not duplicate an earlier initiative, the *Mental Health Gap Action Program (mhGAP)*, which is ongoing. Taking account of the limited mental health budgets of most countries, mhGAP was launched in 2010 to counter the idea that improvements in mental health require sophisticated and expensive technologies and highly specialized staff. Because about three-quarters of people suffering mental health problems are in developing countries, mhGAP is particularly focused on strategies for expanding mental health services in low resource settings. A guide, the *mhGAP Intervention Guide for Mental, Neurological and Substance Use Disorders in Non-Specialized Health Settings (mhGAP-IG)* (http://www.ncbi.nlm.nih.gov/books/NBK138690/), has been developed as a technical tool for facilitating delivery of mental health services by nonspecialist health-care personnel.

mhGAP-IG aims to encourage access to psychosocial intervention and adjunctive medication for tens of millions of people suffering from depression, psychosis, bipolar disorders, epilepsy, developmental and behavioral disorders in children and adolescents (e.g., autism), dementia, alcohol and drug use disorders, self-harm and suicide, and common emotional problems. In addition, mhGAP-IG promotes the use of "advanced" psychosocial interventions "that take more than a few hours" of a health-care provider's time to learn and typically "more than a few hours to implement" (see Box 15.1). Advanced psychosocial interventions are intended to be suitable for delivery in nonspecialized care settings, but "only when sufficient human resource time is made available."

### 15.2.3   Mental Health Interventions: Packages or Processes?

Although emphasis on psychosocial determinants contrasts sharply with the biological emphasis of much contemporary psychiatric practice, the development and practice of psychosocial interventions has nevertheless been substantially influenced by the biomedical model (Deacon, 2013). This is especially evident in relation to the methods used to examine the efficacy of interventions. It will be recalled from Chapters 10 and 11 that the efficacy claims of contemporary biomedical practice are based primarily on results from randomized controlled trials in which patients are assigned to conditions representing active drug and a control consisting of alternative drug or placebo. The same basic approach has in recent years dominated efforts to establish the efficacy of psychosocial interventions. Typically, active psychosocial interventions in such trials are delivered in a standardized

**BOX 15.1  Summary of the World Health Organization's. Intervention Guide for Mental Health Problems in Non-Specialized Health Settings (mhGAP-IG)**

Interventions recommended by mhGAP-IG cover both psychological and social domains requiring substantial dedicated time, which may be offered by trained and supervised nonspecialized health-care personnel. Psychological interventions are usually provided on a weekly basis over a number of months in either individual or group format. Some of the interventions (e.g., cognitive behavioral therapy and interpersonal psychotherapy) have been implemented successfully by health-care personnel in low-income countries as part of research initiatives (WHO, 2010). It is believed that with systematic use of detailed intervention protocols and training manuals scaling-up of interventions is feasible for nonspecialized health-care settings involving nonspecialized personnel under supervision.

| | |
|---|---|
| Behavioral activation | Derived from principles of behavior analysis and a component of behavior therapy and cognitive behavioral therapy for depression. Intervention focuses on activity scheduling to encourage involvement of the person in life activities that are personally rewarding. Recommended as an intervention option for depression, including bipolar depression, and other significant emotional or medically unexplained complaints |
| Cognitive behavioral therapy (CBT) | Based on the idea that feelings are affected by thinking and behavior. People with mental disorder tend to have unrealistic or distorted thoughts, which if unchecked can lead to unhelpful behavior. CBT typically has a cognitive component (helping the person to develop the ability to identify and challenge unrealistic negative thoughts) and a behavioral component (behavioral activation). Recommended as an intervention option for depression, including bipolar depression, behavioral disorders, alcohol use or drug use disorders, and as an intervention option for psychosis after the acute phase |
| Contingency management therapy | A structured method of rewarding specified desired behaviors, such as attending treatment sessions, active participation in treatment, and avoiding harmful substance use. Introduced rewards for desired behavior are reduced over time as natural rewards, especially from the social environment, become established. Recommended as an intervention for people with alcohol use or drug use disorders |
| Family counseling or therapy | Involves multiple planned sessions, usually more than six, over a period of months. Can be delivered to individual families or groups of families. It has supportive, educational, |

*Continued*

**BOX 15.1 Summary of the World Health Organization's. Intervention Guide for Mental Health Problems in Non-Specialized Health Settings (mhGAP-IG)—cont'd**

| | |
|---|---|
| | and treatment functions. It often includes negotiated problem solving or crisis management. Recommended as an intervention for families of people with psychosis, and alcohol use and drug use disorders |
| Interpersonal psychotherapy (IPT) | Designed to help a person identify and address problems in their relationships with family, friends, partners, and other people. Recommended as an intervention option for depression, including bipolar depression |
| Motivational enhancement therapy | A structured therapy, typically lasting four sessions or less. Uses motivational interviewing, which is a goal-directed client-centered method for encouraging behavior change by helping the person to resolve ambivalence regarding reasons for change. Recommended as an intervention to help people who are dependent on substances |
| Skills training for parents of children with behavior problems | Focuses on positive parent-child interactions, the importance of parenting consistency, discouraging harsh punishments, and the practicing of new skills by parents with their children during training. The content should be culturally sensitive, but should not allow violation of children's basic human rights according to internationally endorsed principles |
| Skills training for parents of children and adolescents with developmental disorders | Involves culturally appropriate training to improve development, functioning, and participation of the child within families and communities. Derived from behavioral principles, applied behavior analysis is used to teach specific social, communicative, and behavioral skills (e.g., by rewarding appropriate behavior and carefully analyzing aspects of the person's environment that trigger problem behavior). Invention involves support for parents (during training and ongoing) of children with different levels of intellectual disability and different specific behavior problems |
| Problem-solving counseling or therapy | Involves the counselor and client working collaboratively to identify and isolate key factors that might be contributing to the person's mental health problems, to separate these into specific manageable tasks, and to develop strategies for solving or coping with each problem. Recommended as a treatment option for depression, including bipolar depression; as an intervention option for alcohol use |

**BOX 15.1 Summary of the World Health Organization's. Intervention Guide for Mental Health Problems in Non-Specialized Health Settings (mhGAP-IG)—cont'd**

| | disorders, drug use disorders, self-harm, and other significant emotional or medically unexplained complaints; and for parents of children and adolescents with behavioral problems |
|---|---|
| Relaxation training | Breathing exercises and progressive muscle relaxation to elicit the relaxation response. Progressive relaxation teaches how to identify and relax specific muscle groups. Treatment usually consists of daily relaxation exercises for at least 1-2 months. Recommended as a treatment option for problems of anxiety and depression, including bipolar depression, and for other significant emotional or medically unexplained complaints |
| Social skills therapy | Helps to develop skills and coping in social situations to reduce distress in everyday life. Uses role-playing, social tasks, encouragement, and positive social reinforcement to help improve ability in communication and social interactions. Can be done with individuals, families, and groups. Intervention usually consists of 45-90 min sessions once or twice per week for an initial 3 months and then monthly. Recommended as an intervention option for people with psychosis or behavioral disorder |

*Adapted from WHO (2010).*

manner using treatment manuals. Comparison interventions usually consist of an alternative active intervention (e.g., pharmacotherapy) or placebo that simulates the conventional intervention in key respects other than the presumed active element. Research has been extensive and has led to strong claims about the efficacy of specific psychosocial interventions for specific mental disorders.

In fact, psychosocial interventions of demonstrated efficacy now exist for most psychiatric disorders (e.g., Hofmann et al., 2012). However, as discussed in Chapter 10, near-exclusive reliance on the randomized controlled trial, useful for garnering evidence of efficacy (internal validity), has left biomedical practice as delivered under usual circumstances largely bereft of evidence of effectiveness (external validity). Uncritical adoption of the randomized controlled trial has led to similar concerns being raised about the effectiveness of psychosocial interventions (Ruscio and Holohan, 2006). Thus, on one hand, the delivery of fixed numbers of therapy sessions to relatively homogenous patient groups whose management is prescribed step by step in accordance with an intervention manual has been successful in demonstrating efficacy. On the other hand, the effectiveness of the same interventions when used in routine clinical practice is less clear.

Additionally, just as alternative drugs compete with one another for selection as first-line interventions in biomedical practice, the adoption of the randomized controlled trial for establishing the efficacy of psychosocial interventions has contributed to the proliferation of *empirically established treatment packages* that compete with one another as interventions of choice for patients deemed to fit predetermined categories of psychiatric diagnosis. Moreover, the emphasis in randomized controlled trials on evaluation of complete treatment packages has contributed comparatively little to improved understanding of the underlying causal mechanisms responsible for therapeutic change. Paralleling the current situation with biomedical practice, overreliance on efficacy demonstrations of standardized packages of psychosocial interventions for use with patients of designated diagnostic type may leave psychosocial practitioners ill prepared to manage the diversity of problems and challenging circumstances typically encountered in usual clinical practice.

Notably, the treatment-package approach encouraged by randomized controlled trials is not the sole method for amassing empirical evidence of effective psychosocial interventions. There is a long tradition in clinical psychology that emphasizes the development of general scientific principles of behavior change applicable to a great diversity of presenting mental health problems unconstrained by the formal (and arbitrary) diagnostics of psychiatry (e.g., Borkovec and Castonguay, 1998; Shapiro, 1995; Skinner, 1953; Wolpe, 1958), and there is a strong case for returning to those earlier traditions (Rosen and Davison, 2003). There is need, therefore, for the current overemphasis on empirically established treatments (the "cookbook" approach) to be redirected toward greater emphasis on understanding and training in *empirically supported principles* of emotional and behavior change (the practitioner-as-scientist approach). The result is likely to be interventions of demonstrated efficacy being delivered effectively by practitioners with sufficient understanding of mechanisms of psychosocial change to meet the diversity of challenges characteristic of usual clinical practice.

### 15.2.4  Subjective Well-being, Resilience, and Flourishing

Influenced by decades of biomedical dominance of health care, mental health has come to be widely understood as the absence of mental disorder. In contrast, consistent with its earlier general definition of health (WHO, 1948), the WHO (2005) has defined mental health as a positive *state of well-being*. However, only relatively recently, following decades of biomedical emphasis on the treatment of individuals with mental *disorder*, have principles of health promotion and prevention begun to be applied to mental *health*. Reflecting their multidimensional nature, positive states of mental health are referred to by various terms that are not merely overlapping by are sometimes used interchangeably, including *subjective well-being, resilience, flourishing, happiness, quality of life*, and *life satisfaction*. Consistent with Rose's (1992) principles of prevention discussed in Chapter 12 and illustrated in Figure 12.1, population-wide promotion of positive states of well-being has the potential not merely to enhance overall levels of mental health but also to reduce the incidence of mental health problems.

Personal subjective well-being is evidently an important life goal for many people (Tay et al., 2015) and possibly the most important population goal for public policy (Oishi and Diener, 2014). Therefore, good quality measurements of subjective well-being are vital for determining what factors are most important for the promotion of well-being. National surveys of subjective well-being often involve the use of a few brief items, if not a single item, thereby raising questions about the reliability of what could appear to be superficial and cursory data. For example, in response to the question, *All things considered, how satisfied are you with your life?*, householders have been asked to indicate on an 11-point scale whether they are *0—totally dissatisfied* to *10—totally satisfied* (HILDA, 2013). However, despite the global nature of such questions, comparisons between countries and over time show good consistency in responses from nationally representative surveys (Kuppens et al., 2006; Lucas and Donnellan, 2012). Many factors influence subjective well-being, and although the contribution of different factors is only approximately understood there is agreement regarding the particular importance of per capita income, democracy and freedom, and social welfare (Jorm and Ryan, 2014).

In addition to being valued intrinsically, subjective well-being appears to be protective of health and longevity, although proof of causality remains elusive (e.g., Diener and Chan, 2011; Steptoe et al., 2015; Xu and Roberts, 2010). Assuming a causal role, there appear to be two main pathways. The first is psychophysiological, wherein positive emotions such as optimism, hope, and curiosity may enhance resilience to disease and disease processes including hypertension, diabetes, and respiratory tract infections (Boehm and Kubzansky, 2012; Richman et al., 2005) and negative emotions such as anxiety and depression may increase vulnerability to disease, especially cardiovascular disease (Sin et al., 2015; Watkins et al., 2013). The second likely causal mechanism is behavioral, wherein higher subjective well-being may encourage health-enhancing behavior such as better diet, physical activity, and avoidance of tobacco and harmful consumption of alcohol (Blanchflower et al., 2013; Boehm and Kubzansky, 2012; Huang and Humphreys, 2012). With both mechanisms, however, reverse causation and reciprocal causation are also likely. For example, anxiety and depression may contribute to cardiovascular disease and vice versa, and physical activity, for example, may contribute to improved subjective well-being and vice versa.

Governments worldwide are increasingly coming to appreciate the importance of measuring subjective well-being as an indicator of national progress and level of success in public policy. However, considering the multidimensional character of the construct, deciding how subjective well-being should be measured remains a major challenge. One suggestion is to operationalize mental health as a composite of positive feelings and positive functioning in life referred to as *flourishing* (Keyes, 2002). As illustrated in Figure 15.2 (a variant of Figure 12.1), absence of flourishing is *languishing*, with moderate mental health being conceptualized as lying between the two. Manifest mental disorder that meets formal diagnostic criteria is regarded as being a fourth category representing the point on the mental

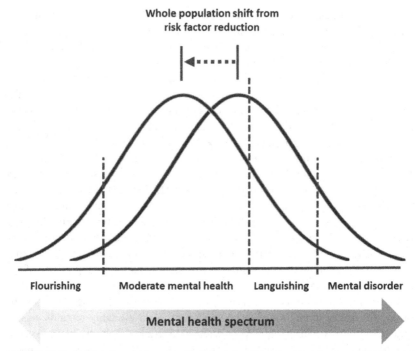

**Whole population shift from risk factor reduction**

Flourishing    Moderate mental health    Languishing    Mental disorder

**Mental health spectrum**

**FIGURE 15.2**  Schematic representation of subjective well-being across the spectrum of mental health. *Note*: Whole-population shifts can markedly alter the size of the area under the curve representing population segments for each of the four states of mental health.

health spectrum that is opposite to flourishing. Figure 15.2 suggests that those who are languishing are at greatest risk of developing mental disorder. Also, as can be seen from the figure, modest whole-population shifts can markedly alter the size of the area under the curve representing population segments for each of the four mental health categories. For example, whole-population shifts to the left in the direction of greater flourishing would be associated with fewer people overall who are languishing or suffering mental disorder.

Recently, the concept of flourishing has been extended to incorporate aspects of most previous measurements of subjective well-being (Huppert, 2009; Huppert and So, 2013). By examining internationally agreed criteria for depression and anxiety, and defining the opposite of each symptom, an operational definition of flourishing has been derived that includes 10 features of subjective well-being. These incorporate positive feeling (*hedonic*), as well as positive function and its associated sense of purpose and meaning in life (*eudemonic*). Using that definition, European Social Survey data were examined for a representative sample of 43,000 people from 23 countries (Huppert and So, 2013). Table 15.1 lists the 10 features, each of which is illustrated by one survey item. Results of the analysis indicated a four-fold difference in rate of flourishing between countries, with Denmark, at 41% of the

**TABLE 15.1** Features of flourishing and illustrative items from the European Social Survey

| Positive feature | A single illustrative survey item |
| --- | --- |
| Competence | Most days I feel a sense of accomplishment from what I do |
| Emotional stability | I felt calm and peaceful (in the past week) |
| Engagement | I love learning new things |
| Meaning | I generally feel that what I do in my life is valuable and worthwhile |
| Optimism | I am always optimistic about my future |
| Positive emotion | Taking all things together, how happy would you say you are? |
| Positive relationships | There are people in my life who really care about me |
| Resilience | When things go wrong in my life it generally takes me a long time to get back to normal (Reverse score) |
| Self-esteem | In general, I feel very positive about myself |
| Vitality | I had a lot of energy (in the past week) |

Adapted from Huppert and So (2013).

population assessed to be flourishing, having the highest rate among the 23 countries surveyed and three countries, Slovakia, Russia, and Portugal, having rates less than 10%. Besides overall rate of flourishing, variability between country profiles for the 10 features of flourishing provides a basis for identifying specific policy targets of particular relevance for each country.

Findings to date provide strong support for the growing international impetus for countries to routinely measure subjective well-being in addition to the many indicators of economic and social well-being that are already routinely taken in most countries (e.g., GDP, employment rates, and consumer price index). The routine measurement of subjective well-being should strengthen the ability of governments to develop and adapt public policy to better serve citizens' needs. However, there is one caveat to such developments that to date appears not to have received attention. The subjective nature of well-being means that in principle coordinated action could enable groups within society to make strategic use of self-reported responses in national surveys. For example, by reporting less rather than more satisfaction in a synchronized manner, groups could influence government initiatives to their own advantage and to the potential disadvantage of other groups who may have equal call on limited resources. Nevertheless, the strong correlation between subjective well-being and observable outcomes, including those of health and longevity, should limit such distortions. Overall, it is to be anticipated that by routinely measuring subjective well-being, governments will be better equipped to develop policies that foster population well-being, promote mental health, and prevent mental disorders.

# REFERENCES

Allison, L., Moncrieff, J., 2014. "Rapid tranquillisation": an historical perspective on its emergence in the context of the development of antipsychotic medications. Hist. Psychiatry 25, 57–69.

American Psychiatric Association, 2013. Diagnostic and Statistical Manual, DSM-5. American Psychiatric Association, Washington, DC.

Anderson, H.D., Pace, W.D., Libby, A.M., et al., 2012. Rates of 5 common antidepressant side effects among new adult and adolescent cases of depression: a retrospective US claims study. Clin. Ther. 34, 113–123.

Antonuccio, D., Healy, D., 2012. Relabeling the medications we call antidepressants. Scientifica 2012, 1–6. http://dx.doi.org/10.6064/2012/965908.

Ashton, H., 2005. The diagnosis and management of benzodiazepine dependence. Curr. Opin. Psychiatry 18, 249–255.

Baker, T.B., McFall, R.M., Shoham, V., 2009. Current status and future prospects of clinical psychology toward a scientifically principled approach to mental and behavioral health care. Psychol. Sci. Public Interest 9, 67–103.

Baldwin, D.S., Kosky, N., 2007. Off-label prescribing in psychiatric practice. Adv. Psychiatr. Treat. 13, 414–422.

Baldwin, D.S., Anderson, I.M., Nutt, D.J., et al., 2005. Evidence-based guidelines for the pharmacological treatment of anxiety disorders: recommendations from the British Association for Psychopharmacology. J. Psychopharmacol. 19, 567–596.

Ballon, J., Stroup, T.S., 2013. Polypharmacy for schizophrenia. Curr. Opin. Psychiatry 26, 208–213.

Barlow, D.H., 2004. Psychological treatments. Am. Psychol. 59, 869–879.

Barry, M.M., Clarke, A.M., Jenkins, R., Patel, V., 2013. A systematic review of the effectiveness of mental health promotion interventions for young people in low and middle income countries. BMC Public Health 13, 1–20. http://dx.doi.org/10.1186/1471-2458-13-835.

Belfer, M.L., 2008. Child and adolescent mental disorders: the magnitude of the problem across the globe. J. Child Psychol. Psychiatry 49, 226–236.

Bet, P.M., Hugtenburg, J.G., Penninx, B.W., Hoogendijk, W.J., 2013. Side effects of antidepressants during long-term use in a naturalistic setting. Eur. Neuropsychopharmacol. 23, 1443–1451.

Biederman, J., Faraone, S.V., 2005. Attention-deficit hyperactivity disorder. Lancet 366, 237–248.

Bijl, R.V., de Graaf, R., Hiripi, E., et al., 2003. The prevalence of treated and untreated mental disorders in five countries. Health Aff. 22, 122–133.

Black, K., Shea, C., Dursun, S., Kutcher, S., 2000. Selective serotonin reuptake inhibitor discontinuation syndrome: proposed diagnostic criteria. J. Psychiatry Neurosci. 25, 255–261.

Blanchflower, D.G., Oswald, A.J., Stewart-Brown, S., 2013. Is psychological well-being linked to the consumption of fruit and vegetables? Soc. Indic. Res. 114, 785–801.

Boehm, J.K., Kubzansky, L.D., 2012. The heart's content: the association between positive psychological well-being and cardiovascular health. Psychol. Bull. 138, 655–691.

Borkovec, T.D., Castonguay, L.G., 1998. What is the scientific meaning of empirically supported therapy? J. Consult. Clin. Psychol. 66, 136–142.

Bushe, C.J., Savill, N.C., 2014. Systematic review of atomoxetine data in childhood and adolescent attention-deficit hyperactivity disorder 2009–2011: focus on clinical efficacy and safety. J. Psychopharmacol. 28, 204–211.

Bystritsky, A., Khalsa, S.S., Cameron, M.E., Schiffman, J., 2013. Current diagnosis and treatment of anxiety disorders. Pharm. Ther. 38, 30–57.

CAMH, 2014. Best Practice Guidelines for Mental Health Promotion Programs: Children (7–12) & Youth (13–19). Centre for Addiction and Mental Health, Toronto, Canada.

Catalano, R.F., Fagan, A.A., Gavin, L.E., et al., 2012. Worldwide application of prevention science in adolescent health. Lancet 379, 1653–1664.

Collins, P.Y., Insel, T.R., Chockalingam, A., et al., 2013. Grand challenges in global mental health: integration in research, policy, and practice. PLoS Med. 10, 1–7. http://dx.doi.org/10.1371/journal.pmed.1001434.

Cosgrove, L., Krimsky, S., 2012. A comparison of DSM-IV and DSM-5 panel members' financial associations with industry: a pernicious problem persists. PLoS Med. 9, 1–5. http://dx.doi.org/10.1371/journal.pmed.1001190.

Cosgrove, L., Wheeler, E.E., 2013. Drug firms, the codification of diagnostic categories, and bias in clinical guidelines. J. Law Med. Ethics 41, 644–653.

Cosgrove, L., Krimsky, S., Vijayaraghavan, M., Schneider, L., 2006. Financial ties between DSM-IV panel members and the pharmaceutical industry. Psychother. Psychosom. 75, 154–160.

Cuijpers, P., 2014. Examining the effects of prevention programs on the incidence of new cases of mental disorders: the lack of statistical power. Am. J. Psychiatr. 160, 1385–1391.

Deacon, B.J., 2013. The biomedical model of mental disorder: a critical analysis of its validity, utility, and effects on psychotherapy research. Clin. Psychol. Rev. 33, 846–861.

Diener, E., Chan, M.Y., 2011. Happy people live longer: subjective well-being contributes to health and longevity. Appl. Psychol.: Health Well-Being 3, 1–43.

Eaton, J., Kakuma, R., Wright, A., Minas, H., 2014. A position statement on mental health in the post-2015 development agenda. Int. J. Ment. Heal. Syst. 8, 1–5. http://dx.doi.org/10.1186/1752-4458-8-28.

Engel, G.L., 1977. The need for a new medical model: a challenge for biomedicine. Science 196, 129–136.

Eyre, H.A., Air, T., Proctor, S., et al., 2015. A critical review of the efficacy of non-steroidal anti-inflammatory drugs in depression. Prog. Neuro-Psychopharmacol. Biol. Psychiatry 57, 11–16.

Fabiano, G.A., Schatz, N.K., Aloe, A.M., et al., 2015. A systematic review of meta-analyses of psychosocial treatment for attention-deficit/hyperactivity disorder. Clin. Child. Fam. Psychol. Rev. 18, 77–97. http://dx.doi.org/10.1007/s10567-015-0178-6.

Faraone, S.V., Buitelaar, J., 2010. Comparing the efficacy of stimulants for ADHD in children and adolescents using meta-analysis. Eur. Child Adolesc. Psychiatry 19, 353–364.

Fazel, S., Zetterqvist, J., Larsson, H., et al., 2014. Antipsychotics, mood stabilisers, and risk of violent crime. Lancet 384, 1206–1214.

Fergusson, D., Doucette, S., Glass, K.C., et al., 2005. Association between suicide attempts and selective serotonin reuptake inhibitors: systematic review of randomised controlled trials. Br. Med. J. 330, 396–399.

Gallego, J.A., Nielsen, J., De Hert, M., et al., 2012. Safety and tolerability of antipsychotic polypharmacy. Expert Opin. Drug Saf. 11, 527–542.

Geddes, J.R., Miklowitz, D.J., 2013. Treatment of bipolar disorder. Lancet 381, 1672–1682.

Hanlon, C., Luitel, N.P., Kathree, T., et al., 2014. Challenges and opportunities for implementing integrated mental health care: a district level situation analysis from five low- and middle-income countries. PLoS One 9, 1–12. http://dx.doi.org/10.1371/journal.pone.0088437.

Harrow, M., Jobe, T.H., Faull, R.N., 2014. Does treatment of schizophrenia with antipsychotic medications eliminate or reduce psychosis? A 20-year multi-follow-up study. Psychol. Med. 44, 3007–3016.

Haw, C., Stubbs, J., 2005. A survey of the off-label use of mood stabilizers in a large psychiatric hospital. J. Psychopharmacol. 19, 402–407.

Mental Health Strategic Partnership, 2014. No Health Without Mental Health: A Guide for Local Healthwatch Organisations. Mind, London. Retrieved on 15 February 2015 from, http://www.mind.org.uk/media/343113/No_Health_Without_Mental_Health_Local_Health_Watch_organisations.pdf.

Heuzenroeder, L., Donnelly, M., Haby, M.M., et al., 2004. Cost-effectiveness of psychological and pharmacological interventions for generalized anxiety disorder and panic disorder. Aust. N. Z. J. Psychiatry 38, 602–612.

HILDA, 2013. HILDA (Living in Australia) Continuing Person Questionnaire (CPQ) W15DR. Melbourne Institute of Applied Economic and Social Research, Melbourne, Australia.

Hofmann, S.G., Asnaani, A., Vonk, I.J., et al., 2012. The efficacy of cognitive behavioral therapy: a review of meta-analyses. Cogn. Ther. Res. 36, 427–440.

Honyashiki, M., Furukawa, T.A., Noma, H., et al., 2014. Specificity of CBT for depression: a contribution from multiple treatments meta-analyses. Cogn. Ther. Res. 38, 249–260.

Huang, H., Humphreys, B.R., 2012. Sports participation and happiness: evidence from US microdata. J. Econ. Psychol. 33, 776–793.

Huppert, F.A., 2009. A new approach to reducing disorder and improving well-being. Perspect. Psychol. Sci. 4, 108–111.

Huppert, F.A., So, T.T., 2013. Flourishing across Europe: application of a new conceptual framework for defining well-being. Soc. Indic. Res. 110, 837–861.

Jääskeläinen, E., Juola, P., Hirvonen, N., et al., 2013. A systematic review and meta-analysis of recovery in schizophrenia. Schizophr. Bull. 39, 1296–1306.

James, A., Hoang, U., Seagroatt, V., et al., 2014. A comparison of American and English hospital discharge rates for pediatric bipolar disorder, 2000 to 2010. J. Am. Acad. Child Adolesc. Psychiatry 53, 614–624.

Johnston, C., Park, J.L., 2015. Interventions for attention-deficit hyperactivity disorder: a year in review. Curr. Dev. Disord. Rep. 2, 38–45.

Jorm, A.F., Ryan, S.M., 2014. Cross-national and historical differences in subjective well-being. Int. J. Epidemiol. 43, 330–340.

Kaaya, S., Eustache, E., Lapidos-Salaiz, I., Musisi, S., Psaros, C., Wissow, L., 2013. Grand challenges: improving HIV treatment outcomes by integrating interventions for co-morbid mental illness. PLoS Med. 10, 1–6. http://dx.doi.org/10.1371/journal.pmed.1001447.

Kessler, D., Bruffaerts, K., de Girolamo, M., et al., 2004. Prevalence, severity, and unmet need for treatment of mental disorders in the World Health Organization World Mental Health Surveys. Lancet 291, 2581–2590.

Kessler, R.C., Berglund, P., Demler, O., et al., 2005. Lifetime prevalence and age-of-onset distributions of DSM-IV disorders in the National Comorbidity Survey Replication. Arch. Gen. Psychiatry 62, 593–602.

Kessler, R.C., Angermeyer, M., Anthony, J.C., et al., 2007. Lifetime prevalence and age-of-onset distributions of mental disorders in the World Health Organization's World Mental Health Survey Initiative. World Psychiatry 6, 168–176.

Keyes, C.L., 2002. The mental health continuum: from languishing to flourishing in life. J. Health Soc. Behav. 43, 207–222.

Kornfield, R., Watson, S., Higashi, A.S., et al., 2013. Effects of FDA advisories on the pharmacologic treatment of ADHD, 2004–2008. Psychiatr. Serv. 64, 339–346.

Kravitz, R.L., Epstein, R.M., Feldman, M.D., et al., 2005. Influence of patients' requests for direct-to-consumer advertised antidepressants: a randomized controlled trial. J. Am. Med. Assoc. 293, 1995–2002.

Kuppens, P., Ceulemans, E., Timmerman, M.E., Diener, E., Kim-Prieto, C., 2006. Universal intra-cultural and intercultural dimensions of the recalled frequency of emotional experience. J. Cross-Cult. Psychol. 37, 491–515.

Lebowitz, M.S., Ahn, W.K., 2014. Effects of biological explanations for mental disorders on clinicians' empathy. Proc. Natl. Acad. Sci. 111, 17786–17790.

Leucht, S., Komossa, K., Rummel-Kluge, C., et al., 2009. A meta-analysis of head-to-head comparisons of second-generation antipsychotics in the treatment of schizophrenia. Am. J. Psychiatr. 166, 152–163.

Lucas, R.E., Donnellan, M.B., 2012. Estimating the reliability of single-item life satisfaction measures: results from four national panel studies. Soc. Indic. Res. 105, 323–331.

McCarthy, S., Asherson, P., Coghill, D., et al., 2009. Attention-deficit hyperactivity disorder: treatment discontinuation in adolescents and young adults. Br. J. Psychiatry 194, 273–277.

McHugh, R.K., Nielsen, S., Weiss, R.D., 2014. Prescription drug abuse: from epidemiology to public policy. J. Subst. Abus. Treat. 47, 160–167. http://dx.doi.org/10.1016/j.jsat.2014.08.004.

Mehta, S., Farina, A., 1997. Is being "sick" really better? Effect of the disease view of mental disorder on stigma. J. Soc. Clin. Psychol. 16, 405–419.

Miklowitz, D.J., Scott, J., 2009. Psychosocial treatments for bipolar disorder: cost-effectiveness, mediating mechanisms, and future directions. Bipolar Disord. 11 (Suppl. 2), 110–122.

Miller, G.A., 2010. Mistreating psychology in the decades of the brain. Perspect. Psychol. Sci. 5, 716–743.

Mohr, D.C., Burns, M.N., Schueller, S.M., et al., 2013. Behavioral intervention technologies: evidence review and recommendations for future research in mental health. Gen. Hosp. Psychiatry 35, 332–338.

Molina, B.S., Hinshaw, S.P., Swanson, J.M., et al., 2009. The MTA at 8 years: prospective follow-up of children treated for combined-type ADHD in a multisite study. J. Am. Acad. Child Adolesc. Psychiatry 48, 484–500.

Moncrieff, J., 2007. Co-opting psychiatry: the alliance between academic psychiatry and the pharmaceutical industry. Epidemiol. Psichiatr. Soc. 16, 192–196.

Moncrieff, J., 2014. The medicalization of "ups and downs": the marketing of the new bipolar disorder. Transcult. Psychiatry 51, 581–598. http://dx.doi.org/10.1177/1363461514530024.

Moncrieff, J., Kirsch, I., 2005. Efficacy of antidepressants in adults. Br. Med. J. 331, 155–157.

Morris, J., Belfer, M., Daniels, A., et al., 2011. Treated prevalence of and mental health services received by children and adolescents in 42 low-and-middle-income countries. J. Child Psychol. Psychiatry 52, 1239–1246.

Newcorn, J.H., Kratochvil, C.J., Allen, A.J., et al., 2008. Atomoxetine and osmotically released methylphenidate for the treatment of attention deficit hyperactivity disorder: acute comparison and differential response. Am. J. Psychiatr. 16, 5721–5730.

Ngo, V.K., Rubinstein, A., Ganju, V., Kanellis, P., Loza, N., Rabadan-Diehl, C., Daar, A.S., 2013. Grand challenges: integrating mental health care into the non-communicable disease agenda. PLoS Med. 10, 1–5. http://dx.doi.org/10.1371/journal.pmed.1001443.

NICE, 2009. Attention deficit hyperactivity disorder: diagnosis and management of ADHD in children, young people and adults. NICE Clinical Guideline 72. Retrieved 20 February 2015 from http://www.nice.org.uk/guidance/cg72.

Nielsen, M., Hansen, E.H., Gøtzsche, P.C., 2012. What is the difference between dependence and withdrawal reactions? A comparison of benzodiazepines and selective serotonin re-uptake inhibitors. Addiction 107, 900–908.

O'Connell, M.E., Boat, T., Warner, K.E. (Eds.), 2009. Preventing Mental, Emotional, and Behavioral Disorders Among Young People: Progress and Possibilities. National Academies Press, Washinton, DC.

Oishi, S., Diener, E., 2014. Can and should happiness be a policy goal? Policy Insights Behav. Brain Sci. 1, 195–203.

Patel, V., Belkin, G.S., Chockalingam, A., Cooper, J., Saxena, S., Unützer, J., 2013. Grand challenges: integrating mental health services into priority health care platforms. PLoS Med. 10, 1–6. http://dx.doi.org/10.1371/journal.pmed.1001448.

Pescosolido, B.A., Martin, J.K., Long, J.S., et al., 2010. "A disease like any other"? A decade of change in public reactions to schizophrenia, depression, and alcohol dependence. Am. J. Psychiatr. 167, 1321–1330.

Pilecki, B.C., Clegg, J.W., McKay, D., 2011. The influence of corporate and political interests on models of illness in the evolution of the DSM. Eur. Psychiatry 26, 194–200.

Pilgrim, D., 2014. Historical resonances of the DSM-5 dispute American exceptionalism or eurocentrism? Hist. Hum. Sci. 27, 97–117.

Polanczyk, G.V., Willcutt, E.G., Salum, G.A., et al., 2014. ADHD prevalence estimates across three decades: an updated systematic review and meta-regression analysis. Int. J. Epidemiol. 43, 434–442.

Polanczyk, G.V., Salum, G.A., Sugaya, L.S., et al., 2015. Annual Research Review: a meta-analysis of the worldwide prevalence of mental disorders in children and adolescents. J. Child Psychol. Psychiatry 56, 345–365. http://dx.doi.org/10.1111/jcpp.12381.

Rahman, A., Surkan, P.J., Cayetano, C.E., et al., 2013. Grand challenges: integrating maternal mental health into maternal and child health programmes. PLoS Med. 10, 1–7. http://dx.doi.org/10.1371/journal.pmed.1001442.

Reinares, M., Sánchez-Moreno, J., Fountoulakis, K.N., 2014. Psychosocial interventions in bipolar disorder: what, for whom, and when. J. Affect. Disord. 156, 46–55.

Reininghaus, U., Dutta, R., Dazzan, P., et al., 2014. Mortality in schizophrenia and other psychoses: a 10-year follow-up of the SOP First-Episode Cohort. Schizophr. Bull. 41, 664–673. http://dx.doi.org/10.1093/schbul/sbu138.

Richman, L.S., Kubzansky, L., Maselko, J., et al., 2005. Positive emotion and health: going beyond the negative. Health Psychol. 24, 422–429.

Rose, G., 1992. The Strategy of Preventive Medicine. Oxford University Press, Oxford, UK.

Rosen, G., Davison, G., 2003. Psychology should list empirically supported principles of change (ESPs) and not credential trademarked therapies or other treatment packages. Behav. Modif. 27, 300–312.

Ruscio, A.M., Holohan, D.R., 2006. Applying empirically supported treatments to complex cases: ethical, empirical, and practical considerations. Clin. Psychol. Sci. Pract. 13, 146–162.

Saha, S., Chant, D., McGrath, J., 2007. A systematic review of mortality in schizophrenia: is the differential mortality gap worsening over time? Arch. Gen. Psychiatry 64, 1123–1131.

Salomon, J.A., et al., 2012. Common values in assessing health outcomes from disease and injury: disability weights measurement study for the Global Burden of Disease Study 2010. Lancet 380, 2129–2143.

Sandler, I., Wolchik, S.A., Cruden, G., et al., 2014. Overview of meta-analyses of the prevention of mental health, substance use, and conduct problems. Annu. Rev. Clin. Psychol. 10, 243–273.

Schomerus, G., Schwahn, C., Holzinger, A., Corrigan, P.W., et al., 2012. Evolution of public attitudes about mental illness: a systematic review and meta-analysis. Acta Psychiatr. Scand. 125, 440–452.

Shapiro, D.A., 1995. Finding out how psychotherapies help people change. Psychother. Res. 5, 1–21.

Shiffman, J., Smith, S., 2007. Generation of political priority for global health initiatives: a framework and case study of maternal mortality. Lancet 370, 1370–1379.

Sibley, M.H., Kuriyan, A.B., Evans, S.W., et al., 2014. Pharmacological and psychosocial treatments for adolescents with ADHD: an updated systematic review of the literature. Clin. Psychol. Rev. 34, 218–232.

Sin, N.L., Yaffe, K., Whooley, M.A., 2015. Depressive symptoms, cardiovascular disease severity, and functional status in older adults with coronary heart disease: the heart and soul study. J. Am. Geriatr. Soc. 63, 8–15.

Skinner, B.F., 1953. Science and Human Behavior. Free Press, New York.

Steel, Z., Marnane, C., Iranpour, C., et al., 2014. The global prevalence of common mental disorders: a systematic review and meta-analysis 1980–2013. Int. J. Epidemiol. 43, 476–493, dyu038.

Steptoe, A., Deaton, A., Stone, A.A., 2015. Subjective wellbeing, health, and ageing. Lancet 385, 640–648.

Suokas, J.T., Suvisaari, J.M., Haukka, J., et al., 2013. Description of long-term polypharmacy among schizophrenia outpatients. Soc. Psychiatry Psychiatr. Epidemiol. 48, 631–638.

Tay, L., Kuykendall, L., Diener, E., 2015. Satisfaction and happiness: the bright side of quality of life. In: Glatzer, W., et al., (Eds.), Global Handbook of Quality of Life. Springer, Dordrecht, Netherlands, pp. 839–853.

Thapar, A., Cooper, M., Eyre, O., Langley, K., 2013. Practitioner review: what have we learnt about the causes of ADHD? J. Child Psychol. Psychiatry 54, 3–16.

Tomlinson, M., Lund, C., 2012. Why does mental health not get the attention it deserves? An application of the Shiffman and Smith framework. PLoS Med. 9, 1–4. http://dx.doi.org/10.1371/journal.pmed.1001178.

Turner, E.H., Matthews, A.M., Linardatos, E., et al., 2008. Selective publication of antidepressant trials and its influence on apparent efficacy. N. Engl. J. Med. 358, 252–260.

Tyrer, P., Kendall, T., 2009. The spurious advance of antipsychotic drug therapy. Lancet 373, 4–5.

UK Department of Health, 2011. No Health Without Mental Health: A Cross-Government Mental Health Outcomes Strategy for People of All Ages. UK Government, London. Retrieved on 15 February 2015 from, https://www.gov.uk/government/uploads/system/uploads/attachment_data/file/213761/dh_124058.pdf.

Visser, S.N., Danielson, M.L., Bitsko, R.H., et al., 2014. Trends in the parent-report of health care provider-diagnosed and medicated attention-deficit/hyperactivity disorder: United States, 2003–2011. J. Am. Acad. Child Adolesc. Psychiatry 53, 34–46.

Waddington, J.L., Youssef, H.A., Kinsella, A., 1998. Mortality in schizophrenia. Antipsychotic polypharmacy and absence of adjunctive anticholinergics over the course of a 10-year prospective study. Br. J. Psychiatry 173, 325–329.

Watkins, L.L., Koch, G.G., Sherwood, A., et al., 2013. Association of anxiety and depression with all-cause mortality in individuals with coronary heart disease. J. Am. Heart Assoc. 2, 1–10. http://dx.doi.org/10.1161/JAHA.112.000068.

Watson, A.C., Otey, E., Westbrook, A.L., et al., 2004. Changing middle schoolers' attitudes about mental illness through education. Schizophr. Bull. 30, 563–572.

Watson, G.L., Arcona, A.P., Antonuccio, D.O., Healy, D., 2013. Shooting the messenger: the case of ADHD. J. Contemp. Psychother. 44, 43–52.

Watson, S.M.R., Richels, C., Michalek, A.P., Raymer, A., 2015. Psychosocial treatments for ADHD a systematic appraisal of the evidence. J. Atten. Disord. 19, 3–10.

Weinmann, S., Read, J., Aderhold, V., 2009. Influence of antipsychotics on mortality in schizophrenia: systematic review. Schizophr. Res. 113, 1–11.

Whiteford, H.A., Degenhardt, L., Rehm, J., et al., 2013. Global burden of disease attributable to mental and substance use disorders: findings from the Global Burden of Disease Study 2010. Lancet 382, 1575–1586.

WHO, 1948. Preamble to the Constitution of the World Health Organization. Adopted by the International Health Conference, New York, 19–22 June, 1946; signed on 22 July 1946 by the representatives of 61 States (Official Records of the World Health Organization, no. 2, p. 100), and entered into force on 7 April 1948.

WHO, 2005. Promoting Mental Health: Concepts, Emerging Evidence, Practice. World Health Organization, Geneva. Retrieved 26 April 2014 from, http://www.who.int/mental_health/evidence/MH_Promotion_Book.pdf.

WHO, 2008. Global Burden of Disease 2004 Update. World Health Organisation, Geneva. Retrieved on 24 April 2014 from, http://www.who.int/healthinfo/global_burden_disease/2004_report_update/en/.

WHO, 2010. mhGAP Intervention Guide for Mental, Neurological and Substance Use Disorders in Non-Specialized Health Settings. World Health Organisation, Geneva. Retrieved on 28 April 2014 from, http://whqlibdoc.who.int/publications/2010/9789241548069_eng.pdf.

WHO, 2013. Mental Health Action Plan 2013–2020. World Health Organization, Geneva. Retrieved on 24 April 2014 from, http://www.who.int/mental_health/publications/action_plan/en/.

WHO, 2014. Health in All Policies Framework for Country Action. World Health Organization, Geneva. Retrieved 26 April 2014 from, http://www.who.int/cardiovascular_diseases/140120HPRHiAPFramework.pdf?ua=1.

Wolpe, J., 1958. Psychotherapy by Reciprocal Inhibition. Stanford University Press, Palo Alto, CA.

Xu, J., Roberts, R.E., 2010. The power of positive emotions: it's a matter of life or death. Subjective well-being and longevity over 28 years in a general population. Health Psychol. 29, 9–19.

# Epilogue

Contemporary biomedical healthcare struggles to control disease after it has occurred, and general lack of success in that regard is evidenced by the worsening global burden of disease. Reducing the incidence of disease and injury through preventive interventions that reduce exposure to risk factors is the only viable means for optimizing the health of populations. Optimal healthcare addresses underlying social determinants of health, and has as a first priority the promotion of ways of living that maximize wellbeing and minimize exposure to disease and injury risk factors.

Commitment to preventive action is undermined by the allure of the technology of biomedicine, with its pills, advanced procedures, scanning devices, and other diverse paraphernalia. Those interventions promise solutions in return for little personal effort and at the cost of surrendering personal control. Moreover, the promise of biomedicine is answered with infrequent cure, immense harm, and high and rising financial burden. The broad summation of outcomes shows that a vision of healthcare in which biomedicine offers comprehensive solutions to leading health problems is a mirage. Benefits from the cumulative panoply of biomedical interventions pales in comparison to that from population-wide life-course reduction in exposure to disease and injury risk factors. In the pursuit of population health, conventional biomedicine should be regarded adjunctive rather than first-line intervention in optimal healthcare.

In essence, risk factor reduction is tantamount to making the effort to stay out of harm's way. To some, that may appear too simple a proposition for resisting the entreaties and dazzling artifices of the scientific–professional–corporate complex that comprises contemporary biomedical healthcare. However, as claimed in the first chapter and repeated thereafter, the foundations of health are prosaic. They are the familiar habits and habitats of everyday human existence. By addressing ways of living, individuals, communities, and nations together possess the means to optimize the health of populations.

# List of Acronyms

| | |
|---|---|
| ADE | adverse drug event |
| ADHD | attention deficit hyperactivity disorder |
| ADI | adverse drug interaction |
| ADR | adverse drug reaction |
| AHRQ | (American) Agency for Healthcare Research and Quality |
| BMI | body-mass index, $\dfrac{\text{height}(m)}{\text{weight}(kg^2)}$ |
| CAM | complementary and alternative medicine |
| CEO | Chief Executive Officer |
| *C. difficile* | *Clostridium difficile* |
| CDC | (American) Centers for Disease Control |
| coxibs | Cox-2 inhibitors |
| DALE | disability-adjusted life expectancy |
| DALY | disability-adjusted life years |
| DNA | deoxyribonucleic acid |
| DSM | *Diagnostic and Statistical Manual* of the American Psychiatric Association |
| DSM-5 | fifth edition (2013) of the *Diagnostic and Statistical Manual* of the American Psychiatric Association |
| DTaP | three-in-one vaccine against diphtheria, tetanus, and pertussis (whooping cough) |
| DTaP/IPV/Hib | five-in-one which combines *DTaP* with vaccines against polio and *Haemophilus influenzae* type b (*Hib*) |
| *E. coli* | *Escherichia coli* |
| FDA | (American) Food and Drug Administration |
| fMRI | functional magnetic resonance imaging |
| GDP | gross domestic product |
| GNH | gross national happiness |
| GWAS | genome-wide association studies |
| HAI (or HCAI) | healthcare-associated infection; hospital-acquired infection |
| HALE | health-adjusted life expectancy |
| HALY | health-adjusted life years |
| Hib | *Haemophilus influenzae* type b |

| | |
|---|---|
| HIV/AIDS | human immunodeficiency virus infection and acquired immunodeficiency syndrome |
| HBV | hepatitis B virus |
| HPV | human papillomavirus |
| HRT | hormone replacement therapy (now, menopausal hormone therapy; HT) |
| IHI | (American) Institute for Healthcare Improvement |
| IOM | (American) Institute of Medicine |
| MDR | multiple drug (or multidrug) resistant |
| MET | metabolic, as in metabolic equivalent of energy spent during physical activity |
| mhGAP | Mental Health Gap Action Programme (World Health Organisation) |
| mhGAP-IG | Intervention Guide for Mental, Neurological and Substance Use Disorders in Non-Specialised Health Settings (World Health Organisation) |
| mRNA | messenger RNA |
| MMR | vaccine against measles, mumps, and rubella |
| MRSA | methicillin-resistant *Staphylococcus aureus* |
| NSAIDs | nonsteroidal anti-inflammatory drugs |
| NHS | (British) National Health Service |
| NDM-1 | New Delhi metallo-ß-lactamase-1 |
| NME | new molecular entity |
| P4 medicine | biomedical health care that is alleged to be predictive, personalized, preventive, and participatory |
| P5 medicine | P4 medicine that is alleged to have a population perspective |
| PMDD | premenstrual dysphoric disorder |
| QALE | quality-adjusted life expectancy |
| QALY | quality-adjusted life years |
| RNA | ribonucleic acid |
| SARS | severe acute respiratory syndrome |
| SSRIs | selective serotonin reuptake inhibitors |
| VRE | vancomycin-resistant enterococci |
| WHO | World Health Organisation |

# Glossary[1]

*Adverse drug event*  The term used to refer to all harmful effects of medication, inclusive of both *adverse drug reactions* and *adverse drug interactions*.

*Adverse drug interaction*  An unintended harmful effect of medication (*adverse drug reaction*) when taken simultaneously with one or more other drugs, whether prescribed or not and whether licit or illicit

*Adverse drug reaction*  An unintended harmful effect of medication.

*Agricultural revolution* (or *Agrarian Revolution*)  The transition from preagricultural subsistence to settled agriculture characterized by a diet of cultivated foods. The first such transition is referred to as the *Neolithic Revolution* or *First Agricultural Revolution*.

*Alleles*  Alternative forms of the same gene.

*Antibiotic stewardship*  A coordinated program to promote the appropriate use of antimicrobial agents, including antibiotics. A main objective is to slow the development of microbial resistance and decrease the spread of infections caused by multidrug-resistant organisms.

*Antibiotics*  A subclass of *antimicrobials* for destroying bacteria and inhibiting bacterial growth.

*Antimicrobials* (see *antibiotics*)  Agents that kill microorganisms or inhibit their growth across a wide range including bacteria, viruses, and fungi. Includes formulations for cleaning surfaces (e.g., disinfectants) as well as formulations that are ingested to destroy microorganisms in the body.

*Bacteria* (plural of *bacterium*)  A large group of typically single-celled microorganisms present in most habitats on earth that can exist either as independent (free-living) organisms or in close mutual (symbiotic) and nonmutual (parasitic) relationship with other life forms.

*Big Pharma*  Usually, pharmaceutical companies collectively, as a sector of industry, but occasionally, a large individual company in that sector.

*Biomedicalization* (see also *medicalization* and *pharmaceuticalization*)  Largely synonymous with *medicalization*, but with greater emphasis on the role of technology in diagnosis and treatment.

*Biopsychosocial model* (see also *medical model*)  The general proposition that rather than being viewed as the absence of dysfunction in biological processes health is better understood as involving the interplay between biological, psychological (behavioral, cognitive, and emotional), and social (interpersonal, societal, and cultural) variables.

*Bipolar disorder* (also *bipolar affective disorder*, formerly *manic depression*)  A mental disorder characterized by extreme shifts from periods of elevated mood and activity to periods when mood and activity are depressed, with both periods characterized by impeded ability to perform daily tasks.

---

1. Terms in bold italics are defined in the Glossary.

*Body mass index (BMI)* A measure of body composition based on weight (mass) and height, determined by the formula: $\frac{\text{weight (kg)}}{\text{height (m)}^2}$, expressed as kg/m$^2$.

*Burden of disease* Refers to the aggregate of population mortality and morbidity from specific diseases and injuries or all disease and injury combined.

*Clinical practice guidelines* Sets of systematically developed recommendations based on scientific evidence to guide optimal patient care for specified diagnoses.

*Cognitive behavioral therapy* A type of psychotherapy comprising a broad range of behavioral and cognitive interventions that, as a group, do not submit to precise definition. Interventions tend to be problem-focussed, systematic, and time-limited, frequently involving graduated goal-oriented assignments intended to facilitate cognitive and/or behavior change.

*Common complex diseases* (see also *noncommunicable diseases*) Noninfectious and non-transmissible between persons, typified by slow progression (incubation) and long duration, including cardiovascular disease (primarily, coronary heart disease and stroke), cancer, chronic respiratory disease, and diabetes.

*Complementary and alternative medicine (CAM)* Refers to healthcare products and practices that are not part of conventional biomedical care. *Complementary* typically refers to interventions used together with conventional medical care (e.g., some uses of acupuncture), whereas *alternative* refers to interventions used in place of conventional medical care (e.g., homeopathy). Classifications are not rigid, with practices once considered alternative or complementary coming later to be regarded as part of conventional medical care (e.g., guided imagery and massage for managing pain).

*Conflict of interest* A set of circumstances that creates a risk that professional judgment or actions regarding a primary interest (e.g., patient care) will be unduly influenced by a secondary interest (e.g., benefit to the practitioner).

*Continuing medical education* Continuing education for medical professionals to maintain competence and to learn about new developments.

*Continuing professional development* (or *continuing professional education*) Profession-related learning that usually occurs after completion of formal training, and may be required to maintain professional certification or licensure. It may include learning activities that are structured (e.g., attendance at courses, conferences, and workshops) as well as informal (e.g., self-directed reading).

*Delusion* A firm conviction or belief that is held despite strong invalidating evidence. Bizarre and florid delusions are a sign of mental disorder.

*Demographic transition* (see also *epidemiologic transition*) The transition from high to low rates of births and deaths characteristic of economic change from preindustrial to industrial.

*Designer drug* A synthetic drug analog devised to circumvent laws controlling the use a legally restricted or prohibited drug.

*Disease mongering* The practice of widening the diagnostic boundaries of disease to expand markets for those who sell and deliver medical interventions. Also known as *diagnostic creep* and the *selling of sickness*, it is an aspect of the more general processes of *medicalization*, *biomedicalization*, and *pharmaceuticalization* of health.

*Distal* (see also *proximal*) Being away or removed from, in space or time, a point of reference.

*DNA* A molecule comprised of two strands coiled around each other to form a double helix. Carries the genetic information in the cell, and capable of self-replication and synthesis from *RNA*.

*Dose-response relationship* Describes the change in effect caused by varying levels of exposure. For example, incidence of disease (e.g., lung cancer) may increase in proportion to greater exposure to a behavioral risk factor (e.g., cigarette smoking).

*Empirical evidence* The term derives from the Greek word for experience, and refers to knowledge acquired by means of observation or experimentation in contrast to reason and reflection. In practice, reason and reflection (often incorporating prior observation) are used to propose hypotheses that are then tested using empirical methods.

*Endoscope* An instrument, often consisting of a tube, light source, and lens, that can be inserted into a cavity in the body or the interior of a hollow organ for the purpose of relaying images to a viewer (e.g., surgeon). The device may also include a channel for conveying medical instruments or manipulators.

*Endoscopy* Typically, looking inside the body using an *endoscope*.

*Epidemiologic transition* (see *demographic transition*) Usually refers to the transition from high to low rates of births and deaths, characterized by a transition from high mortality due to acute infectious diseases largely among the young to lower overall mortality but higher morbidity due to chronic degenerative diseases in adulthood.

*External validity* (see *internal validity*) Refers to the level of confidence with which conclusions about causal relationships inferred from a scientific study can be generalized to circumstances not identical to those of the reference study.

*Free sugar* Usually, refers to sugar added to foods by the consumer, during cooking, or by manufacturers during the production of processed foods. It can also refer to sugars naturally present in honey, syrups, and fruit juices.

*Generic drugs* (or *generics*) (1) A drug product that is identical or similar to a branded product, especially with respect to key active ingredients. They are usually brought to the market with little advertising and are typically lower in price than branded formulations. (2) The chemical name of a drug as distinct from its brand name.

*Genome* The complete set of genes or genetic material present in a cell or an organism.

*Genomics* The study of genomes.

*Germ theory* The theory that microorganisms (pathogens), too small to be seen without magnification, rather than *miasma* emanating from rotting organic matter, cause epidemic infectious diseases such as cholera and bubonic plague (Black Death).

*Global Trigger Tool* A protocol for measuring medical harm in which an expert team systematically reviews patient records using defined criteria and procedures.

*Gross national income* The total domestic and foreign output of a country, consisting of gross domestic product plus income earned by foreign residents minus income earned in the domestic economy by non-residents.

*Habit* A routine or persistent pattern of behavior.

*Habitat* A given species' characteristic milieu, inclusive of the physical and social features of that environment.

*Hallucination* A vivid perception (visual, auditory, tactile, olfactory, or gustatory) that is experienced and held to be real despite the absence of an external stimulus. Usually drug induced or due to mental disorder.

*Healthcare-associated infection* Infection not present until such time as a patient comes into contact with a healthcare setting, most often a hospital.

*Hyperglycemia* An excess of glucose in the bloodstream, often associated with diabetes mellitus.

*Identifiable victim effect* Refers to the tendency for identified victims of misfortune to attract more attention and support than that given to preventing a larger number of unidentified people from becoming victims in the first place.

*Impact factor* An index reflecting the reputed importance or influence of an academic journal within its field based on the number of citations articles receive.

*Impaired glucose tolerance* Persistently raised blood glucose to levels that are higher than normal but less than the level required for a diagnosis of diabetes.

*Incidence* (see *prevalence*) A measure of disease frequency or risk indicated by the number of new cases in a defined population within a specified time period. Assume, for example, a healthy population of 100,000 people and 5000 develop disease over the course of 12 months. The incidence of disease is 5000 per 1000,000 (or 5%) per year.

*Informed consent* Permission given by a person for the conduct of a healthcare intervention based on understanding of all relevant facts concerning the implications and consequences of the intervention.

*Insulin* A hormone produced in the pancreas, which regulates the amount of glucose in the blood.

*Insulin resistance* A physiological condition in which cells fail to respond normally to the hormone *insulin* that is produced by the body such that the insulin is unable to be used effectively, leading to *hyperglycemia*.

*Internal validity* (see *external validity*) Refers to the level of confidence with which conclusions about causal relationships inferred from a scientific study can be accepted over alternative explanations.

*Laparotomy* A surgical procedure involving an incision through the abdominal wall to gain access to the abdominal cavity, often for exploratory purposes.

*Late Pleistocene* Period of the Earth's history incorporating the last glacial period (or *Ice Age*). Its ending, about 12,000 years ago, coincides roughly with the advent of settled agriculture.

*Medical model* (*medical paradigm* or *biomedical model*, see also *biopsychosocial model*) The general proposition that biology provides a complete understanding of human health, wherein health is equated with the absence of disease.

*Medicalization* (see also *biomedicalization* and *pharmaceuticalization*) The process of defining and treating human problems, especially aspects of life previously outside the jurisdiction of medicine, as medical conditions or diseases.

*Metabolic syndrome* A disorder of energy utilization and storage, diagnosed by the co-occurrence of three or more of the following: abdominal ("central") obesity, raised blood pressure, raised fasting plasma glucose, high serum triglycerides, and low high-density cholesterol (HDL) levels.

*Meta-analysis* A method for statistically combining the results of similar studies with the aim of improving the precision of estimated effects.

*Miasma theory* (*miasmatic theory*) The theory that epidemic infectious diseases such as cholera and bubonic plague (Black Death) are caused by *miasma* (Ancient Greek for *pollution*) comprising poisonous vapor, *bad air*, or *night air* that emanated from decomposed organic matter. With the discovery of microorganisms in the nineteenth century, the miasma theory was displaced by the *germ theory*.

*Monogenic disorder* An inherited disorder controlled by a single gene.

*Moral hazard* Refers to the tendency of decision makers to tolerate greater risk when the consequences are borne by others.

*mRNA* A large family of *RNA* molecules that mediate the transfer of genetic information from *DNA* in the cell nucleus to ribosomes (cites of protein synthesis) in the cytoplasm.

*Natural selection*  The process whereby natural variations in phenotype (observable traits) lead to some traits being better adapted to particular features of the environment (ecological niche) and thereby confer a reproductive advantage for the individuals possessing that phenotype. This, in turn, may lead to a gradual increase in the population frequency of the genotype (genetic complement) responsible for the successful phenotype. Successive adaptations may, over time, lead to the emergence of new species.

*Neolithic Revolution* (or *First Agricultural Revolution* or *Neolithic Demographic Transition*) The initial transition from hunting and gathering to settled agriculture that occurred about 12,000 years ago.

*Noncommunicable diseases*  Noninfectious and nontransmissible between persons, includes *common complex diseases* and injuries.

*Odds ratio*  A measure of *association* that compares the odds of an outcome (e.g., disease) for those at risk (e.g., exposed to an environmental pollutant) to the odds of the outcome for those deemed not at risk. An odds ratio that approximates unity (1.0) suggests no difference in risk between exposed and not-exposed groups, whereas an odds ratio significantly above (or below) unity for the exposed group would suggest greater (or lower) risk compared to the not-exposed group.

*Off-label use*  Use of a drug or medical device for a purpose other than that for which it has been approved, including prescribing a drug for a child when that drug has been approved for use only for adults, prescribing a drug for a condition other than that for which it has been approved, and prescribing a drug at an unapproved dosage.

*Operational definition*  Specifies empirical (i.e., observable) properties of a variable (term or object) in order that the variable can be reliably observed or measured by others.

*Opioid analgesic*  A drug for pain relief containing morphine (the main active chemical in opium) or other morphine-like chemical.

*Optimism bias* (or *unrealistic optimism*) The belief that one is at less risk than others from a specified hazard or negative event.

*Paleolithic*  Period of human prehistory to the end of the *Late Pleistocene* characterized by nomadic hunting and gathering, and diverse cultural practices, including art (cave painting, rock art, and jewellery making) and rituals (e.g., burial ceremonies).

*Passive smoking*  The inhalation by nonsmokers of "secondhand" or environmental tobacco smoke from active smokers (see *third-hand smoke*).

*Peer review*  A form of self-regulated quality control wherein work is evaluated by one or more people of similar competence (peers) to those who produced the work. It is often used to determine the suitability of a scientific report for publication.

*Pharmaceuticalization* (see also *medicalization* and *biomedicalization*) The process of defining and redefining human problems, including problems of everyday living, as having pharmaceutical solutions.

*Physical dependence* (or *dependence*) A state caused by repeated use of a drug and indicated by negative physical symptoms (*withdrawal effects* or *withdrawal syndrome*) following abrupt discontinuation or reduction in dose. Symptom relief (*withdrawal reversal*) follows resumption of intake or increase in dose.

*Placebo*  As typically defined, a placebo is an inert (i.e., inactive) agent (e.g., "sugar pill") or procedure that produces a therapeutic effect. However, this definition is either contradictory or incomplete because it does not explain how an inactive agent can have active therapeutic effects (see the text for further discussion).

*Prevalence* (see also *incidence*) A measurement of disease frequency indicated by the number of cases as a proportion of the population. Usually expressed as a fraction or a percentage (e.g., 5000 cases per 100,000 of population or 5%).

***Product placement*** A marketing strategy whereby a marketing message, brand logo, or product is embedded in a visual or graphic medium of entertainment, including television, film, music, and video games (e.g., showing the lead character in a film drinking soda with the brand name displayed).

***Proximal*** (see also ***distal***) Being near in space or time to a point of reference.

***Psychosis*** A mental disorder characterized by a "loss of contact with reality", indicated by ***thought disorder***, ***delusions***, or ***hallucinations***.

***Puerperal fever*** (*childbed fever, the doctors' plague*) A bacterial infection contracted by women during childbirth or miscarriage. It can develop into puerperal sepsis, a serious and potentially fatal form of septicemia.

***Quasi-experiment*** While allowing comparison of groups measured on the same outcome variables, lacks the element of random assignment to groups that characterizes traditional experiments. Accordingly, inferences concerning causal attribution, though permissible, tend to be less strong than in a traditional experiment.

***Randomized controlled trial*** (*clinical trial*) A scientific experiment in which people are allocated at random to receive one of two or more alternative interventions, thereby permitting a fair comparison to be made of the effects of different interventions.

***Regression toward the mean*** All distributions of values contain natural variation, and it is a property of distributions that values that appear distant from the mean (i.e., extreme values or *outliers*) when a variable is measured first will tend to be closer to the mean when a second measurement is taken.

***Risk compensation*** The tendency for people to adjust their behavior according to perceived level of risk, wherein higher perceived risk encourages more carefulness and lower perceived risk encourages less carefulness.

***Risk factors*** Variables associated with increased incidence of disease. Since statistical *association* alone does not establish *causation*, the term *risk marker* is sometimes used to label disease-associated variables, while requiring additional evidence of a causal contribution to disease before labeling the variable a *risk factor*.

***RNA*** A molecule comprised of a single strand that is a constituent of all living cells and many viruses. Important in protein synthesis and in the transmission of genetic information transcribed from ***DNA***.

***Schizophrenia*** A mental disorder characterized by symptoms of ***psychosis***, including ***thought disorder***, ***delusions***, and ***hallucinations***.

***Secular*** Refers, in the present context, to changes in health and health-related indicators that are not due to or derived from biomedical healthcare.

***Self-serving bias*** A cognitive bias in which people tend to judge, reason, and evaluate life events in ways that are favorable to their own interests, which include their view of themselves, while regarding others' interests relatively less favorably.

***Senescence*** The process of being or of becoming old.

***Spontaneous reporting systems*** Repositories for recording events (e.g., instances of medical harm), usually administered by a central or regional regulatory authority, wherein reporting obligations may be either voluntary or compulsory.

***Staphylococcus aureus*** A bacterium frequently found in nonpathogenic form on human skin and in the respiratory tract. The emergence of antibiotic-resistant forms has become a worldwide problem.

***Surrogate endpoint*** (or *marker*) Used in clinical trials as a substitute measurement (e.g., blood test result) of the effect of an intervention when measuring *clinical endpoints* (manifest disease) is impractical. However, the relationship between surrogate and clinical endpoints is not guaranteed, and this can lead to invalid inferences about the clinical efficacy of an intervention.

**Third-hand smoke** Refers to residual nicotine and other chemicals from tobacco smoke, which are deposited on surfaces whenever smoking occurs within enclosed spaces, being subsequently inhaled by occupants of those spaces (see also *passive smoking*).

**Thought disorder** Disorganized thinking evidenced by disorganized speech, including gibberish, illogicality, and ideas of reference (expressed beliefs of being controlled by things completely unconnected to oneself, such as real or imagined voices emanating from radio or TV).

**Transcription** (see also *translation*) In genetics, the transfer of genetic information involving the synthesis of *mRNA* from *DNA*.

**Translation** (see also *transcription*) In genetics, the synthesis of protein from *mRNA*.

**Translational research** "Bench-to-bedside" transfer wherein knowledge from basic laboratory research is "translated" to useful practical applications that are tested in clinical trials and eventually incorporated into routine healthcare.

**Type 1 error** A spurious finding of difference when none exists.

**Type 2 error** A spurious finding of no difference when a difference exists.

**Viral marketing** Building brand awareness and promoting purchases by encouraging the distribution of a marketing message between people in a target audience, often through electronic or digital platforms.

**Virus** Submicroscopic structures consisting of a core of *RNA* or *DNA* surrounded by a protein coat. Replicating only within the cells of living hosts, including bacteria, plants, and animals, viruses are typically not considered to be living organisms.

**Whistleblower** A person who informs about unlawful or immoral conduct by another person or organization.

# Subject Index

Note: Page numbers followed by *b* indicate boxes, *f* indicate figures and *t* indicate tables.

# Author Index

Note: Page numbers followed by *f* indicate figures, *b* indicate boxes, *np* indicate footnotes and *t* indicate tables.

494    Author Index

Ganju, V., 448
Garatachea, N., 273
Garattini, S., 197
García-Franco, A.L., 270
García-Muse, T., 263
Gardiner, E., 13
Garin, N., 341–342
Garnatz, J., 142
Garrouste-Orgeas, M., 164
Garry, D.A., 123
Gautam, O.P., 20
Gavin, L.E., 447–448
Gawande, A.A., 141–142
Geddes, J.R., 438, 445–446
Geels, F., 20
Gelders, S.F., 138, 414–415
Gelijns, A., 141–143
Gellad, W.F., 213–214
Geoghegan-Quinn, M., 167
George, J.N., 185
Gerstel, E., 341–342
Gérvas, J., 335–337
Getz, L., 184–185
Ghafur, A.K., 155, 158–159
Giampaoli, S., 74
Gibberd, R.W., 115–117
Gibbons, R.V., 193–194
Gibbs, V.C., 142f
Gibson, N., 388
Gillings, M.R., 166
Gilpin, E.A., 389
Giovino, G., 389
Giske, C.G., 158
Glantz, S.A., 341–342, 388–389, 391
Glanz, K., 352–353
Glasgow, R.E., 3–5
Glass, K.C., 439–440
Glatzer, W., 455
Gledhill, N., 404
Gluckman, P.D., 37
Gluud, C., 216–218, 217f, 231
Godfrey, K.M., 37
Godlee, F., 271
Gohlke, H., 205, 397–398
Gold, M.R., 64
Golder, S., 233
Goldstein, H., 343, 387
Golomb, B.A., 184
Gonzalez-Garay, M.L., 262–263
Good, C.B., 213–214
Goodman, B., 190–191
Goodman, J.C., 117–118
Goodman, S.H., 344

Goodney, P.P., 126
Gordon, A., 152
Gordon-Dseagu, V., 399–400
Gorski, D.H., 326
Gortmaker, S.L., 411
Gottlieb, L., 263, 356
Gøtzsche, P.C., 186, 218–226, 233–234, 253, 256–260, 308–309, 316, 347, 442
Govoni, R., 392
Goyal, J.P., 272
Gradmann, C., 11–12
Grady, D., 269–270, 300
Graham, R., 186–187
Grana, R., 391
Grande, D., 202
Grant, R.W., 252–253
Graves, J., 193
Gray, J.A., 290, 292–293
Green, J.L., 137
Green, M.J., 202
Greenberg, C.C., 141–142
Greenberg, R.P., 223
Greene, L.S., 36
Greenfield, S., 324
Greenhalgh, T., 293–294, 388
Greenwood, D.C., 343
Griffin, F.A., 108, 118–119, 120t, 121–122, 121f, 124, 134–135
Griffin, M., 388
Grimshaw, J.M., 237
Griskeviciene, J., 147–149, 149t
Grobbee, D., 394
Groene, O., 145
Grosse, S.D., 246
Grover, S.A., 98
Grucza, R.A., 392
Gruen, R.L., 189–190, 190f
Gruenberg, E.M., 72, 75, 367, 404–405
Gruskin, S., 355
Guedes, I.N., 12–13
Gulbrandsen, P., 199–200
Gulliford, M.C., 346
Gundel, L.A., 390
Guthrie, R., 261–262
Guyatt, G.H., 231, 270, 289–290
Gwin, M.L., 3–5
Gwinn, M., 261

**H**
Haby, M.M., 442–443
Hackbarth, A.D., 108, 121, 124
Haglund, A.B., 115–116
Haglund, M., 198

**504** Author Index

Printed in the United States
By Bookmasters